Advances in Olericulture

Series Editor

Silvana Nicola, Department of Agricultural, Forest and Food Sciences, DISAFA, Vegetable Crops & Medicinal and Aromatic Plants, VEGMAP
University of Turin
Grugliasco (Turin), Italy

The book series Advances in Olericulture provides a state-of-the-art account of research in olericulture, the applied life science of production and utilization of vegetable crops. The series focuses on various aspects of vegetable science and technology covering primarily but not exclusively species where the vegetative organ is the economically important component. The series of books spans current topics from sustainable fertilization to organic production; from open field cultivation to advanced soilless growing techniques; from vegetable seed and seedling physiology to vegetable quality and safety; from environmental stresses to phyllosphere communities interaction with vegetables; from postharvest biology and technology to minimally processing of vegetables. The series is designed to present the most advanced scientific information available linking basic and applied research for serving olericulturists, research workers, teachers and advanced students.

Shashank Shekhar Solankey
Editor

Advances in Research on Potato Production

Springer

Editor
Shashank Shekhar Solankey
Department of Horticulture (Vegetable Science), Nalanda College of Horticulture, Noorsarai, Nalanda
Bihar Agricultural University, Sabour
Bhagalpur, Bihar, India

ISSN 2367-4083 ISSN 2367-4091 (electronic)
Advances in Olericulture
ISBN 978-3-031-82709-9 ISBN 978-3-031-82710-5 (eBook)
https://doi.org/10.1007/978-3-031-82710-5

This Springer imprint is published by the registered company Springer Nature Switzerland AG
The registered company address is: Gewerbestrasse 11, 6330 Cham, Switzerland

Preface

The potato (*Solanum tuberosum L.*), often referred to as the *King of Vegetables*, is consumed by over one billion people worldwide and is considered the world's most important non-grain food crop, playing a significant role in global food security. Globally, potatoes are crucial for both smallholder farmers as well as international markets, and they nourish populations and bolster agricultural sustainability. Nutritionally, potatoes are exceptional as they are nearly fat-free, gluten-free, and rich in vitamin C, B_6, phenols, iron, potassium, phosphorus, magnesium and protein comparable to cereals. Their high energy density surpasses that of many other vegetables, making potatoes a vital ally in combating hidden hunger, a pressing global health issue.

Potatoes are not just a staple food but a symbol of resilience and adaptability; therefore, it was recognized by the United Nations in 2008 as the *International Year of the Potato* due to its potential to address hunger and poverty. In 2024, the FAO honored the potato by designating 30th May of every year as the *International Day of Potato*, celebrated for the first time with the theme "Harvesting Diversity, Feeding Hope." This theme emphasizes potato's role in global food security and its potential to enhance resilience against climate change. In contrast, *National Potato Day* is celebrated annually on 19th August. Both initiatives highlight the critical connection between poverty, food security, and the potato's role in addressing these issues. Sustainable potato production is essential for ensuring food security and social welfare, particularly in the context of future climate change.

Despite their importance, potatoes are highly sensitive to environmental conditions due to their shallow root system. Projected climate impacts, such as rising temperatures and increased drought, could result in a 32% reduction in potato yields by 2050. To address these challenges, future potato breeding programs must prioritize enhancing tolerance to both abiotic and biotic stresses. Utilizing conserved natural germplasm from gene banks and implementing climate-smart agricultural practices will be crucial. Advances in molecular techniques such as marker-assisted breeding, genome-wide association studies, functional genomics and genetic engineering offer promising pathways for developing resilient potato varieties.

Disease-free potato seed production is another critical area, achieved through tissue culture techniques and the production of mini-tubers using aeroponics or root culture methods. Ongoing research focuses on improving potato production, tuber quality, and stress management, addressing both biotic and abiotic factors. As a staple food for millions, the potato not only has a rich history but also a promising future, driven by increasing global demand and continued research into potato production, economics and marketing strategies, particularly within the Indian context.

This book, *Advances in Research on Potato Production*, aims to illuminate the extensive scientific progress made in this field, tracing both the historical development and future prospects of potato cultivation. It is intended for professionals, researchers, and postgraduate students engaged in advanced potato production, breeding, seed production, stress management, postharvest technologies, and marketing strategies, with a particular focus on the Indian potato sector. This comprehensive resource encompasses 18 chapters that cover various aspects such as future trends and goals, genotypic selection, climate change impacts and mitigation strategies, advancements in conventional and modern breeding techniques (including molecular breeding, gene editing and genetic transformation), disease-free seed production, aeroponics and root culture techniques, agronomic management practices, organic cultivation, soil and nutrient management, geospatial technologies, quality improvement, postharvest management and storage and stress management. It also explores the Indian potato sector's economics and marketing practices.

Each chapter of this book offers a comprehensive overview and specific insights into recent breakthroughs, addressing the current state of research while proposing critical questions and future directions. The main aim is to bridge the gap between scientific discovery and practical application, enhancing our understanding of how research advancements can translate into tangible benefits for farmers and consumers alike.

I extend my heartfelt gratitude to the dedicated researchers, practitioners, and institutions whose efforts have made this book possible. Their unwavering commitment to advancing knowledge and improving agricultural practices has been instrumental in driving these innovations. I hope this volume will serve as a valuable resource for researchers, policymakers, and agricultural professionals, inspiring continued progress in potato production.

As we navigate the complexities of modern agriculture, it is crucial to remember that potato research is not merely about increasing yields but also about fostering a sustainable and equitable food system for future generations. I invite you to explore the chapters that follow, discover the advancements made and join me in my ongoing quest to unlock the full potential of this remarkable crop.

First and foremost, I wish to express my profound gratitude to the *Almighty God* for providing me the courage, ideas, and inspiration to complete this book. I am deeply grateful to my family, my mother, Smt. Prem Shila Singh; my wife, Dr. Reshma Singh; and my daughters, Arya and Aadya for their unwavering love, support and encouragement throughout this journey.

I am also thankful to all the subject matter experts for their valuable contributions, providing innovative information and detailed scripts that made this book possible. I also acknowledge my organization Bihar Agricultural University, Sabour, Bhagalpur, Bihar, India for their generous support and cooperation.

Special thanks go to the Springer Nature academic staff, particularly Series Editor Dr. Silvana Nicola (Advances in Olericulture), Ms. Deepthi Vasudevan, Production Editor (Books) and Ms. Thirumangai Thamizhmani, Production Editor (Books) for their generous support in compiling and completing this manuscript. I value the direct and indirect help, as well as the insightful recommendations from professors, colleagues, and friends, throughout the preparation of this book. I welcome any suggestions from readers that could enhance the book's value and acknowledge the support received from various experts, books, journals, writers, publishers and online platforms consulted during its preparation. We deeply appreciate Springer Nature for publishing this book with such meticulous care.

NCOH, Noorsarai, Nalanda, Bihar, India
May, 2025

Shashank Shekhar Solankey

Contents

Editors and Contributors

About the Editor

Shashank Shekhar Solankey is presently working as an Associate Professor–cum–Senior Scientist (Vegetable Science) at Nalanda College of Horticulture, Noorsarai, Nalanda (Bihar Agricultural University, Sabour, Bhagalpur), Bihar, India. He obtained his master's degree in Vegetable Science from Acharya Narendra Deva University of Agriculture and Technology, Kumarganj, Ayodhya, UP, India, in 2006, and doctorate in Horticulture from Banaras Hindu University, Varanasi, UP, India, in 2010. From 2010 to 2012, Dr. Solankey served as a Senior Research Fellow and Research Associate at ICAR-Indian Institute of Vegetable Research (IIVR), Varanasi, Uttar Pradesh. He joined Bihar Agricultural University, Sabour, as an Assistant Professor–cum–Junior Scientist (Vegetable Science) on September 17, 2012. He has been actively involved in teaching, research, extension and training activities as per the mandate of university. Thereafter, he was deputed at Dr. Kalam Agricultural College, Kishanganj, under the umbrella of BAU, Sabour, on September 7, 2015 and acted as Nodal Officer of the newly established Horticulture Research Centre, Kishanganj, Bihar, from May 2021 to June 2022. He has handled four research projects on vegetable crops as PI/Co-PI with the objective of biotic and abiotic stress management as well as quality improvement in solanaceous vegetables and okra at BAU, Sabour. He has contributed to the development of two brinjal varieties through CVRC during 2024, namely "Sabour Sadabahar" and "Sabour Krishnakali" and one technology on "Management of Sucking Pests in Okra." In 2024, he collaboratively registered a patent for the "Smart Horticultural Spreader Machine" with the Government of India. At present, he is handling two seed-money research projects entitled "Improvement of Okra Genotypes for YVMV Tolerance" and "Collection, Evaluation and Assessment of the Feasibility of Promising Vegetables (Garden Pea and Cowpea) for Bihar," and he is serving as Co-P.I. in one CRA program on potato, as well as being involved in several other research projects at NCOH, Noorsarai. Dr. Solankey has supervised five M.Sc. students and served as a member of advisory committee for ten M.Sc. and three Ph.D. students. He has served as content creator of two postgraduate courses at national level under ICAR-NAHEP Call-3: "Production Technology for Cool Season Vegetable Crops (VSC-501: Unit-IV)" and "Production Technology for Warm Season Vegetable Crops (VSC-502: Unit-II)." He has published 66 research papers, 10 review papers, 2 souvenir papers, 9 edited books, 1 authored book, 59 book chapters, 6 technical bulletins, and 50 popular articles. Dr. Solankey is a life member of professional societies, including Indian Society of Vegetable Science, ICAR-IIVR, Varanasi; Indian Academy of Horticultural Sciences (IAHS), Pusa, New Delhi; International Society for Noni Science (ISNS), Perungudi, Chennai; Society for Scientific Development in Agriculture and Technology, Meerut; and Bihar Horticulture Society, BAU, Sabour, Bihar. He also serves as a reviewer for *The Indian Journal of Agricultural Sciences*, *Scientia Horticulturae*, *Vegetos* and *International Journal of Plant & Soil Science.* In recognition of his contributions, Dr. Solankey was awarded with Best Teacher Award and Best Researcher Award by Bihar Agricultural University in 2016. He received Fellowship of the International Society for Noni Science (ISNS), Perungudi, Chennai, in 2022. Besides these, he has been the recipient of 18 other reputed awards and recognitions.

Contributors

Anupam Adarsh Department of Horticulture (Vegetable Science), Nalanda College of Horticulture, Noorsarai, Nalanda, Bihar Agricultural University, Sabour, Bhagalpur, Bihar, India

Shirin Akhtar Department of Horticulture (Vegetable Science), Bihar Agricultural College, Bihar Agricultural University, Sabour, Bhagalpur, Bihar, India

B. C. Anu Krishi Vigyan Kendra, Turki, Muzaffarpur, Dr. Rajendra Prasad Central Agricultural University, Pusa, Samastipur, Bihar, India

Satyaprakash Barik Horticulture, Department of Agriculture and Allied Sciences, CV Ramon Global University, Bhubaneswar, Odisha, India

Papia Biswas Department of Agronomy, Nalanda College of Horticulture, Noorsarai, Nalanda, Bihar Agricultural University, Sabour, Bhagalpur, Bihar, India

Anukul Chandra Department of Extension Education, Bihar Agricultural College, Bihar Agricultural University, Sabour, Bhagalpur, Bihar, India

Ram Chandra Department of Mycology and Plant Pathology, Institute of Agricultural Sciences, Banaras Hindu University, Varanasi, Uttar Pradesh, India

V. Chandra Department of Horticulture (Fruits & Fruit Tech.), Bihar Agricultural College, Bihar Agricultural University, Sabour, Bhagalpur, Bihar, India

Kevin Christopher Department of Extension Education, Dr. Kalam Agricultural College, Kishanganj, Bihar Agricultural University, Sabour, Bhagalpur, Bihar, India

Debashis Dash Department of Agriculture and Allied Sciences, CV Raman Global University, Bhubaneswar, Odisha, India

Mahendra Prasad Das Department of Vegetable Science, Odisha University of Agriculture and Technology, Bhubaneswar, Odisha, India

Sunanya Das Department of Botany, School of Applied Sciences, Centurion University of Technology and Management, Bhubaneswar, Odisha, India

Rajkumari Asha Devi Department of Horticulture (Vegetable Science), Bihar Agricultural College, Bihar Agricultural University, Sabour, Bhagalpur, Bihar, India

Swaraj Kumar Dutta Department of Agronomy, Dr. Kalam Agricultural College, Kishanganj, Bihar Agricultural University, Sabour, Bhagalpur, Bihar, India

Ayesha Fatima Department of Agronomy, Dr. Kalam Agricultural College, Kishanganj, Bihar Agricultural University, Sabour, Bhagalpur, Bihar, India

Himansu Sekhar Gouda Department of Agriculture and Allied Sciences, CV Raman Global University, Bhubaneswar, Odisha, India

Neeharika Kanth Department of Horticulture, Pt. Deen Dayal Upadhyay College of Horticulture and Forestry, Piprakothi, Dr. Rajendra Prasad Central Agricultural University, Pusa, Samastipur, Bihar, India

Krishna D. K. Department of Extension Education, Advance Centre on Sericulture, Kishanganj, Bihar Agricultural University, Sabour, Bhagalpur, Bihar, India

Ajeet Kumar Department of Soil Science and Agricultural Chemistry, Nalanda College of Horticulture, Noorsarai, Nalanda, Bihar Agricultural University, Sabour, Bhagalpur, Bihar, India

Anand Kumar Department of Soil Science and Agricultural Chemistry, Nalanda College of Horticulture, Noorsarai, Nalanda, Bihar Agricultural University, Sabour, Bhagalpur, Bihar, India

Avinash Kumar Department of Agricultural Economics, Bihar Agricultural College, Bihar Agricultural University, Sabour, Bhagalpur, Bhagalpur, Bihar, India

Alpana Kumari Climate Resilient Agriculture Programme on Potato, Department of Horticulture (Vegetable Science), Nalanda College of Horticulture, Noorsarai, Nalanda, Bihar Agricultural University, Sabour, Bhagalpur, Bihar, India

Dhyanananda Kumari Assistant Professor–cum–Junior Scientist, Department of Horticulture (Fruits & Fruit Technology), Nalanda College of Horticulture, Noorsarai, Nalanda, Bihar Agricultural University, Sabour, Bhagalpur, Bihar, India

Meenakshi Kumari Department of Vegetable Science, Faculty of Agricultural Sciences, SGT University, Gurugram, Haryana, India

Randhir Kumar Department of Horticulture (Vegetable Science), Nalanda College of Horticulture, Noorsarai, Nalanda, Bihar Agricultural University, Sabour, Bhagalpur, Bihar, India

Santosh Kumar Department of Mycology and Plant Pathology, Institute of Agricultural Sciences, Banaras Hindu University, Varanasi, Uttar Pradesh, India

V. Kumar Department of Soil Science and Agriculture Chemistry, Bihar Agricultural College, Bihar Agricultural University, Sabour, Bhagalpur, Bihar, India

Milan Kumar Lal Division of Crop Physiology, Biochemistry and Postharvest Technology, ICAR–Central Potato Research Institute, Shimla, Himachal Pradesh, India

Shubham Maurya Department of Horticulture, Post–Graduate College of Agriculture (PGCA), Dr. Rajendra Prasad Central Agricultural University (RPCAU), Pusa, Samastipur, Bihar, India

Moti Lal Meena Krishi Vigyan Kendra, Turki, Muzaffarpur, Dr. Rajendra Prasad Central Agricultural University, Pusa, Samastipur, Bihar, India

Arindam Nag Department of Extension Education, Dr. Kalam Agricultural College, Kishanganj, Bihar Agricultural University, Sabour, Bhagalpur, Bihar, India

P. Nagarjun Department of Agronomy, Dr. Kalam Agricultural College, Kishanganj, Bihar Agricultural University, Sabour, Bhagalpur, Bihar, India

Hena Parween Department of Soil Science and Agricultural Chemistry, Dr. Kalam Agricultural College, Kishanganj, Bihar Agricultural University, Sabour, Bhagalpur, Bihar, India

C. Patel Department of Soil Science and Agriculture Chemistry, Bihar Agricultural College, Bihar Agricultural University, Sabour, Bhagalpur, Bihar, India

Pushpam Patel Krishi Vigyan Kendra, Barh, Patna, Bihar Agricultural University, Sabour, Bhagalpur, Bihar, India

V. K. Patel Department of Horticulture, Sri. Durga Ji PG College, Chandeshwar, Azamgarh, Uttar Pradesh, India

Menka Pathak Department of Vegetable Science, Odisha University of Agriculture and Technology, Bhubaneswar, Odisha, India

Madhuri Pradhan Department of Agriculture and Allied Sciences, CV Raman Global University, Bhubaneswar, Odisha, India

Kartik Pramanik Department of Vegetable Science, MS Swaminathan School of Agriculture, Centurion University of Technology and Management, Parelakhemundi, Odisha, India

K. Prasad Department of Horticulture, Tirhut College of Agriculture (TCA), Dholi, Dr. Rajendra Prasad Central Agricultural University, Pusa, Samastipur, Bihar, India

Pankaj Kumar Ray Krishi Vigyan Kendra, Saharsha, Bihar Agricultural University, Sabour, Bhagalpur, Bihar, India

Rina Roy Department of Agronomy, Dr. Kalam Agricultural College, Kishanganj, Bihar Agricultural University, Sabour, Bhagalpur, Bihar, India

Bholanath Saha Department of Soil Science, Dr. Kalam Agricultural College, Kishanganj, Bihar Agricultural University, Sabour, Bhagalpur, Bihar, India

Jyoti Prakash Sahoo Agricultural Biotechnology, Department of Agriculture and Allied Sciences, CV Raman Global University, Bhubaneswar, Odisha, India

Nandini Sahu Agricultural Entomology, Department of Agriculture and Allied Sciences, CV Raman Global University, Bhubaneswar, Odisha, India

Neetu Saroj Department of Horticulture, Post–Graduate College of Agriculture (PGCA), Dr. Rajendra Prasad Central Agricultural University (RPCAU), Pusa, Samastipur, Bihar, India

A. K. Singh Bihar Agricultural University, Sabour, Bhagalpur, Bihar, India

Aniruddh Pratap Singh Department of Horticulture (Vegetable Science), Nalanda College of Horticulture, Noorsarai, Nalanda, Bihar Agricultural University, Sabour, Bhagalpur, Bihar, India

Hemant Kumar Singh School of Agriculture, Central University of South Bihar, Gaya, Bihar, India

Kamlesh Kumar Singh Department of Soil Science, Tirhut College of Agriculture (TCA), Dholi, Dr. Rajendra Prasad Central Agricultural University, Pusa, Samastipur, Bihar, India

Mahendra Singh Department of Soil Science and Agriculture Chemistry, Acharya Narendra Deva University of Agriculture & Technology, Kumarganj, Ayodhya, Uttar Pradesh, India

Vinit Pratap Singh College of Agriculture Campus, Kotwa, Azamgarh, ANDUA&T, Kumarganj, Ayodhya, Uttar Pradesh, India

Aditya Sinha Department of Extension Education, Bihar Agricultural College, Bihar Agricultural University, Sabour, Bhagalpur, Sabour, Bihar, India

Anuradha Sinha Krishi Vigyan Kendra, Aurangabad, Bihar Agricultural University, Sabour, Bhagalpur, Bihar, India

Shashank Shekhar Solankey Department of Horticulture (Vegetable Science), Nalanda College of Horticulture, Noorsarai, Nalanda, Bihar Agricultural University, Sabour, Bhagalpur, Bihar, India

Swapnil Department of Genetics and Plant Breeding, Centurion University of Technology and Management, Paralakhemundi, Odisha, India

Barsa Tripathy Department of Vegetable Science, Institute of Agricultural Sciences, Siksha 'O' Anusandhan Deemed to be University, Bhubaneswar, Odisha, India

D. K. Verma Department of Soil Science and Agriculture Chemistry, Bihar Agricultural College, Bihar Agricultural University, Sabour, Bhagalpur, Bihar, India

S. Vinod Kumar Department of Mycology and Plant Pathology, Institute of Agricultural Sciences, Banaras Hindu University, Varanasi, Uttar Pradesh, India

Abbreviations

%	Percent
@	at the rate of
°C	Degree Celsius
°F	Degree Fahrenheit
2,4,5-T	2,4,5-Trichlorophenoxy acetic acid
2,4-D	2-4 Dichlorophenoxy acetic acid
a.i.	Active ingredient
ABA	Abscisic acid
ADP	Adenosine diphosphate
AFLP	Amplified fragment length polymorphism
Al	Aluminum
ALS	Acetolactate synthase
APX	Ascorbate peroxidase
ARC	Apical rooted cutting
AS	Ammonium sulfate
ATP	Adenosine triphosphate
B	Boron
BMPs	Best management practices
Bt	*Bacillus thuringiensis*
C	Carbon
Ca	Calcium
CAASPS	Computer Aided Advisory System for Potato Crop Scheduling
CAN	Calcium ammonium nitrate
CAS	Controlled atmosphere storage
CAT	Catalase
CH_4	Methane
CI	Chlorophyll index
CIP	International Potato Center
CIS	Cold-induced sweetening
cm	Centimeter
CNNs	Convolutional neural networks

CO_2	Carbon dioxide
CPB	Colorado potato beetle
CPE	Cumulative pan evaporation
CPRI	Central Potato Research Institute, Kufri, Shimla
CQAs	Caffeoyl quinic acids
CQAs	Caffeoylquinic acids
CRF	Controlled release formulations
CRISPR	Clustered regularly interspaced short palindromic repeats
D/N	Day/night
DAP	Days after planting
DMN	1,4-Dimethyl naphthalene
DNA	Deoxyribonucleic acid
DRIS	Diagnosis and recommendation-integrated system
DSBs	Double-strand breaks
DTPA	Diethylenetriaminepentaacetic acid
EAPR	European Association for Potato Research
EC	Electrical conductivity
ELISA	Enzyme-linked immunosorbent assay
EPNs	Entomopathogenic nematodes
FAO	Food and Agriculture Organization, Rome, Italy
Fe	Iron
FPOs	Farmer producer organizations
FYM	Farmyard manure
GAME4	Glycoalkaloid metabolism 4
GAP	Good Agricultural Practices
GDP	Gross Domestic Product
GE	Genetic engineering
GE	Genome editing
GEBVs	Genomic estimated breeding values
GHI	Global Hunger Index
GI	Genetic improvement
GIS	Geographic information system
GM	Genetically modified
GMOs	Genetically modified organisms
GPX	Guaiacol peroxidase
GS	Genomic selection
GWAS	Genome-wide association studies
H_2O_2	Hydrogen peroxide
H_2PO_4	Dihydrogen phosphate
H_2SO_4	Sulfuric acid
HCl	Hydrogen chloride
IAA	Indole acetic acid
IARI	Indian Agricultural Research Institute, New Delhi
ICAR	Indian Council of Agricultural Research, New Delhi
IDM	Integrated Disease Management

INM	Integrated Nutrient Management
IPCC	Intergovernmental Panel on Climate Change, Geneva, Switzerland
IPM	Integrated Pest Management
IT	Information technology
IWM	Integrated weed management
IWUE	Irrigation water use efficiency
$K_2Cr_2O_7$	Potassium dichromate
K_2SO_4	Potassium sulfate
K_3O_3P	Potassium phosphite
KCl	Potassium chloride
kg	Kilogram
$KMnO_4$	Potassium Permanganate
LoQ	Level of quantification
MAB	Marker-assisted breeding
MAP	Modified atmosphere packaging
MAS	Marker-assisted selection
MDA	Malondialdehyde
Mg	Magnesium
MLAs	Machine learning algorithms
mRNA	Messenger ribonucleic acid
N	Nitrogen
N_2O	Nitrous oxide
Na	Sodium
NAA	Naphthalene acetic acid
NAAS	National Academy of Agricultural Sciences
NAB	National Accreditation Body
NADPH	Nicotinamide Adenine Dinucleotide Phosphate
$NaHCO_3$	Sodium bicarbonate
NaOH	Sodium hydroxide
NBTs	New breeding techniques
NCAR	Net carbon assimilation rate
NDVI	Normalized Difference Vegetation Index
NH_4^+	Ammonium ion
NH_4F	Ammonium fluoride
NH_4HCO_3	Ammonium bicarbonate
NO_3-	Nitrate
NOTS	Naturally occurring toxic substances
NPK	Nitrogen, phosphorus, and potassium
NPOP	National Programme for Organic Production
NSOP	National Standards for Organic Production
O_2	Oxygen
O_3	Ozone
PAR	Photosynthetically active radiation
PATs	Precision agricultural technologies
PCR	Polymerase chain reaction

PGPR	Plant growth-promoting rhizobacteria
PGPYC	Potato Growing Period and Yield Calculator
PGSD	Potato Growing Season Descriptor
pH	Potential of hydrogen
PLRV	Potato leafroll virus
POD	Peroxidase
PPF	Photosynthetic photon flux
ppm	Parts per million
PPM	Potato Pest Manager
PTGS	Post-transcriptional gene silencing
PVA	Potato virus A
PVM	Potato virus M
PVX	Potato virus X
PVY	Potato virus Y
PWM	Potato Weed Manager
QTL	Quantitative trait locus
QUEFT	Quantitative Evaluation of the Fertility of Tropical Soils
RAPD	Random amplified polymorphic DNA
RDF	Recommended dose of fertilizers
RFID	Radio frequency identification
RFLP	Restriction fragment length polymorphism
RNA	Ribonucleic acid
RNAi	RNA interference
ROS	Reactive oxygen species
RS	Remote sensing
RTNM	Real-time nitrogen management
RuBisCO	Ribulose bisphosphate carboxylase
RWC	Relative water content
S	Sulfur
S.L.	Soluble (liquid) concentrate
SC	Suspension concentrate
SDA	Steroidal glycoalkaloid
SDI	Subsurface drip irrigation
SGAs	Steroidal glycoalkaloids
SNC	Single node cutting
SOD	Superoxide dismutase
SPT	Seed plot technique
SSCP	Single-strand conformation polymorphism
SSNM	Site-Specific Nutrient Management
SSR	Simple Sequence Repeat
STCR	Soil test crop response
STV	Soil test values
SVMs	Support vector machines
TAG	Triacylglycerol
TALENs	Transcription activator-like effector nucleases

TPS	True potato seeds
TYE	Targeted yield equations
VIGS	Virus-induced gene silencing
VPD	Vapor pressure deficit
VRAT	Variable rate application technology
WHO	World Health Organization, Geneva, Switzerland
WUE	Water usage efficiency
ZFNs	Zinc-finger nucleases
Zn	Zinc

The Present Advances in Research for Potato Production: Future Trends and Goals

Shashank Shekhar Solankey, Alpana Kumari, Shirin Akhtar, Anupam Adarsh, Anuradha Sinha, and Meenakshi Kumari

Abstract An overview of recently developed technological options for raising crop yields and enhancing yield stability in the production of potatoes is given in this chapter. The detailed discussion of biotic and abiotic restrictions under the present climate change scenario on crop production includes technology like seed plot techniques for controlling viral disease to address these constraints. This chapter provides a thorough rundown of the technology available for producing potatoes. It covers the various technologies and tactics aimed at yield enhancement, yield stabilisation, and quality enrichment, with a focus on improved varieties. Scientists working on potato crops are addressing these challenges by developing advanced technologies such as genome sequencing, high-tech seed production methods like aeroponics and ARC, integrated disease and pest management, optimized plant nutrition and water management, and the use of information technology for monitoring soil health. Additionally, they are advancing other innovative research and development technologies to improve potato production.

Keywords Advanced technologies · Aeroponics · ARC · Climate change · Yield · Quality

S. S. Solankey (✉) · A. Kumari · A. Adarsh
Department of Horticulture (Vegetable Science), Nalanda College of Horticulture, Noorsarai, Nalanda, Bihar Agricultural University, Sabour, Bhagalpur, Bihar, India

S. Akhtar
Department of Horticulture (Vegetable Science), Bihar Agricultural College, Bihar Agricultural University, Sabour, Bhagalpur, Bihar, India

A. Sinha
Krishi Vigyan Kendra, Aurangabad, Bihar Agricultural University, Sabour, Bhagalpur, Bihar, India

M. Kumari
Department of Vegetable Science, Faculty of Agricultural Sciences, SGT University, Gurugram, Haryana, India

S. S. Solankey (ed.), *Advances in Research on Potato Production*,
Advances in Olericulture, https://doi.org/10.1007/978-3-031-82710-5_1

1 Introduction

Potato, botanically known as *Solanum tuberosum* ssp. *tuberosum*, has much more characteristics of an ideal food crop. When compared with other crops, its yields of edible energy, protein and dry matter are far higher per unit of land or time. Potato tubers are a highly healthy and wholesome source of food. The potato tubers contain carbohydrates, proteins, minerals, vitamin C, several complex vitamins and high-quality dietary fibre (Jaiswal et al. 2023; Kumar et al. 2023). Potato is one of the most demanding and consumed vegetable in India. Moreover, it is a primary food source in various European and Latin American nations. Contrarily, it continues to be utilised as a nutritional supplement in our nation. In India, the average individual consumes a meagre 15 kg of fresh potatoes per year, but in Denmark and The Netherlands, the figures are far higher at 193 and 126 kg per year, respectively. Hence, there exists a significant opportunity to augment both the consumption and production of potatoes within the nation. Most of the earlier varieties were either developed at ICAR-CPRI, Shimla or evolved from selections from improved varieties such as Satha, Gola, Up-to-date, which were replaced by new high yielding and disease resistant varieties. Potato flowers are hermaphrodite and self-pollinated. *S. tuberosum* ssp. *tuberosum* produces tubers on both long and short days in tropical regions or at lower altitudes between 500 and 2000 meters above sea level. In contrast, *S. tuberosum* ssp. *andigena* produces tubers only at high altitudes (above 2000 m) and under short-day conditions.

As per the FAOSTAT database of 2022–2023, around 375 million tonnes of potatoes were produced globally in 2022; China contributing 95.5 million tonnes followed by India with 56 million tonnes of potato production are the top producers in the world.

2 The History of Potato in India

Potato cultivation in the Indian subcontinent began around four centuries ago, introduced by European seafarers. The domestication of this New World crop began approximately 8000 years ago in the Andes Mountains of South America, specifically in the region shared by Bolivia and Peru, where ancient societies started cultivating it as a staple food. The Spanish introduced the potato to Spain from Peru in 1565 and Sir Francis Drake is believed to have brought it to England around 1586. The introduction of potatoes to India is most probable to have occurred during the late sixteenth to early seventeenth century by Portuguese traders. The earliest recorded reference to potatoes in India can be found in Edward Terry's description of a grand feast organised by Abdul Hassan Asaf Khan, the Governor of Punjab under Mughal Emperor Jehangir. This banquet, held in 1615 at Ajmer, was held in honouring Sir Thomas Roe, the British Ambassador. Although the potato was initially introduced to the Southern coast of India, it is unlikely that the local farmers planted it. Instead, it swiftly became entrenched in the gardens of Surat and Karnataka. Over the next 250 years, agricultural and horticultural groups, as well as botanical gardens, actively worked to encourage the cultivation of this plant across India. The Royal Agri-Horticultural Society, established in 1820 in Calcutta by the renowned missionary scholar William Carrey, had a significant role in promoting potato production in eastern India. Captain Trichmond cultivated potatoes in 1832 using a variety that was brought from England and distributed the Royal Agri-Horticultural Society. It is noted the size and weight of the England potatoes were far better than those of the regional forms. Thc first cultivation of these new English potato types took place in hilly locations such as Naini Tal and Shillong. Meanwhile, some military officials displayed a strong enthusiasm for expanding and establishing potato farming in the hilly regions of India. The cultivation of this crop in the hills located to the north of Dehradun was initially initiated by Major Young and then strengthened by Captain Townsend. Captain Mundi, a passionate military officer, showed great enthusiasm in introducing potato cultivation to the Simla highlands by 1828. In 1830, Mr. Sullivan brought potatoes to the Nilgiri hills. The Government Botanic Garden, established in 1848 in Ootacamund, expanded potato cultivation across the Nilgiri hills. The imported European potato types thrived in the hills due to the favourable and mild climate. By the year 1840, potato cultivation had become firmly entrenched in both plains and hills of the Indian sub-continent. Initially, at the introduction of the potato, the orthodox Brahmins expressed their disapproval due to its absence as a recognised food item in the Hindu Puranas. The prejudice gradually dissipated, and potatoes were included as a prominent vegetable supplement, primarily due to their mild flavour, which easily complements other cuisines. During fasting days, Hindus in certain regions of India now consume significant amounts of potatoes, as they are considered permissible root crops to be eaten during fasting. By 1941, potato cultivation in the country was limited, with the entire Indian sub-continent contributing less than 1% of global potato acreage and production. The trend of potato production is expected to continue for the next 100 years. The main reason for this was the absence of locally adapted potato varieties and technologies suitable for cultivation in sub-tropical climates. An organised research programme on potatoes was launched on 1st April 1935 (Fig. 1). This initiative established three breeding and seed production stations, two at Kufri, Shimla (both located in the Shimla hills) and one at Bhowali (located in the Kumaon hills), all under the Indian Council of Agricultural Research, New Delhi. With the help of Sir Herbert Steward, the Government of India's Agriculture Advisor, and Sir Pheroz M. Kharegat, the Secretary of the Ministry of Agriculture, a proposal to establish the CPRI was formulated in 1945. Dr Benjamin Peary Pal, Dr Srinivasa Ramanujam, Dr Pushkarnath and Dr Raghunath Sahai Vasudeva played a key role in formulating the strategy and establishing the institute. In order to oversee the project's execution, Dr S. Ramanujam, Second Economic Botanist at IARI, was given the position of Officer on Special Duty in 1946.

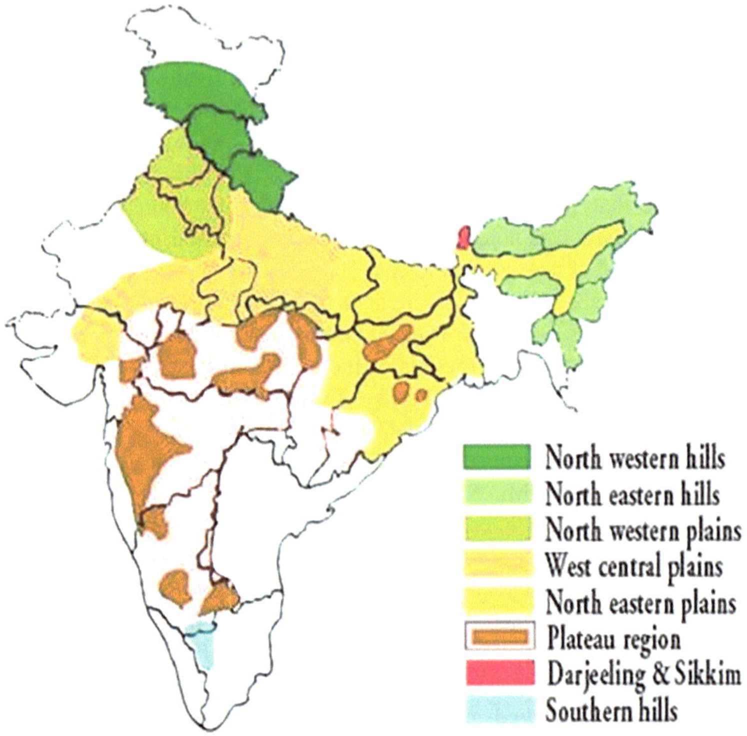

Fig. 1 Potato growing regions in India

3 Advances in Research for Potato

The International Potato Centre (CIP) in Lima, Peru, pursues two distinct potato production-related goals through technological advancements. The primary goals are to reduce negative impacts on climate change and public health while simultaneously increasing or maintaining the production of the potato crop in challenging circumstances without the need for expensive outside resources. Moreover, technical developments strive for quality improvements linked to market value or consumption.

In recent decades, aeroponics as well as ARC technologies, seed plot technologies, development of new varieties and improved production system are the four pillars of potato production system. Under crop improvement practices, a variety of new technologies are being used including genetic modification of potatoes. Technological developments aimed at improving planting material availability, quality and productivity that are referred to as seed technologies. The aeroponics method involves hanging plants in large containers and providing them with spraying of water and nutrients solutions to help them flourish. A thorough understanding of advanced potato production systems is necessary in present days for gaining higher productivity.

(i) *Use of Drones for Transforming Potato Farming*

Advanced crop monitoring and management system includes invisible details captured by drones fitted with multispectral sensors and high-resolution cameras soar above potato fields. By showing problems like nutrient shortages, disease outbreaks and water stress, these photos offer vital information of the plant health. Irrigation optimisation drones identify varying levels of soil moisture across the potato field. This information is very useful for effective water management and enabling farmers to apply the right amount of water at the right time. By preventing over- or under-watering, drones help in conserving water resources, an increasingly important factor in sustainable agriculture, especially in regions facing water scarcity. Besides these, drones effectively identify the pest and disease incidence at early stages and allow for timely and targeted management.

(ii) *Changing the Yield Frontier and Increasing Yield Stability*

The potato production can be enhanced by advancing the yield frontier and optimising yield stability technology to achieve higher yields. Adjusting the genetic makeup of potato varieties to the length of the photoperiods can contribute to achieving improved yield potential in potato production. Generally, the duration of sunlight has a significant impact on tuberisation, which is typically stimulated by shorter periods of daylight. There is a short crucial photoperiod for the tuberisation process in *Solanum tuberosum* spp. *andigena*. Conversely, *S. tuberosum* spp. *tuberosum* exhibit a longer critical photoperiod while some genotypes show neutral response against photoperiodism. Genes related to photoperiod show variation within and between gene pools. Clones have been produced from *S. andigena* populations that can grow tubers under long-day conditions, akin to *S. tuberosum*, by using a recurrent selection strategy (Luo 2021). Flowers are hermaphrodite and self-pollinated.

4 Advances in Crop Improvement of Potato

The implementation of crop improvement programmes has led to increased productivity in several crops through the development and selection of superior-quality hybrids that surpass their parental lines. Heterosis refers to the enhanced performance of hybrids (F_1s) compared to their parent lines, which may manifest as increased biomass, yield, growth rate, reproduction rate or tolerance to biotic and abiotic stresses. In context to maximising adaptation and yield to maintain produce quality, the main breeding goals are to increase heat and drought tolerance, resistance against viruses and late blight, mitigate gradual degeneration and optimise the response to photoperiod (Moldovan et al. 2011).

(A) *Genetic Improvement for Abiotic Stress Tolerance*

Drought is the primary abiotic factor that significantly impacts potato productivity. The potato breeding plan entails the assessment of several biological and physical parameters. Measuring changes in the timing of growth initiation among clones is crucial to avoid confusing drought tolerance with drought susceptibility. Relative leaf water content quantifies variations in leaf tissue hydration and serves as a gauge of the plant's ability to sustain its water levels under moisture stress condition. Leaf area measurement is also important because of its direct correlation with photosynthesis and transpiration. Water use efficiency refers to the amount of water required by a plant to produce a specific amount of biomass or yield. Plants that have a high water use efficiency are preferred for breeding drought-tolerant varieties. However, there are other features that contribute to water usage efficiency, which include transpiration efficiency, growth regulation, carbon allocation and gas exchange efficiency (Tournex et al. 2003). The production of plants relies on the effectiveness of photosynthesis. One way to assess the state of photosynthesis is to measure the fluorescence of chlorophyll. During drought conditions, this approach aids in the identification of many genotypes with differing levels of photosynthetic efficiency (Tournex et al. 2003). The drought-tolerant wild species, such as *S. stenotonum, S. demissum, S. acaule*, *S. ajanhuiri, S. xjuzepczukii* and *S. curtilobum*, and varieties like Alpha, Bintje and Kufri Sheetman are widely used as a prominent donor for development of the moisture stress tolerant varieties of potatoes (Kumar et al. 2012). Besides these, recently few new potato varieties against moisture stress have been released by ICAR-CPRI, Shimla, such as Kufri Thar-1, Kuri Thar-2, Kufri Thar-3 and Kufri Daksh.

Temperature is another significant abiotic factor that influences the stability of potato yield and production. Elevated temperature impacts the rates of photosynthesis and respiration, causing a decrease in the former and an increase in the latter. Temperatures over 20 °C lead to a decrease of around 25% in the photosynthetic rate. The ability to withstand heat stress may be more related to variations in respiration rather than photosynthesis (Levy and Veilleux 2007). Tuber growth and tuber initiation are hindered when the temperature exceeds 25 °C, resulting in delays in the process of tuberisation. Elevations in either diurnal or midnight temperature over the optimum range (18–20 °C) result in diminished tuber production, with elevated nocturnal temperature having a detrimental effect on tuber enlargement and the buildup of dry matter. Elevated temperatures can induce physiological disorders in plants, including irregular form, premature sprouting, cracking and increased formulation of glycoalkaloids (α-solanine, α-chaconine) in tubers. This might result in the production of bitter tubers that may be hazardous for health (Levy and Veilleux 2007).

(B) *Improvement for Biotic Stresses in Potato*

Potato virus Y (PVY) is the largest group of plant viruses, with more than 111 identified and 86 tentative species across 30 plant families. It is distributed globally, with a particularly high prevalence in temperate and subtropical regions where potatoes are extensively cultivated (Mostafa and Hossain 2024). The process of breeding for late blight resistance involves selecting traits that either entirely inhibit the establishment of the pathogen or decrease the rate at which the disease develops in the field. The time interval before infection occurs, the rate at which disease lesions expand and the ability of the pathogen to release spores or reproduce on the host plants, all contribute to the resistance that reduces the severity of the disease. The potato germplasm has varied features that can be integrated into new varieties. These varieties can then be used in management approaches that need minimal or no use of fungicides to control the diseases.

Among viruses, potato leafroll virus (PLRV), potato virus X (PVX), potato virus Y (PVY), potato virus M (PVM) and potato virus A (PVA) are the most important viruses in terms of their prevalence and economic impact. Chlorosis, mosaic patterns, upright growth, curling of the leaves, changes in the leaf colour and appearance and necrotic lesions on the leaves' surface and stems are the signs of vital viral infection in potato crop. Moreover, infections may have a direct effect on tubers' size and numbers. Plant virus infections reduce plant vigour, tuber quality as well as yields, even though they rarely cause plant mortality (Stevenson et al. 2001). Viral infections have been shown to have a wide range of effects on potato yield, with confirmed decreases of up to 90% (Valkonen et al. 2017). Climatic factors such as temperature, CO_2 concentration, humidity and precipitation are key factors in the epidemiology of PVY viruses, influencing their transmission, replication and spread (Mostafa and Hossain 2024). The wild species *S. acaule, S. berthaultii, S. tuberosum* subsp. *Andigena* are resistant sources of PVX, while the species (Table 1) *S. phureja, S. demissum, S. stoloniferum* are resistant donors for PVY; moreover, the species *S. acaule, S. demissum, S. tuberosum* subsp. *andigena* are resistant sources of PLRV (Hazra and Som 2015).

5 Enhancement of the Potato Crop Through Genetics and Advanced Clones and Varieties

The International Potato Center (CIP) is crucial for producing and supplying innovative potato varieties and clones for the tropical, subtropical and temperate agroecologies of poor countries, claim Thiele et al. (2008). In order to create candidate varieties that farmers can easily adopt to increase productivity and production and seize opportunities for income generation once they are released as varieties in developing countries, the CIP's potato breeding programme aims to produce improved populations and clones that are resistant or tolerant to biotic and abiotic stresses.

(A) *Gastelo Experiment*

The International Potato Center (CIP) has utilised a diverse range of genetic resources, such as wild species, landrace and genetically modified genotypes, to

Table 1 Wild species of potatoes with various important genes

Source of genes	Wild species
Viruses—PVX	*S. acaule, S. berthaultii, S. tuberosum* subsp. *andigena*
PVY	*S. phureja, S. demissum, S. stoloniferum*
PLRV	*S. acaule, S. demissum, S. tuberosum* subsp. *andigena*
Late blight—Vertical resistance	*S. demissum, S. verrucosum, S. stoloniferum*
Late blight—Horizontal resistance	*S. berthaultii, S. chacoense, S. microdontum, S. vernei*
Wart	*S. acaule, S. berthaultii.*
Common scab	*S. chacoense, S. tuberosum* ssp. *andigena*
Bacterial wilt	*S. Chacoense, S. microdontum,*
Cyst nematodes	*S. tuberosum* subsp. *andigena, S. berthaultii. S. vernei*
Root-knot nematode	*S. spegazzinii*
Aphids	*S. berthaultii.*
Frost	*S. acaule, S. ajanhuiri*
Heat tolerance	*S. chacoense, S. commersonii*
High protein content	*S. phureja*

Source: Hazra and Som (2015)

create enhanced populations that are well-suited to the challenging climatic conditions. So far, these efforts have led to the creation of two advanced populations: (i) Population B, which is resistant to late blight and adapted to highland-tropical conditions, and (ii) Population LTVR, which is resistant to viruses and adapted to lowland sub-tropical conditions (Fig. 2).

Population is being enhanced to possess strong resistance to late blight as well as desirable traits such as high tuber yield, quality suitable for both table and industrial use and the ability to adapt to various environments. Moreover, it is being developed to tolerate other stresses caused by living organisms or environmental factors that are commonly found at testing sites. Population B is characterised by enhanced adaptation to warm climates, increased resistance against late blight and medium maturity, which allows it to grow for 90 days even under short day length conditions. This population is referred to as LBHT (late blight and heat tolerance). Clone classifications are divided into two groups. The first group is relatively early, heat tolerant, resistant to late blight and PVY. These clones would be ideally adapted in Asia and Sub-Saharan Africa. The second group is very beneficial in highland areas because of their considerable resistance to late blight and relatively early maturity, even though they do not possess PVY resistance.

The main characteristic of the LTVR population that makes it notable is its resistance to the three main viral infections that harm potatoes: PVY, PVX and PLRV. Moreover, it demonstrates tolerance to warm, dry conditions as well as early tuberisation on short days and mid-maturity in long days. While the incorporation

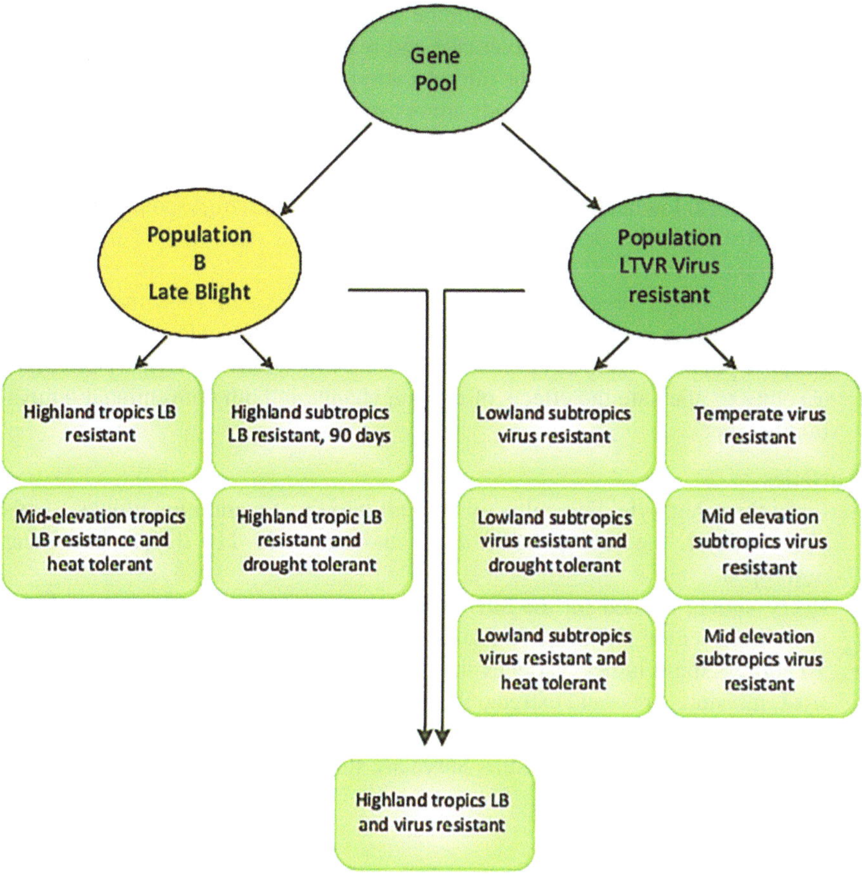

Fig. 2 Advanced populations with specific adaptation to the most important agroecologies of the tropic. (Gastelo et al. 2013)

of late blight resistance has been more emphasised in recent years, the population's early maturity also provides some level of protection against late blight, allowing it to escape its detrimental effects to a certain extent.

(B) *Advanced Clones and Varieties*

From the hot -arid coast to the cold - humid mountains and everywhere in between the experimental site of CIP's headquarters, Peru, based on the requirements of various potato agro-ecologies, the advanced clones that CIP nurseries offer incorporate unique tuber qualities. To facilitate cross-border exchange and assessment of CIP's advanced bred clones, 12 sets of clones are available as nurseries. Agronomic features, biotic and abiotic constraints, tolerance and resistance to these conditions and trait combinations required for productivity in a range of agro-ecological environments tropical, subtropical and temperate regions at different elevations are all provided by the nurseries (Thiele et al. 2008).

- Tropical highland conditions combine drought tolerance with resilience to viruses and late blight. Heat tolerance and resistance to late blight are common traits of the clones grown in the mid-elevation of tropics.
- Clones from subtropical highlands are resistant to late blight and have a short to medium growth cycle. The resistance of clones to PVY and PVX is high in subtropical and temperate regions in mid-elevation regions.
- In subtropical lowlands, viral tolerance is combined with high temperature tolerance. Although the desired cultivar can withstand heat, it is vulnerable to late blight.
- When late blight pressure is great, advanced clones can yield more than 20 tonnes of tubers per hectare without the need for any fungicide protection.
- Because potato cultivation is spreading into new locations, all clones possess the capacity to alleviate the effects of climate change in both conventional cultivation practices.

Varieties

- Forty-three high-yielding cultivars that fall into three maturation groups—early (70–80 days), medium (90–100 days) and late (100–110 days)—have been developed.
- Every variety created for the hills and the most recent medium-duration types introduced for the plains are resistant to late blight.
- To address these issues, varieties that are resistant or tolerant to warts, viruses, frost, moisture stress and extreme thermoperiods have also been bred.
- The varieties Kufri Himsona, Kufri Lauvkar, Kufri Chipsona-1, 2, 3, Kufri Frysona, Kufri Sangam and Kufri Surya are suitable for processing.

6 Production of Potato Under Aeroponic System

Aeroponics is a technique that allows plants to grow in a completely humid environment without the use of soil. Plants do not require water, soil or sunlight to flourish. In a closed box, nutrient solution is sprayed on the roots of plants and mini-tubers can be picked 10–15 times at regular intervals. The discovery and eradication of viruses usually found in potato tubers as well as rodent multiplication of four generations of disease-free tubers are the foundations of this unique healthy seed production technique. At present aeroponics is one of the most rapid methods of propagation useful in seed potato production (Buckseth et al. 2016).

Benefit

- Forty to fifty mini-tubers are produced by each plant and plucked at every 5 days intervals.
- It is also easy to grow them uniformly, and there is no possibility of occurrence of soil-borne diseases because plants grow in the air under aeroponic system.
- Harvesting, handling and storage of mini-tubers are easy and they can be comfortably transported from one place to another.

- In aeroponic system, plants are grown on 20 mm on the upper surface of the specially made boxes. Round holes are made at prescribed or required distance and one tissue cultured plant is planted in each hole. The roots are kept in the boxes to protect them from the sunlight.
- One time investment in this technology is required and this setup lasts for a long time.

Methods

There are 13 macro- and micro-nutrients in the nutrient solution. The pH of the nutrients is around 6.0 and the top section is still exposed to sunlight. The nutrient solution continues to circulate. Potato mini-tubers begin to form after the establishment of lengthy roots. Potato mini-tubers weighing three to four grammes are harvested for the first time after 45–50 days after transplanting. The nutrient solution is replaced on a regular basis to avoid any kind of infection. When compared to traditional approaches, the aeroponic technology can produce 10–12 times the number of seed tubers per plant and is a brilliant idea for producing high-quality disease seed potatoes in greater quantities in less time in high populous country like India.

7 ARCs—Apical Rooted Cuttings

ARCs are transplants produced in a screenhouse/nethouse from tissue culture plantlets. Instead of allowing these plantlets to mature and form mini-tubers, cuttings are taken from them. Once rooted, these cuttings are transplanted into the field to produce seed tubers. Apical rooted cuttings offer an alternative to mini-tubers in potato seed production systems. While mini-tubers provide greater flexibility as being storable until planting in a small cold storage and easier to transport, however the the apical rooted cuttings surpass mini-tubers in terms of productivity for generating seed tubers. Moreover, cuttings become profitable after just two seasons of multiplication, whereas mini-tubers typically require 3–4 seasons to achieve profitability. ARCs are produced in a screenhouse from tissue culture plantlets. Instead of allowing these plantlets to mature and produce mini-tubers, cuttings are taken from them. Once the cuttings are rooted, they are transplanted into the field to generate seed tubers. ARCs offer an alternative to mini-tubers in the present potato seed production systems. While, mini-tubers are more versatile because they can be stored and easily transported, the productivity of cuttings for seed tuber production outpaces that of mini-tubers, becoming profitable after two seasons of multiplication, whereas mini-tubers require 3–4 seasons to achieve profitability (Reddy et al. 2021).

(A) *Seed Plot Technique*

The unique approach of producing healthy disease-free seed tubers in India's subtropical plains is known as "Seed Plot Technique". It is the third most significant technological advancement that has transformed potato farming in India (Singh and Buckseth 2024).

The planting material of potatoes deteriorates with time because of viral infections and other degenerating factors that are inherited from previous generations because potatoes are clonally propagated crop. The utilisation of substandard planting material diminishes both tuber quality and yields, contributing to yield reductions on around 5 million hectares of potato. The foundational idea of the Seed Plot Technique was developed by Dr Puskarnath in 1967. In this technique the seed crop cultivation should have a low aphid period (20 aphids/100 compound leaves) of at least 70–75 days. The requirements for the success of this technology were:

(i) Serious illnesses such as warts, cyst nematodes, black scurf, common scab, fungal and bacterial wilts should not be present in the soil.
(ii) There should be a low aphid or aphid-free period of at least 70–75 days following tubers planting. During crop season, the maximum and minimum temperatures should be between 8 and 28 °C
(iii) The used varieties should degenerate slowly. For that irrigation is restricted during mid-December thereafter, crop dehaulming (cutting above-ground foliage) should be done at the end of December, or when the critical threshold of 20 aphids per 100 compound leaves is observed. Two agricultural field inspections are conducted in order to eliminate virus-infected plants and off-type plants.

When compared to traditionally grown seeds in hilly areas, seeds generated during low aphid conditions sprouted more due to good physical condition of tubers after 6–7 months of storage at low temperature (2–3 °C).

The technique was further enhanced in response to the appearance and subsequent buildup of aphids in the Indo-Gangetic plains, spanning from northwestern to northeastern regions. In areas where aphids appear early in the crop growing cycle such as Madhya Pradesh, Gujarat and Maharashtra, the spraying of systemic insecticides has proven successful in reducing vector populations and facilitating the production of high-quality seed tubers. Agrotechnological methods for seed production using the seed plot technique have been improved and standardised over time. Here are the many advanced agro-techniques applied in healthy seed production.

Field Selection Field should be free from perennial weeds, soil-borne disease and insect-pests. Crop rotations of 2–3 years, particularly with cereal crops should be practiced. It is preferable to have a light-textured, sandy loam soil that drains readily and has a pH range of neutral to slightly alkaline, i.e. 7.4–7.8.

The goal of hot season agriculture is achieved by deep tillage, deep ploughing the field at the end of April and leaving it open in May and June with one or two further ploughings.

Green Farming It is useful to minimise the occurrence of insect-pests and diseases to use 45–55 days old green manuring crops like Sunhemp or Dhaincha for at least 1 month prior to planting potatoes during the wet season for optimal decomposition. It also increases the soil fertility and water-holding capacity, which is beneficial to the growth and production of potatoes.

Tillage There is reduction in the amount of fuel, time and money used on cultivation by applying minimal tillage. The optimum method for producing seeds is to combine chemical weed control with low tillage, as it ensures minimal interaction with standing crops.

Seed Plot Isolation To stop viral illnesses from mixing and spreading, 5-m isolation distance is recommended.

Healthy Seed Source Seeds should always be procured from a reliable source, particularly from government organisations, ensuring they should be disease- and pest-free, healthy and true to type. Initially, use of foundation or certified seed tubers should be done and seeds should be replaced every 3–4 years.

Seed Tubers Treatment Treating seed tubers to promote the growth of many strong, healthy sprouts before planting by ensuring uniform stand, early maturity and rapid emergence of the crop should be done. Additionally, it increases tuber number, size and ultimately tuber yield. The seed tubers should be taken out 10 days prior to planting from the cold storage and stored for a full day in the cold store's pre-cooling chamber. Further, the seed tubers are spread thinly on a cemented floor in a well-ventilated room with soft lighting. Sort the tubers that have sprouted hairs, blindness, rotting or are unhealthy appearance. In the field, plant the tubers that have sprouted in the trays.

The ideal day and night temperature for potato planting is from 30–32 °C to 18–20 °C, respectively, and growing period must be free from aphids. In the Northwestern, West-central and Eastern Plains of India, the first, second, third and fourth weeks of October, respectively, are the ideal times to plant the seed crops.

Planting Methodology and Seed Quantity Whole or cut tubers weighing 25–30 g are planted at the centre of the ridge, spaced 15–20 cm apart and 5–7 cm deep, then covered with pulverised soil. The seed rate is 1.5–1.8 tonnes per hectare for round varieties and 2.0–2.5 tonnes per hectare for oval varieties (Swarup 2006).

Seed Treatment The whole tuber should be treated with ethylene chlorohydrin vapour @ 1 l/20 quintals seed in an airtight chamber for 4 h. Then the tubers should be cut in pieces and dipped in 1% solution of thiourea (1 kg/10quitals seed) for 1 h. Finally, the cut pieces are immersed in 1 ppm solution of gibberellic acid (GA_3) for 10 min to break the dormancy of potato tubers. The sequential treatment of tubers is known as *Triple Treatment* and has been developed and recommended by ICAR-CPRI, Shimla.

Manure and Fertilisers Green manure is the best for potato crop. Before planting, the risk of diseases, such as scab, is very high. A fertiliser dose of 180–240 kg N, 60–90 kg P_2O_5 and 85–130 kg K_2O per hectare is recommended for potato crop (Singh and Buckseth 2024). This eliminates the need for expensive nitrogenous fertilisers such as calcium ammonium nitrate (CAN) and ammonium sulphate (AS).

Weed Control and Intercultural Operations Hand weeding and earthing up are the essential intercultural operations that are used to get weed-free crop, keep the soil loose for a healthy tuber development and cover the tubers 20–25 cm deep with the help of loose soil. Two to three days after planting is the best time to use any of the pre-emergence herbicide in the moist soil, such as Metribuzin (0.75 kg/ha), Linuron (0.5 kg/ha), Oxyfluorfen (0.15 kg/ha), Alachlor (1.5 kg/ha) or Isoproturon (0.75 kg/ha).

Water Management Pre-sowing irrigation before land preparation encourages early and uniform sprouting. There should be 5–7 days interval between the first and second irrigations. Heavy irrigation prior to emergence causes anaerobic conditions, resulting in seed tuber rotting, uneven emergence and lower tuber yields. A week following the initial irrigation, a second irrigation is applied. Six to eight irrigations are required, with the last irrigation occurring approximately 10 days before dehaulming in light sandy soils while 15 days in heavy clay soils. Besides furrow irrigation, the potato crop can be irrigated by drip or sprinkler method.

Plant Protection In the seed crop, plant protection against aphids and other virus-transmitting vectors is essential. The use of systemic insecticide Fipronil 5% SC @ 75 g a.i./ha at the time of earthing up controls jassids, leafhoppers and white flies in their early stages of development for up to 30–35 days. Depending on the length of the crop and the level of infestation, two applications of imidacloprid 17.8% S.L. @ 0.002% may be made after aphid development at intervals of 12–15 days.

Cutworm attack is effectively reduced during the early phases of crop development by drenching the ridges with Chlorocyper (Chlorpyriphos 50% + Cypermethrin 5% EC) @ 2.0 l/ha prior to final earthing up.

To avoid early and late blight, a single preventive spray of Mancozeb @ 0.2% is useful; depending on the weather, it may be repeated every 7–14 days intervals. It will also treat *Phoma* blight and other foliar diseases. If there is an extensive and persistent late blight infestation, systemic fungicides like metalaxyl 8% + mancozeb 64% WP @ 2.5 kg/ha or Cymoxanil 8% + Mancozeb 64% WP@ 7 kg/ha can be sprayed and if needed; the treatment can be repeated every 7–10 days intervals.

Crop Inspection and Rouging The potato crop is examined at least twice a season. The first inspection should be done before flowering, and the second inspection will be done at the time of harvesting. The initial rouging is completed 25–30 days following planting before earthing up. Three to four days prior to dehaulming, the third rouging takes place, and the second takes place 40–45 days after planting.

Dehaulming When the aphid population in the plains hits the critical threshold of 20 aphids/100 compound leaves in the nearby area, haulms of the seed crop must be removed between 5 and 15 January. Dehaulming is achieved by either spraying the non-selective herbicide Paraquat at 0.5 kg/ha or by cutting the haulms 10–15 cm above the ground with the help of sickle. In order to shield exposed tubers from direct sunlight, it is best to leave the haulms on the ridges to cover the exposed tubers. Any regeneration of leaves is eliminated after a week of dehaulming because aphid vectors find delicate and succulent leaves more enticing and spreading virus very quickly (Kumar et al. 2007).

Curing and Harvesting Depending on the location, soil type and variety, the best time to harvest potatoes is 75–120 days after planting. After harvesting, the potatoes are sun-dried for 1 day in the field and thereafter, kept in heaps for 10–15 days to cure the skin under shade. The tubers should not be exposed to direct sunlight until they turn green.

Seed Treatment and Grading The potato seed tubers are categorised into four common grades: small (25 g), medium (25–50 g), large (50–100 g) and extra-large (>100 g). For 30 min, seed tubers should be submerged in a boric acid (3%) solution to protect against tuber-borne diseases such as black scurf and common scab. Dip treatment with fungicide such as Emisan-6 @ 0.25%, an organo-mercurial compound for 20 min, is also effective but is regarded toxic and should be avoided. Before bagging and storing the treated tubers, they should be properly dried in the shade.

Packing and Storage By the end of February, the 50 kg bags containing the seed tubers are kept in a cold store at 40 °C. The treated seed sacks are carefully wrapped and labelled to prevent human consumption of poisonous tubers.

Adoption and Impact The technology for producing healthy seed in subtropical plains was primarily responsible for India's remarkable expansion in potato acreage, production and productivity following independence. The seed plot technique alone resulted in an increase of over 62.7% in area and 51.2% increase in yield. By using this method, a robust national seed production system was established to guarantee the superior quality of nucleus and breeders' seed for improved varieties. Numerous farmers were also able to contribute to the production of high-quality potato seeds because of this technology.

The combination of seed plot technology with enhanced plant protection measures, virus testing capabilities and agronomic practices improves the potato productivity worldwide. New generation, short- to medium-duration cultivars including seed plot technique have increased the area expansion and advancement of high-yielding potato varieties such as Kufri Chandramukhi, Kufri Bahar, Kufri Jyoti, Kufri Badshah, Kufri Pukhraj, Kufri Khyati, Kufri Sindhuri, Kufri Neelkanth, Kufri Uday, Kufri Mohan and Kufri Lalima.

8 Recent Approaches in Nutrient Management of Potato

The complicated dynamic of combining soil, crop, weather and modern nutrient management deals with the best balance between resource conservation and profit maximisation. Extra efforts should be made to reduce off-site nutrient movement that could be harmful to the environment while enhancing crop quality, yields, economic returns and nitrogen use efficiency. In an effort to achieve smart soil nutrient management, a number of strategies are gaining traction, including soil test crop response (STCR), targeted yield equations (TYE), site-specific nutrient management (SSNM), real-time nitrogen management (RTNM) utilising the chlorophyll index (CI), diagnosis and recommendation integrated system (DRIS), variable rate application technology (VRAT), remote sensing (RS), use of geographic information systems (GIS) and controlled release formulations (CRF) for input delivery that is targeted using nanotechnology.

Potato is a heavy nutrient feeder crop; the optimal rate of NPK fertilisation in various zones is determined by tuber yield response in trials conducted in different agro-climatic zones of India that are suitable for potato growth. Fertiliser management for potato crop is generally based on area averages, or "package of practices". These are suggestions considering the general growth conditions of the areas, including the quality of the soil, the local climate and weather and the cultivars to be grown. Such an approach results in the application of excessive or suboptimal inputs, which deteriorates the environment and inefficiently uses costly fertilisers.

Soil Testing In general, soil testing is an important indicator of the suitability of the soil for a particular crop that can be obtained by testing its pH, EC, organic carbon and texture. Organic matter in the soil, readily available nitrogen, potassium, phosphorus and secondary nutrients (Ca, Mg, S) as well as micronutrients are significant parameters for determining nutrition supplying capacity in potato crops.

Organic Matter Content The significance of organic matter lies in its ability to improve soil fertility or the availability of nutrients and in its profound influence on the soil's physical environment, microflora and fauna populations. It is also well recognised that soils rich in organic matter have significant concentrations of micronutrients like zinc, copper, iron and manganese. In soil testing labs, the method which involves oxidising organic matter with chromic acid and sulfuric acid is frequently utilised to determine the amount of organic matter present in the soil sample.

N_2 Organic carbon has proven to be a very powerful indicator of available nitrogen since most nitrogen in soil is contained in organic form and must be mineralised in NH_4^+ and NO_3^- before being utilised by potato plants. The rate at which organic N is mineralised for plant absorption is indicated by the presence of NO_3-N, a small fraction of total N. Potatoes can utilise total N, $KMnO_4$-oxidisable N, $K_2Cr_2O_7$-oxidisable N and K_2O_4-oxidisable N.

Organic Materials The oxidation of soil organic matter in an alkaline $KMnO_4$ medium (NaOH) is the basis of the $KMnO_4$-oxidisable-N (Subbiah and Asija 1956) method, which is extensively employed in soil testing facilities across the nation. $K_2Cr_2O_7$-oxidisable N works with all types of soil and is based on the oxidation of organic matter in acidic conditions using H_2SO_4. The content of NO_3^- in the soil was substantially correlated with the potato and response to N in heavy soils of Patna and Nalanda (which have high levels of clay and organic matter) in the eastern Indo-Gangetic plains of India.

Phosphorus $H_2PO_4^-$ ions in acidic soils and HPO_4^- and $H_2PO_4^-$ ions in neutral and alkaline soils mostly meet plant requirements. Because it can differentiate between absorbed and acid-soluble forms of P, the Bray P (0.03 N NH4F in 0.25 N HCl) method is suitable for soils containing significant concentrations of Al and Fe compounds. The low Ca activity prevents the precipitation of liberated phosphate as calcium phosphate and Olsen's technique (0.5 M $NaHCO_3$, pH 8.3) only partially recovers phosphate from Fe and Al. It has been found to be a reliable indication of P availability in black and alluvial soils that contain P in the forms of Ca and Mg. In acidic hill soils, the organic carbon test predicts potato P and K availability just as well.

Potassium At any given time, crop species, soil type, soil capacity to release K, ability to use K from soil pool and fertiliser as well as manure influence the methodology used to evaluate the availability of K in the soil. In soil testing facilities, a variety of chemical extractants have been used, with each method having its own advantages. Each test's validity is also determined by its ability to correlate crop response to the amount of K recovered by that approach.

Di-ethylene triamine penta-acetic acid (DTPA) test for soil micronutrient availability, specifically Zn, Fe, Mn and Cu, is commonly used in potatoes. In addition to other methods, 0.1 N HCl and 0.05 N HCl + 0.025 N H_2SO_4 have been used to quantify the amounts of Zn and Cu retained by soil organic matter. Malathi and Stalin (2018) found that the NH_4HCO_3-DTPA test has an advantage over other methods because it can estimate micronutrients as well as K, Ca and Mg in the same extract.

Soil Test Value Interpretation Based on the results of the soil test, a moderate rate of fertilisation can be achieved (STV). Fertiliser recommendations based on soil test findings promote tuber yield and promote fertiliser economy through balanced application. A fixed value below which the crop is likely to respond to nutrient delivery is identified in the interpretation of soil test results. They are referred to as crucial constraints.

The target yield concept strikes a balance between crop requirements that change throughout the field and the amount of fertiliser applied. Compared to recommended potato fertilisation, precision fertilisation, using soil tests and yield-based prescription equations, improves fertiliser efficiency by aligning applications with soil needs and target yields. Site-specific nutrient management (SSNM) entails applying fertiliser, particularly N, in a greater number of doses to provide adequate N supply in synchrony with crop needs while minimising environmental impact.

Where fertiliser P_2O_5 and K_2O rates are calculated based on soil fertility and yield potential of the variety. One plot is grown with abundant fertiliser additions

(NPK plot or NPK plus micronutrient plot) in the nutrient omission plot technique and the yield obtained is utilised to compute the total requirement for P_2O_5 and K_2O.

Approaches Based on Models One recent method for recommending fertiliser is the Quantitative Evaluation of the Fertility of Tropical Soils (QUEFT) model which served as the foundation for the potato nutrient application. For NPK SSNM (site-specific nutrient management) in potatoes, the QUEFTS model was calibrated (Choudhary et al. 2023).

Real-Time N Management Real-time N management estimates plant N status and is used to improve N management by estimating crop fertiliser needs before sowing and providing fertiliser during the cropping season based on crop requirement. It also aids in maintaining the soil's nutrient balance and reducing nutrient losses.

Precision Farming, Remote Sensing and GIS The application of GIS-based technology and remote sensing in sustainable agricultural growth and management is gaining popularity day-by-day. Technological developments in precision farming are also very intriguing. Soil mapping, crop stress monitoring, yield mapping, terrain analysis and soil organic matter estimation have all benefited from the use of remote sensing. Many portions of India have extremely good spatial mapping of available nutrients in potato growing pockets utilising remote sensing and GIS methods.

Precision farming entails applying input variables such as fertiliser in the appropriate amount, instead of applying a preventive measure; variable rate application technology (VRAT) applies the appropriate amount, at the correct place and at the right time based on crop need. An input application called variable rate technology (VRAT) accounts for the geographical heterogeneity of soil to ensure that no area of the field receives excessive or insufficient fertiliser. VRAT comes in a variety of forms that are generally categorised as sensor-based or GPS/ map-based (Naithani 2023).

9 Impact of Climate Change in Potato Production

Global potato production is predicted to be significantly impacted by climate change. The distribution and population of potato growing areas will be significantly impacted by climate change, as well the distribution and population of numerous pests and diseases that affect potatoes. Potatoes are prone to factors such as biotic, including nematodes, fungi, bacteria, pathogens, pests and disease damages, and abiotic including sunlight, drought, humidity, precipitation, frost, salinity and temperature changes and intense weather variation (Adekanmbi et al. 2023). As potato is the most important horticultural crop in the world so far, potato cultivation needs to adjust to climate change in order to prevent crop full fill the demand of future population.

Carbon Dioxide The potato tubers have high amount of carbohydrates in the form of starch; hence, higher atmospheric carbon dioxide results in higher tuber yields. Higher amounts of atmospheric carbon dioxide cause potatoes to open their stomata less to take up a same amount of carbon dioxide for photosynthesis, resulting in reduced water loss through stomatal transpiration. As a result, potato plant water use efficiency (the quantity of carbon assimilated per unit water lost) is expected to improve.

Temperature Potato is basically a cool season crop so far; temperate climates are ideal for potato production. Variations in temperature beyond 5–30 °C have a severe impact on tuber growth and yield. According to the Intergovernmental Panel on Climate Change (IPCC), by 2100, there will be 6.4 °C increase in the global temperatures. Depending on the present temperature in that location, variable temperatures have diverse effects on potato yield. Over 30 °C temperatures can cause negatively impact on potato growth in various ways, including:

- Slowing down of tuber growth and commencement.
- Less starch partitioning to tubers.
- Tuber physiological injuries (e.g. browning).
- Reduced/no tuber dormancy, causing tubers to grow too early.
- Insect-pests, for example potato tuber moth and Colorado potato beetle, which are expected to move into tropical to temperate places.
- Aphids, which serve as vectors for several potato viruses and will be able to spread when temperature increases.
- At higher temperatures, pathogens that cause potato blackleg disease (such as Dickeya) can emerge and spread very quickly, which means that they will probably become more common in temperate regions also.
- Higher temperatures are predicted to boost bacterial diseases like *Ralstonia solanacearum* and hasten their spread through flash flooding in hilly areas.
- Late blight benefits from warmer temperatures and more precipitation. It is anticipated that late blight would become less in some places like Finland and more destructive in the United Kingdom.
- Reduced seed germination and seedling growth have an impact on crop growth and yield.
- Reduced leaf expansion, reduced shoot and root growth, early maturity, reduced plant height, photosynthesis disturbance, poor nutrient intake, etc. directly influence the tuber yield and quality of potatoes. Crop yield fluctuations are expected to create alterations in the locations where potato crops can be produced profitably. Shifting potato growing regions in some nations could help to minimise yield losses due to rising temperatures and limited water availability. Competition for land among potato crops, other crops and other land uses is a potential issue in changing potato production.

(A) *Improving Irrigation Technology and Water Use Efficiency*

Improving irrigation technology and water use efficiency are required to sustain potato crop yields without overburdening water supplies. Sub-surface drip irrigation system is the best example of efficient water-saving technology that has been tested in potato crops. Few potato varieties such as Kufri Thar-1, 2, 3 and Kufri Daksh somehow have better water use efficiency.

10 Breeding Strategies for Development of Resistant Varieties

In potato, late blight and various viruses caused severe loss and its proper management requires the most extensive and expensive pest management practices during the crop life cycle. Introducing resistance against *P. infestans*, PLRV, PVX and PVY are therefore often high on the wish list in potato breeding. The different breeding strategies and potential future strategies to increase resistance, like traditional cross-breeding, traditional genetic modification and the evolving genome editing methods and strategies are shown in Fig. 3.

However, till date a small number of potato cultivars, hybrids and related species has been released in respect to biotic and abiotic stress tolerance as below:

- Zahov: a heat-tolerant polyploid potato hybrid derived from many wild potato species. Several wild potato species are more heat tolerant than *Solanum tuberosum*.
- Kufri Surya: is an Indian potato cultivar that is heat tolerant and early maturing.
- *Solanum verrucosum*: a wild potato species with genes that confer late blight resistance.

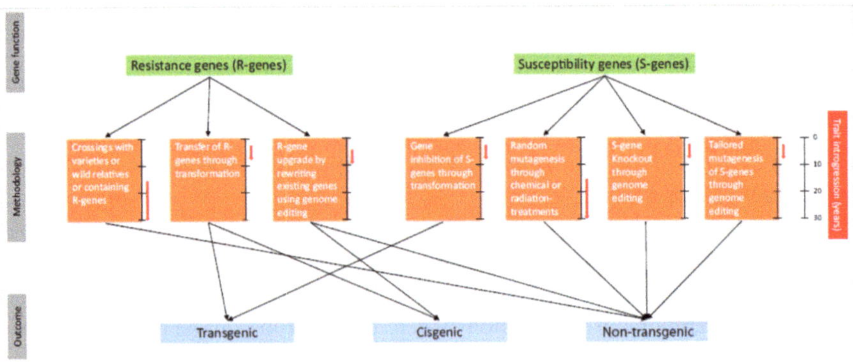

Fig. 3 Examples of breeding strategies to increase late blight resistance in potato. Two main target strategies are displayed—adding or modulating resistance genes (R-genes) and knocking out, inhibiting or modulating susceptibility genes (S-genes)—that might be used both individually and in combination to make the plants more resistant to the disease. Based on the breeding method used, the times for trait introgression differ substantially and the outcome will be a transgenic, non-transgenic or cisgenic plant that might be differently regulated based on the technique and where in the world it will be cultivated. (Source: Hofvander et al. 2021)

- Sullu and an Andean potato is improved drought resistance varieties.
- SpuntaG2: A genetically engineered Spunta of potatoes cultivar that is resistant to Potato Tuber Moth because it carries the *Bacillus thuringiensis cry1Ia1* gene.

11 Varietal Development

Many improved disease-free table purpose and processing varieties of the potato have been developed by ICAR-CPRI, Shimla and PAU, Ludhiana (Table 2).

12 Genetic Advancement of Potatoes Through Molecular Breeding and Biotechnology

The potato is an ideal crop for introducing new traits through molecular breeding and biotechnological tools. It was one of the first crops to be genetically modified, with Monsanto's NewLeaf™ being commercially grown in 1995 (Dennis et al. 2016). Traditional potato breeding remains inefficient and slow, having seen little change over the past century. Potatoes demand significantly higher amounts of nutrients, pesticides, and water to achieve optimal yield, quality and protection from pests and diseases. Combining tuber quality traits along with disease resistance is desired by potato processors, i.e. most significant challenge in potato breeding. The extensive genetic diversity in both wild and cultivated potato relatives enables the relatively straightforward identification, isolation and introduction of new genes for specific traits using biotechnological tools. For example, genes from wild potato relatives can enhance resistance to diseases like late blight and *Verticillium* wilt, as well as to pests and environmental stresses such as water scarcity and cold-induced sweetening in cultivated species, *S. tuberosum*. Unlike traditional transgenic plants that incorporate DNA from bacteria, viruses or other organisms, these improvements are achieved by inserting potato DNA derived from wild relatives into cultivated varieties. The transgenic potatoes introduced in the mid-1990s were technologically successful and provided notable benefits to producers, consumers and the environment. Despite these advantages, safety concerns about GMOs led to their withdrawal from the market in 2002 and they remained absent for over a decade. With increasing acceptance of transgenic crops, there has been renewed interest in developing genetically modified potatoes. This resurgence led to their reintroduction to the market in 2015 (Dennis et al. 2016).

(A) *Utilisation of Tissue Culture Plants in Potato Seed Production*

The successful establishment of tissue culture plants from potato tubers was first reported in 1951 (Srivastava et al. 2012). Since then, in vitro cultures of potatoes have been developed from various plant parts, including petioles, ovaries, anthers,

Table 2 Important characters of promising Indian potato varieties

Varieties	Tuber	Maturity	Yield (t/ ha)	Special features
Kufri Chandramukhi	White-cream, ovoid tubers with shallow eyes and white flesh	Early	20–25	Attractive tubers with excellent flavour and taste
Kufri Lauvkar	White-cream, round tubers with medium deep eyes and cream flesh	Early	25–30	Special features: Good for processing when grown in warmer areas
Kufri Pukhraj	Yellow, ovoid tubers with shallow-medium eyes and yellow flesh	Early to Medium	35–40	Special features: Early bulker, suitable for low input eco-system
Kufri Uday	Red, ovoid tubers with shallow eyes and yellow flesh	Early to medium	36–38	Special features: Field resistance to late blight
Kufri Thar-1, 2 and 3	tubers light yellow ovoid, eyes shallow, flesh light yellow;	Medium	30–35	Good keeping quality, good taste, typical potato flavour, mealy texture, tolerant to moisture stress
Kufri Bahar	White-cream, ovoid tubers with medium eyes, white flesh	Medium	30–35	Early bulker; suitable for long-distance transportation
Kufri Daksh (2023)	White-cream, ovoid tubers with shallow-medium eyes, creamy flesh	Medium	30–35	Water and heat stress tolerance; moderately resistant to late blight
Kufri Chipsona-3	White-cream, ovoid tubers with shallow eyes and white flesh	Medium	30–35	Suitable for making chips and French fries
Kufri Chipsona-5	White-cream, ovoid tubers with shallow eyes and cream flesh	Medium	30–35	Moderately resistant to late blight
Kufri FryOm	White oblong tubers with shallow eyes and white flesh	Medium	30–35	Field resistance to late blight and potato virus Y
Kufri Frysona	White-cream, long-oblong tubers with shallow eyes and white flesh	Medium	30–35	Special features: Suitable for making French fries
Kufri Himalini	White-cream, ovoid tubers with medium-deep eyes and cream flesh	Medium	30–35	Special features: Yield good in hills and plains, possess day-neutrality feature

(continued)

Table 2 (continued)

Varieties	Tuber	Maturity	Yield (t/ ha)	Special features
Kufri Jyoti	White-cream, ovoid tubers with shallow eyes and cream flesh	Medium	33–35	Special features: Early bulker with slow rate of degeneration and wider adaptability
Kufri Ganga	White-cream, ovoid tubers with shallow eyes and cream flesh	Medium	35–40	Special features Tolerant to moderate drought conditions
Kufri Lohit	Attractive red, ovoid tubers with shallow eyes, cream flesh	Medium	36–38	Special features: Field resistance to late blight
Kufri Manik	Tubers red round, eyes medium-deep, flesh yellow; moderately resistant to late blight	Medium	22–25	Possess anthocyanin and carotenoids
Kufri Sangam	White-cream, ovoid tubers with shallow eyes and white flesh	Medium	35–40	Dual-purpose variety as suitable for table as well as for processing purpose
Punjab Potato 101 and Punjab Potato 102	Tubers are light brown, oval round, uniform with shallow eyes, and creamy white flesh	Medium	25–30	Table purpose variety
Kufri Neelkanth	Purple, ovoid tubers with shallow eyes, yellow flesh	Medium	35–38	Special features: Antioxidant-rich specialty potato variety
Kufri Jamunia	Dark purple, oblong tuber with shallow eyes	Medium	32–35	Highly anthocyanin content, high yielding and late blight resistant
Kufri Sindhuri	Red, round tubers with deep eyes and cream flesh	Late	30–35	Special features: Suitable for low input eco-system
Kufri Badshah	White-cream, ovoid tubers with shallow eyes and cream flesh	Late	30–35	Resistant to early blight, late blight and PVX
Kufri Karn	Tubers white-cream ovoid, eyes shallow, flesh white	Late	25–30	Resistant to late blight, APCLV, PVX, Y, S, A, M, PLRV; moderately resistant to PCN
Kufri Bhaskar	White-cream, ovoid tubers that have shallow eyes and a mealy texture	Early–medium	30–35	Resistant to hopper and mite infestations, which contributes to better plant health and improved yields

stems, roots and meristem shoot tips. In vitro technology produced, disease-free potato clones combined with conventional multiplication methods, have become a crucial component of seed production in many potato growing countries (Naik and Sarkar 2000).

(B) *Meristem Shoot Tip Culture for Disease-Free Plantlets Production*

As represented in Fig. 4, the following steps are involved in the meristem shoot tip culture:

1. Select apparently healthy plants from the field or sample tubers.
2. Test these plants/tubers for freedom from viruses using ELISA test.
3. If no plants/tuber is found free from all the viruses, then one must resort to meristem shoot tip culture.
4. Select a plant/tuber that is infected with minimum (one or two) viruses for use in meristem shoot tip culture.

(C) *Micropropagation*

Micropropagation involves culturing disease-free micro-plants' nodal explants on semi-solid (agarified) or liquid culture media, while semi-solid medium is used for initial nodal segment propagation while, liquid medium promotes faster growth of potato micro-shoots. In vitro-derived micro-plants can be used for:

1. Micro-Tuber Production: To produce micro-tubers in vitro.
2. Greenhouse Transplants: For the production of mini-tubers.
3. Further In Vitro Multiplication: Using single node cuttings (SNCs) or apical rooted cuttings (ARCs).
4. Synthetic Seed Production: As source material for synthetic seed production.

(D) *Genetic Engineering for Potato Improvement*

Genetic engineering (GE), also known as genetic modification (GM) or genetic improvement (GI), involves transferring genes from one organism to another to introduce desirable traits, such as disease resistance, into plants, animals or microorganisms. Potatoes were among the first crop plants to be successfully regenerated

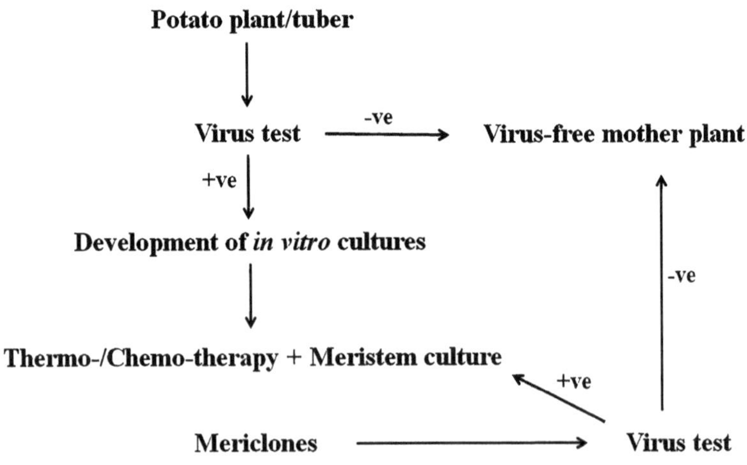

Fig. 4 An outline of major steps involved in potato meristem culture. (Source: Tiwari et al. 2021)

using recombinant gene technology, allowing the introduction of new traits through gene substitution. Kanamycin resistance has often been used as a selectable marker gene for transformed potato events (Mathur et al. 2017).

(E) *Marker-Assisted Breeding (MAB) in Potatoes*

The implementation of molecular markers in tetraploid potato breeding has been limited due to the high level of natural allelic variation and tetrasomic inheritance. This complexity makes marker transfer across mapping populations challenging. Successful examples of marker-assisted breeding in tetraploid potatoes include:

1. Resistance to Potato Cyst Nematode: Selected by Moloney et al. (2010).
2. Multiple Disease Resistance Traits: Developed by Gebhardt et al. (2006) using PCR-based markers for traits like resistance to potato virus Y (PVY), *Globodera rostochiensis* and potato virus X (PVX).

(F) *Somatic Hybridisation for Disease and Virus Resistance in Potato*

Somatic hybrids have significant potential in advancing potato breeding by transferring durable resistance to late blight and Potato Virus Y (PVY) and by broadening the gene pool of the cultivated potato, *S. tuberosum*. Interspecific potato somatic hybrids were created through protoplast fusion between *S. tuberosum* (C-13) and the wild species *S. pinnatisectum* (CGN No.: 17745), aiming to confer high resistance to late blight disease (Fig. 5). Additionally, protoplast fusion was performed between dihaploid *S. tuberosum* (C-13) and the wild species *S. etuberosum* (CGN No.: 23066) to enhance resistance to Potato Virus Y. Notably, *S. etuberosum* is a non-tuberous species, whereas the somatic hybrids can produce tubers, although they do not resemble the tubers of tetraploid potatoes (Tiwari et al. 2021).

13 The Future of the Potatoes: Opportunities and Challenges in Sustainable Agri-Food Systems

Potatoes, with their adaptability, productivity and nutritional value, have a significant role in addressing food security and improving household incomes, especially as the global population continues to grow. Key concerns identified by the European Association for Potato Research (EAPR) Conference in 2017 (Andrivon 2017) include:

1. *Food Security and Safety*: Meeting the needs of a growing population while ensuring food safety.
2. *Sustainable Production*: Managing natural resources efficiently and utilising new technologies such as advanced breeding techniques, biocontrol, and big data management.
3. *Innovation in Agri-food Systems*: Converting scientific research into practical products and processes to enhance agri-food systems.

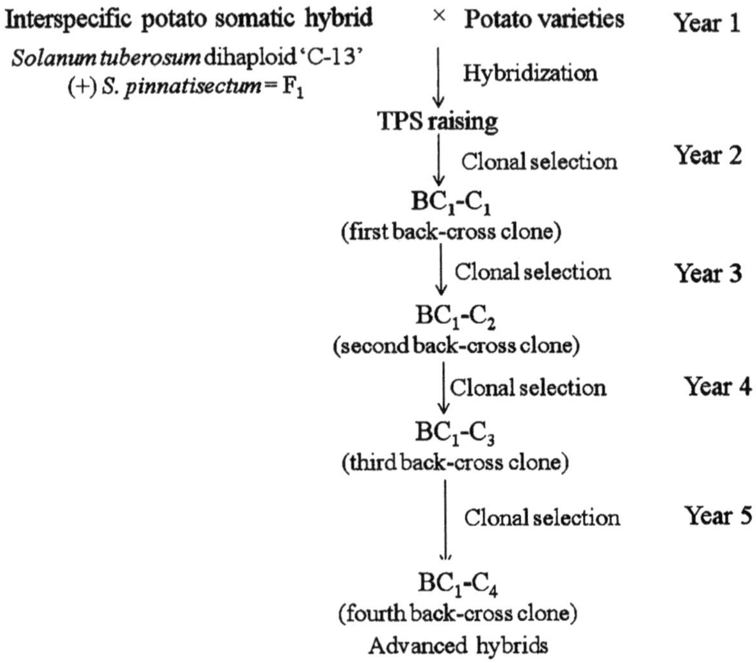

Fig. 5 Development of advance breeding lines of somatic hybrids in potato. (Source: Tiwari et al. 2021)

According to the Global Nutrition Report (IFPRI 2015), food system evolution progresses through five stages:

1. *Rural Food Systems*: Low agricultural productivity with high reliance on staple crops (e.g. Bangladesh, Ethiopia, Bolivia).
2. *Emerging Food Systems*: More urbanised but still reliant on staples (e.g. Pakistan, Thailand, Peru).
3. *Transitioning Food Systems*: Countries like Brazil and Malaysia.
4. *Mixed Food Systems*: Moderate productivity, urbanisation and lower dependence on staples (e.g. Germany, Italy).
5. *Industrial Food Systems*: Highly urbanised with minimal dependence on staple crops (e.g. USA, NW Europe).

(B) *Potato Distribution and Consumption Worldwide*

Potato is the world's third most important food crop by human consumption, following wheat and rice. Despite a significant portion of potato produce being used for seed and animal feed, approximately two-thirds of the harvest is consumed fresh, with around 1.3 billion people eating potatoes as a staple food, including high-population countries like India and China. The trend of potato production around the world is tabulated in Table 3.

Table 3 Potato production and demand trends around the world

Region	2016–2018			Average annual growth rate								
	Production (000 t)	Area (000 ha)	Yield (t/ha)	Production (%) area (%) 1961–1963 vs 1998–1990		Yield (%)	Production (%) area (%) 1988–1990 vs 2016–2018		Yield (%)	Production (%) area (%) 1961–1963 vs 2016–2018		Yield (%)
Africa	25,026	1828	13.7	4.9	3.7	1.1	4.2	3.3	0.9	4.5	3.5	1.0
Asia[a]	184,937	9325	19.8	3.7	2.4	1.3	3.9	2.4	1.4	3.8	2.4	1.3
Europe	109,783	4823	22.8	−1.0	−1.8	0.8	−1.5	−2.8	1.4	−1.2	−2.3	1.1
North America	26,390	5550	47.5	1.15	−0.2	1.4	1.0	−0.5	1.5	1.1	−0.4	1.4
LAC	20,161	1051	19.2	2.2	0.0	2.2	1.7	0.2	1.5	2.0	0.1	1.9
World	366,298	17,584	20.8	0.1	−0.8	0.9	1.1	−0.1	1.2	0.6	−0.4	1.0

[a]= Asia + Ocean

Source: FAOSTAT Crops, last Update June 15, 2020

14 Research and Development on Potato Using Information Technology

The modern world's credo is information technology (IT). These days, IT is present in every aspect of human endeavour, even in agriculture. Due to a shortage of IT-trained professionals, its adoption in agriculture is substantially slower than in the fields of health, engineering and commerce sectors (Rawat 2019).

The ICAR-Central Potato Research Institute, Shimla created Potato Weed Manager (PWM), VarTRAC, Computer Aided Advisory System for Potato Crop Scheduling (CAASPS), Additional decision support tools include the Potato Pest Manager (PPM), Potato Growing Season Descriptor (PGSD), Potato Growing Period and Yield Calculator (PGPYC), etc. At the ICAR-CPRI, Shimla, an electronic book and a picture library of potatoes are also being established to understand more about potato crop easily.

(A) *Decision-Making Aids*

1. An online model called INDO-BLIGHTCAST is used throughout India to predict potato late blight. In this direction ICAR-CPRI, Shimla developed INDO-BLIGHTCAST, a web-based forecasting programme that uses daily weather data from meteorological stations to determine when the late blight disease will start. This model performs better than the JHULSACAST model, which depends on daily rainfall, relative humidity and hourly temperature data. The location-specific calibration and large data needs of JHULSACAST posed major obstacles to its widespread use. The INDO-BLIGHTCAST is relevant throughout India because of its web-based and simple format. It does not require local calibration for various places; it only requires the daily weather. It is therefore more reliable and produces more accurate projections as compared to other software.
2. Two modules make up INDO-BLIGHTCAST: one is for data entry and the other allows users to view the status of late blight forecasts.

(B) *PPM (Potato Pest Manager)*

Step 1: Photographs depicting disease/ pest symptoms are grouped in a photo gallery and exhibited in sequence.

Step 2: For a correct diagnosis, information on the biotic/ abiotic variables presents, as well as the symptoms, is required. This is accomplished by a series of confirming inquiries.

Step 3: The responses to the confirmatory questions concerning pest damage and illness symptoms are used to calculate the confidence percentage.

Step 4: The preventive actions are the behaviours that would have prevented/ mitigated the occurrence of the insect-pests and disease.

Stage 5: In this stage, information needed to suggest control measures for the standing crop, i.e. gathered through a series of additional inquiries. For example, information on the severity of disease/ pest damage, crop age and so on is generally required when choosing on the chemicals to be used, their dosage, number of sprays, etc.

Step 6: Based on the options selected, this step provides the reliable recommendations. This web-based tool/ DSS is written in ASP.NET and the database is written in Microsoft SQL.

3. Potato Growing Season Descriptor (PGSD): The PGSD is a database that contains daily weather information for various locations that is produced by the MARKSIM weather generator. During the fall growing season, calculate the cumulative growth degree days, accumulated physiological days, mean day temperature, mean night temperature and length of the growing season. In this programme, daily meteorological data is analysed using algorithms. In a constraint-free environment, the programme also uses a summary model to predict the production of the autumn crop.
4. Potato Yield and Growing Period Calculator (PYGPC): Potatoes can be cultivated in a range of environments. It can be planted early in the season, at the main season or in the late season, making it incredibly flexible in terms of planting schedule. It is a great option to grow it in a range of cropping system due to its wider adaptability. The growing season(s) and duration of each growing season(s) must be determined to efficiently utilise this flexibility at any particular place. The primary factor influencing potato growth in India's subtropical environment is temperature. The minimum temperature must be less than 21 °C for at least three weeks after the maximum temperature suitability criterion is met, the maximum temperature must not exceed above 35 °C and the temperature must be higher than 2 °C to guarantee a frost-free growing season. The estimated potential yield was calculated by building macros for predicting gross photosynthesis (GPHOT), maintenance respiration, linear growth rate and other factors using the outputs of the database screening macros. An interface, or graphic user interface, was designed in Microsoft Access/ Visual Basic. Fields were made to show the multiple outputs, the position and its spatial characteristics.
5. Computer-Aided Advisory System for Potato Crop Scheduling (CAASPS): These objectives guided the development of DSS's CAASPS:

   (a) To provide information on the predicted yields of different kinds planted at different periods, allowing farmers to choose the best variety for their specific field.
   (b) To assist in determining harvest dates based on yield at 60, 70, 80 and 90 days after planting, respectively.
   (c) To demonstrate varietal performance under various planting dates and crop durations, so assisting in the selection of the suitable variety. A database plus a user interface comprises this DSS. The database includes state, district and location names, as well as yield outputs from the Infocrop-potato model.

(C) *The Following Are the Model Outputs*

1. Using the MARKSIM weather generator, a weather database for major areas in India was built.

2. The location's appropriate thermal window was identified by specifying the minimum temperature (21 °C) and maximum temperature (35 °C) for the screening parameters.
3. Five planting scenarios were tested using the Infocrop-potato model. The planting scenarios were staggered 10 days apart and began 10 days prior to the acceptable temperature window that was indicated by the screening rules.
4. The model was run for ten kinds under potential scenarios for each planting date, and 80% of the potential yield was considered as attainable yield.
5. Each variety's yield output at 60, 70, 80 and 90 days after planting was connected to the relevant spatial data in MS Access, including the place names, state and district.
6. Potato Weed Manager (PWM): Weeds have been reported to reduce potato productivity by up to 80%. For suppressing weeds in potato crops, a variety of cultural, mechanical and chemical approaches are available. Herbicides are available to manage various types of weeds at various stages. To address this issue, the Potato Weed Manager decision support tool was created. The software was created using DOT NET technology and the database is SQL. This software runs on the ICAR-CPRI application server and is linked to the ICAR-CPRI website. This app includes photos of weeds for the user's identification.
7. VarTRAC: A tool based in bioinformatics used to determinate potato varieties.

Plant breeders, authorities in charge of variety registration and certification, seed producers, retailers, farmers, growers, processors and other final consumers, all depend on accurate identification of potato cultivars. Nowadays, morphological descriptors (DUS guidelines) are used worldwide for variety identification. In the near future, DNA fingerprint data will be used to supplement morphological characteristics. ICAR-CPRI created VarTRAC computer programme for rapid identification of a variety using information from DNA fingerprints and morphology.

Every morphological character in the database has a field allocated to it when it was created in MS Access. Every feature needed to distinguish a potato variety has been presented. Users only need to choose the appropriate score for each character from a drop-down menu that assigns scores to each character. Furthermore, support has been provided to ensure precise scoring. Regarding DNA fingerprints, information has been gathered on 127 alleles, with each allele lacking nothing compared to those that are present.

15 Conclusion and Future Prospects

Unique features of potato present advantages as well as challenges to the plant breeders. At present advanced techniques are used to create high-quality potato seeds by using tissue culture techniques in combination with aeroponic systems or apical rooted cuttings, which have been tried to lessen the deficiency of high-quality seeds. To raise potato yields, these technologies ought to be promoted in the majority of emerging nations. Potatoes have a high degree of heterozygosity, tetrasomic inheritance and incompatibility barriers that impede their genetic progress. Moreover, modern molecular breeding aids in the quick identification of desirable genes and the production of high-quality traits like potatoes with increased nutritional content and starches with modified amylose-to-amylopectin ratios. The potato is an ideal crop for introducing genetic modifications. The biggest obstacle in potato breeding continues to be combining the desirable qualities of tubers for consumers and processors with the agronomic performance and disease resistance. Fortunately, identification, isolation and biotechnology are made relatively simple by the enormous genetic variety found in both wild and cultivated relatives of the introduction of novel genes for a particular feature in potatoes. Molecular marker-assisted breeding, genetic engineering and tissue culture are the major biotechnological techniques utilised to enhance selection aided by molecular markers in potatoes genetic improvement. Identifying successful potato-growing practices in various countries could aid replication in other regions and be integrated into the CIP/ ICAR-CPRI database. The techniques discussed here are just the beginning of how they can enhance safety and sustainability in global potato production.

However, Indian potato varieties have a limited genetic base. Over the past 60 years, only 59 varieties have been developed, all of which can be traced back to just 91 parents. Most of these parents are genetically tetraploid cultivars with minimal representation of *S. tuberosum* subsp. *andigena* and other wild species. There is a need to better utilise *S. tuberosum* subsp. *andigena*, which thrives in short days and suits well under sub-tropical plains. However, as of the latest data, India has released over 250 potato varieties to address diverse agricultural conditions and enhance traits such as yield, disease resistance and better-quality characters. The underutilisation of wild and cultivated species in varietal improvement is due to the time required to develop desirable parental lines. Therefore, gene bank collections should be transformed into active, working collections for development of genetically superior varieties under climate change scenario.

References

Adekanmbi T, Wang X, Basheer S, Nawaz RA, Pang T, Hu Y, Liu S (2023) Assessing future climate change impacts on potato yields—a case study for prince Edward Island. Canada Foods 12:1176

Andrivon D (2017) Potato facing global challenges: how, how much, how well? Potato Res 60:389–400. https://doi.org/10.1007/s11540-018-9386-z

Buckseth T, Sharma AK, Pandey KK, Singh BP, Muthuraj R (2016) Methods of pre-basic seed potato production with special reference to aeroponics—a review. Sci Hortic 204:79–87

Choudhary B, Dongre R, Sharma SK, Bakode S (2023) Nutrient management in potato (Solanum tuberosum L.) through evaluation of the fertility status of soils in Satpura Zone of Madhya Pradesh. Biol Forum 15(9):99–103

Climate change and potatoes: the risks, impacts and opportunities for UK potato production. Cranfield Water Science Institute. http://www.potato.org.uk/sites/default/files/%5Bcurrentpa

ge%3Aarg%3A%3F%5D/CC%20impacts%20potatoes_Final_20Sept2011.pdf. Retrieved 7 Nov 2012

Dennis H, Guenthner J, Collinge S, Butler N, Douches D (2016) Biotech potatoes in the 21st century: 20 years since the first biotech potato. Am J Potato Res 93(1):1–20

Gastelo M, Diaz L, Landeo J, Bonierbale M (2013) New elite potato clones with heat tolerance, late blight and virus resistance to address climate change. In: Transforming potato & sweet potato value chains for food and nutrition security. African Potato Association Conference 2013. Naivasha

Gebhardt C, Bellin D, Henselewski H, Lehmann W, Schwarzfischer J, Valkonen JP (2006) Marker-assisted combination of major genes for pathogen resistance in potato. Theor Appl Genet:1458–1464

Hazra P, Som MG (2015) Production technology of different vegetable crops. Kalyani Publishers, pp 227–228

Hofvander P, Andreasson E, Andersson M (2021) Potato trait development going fast-forward with genome editing. Trends Genet. https://doi.org/10.1016/j.tig.2021.10.004

International Food Policy Research Institute (IFPRI) (2015) Global nutrition report 2015: actions and accountability to advance nutrition and sustainable development

Jaiswal AK, Singh B, Mehta A, Lal M (2023) Post-harvest losses in potatoes from farm to fork. Potato Res 66(1):51–66

Kumar D, Singh SV, Pandey SK, Singh BP, Rawal S (2007) Effect of growing season on chipping quality of potatoes under sub-tropical climates. Potato J 34(3–4):180–186

Kumar R, Solankey SS, Major S (2012) Breeding for drought tolerance in vegetables. Veg Sci 39(1):1–15

Kumar A, Dash GK, Sahoo SK, Lal MK, Sahoo U, Sah RP, Ngangkham U, Kumar S, Baig MJ, Sharma S, Lenka SK (2023) Phytic acid: a reservoir of phosphorus in seeds plays a dynamic role in plant and animal metabolism. Phytochem Rev:1–24

Levy D, Veilleux RE (2007) Adaptation of potato to high temperatures and salinity-a review. Am J Potato Res 84:487–506. https://doi.org/10.1007/BF02987885

Luo SP (2021) Nutrient stoichiometry and microbial community response to biochar addition in karst loess. Southwest University, Chongqing. https://doi.org/10.27684/d.cnki.gxndx.2021.000107

Malathi P, Stalin P (2018) Evaluation of AB—DTPA extractant for multinutrients extraction in soils. Int J Curr Microbiol App Sci 7(03):1192–1205. https://doi.org/10.20546/ijcmas.2018.703.141

Mathur V, Javid L, Kulshrestha S, Mandal A, Reddy AA (2017) World cultivation of genetically modified crops: opportunities and risks. In: Eric L (ed) Sustainable Agriculture Reviews. Springer, Cham, pp 45–87

Moldovan C, Morar G, Vatca S, Todoran F (2011) Research concerning the influence of photoperiod upon stolonization and tuberization. Bull UASVM Agric 68(1):224–230

Moloney C, Griffin D, Jones PW, Bryan GJ, McLean K, Bradshaw JE, Milbourne D (2010) Development of diagnostic markers for use in breeding potatoes resistant to Globodera pallida pathotype Pa2/3 using germplasm derived from Solanum tuberosum ssp andigena CPC 2802. Theor Appl Genet 679–689

Mostafa M, Hossain MM (2024) The impact of climate factors on the epidemiology and management of Potato virus Y. Online J Biol Sci 2024. https://doi.org/10.3844/ojbsci.2024.858.876

Naik PS, Sarkar D (2000) *In vitro* propagation and conservation of genetic resources in potato. In: Chadha KL, Ravindran PN, Leela S (eds) Biotechnology in horticultural and plantation crops. Malhotra Publishing House, New Delhi, pp 369–406

Naithani P (2023) Variable rate technology in Indian agriculture scenario and solutions ahead, pp 74–75

Rawat S (2019) Advancement in potato production technology & its future prospects (Use of information technology for potato research and development). ICAR-Central Potato Research Institute, Shimla, pp 78–84

Reddy R, Dhananjaya BN, Das A, Ranganath GJ, Reddy M (2021) Apical Rooted Cutting (ARC) technology for Rapid local seed potato production. In: International Potato Centre (CIP), Bangalore

Singh RK, Buckseth T (2024) Agro-techniques for quality seed potato production ICAR-Central Potato Research Institute, Regional Station, Modipuram, Meerut, Uttar Pradesh. Indian Farm 74(5):12–15

Srivastava AK, Diengdoh LC, Rai R, and Bag TK (2012) Tissue Culture –Technology Harnessed for Potato Seed Production. Central Potato Research Station, 5th Mile, Upper Shillong, Shillong – 793 009

Stevenson W, Loria R, Franc G, Weingartner DP (eds) (2001) Compendium of potato diseases, 2nd edn. APS Press, St. Paul

Subbiah BV, Asija GL (1956) A rapid procedure for the estimation of available nitrogen in soils. Curr Sci 25:259–260

Swarup V (2006) Vegetable science and technology in India. Kalyani Publishers, pp 598–605

Thiele G, Hareau G, Suárez V, Chujoy E, Bonierbale M, Maldonado L (2008) Varietal change in potatoes in developing countries and the contribution of the International Potato Center: 1972–2007, Social sciences working paper series 2008-6. International Potato Center (CIP), Lima. 46 pp

Tiwari JK, Buckseth T, Dalamu, Luthra SK, Singh RK, Bhardwaj V, Vanishree G, Kardile Hemant B, Vinod K, Kumar M (2021) Plant tissue culture and somatic hybridization in potato, eTechnical bulletin No. 1 (Ref.: No.F.PME/3-3/2021 dated 4.7.2021). ICAR—Central Potato Research Institute, Shimla

Tournex C, Devaux A, Camacho MR, Mamani P, Ledent JF (2003) Effect of water shortage on six potato genotypes in the highlands of Bolivia (II): water relations, physiological parameters. Agronomie 23:181–190

Valkonen JPT, Gebhardt C, Zimnoch-Guzowska E, Watanabe KN (2017) Resistance to Potato virus Y in potato. In: Potato virus Y: biodiversity, pathogenicity, epidemiology and management. Springer, pp 207–241. https://doi.org/10.1007/978-3-319-58860-5_8

Advances in Genotypic Selection for Potato Improvement

Shirin Akhtar, Shashank Shekhar Solankey, Anuradha Sinha, and Pushpam Patel

Abstract Potatoes are one of the most important food crops of the world, being rich in carbohydrates, protein, minerals, and vitamins, and is a staple of masses. Cultivated potato has an extremely narrow genetic base. Therefore, improvement of the crop requires creation of variation. The wild relatives of potato are diverse and sources of resistance and nutritional and bioactive compounds. Several accessions of potato are being conserved in different international centers such as CIP, Lima, Peru, USDA, USA, CPRI, India, etc. The basic potato improvement involves genotypic selection as the initial step. Different conventional breeding approaches and recent biotechnological tools are used for potato improvement.

Keywords Potato · Wild relatives · Genetic improvement · Breeding approaches

1 Introduction

Among the food crops of the world, potato ranks third after rice and wheat considering its importance as a staple food of masses (FAOSTAT 2020). Potato is also a source of high income to the farmers and hence regarded as a cash crop. Potato can be a crop

S. Akhtar (✉)
Department of Horticulture (Vegetable Science), Bihar Agricultural College, Bihar Agricultural University, Sabour, Bhagalpur, Bihar, India

S. S. Solankey
Department of Horticulture (Vegetable Science), Nalanda College of Horticulture, Noorsarai, Nalanda, Bihar Agricultural University, Sabour, Bhagalpur, Bihar, India

A. Sinha
Krishi Vigyan Kendra, Aurangabad, Bihar Agricultural University, Sabour, Bhagalpur, Bihar, India

P. Patel
Krishi Vigyan Kendra, Barh, Patna, Bihar Agricultural University, Sabour, Bhagalpur, Bihar, India

S. S. Solankey (ed.), *Advances in Research on Potato Production*, Advances in Olericulture, https://doi.org/10.1007/978-3-031-82710-5_2

that may cater to nutritional security, having high amount of minerals and vitamins and producing higher energy and protein per unit area when compared to other food crops (Lutaladio et al. 2009), and has a very high harvest index going above 75% (Scott and Suarez 2011; Thiele et al. 2010), which makes it an important crop for checking malnutrition, hunger, and poverty. There has been an increase in global potato production in the last few decades which is chiefly attributed to enhanced area under cultivation (Walker et al. 2011). Higher productivity requires proper genotypic selection and further genotypic development in terms of yield and resistance traits.

2 Taxonomy, Origin, and Evolution

Potatoes including their cultivated and wild forms are tuber-bearing species belonging to the genus *Solanum* which is the largest genus with around 1500–2000 species (Burton 1989). Among these, around 235 species tuberize; they are classified into the section *Petota* which has two sub-sections *Potatoe* and *Estolonifera*, and all the tuber-bearing potato species and wild relatives along with common potato *Solanum tuberosum* belong to the sub-section *Potatoe* (Hawkes 1990).

Solanum tuberosum, the cultivated potato, is further classified into two sub-species, *tuberosum* and *andigena*. The former had its origin in Peru, tuberizes under long days in temperate climatic conditions and short days in the tropical and subtropical regions at lower altitudes (500–2000 m above mean sea level), and is extensively distributed over Europe North America and Asia, while the latter subspecies is a native of the mountainous regions of Andes of South America and tuberizes under short-day conditions at high altitudes (>2000 m above mean sea level) and is grown mostly in the surroundings of Central and South America (Hanneman 1994; Hawkes 1990).

The cultivated potato (*S. tuberosum* subsp. *tuberosum*) is a heterozygous and asexually propagated vegetable crop with chromosome number 2n = 4x = 48. There are theories that potatoes originated about 7000–10,000 years ago from the wild species *Solanum leptophyes*, giving rise to *Solanum stenotomum* which was the first domesticated diploid species, and its evolution led to the modern-day potato. Besides, three other wild species, viz., *Solanum sparsipilum*, *S. acaule*, and *S. megistacrolobum*, played an important role in evolution of the modern potatoes. Previously researchers put forward the theory of *S. tuberosum* being an autotetraploid derived from *S. stenotomum*, but later researchers came to the conclusion that it is an allotetraploid derived through hybridization of *S. stenotomum* and *S. sparsipilum* (Ram 1998).

3 Gene Pool and Polyploidy

Extensive inter- and intra-specific crosses have resulted in the evolution and development of the modern-day potato varieties. A polyploid series with genetic characteristics that promote gene transfer across ploidy levels, ranging from 2n = 2x = 24 to 2n = 6x = 72, makes up the diverse gene pool of potato that is accessible for improvement of the crop. In the Andean center of origin, there is huge diversity in farmers' varieties and landraces found for different traits which can be readily used for cultivation. These landraces, local cultivars, improved varieties and the wild species in the *Petota* section of genus *Solanum* form the gene pool 1 and 2 and are simple to employ in breeding initiatives (Harlan and Wet 1971).

The large number of wild tuber-bearing species (more than 200) of potatoes under section *Petota* produced tubers that were of extremely reduced size, often containing large amounts of toxic glycoalkaloids; besides, in many cases, short-day conditions were needed for tuber formation (Spooner et al. 2014). Domestication of these wild potatoes was carried out in various places of coastal South America besides the Andean region for consumption purpose, the basis of selection for which was enhanced size of tuber and lower glycoalkaloid content (Ugent et al. 1982, 1987), and this marked the initiation of potato breeding.

Potatoes that are diploid are natural outcrossers because gametophytic self-incompatibility is common in them (De Nettancourt 1977; Pushkarnath 1942), thus reducing inbreeding and thereby promoting genetic variation at intraspecific level. However, tetraploid potato has been found to be self-compatible, like in case of many other polyploid plants where gametophytic self-incompatibility system that is observed in the diploids is not reported in their polyploid genotypes as reported by Frankel and Galun (1977) and Levin (1983), but the molecular mechanism behind which still remains obscure (Comai 2005). Selfing in most potatoes lead to high inbreeding depression. Hence, elimination of unfavorable alleles and accumulation of favorable alleles to fixation becomes tough due to this inbreeding depression at both diploid and tetraploid levels coupled with self-incompatibility at the diploid level (Bonierballe et al. 2020).

However, cross-compatibility between the cultivated potato (*Solanum tuberosum* subsp. *tuberosum*) with wild diploid species like *Solanum sparsipilum*, *Solanum berthaultii*, *Solanum spegazzinii*, etc. and other related cultivated species (*S. tuberosum* subsp. *andigena*, *Solanum phureja*, and *Solanum stenotomum*) remains which can create genetic recombination. Wild species have been efficient sources of genes for resistance to diseases and insect pests as well as improved quality attributes and proven to be helpful for transgression of these genes for improvement of the crop.

4 Diversity in Related Potato Species

For the success of any effective plant breeding program, diversity in the genetic material is a necessity. The selection of diverse parental lines for hybridization program is of utmost importance in order to develop improved genotype with high yield and other targeted traits (Hayder et al. 2007).

The cultivated potato has an extremely narrow genetic base due to limited introduction of germplasm from their natural range in South America (Rauf et al. 2010). Mostly the cultivated genotypes are tetraploids (2n = 4x = 48), highly heterozygous, and naturally outcrossers, thereby suffering from inbreeding depression. The basic chromosome number of cultivated and wild potatoes is x = 12 and the species may be diploid (2n = 2x = 24), triploid (2n = 2x = 36), tetraploid (2n = 4x = 48), or pentaploid (2n = 5x = 60). Only 9 species and 141 intra-specific taxa have been identified, of which 7 species are cultivated (Hawkes 1990). The specific traits and distribution of the potato cultivated and wild species have been given in Table 1.

Table 1 Specific traits and distribution of the potato cultivated and wild species

Potato species	Particular traits and description	References
Solanum ajanhuiri (2n = 2x = 24) diploid species	Created by spontaneous hybridization between tetraploid wild species *S. bolivense* (*S. megistacrolobum*) and diploid cultivars of the *S. tuberosum* L. *andigenum* group; exhibits resistance to frost; found in the high Andean Altiplano, between central to northern Bolivia and southern Peru, at elevations of 3700–4100 m	Spooner et al. (2010) and Ovchinnikova et al. (2011)
S. juzepczukii (triploid) (2n = 3x = 36)	Formed via the hybridization of a tetraploid wild species, *S. acaule* Bitter, with a diploid cultivar of *S. tuberosum* L. *andigenum* group; it can grow up to 4000 m in elevation and is found in central Peru and southern Bolivia; includes significant amounts of glycoalkaloids, and the locals use freeze-drying to create detoxified processed potatoes, or "chuño"	Rodríguez et al. (2010), Spooner et al. (2010), and Irikura (1989)
S. chaucha (2n = 3x = 36) triploid species	Originated from natural hybridization between *S. tuberosum* subsp. *andigena* and *S. stenotomum* distributed from 2100 to 4100 msl throughout Peru, with lower frequency in Bolivia, and rarely found in Ecuador and Colombia	Hawkes (1990)
S. tuberosum (2n = 4x = 48)	Most popular cultivated potato in most parts of the world; originated from the coastal regions of South-Central Chile	Kumari et al. (2018)
S. curtilobum (2n = 5x = 60)	Formed by hybridization between tetraploid forms of *S. tuberosum* L. *andigenum* group (synonym for *S. tuberosum* subsp. *andigenum*) and *S. juzepczukii*; cultivated in the Andean Altiplano at an altitude range of approximately 4000 m; tubers are bitter, owing to high glycoalkaloid content, and the species is used to prepare "chuño"	Rodríguez et al. (2010), Spooner et al. (2010), and Irikura (1989)
S. phureja	Cultivated from central Peru to Ecuador, Colombia, and Venezuela since the pre-Spanish era, believed to have originated from *S. stenotomum*; high ascorbic acid, protein, starch and carotenoid content; non-sweetening even at low temperature	Hawkes (1990)
S. stenotomum (2n = 2x = 24) diploid species	Most primitive form of cultivated potato, distributed between Central Peru and Central Bolivia; first domesticated potato derived from diploid wild species; high carotenoid content	Kumari et al. (2018)
S. vernei	High starch and protein content, low reducing sugar content	Kumari et al. (2018)
S. stoloniferum	High ascorbic acid content	Kumari et al. (2018)

Reclassification of the cultivated potatoes was done based on the genotyping of 742 potato landraces by Spooner et al. (2007) using SSR and chloroplast markers. The classification was given as follows:

(a) *Solanum tuberosum*, having two cultivar groups (the Andigenum group containing diploids, triploids and tetraploids and the Chilotanum group containing only tetraploids)
(b) *Solanum x ajanhuiri* (diploid);
(c) *Solanum x juzepczukii* (triploid)
(d) *Solanum x curtilobum* (pentaploid)

5 Conservation of Potato Biodiversity

The cultivated potato is a highly diverse crop with varying levels of ploidy, and various improved varieties, landraces, and wild species belonging to different levels of ploidy from diploid to pentaploid are grown throughout the South American Andean belt. Many genebanks have gathered, preserved, and conserved this diversity and made available for use in breeding and research programs worldwide. Over 30 years ago, the cultivated potatoes were divided into seven species and nine taxa by Hawkes (1990) which are as follows:

(i) *Solanum tuberosum* (C. Linneo) subsp. *andigenum* (Juz. & Bukasov)
(ii) *S. tuberosum* (C. Linneo) subsp. *tuberosum*
(iii) *S. x chaucha* (Juz & Bukasov)
(iv) *S. stenotomum* subsp. *goniocalyx* (Juz. & Bukasov)
(v) *S. stenotomum* subsp. *stenotomum* (Juz. & Bukasov)
(vi) *S. x juzepczukii* (Bukasov)
(vii) *S. phureja* (Juz. & Bukasov)
(viii) *S. x ajanhuiri* (Juz. & Bukasov)
(ix) *S. x curtilobum* (Juz. & Bukasov)

The taxonomy of potatoes has undergone revisions in recent times, leading to the reclassification of cultivated potatoes into four species: *Solanum tuberosum*, *S. x juzepczukii*, *S. x ajanhuiri*, and *S. x curtilobum* (Spooner et al. 2014). The taxa of wild relatives of potatoes have undergone a similar taxonomic revision; Hawkes' (1990) prior taxonomic categorization recognized 228 wild species in 21 series. However, the number of wild potato relatives (*Solanum* section *Petota*) has been decreased to 107 species by the current taxonomic categorization by Spooner et al. (2014).

The diverse potato germplasm is being maintained at several genebanks, some important ones being at the International Potato Center (CIP), Peru and United States Department of Agriculture-Agricultural Research Service (USDA-ARS).

Approximately 1800 potato accessions consisting of traditional cultivars were donated by the Peruvian National Potato Program to the International Potato Center (CIP) genebank with which its journey was initiated and the collections were housed at the Santa Ana-INIA Huancayo CIP Huancayo Experimental Station. Initially, the genebank basically focused on collecting more germplasm, removing mixtures, unusual or off-type plants, and those exhibiting indications of viral symptoms. The number of cultivated potato accessions peaked at 17,347 during the 45-year history of the potato genebank. Following a thorough investigation that included the identification and elimination of duplicates, 4870 accessions—including 4467 traditional landrace cultivars from 17 countries, mostly in the Andes, as well as improved varieties—are currently part of the active cultivated potato collection. There are about 289 accessions of diploid *Solanum stenotomum* subsp. *stenotomum*, 102 accessions of diploid *Solanum stenotomum* subsp. *goniocalyx,* 206 accessions of diploid *Solanum phureja,* 14 accessions of diploid *Solanum* × *ajanhuiri,* 235 accessions of various diploids and tetraploids of *Solanum sp.,* 121 accessions of triploid *Solanum* × *chaucha,* 36 accessions of triploid *Solanum* × *juzepczuckii,* 3233 accessions of tetraploid *Solanum tuberosum* subsp. *andigenum,* 179 accessions of tetraploid *Solanum tuberosum* subsp. *tuberosum,* and 6 accessions of *Solanum* × *curtilobum* (https://cipotato.org/genebankcip/potato-cultivated/ accessed on August 13, 2024).

The whole clonal collection is preserved in vitro and distributed globally when required through tissue culture. The genetic stocks having resistance to diseases (late blight, bacterial wilt), as well as released varieties primarily from Latin American countries, are chiefly conserved. Under the provisions of the International Treaty for Plant Genetic Resources for Food and Agriculture (ITPGRFA), this global collection is disseminated with the Standard Material Transfer Agreement (SMTA) and is preserved in trust. Through ex situ and in situ conservation as well as cryopreservation, all accessions in the collection are preserved and kept available for study, breeding, and training.

6 Basic Potato Improvement Goals

Any breeding program depends on a clear set of prioritized breeding goals. The constraints in production and cropping systems of target population and environment, the needs of stakeholder groups and their preferences, often define the key breeding objectives. To carry on the breeding program, the breeder should be aware of the pros and cons of the leading varieties and have descriptors for the various traits.

The basic goals of potato breeding are aimed at consumer and market use, productivity enhancement, and resistance to diseases and pests and to some extent to abiotic stresses.

The potato improvement goals may be summarized as follows:

- Enhanced yield (>25 tons/ha)
- Earliness (maturity in 70 days)
- Table quality (flavor should be acceptable, dry matter content should be >17%, and glycoalkaloid content should <20 ppm)

- Good storability (measured through days to sprouting, less rotting percentage, lesser moisture loss at any particular storage temperature)
- Resistance to late blight (<6 disease score)
- PVY resistance (extreme resistance to PVY)
- PLRV resistance (high level of resistance)
- PVX resistance (extreme resistance to PVX)
- Bacterial wilt resistance (measured through percentage of wilted plants and degree of wilting as noted by scoring and percent latent infection)
- Heat tolerance (tuberization when night temperature goes to >22 °C; bulking under warm day temperatures)
- Water productivity (measured through fresh tuber yield: applied water expressed as $kgha^{-1}mm^{-1}$)
- Cold chipping ability (>18% dry matter score; after cold storage chip color score of 20% selections should be 1–2)

Developing earliness, resistance to biotic stress like late blight, and abiotic stress like drought, besides high yield, was the most important research arenas for potato improvement (Bonierballe et al. 2020).

7 Potato Breeding: Principles, Methods, and Approaches

Improved variety of potatoes has been successfully bred from largely tetraploid landraces and advanced breeding clones from two gene pools, i.e., Chilotanum that is adapted to long-day conditions and distributed in the lowlands, and Andigenum that is adapted to the short-day conditions and distributed in the uplands (Spooner et al. 2007; Bonierbale et al. 2020). It was the Andigenum group that gave rise to the contemporary Tuberosum group, which is well suited to the Northern hemisphere (Gavrilenko et al. 2013).

Clonal (vegetative, asexual) reproduction is the method used to propagate potato and tubers are used for the purpose. It is feasible to create botanical seed in potatoes, which allows for the creation of novel genotypes which may be diploid, tetraploid, or hexaploid in nature. Once variation has been produced through sexual means of propagation, any seedling can be propagated through vegetative means to establish a new variety. Clonal propagation produces a low propagation coefficient (five- to tenfold every generation), despite the fact that most potatoes generate several fruits per plant and a considerable number of botanical seeds (~200) per fruit. Exposure to diseases and pathogens is facilitated by clonal breeding which aids in the breeding process by removing unwanted genotypes, but it also necessitates precautions against harmful viruses that are spread through seed tuber. In potato, the final harvested produce (tuber) remains below the soil, which restricts the phenotyping of the developmental stages of the tuber.

Potato is an asexually propagated or clonally propagated crop. It has the advantage of fixation of the traits since the clones are true-to-type and therefore the maintenance is simpler compared to sexually propagated crops. However, transmission of diseases through vegetative means of propagation is more which is a disadvantage. Besides, potato propagule is bulky, heavier, and perishable in nature, and production of healthy planting material is a costly affair. Therefore, the aim of potato breeding should be a combination of high-priority traits through crossing of diverse parental genotypes and identification of superior individual plants having many desirable traits and lesser negative traits.

Selection of parental lines becomes very important in such condition. Potatoes typically have heterozygous parental genotypes, with the exception of inbreeding lines produced through doubled monoploid production or self-fertilization. Therefore, the resultant F_1s differ at many loci often. Once these heterozygous genotypes have been clonally propagated and fixed (i.e., stabilized genetically), they are amenable to selection. Finding suitable parents is particularly difficult because of polyploid nature, heterozygous genotypes, and heterosis; nonetheless, the performance of a parental line may be determined by the progeny. Good parental lines can be identified well in advance through genomic selection. A considerable extent of allelic diversity and interactions is observed in heterozygous tetraploid potato genotypes, and this determines their performance.

Breeding of potato involves crossing of one "F_1 clone hybrid" with another "F_1 clone hybrid." In potato breeding program, the parental genotypes used are seldom inbred, and thus the resultant genotype once created cannot be recreated since the recombination is unique. The resultant progeny is highly heterogenous and heterozygous, segregate for diverse traits, and directional breeding can result in accumulation of good traits. Breeding strategies to improve potato has focused on the yield enhancement through maximizing heterozygosity for the trait through crosses (Chase 1963), though use of wild and unadapted parental lines is most unlikely to produce high yielding progeny (Bonierbale et al. 1993). In the cultivated gene pool, two-way crosses involving cultivated and wild species have resulted in marginal heterosis at times with respect to increased yield and vigor, but in three-way crosses involving one cultivated potato and two wild species heterosis has not been noted (Sanford and Hanneman 1982). Therefore, genetic distance or maximum heterozygosity alone is not likely to result in heterotic groups in potato.

The advantage of potato, which is clonally propagated is fixation of the traits when clonal selection is followed and any transgressive and superior segregant can be released as variety. However, clonal selection as a method of potato improvement is often neglected (Gruneberg et al. 2009).

7.1 Importance of Selection in Potato Breeding

Every time a hybrid is created, a significant amount of features segregate because potatoes are highly heterozygous. Each progeny inherits a different mix of favorable alleles at certain sites and unfavorable alleles at others, and is never "perfect." Determining which phenotypes are valuable to retain and which should be eliminated at each level of the selection process is, therefore, of great importance, which depends on the breeding priorities. Early generation selection in potato is only applicable for highly heritable traits. On the other hand, meaningful yield assessments are only possible at a later time. As selection stages advance, the product profile becomes crucial for the breeder and the plants not conforming to the breeders' criteria are eliminated. This means that rather than selecting solely based on individual traits, selection should take into account the overall attributes that make up the final product.

The preferred technique for selecting superior potato lines has been phenotypic recurrent selection, which typically begins with 100,000 seedlings from 200 to 300 crossings and continues for many years using clonal selection (Bradshaw 2017; Bradshaw 2009). There are numerous instances of significant varieties being released following clonal selection and crossing for at least 30 years, while 13–14 years should be the ideal amount of time as per many researchers (Jansky and Spooner 2017; Bradshaw 2009). The primary causes of this lengthy cycle of variety development are the quantitative nature of the majority of significant traits, the quick expression of inbreeding depression, and the lack of strong early-generation selection. Due to their slow rate of multiplication and susceptibility to pathogen contamination, tuber propagation causes additional delays for enough production of high-quality seed tubers for field selection at multiple locations.

Successive cycles of crossover and selection can lead to accomplishment of medium- to long-term genetic gain. Recurrent selection is the process of selecting plants again and again, generation after generation, by inter-plant mating that results in a recombinant population that will be used in the subsequent selection cycle. Improvement of mean performance of plant population is the chief aim, while maintaining the high amount of genetic variability remains a secondary, yet important, objective. Recombination time and again aids in breaking of linkage drags, thus allowing good traits from the non-adapted genotype to be carried forward and undesirable ones discarded. Thereafter, the selected genotypes may be directly used as variety or as pre-breeding lines or as parents in future breeding programs.

Unlike diploid potatoes, polyploid potatoes have the ability to pass on multiple alleles from one locus to the next generation through gametes, and selection may be based on dominance genetic variance till the population attains homozygosity, which is very common in tetraploid potatoes (Wricke and Weber 1986; Gallais 2004). This utilization of dominance variance in population improvement of potatoes leads to creation of more diversity and brings more flexibility in the potato improvement programs (Ortiz 1998).

Any breeding program must include the selection of parents as a crucial first step, particularly for medium- and long-term breeding success. Besides, for success of any breeding program, the breeding goals, the availability of infrastructure and resources, and most importantly the selection of parental lines play a very important rolc.

Due to the highly heterozygous and polyploid nature of potatoes, the clone performance is determined by the dominance and epistatic effects. As a result, until a parent or particular cross combination is tested, little is known about its worth.

Strategically, parental value can be determined by evaluating pedigrees, conducting progeny experiments in suitable breeding schemes, or empirically by observing selection ratios. The best progeny of potato can be traced back to a relatively small number of crosses, even in cases where hundreds are made. Predicting which genotypes will make the greatest parental lines is therefore beneficial, as then efforts may be focused on generating and assessing the most promising cross combinations. It is important to evaluate the performance of all cross combinations. The genetic makeup of any trait donor determines its parental value. The ideal parental lines may be identified through targeted traits and progenies that achieve those targeted traits may be identified through specific combining ability (SCA) and the better parental genotypes through general combining ability (Sprague and Tatum 1941). Parental lines with high GCA are good combiners, while the superior cross combination may be identified through SCA.

7.2 Early and Late Generation Selection

Early generations allow for the selection of highly heritable traits for which a single or small number of plants can be used to accurately assess a genotype. However, for complex traits such as yield, selection can only be possible after replicated trials.

Potato improvement requires a lot of time (at least 8–10 years or more from crossing to variety release), and the quantity of planting material of potato is low due to the low propagation coefficient of the crop. When there is little availability of seeds, evaluation and selection are at an earlier stage. However, for complex traits, replicated trials are necessary and selection at later stage is only possible. Selection at multiple stages may also be done from early to late stages through indices providing weightage to different characters. In actuality, some characters are chosen in a sequential manner, particularly in situations where the lowest possible value is evident (such as in the cases of tuber color, shape or size, and pest and disease resistance), while other characters are chosen all at once by combining them into an index, which is frequently an index that is formed intuitively, like the score values of overall performance.

In potatoes, if large amounts of true seed plants can be cloned and the subsequent population is evaluated properly, it is theoretically possible to identify a superior genotype in the first year itself since highest amount of variation resulting from recombination is created. In actuality, however, in order to allow for sufficient plot size and replication for assessments conducted at a later stage, the individual genotypes are chosen during the initial phases of each clonal selection cycle propagated by tubers.

The susceptibility of clonal planting material to disease should be taken into account in the overall selection system. Prior to their distribution throughout other geographic areas, disease-free tuber production of the superior selected clones should be done in areas that are free from the diseases.

When it comes to early selection, many breeding programs stick to one site and use disease-free seeds (produced at a different location) usually done at later phases of selection, in order to carry out evaluation under ideal conditions of station.

To help maintain seed clean over time, this site has to be kept virus-free as much as possible. If enough clonal seed can be generated, the first clonal generation can be assessed in multiple locations. The method involves transformation of true seed families (TSF) into true tuber families (TF), and at least four clones of each genotype from each family are produced that may be simultaneously evaluated at three locations, enabling a multi-location trial of first-generation selected clones.

7.3 Utilization of Diverse Genotypes Through Traditional and Modern Breeding Approaches

Comparing the genetic base of potato varieties grown in major commercial areas to the accessible gene pool for traditional potato breeding reveals a rather narrow genetic base. This is probably because of the limited genetic base of the original tetraploid sources (Andigenum and Tuberosum) which were employed as breeding stock. Crossability among diverse wild and cultivated species is possible directly, or via any bridge species (Plaisted and Hoopes 1989). Nonetheless, the successful introgression of beneficial genes from wild species into contemporary potato types has been limited. According to Sood et al. (2017), about 40% of wild species have genetic features that are valuable for abiotic stressors, illnesses, and pests. Some major interspecific crosses with wild species have led transfer of major genes for disease resistance, such as *Solanum demissum* and *Solanum bulbocastanum* (also a bridge species) for late blight, *Solanum stoloniferum* for viral diseases, and *Solanum spegazzini* and *Solanum vernei* for nematodes (Finkers-Tomczak et al. 2011; Bradshaw 2009; Bradshaw and Ramsay 2005). Apart from these, the bridge species *Solanum acaule* is also a donor frost resistance, besides viral diseases and late blight (Watanabe et al. 1992), *Solanum phureja* having high protein content (Haverkort et al. 2009). All these wild species have been involved in development of the modern-day potato varieties. Due to the quantitative nature of many significant potato traits and the uncertainty surrounding pedigree information, the actual contribution of wild relatives to present potato cultivars remains underestimated.

Unreduced (2n) gametes in diploid potatoes have aided the introgression of wild species genomes into the cultivated groups; this phenomenon has been found to be strongest in group Tuberosum due to intense breeding efforts utilizing a number of wild species (Plaisted and Hoopes 1989; Hardigan et al. 2017). The idea that there might be an inversely proportional relation between disease resistance and yield stems from the fact that wild species are carriers of genes for wild features that come as linkage drag along with the genes for disease resistance (Ning et al. 2017). The limited use of wild species in potato breeding may have resulted from this notion.

In the realm of traditional breeding, methods such as inter- and intraspecific crosses, backcrossing, recurrent selection, and germplasm evaluation have been employed in Southern and Central American countries to produce more productive cultivars that are resistant to *P. infestans* (Gabriel et al. 2001; Barquero et al. 2005; Solano et al. 2014). Three new varieties of creole potatoes were obtained in Colombia by Rodríguez et al. (2009) and recorded as Criolla Latina, yielding between 18 and 20 t/ha, Criolla Paisa, yielding between 22 and 25 t/ha, and Criolla Colombia, yielding between 13 and 15 t/ha. While the latter was observed to be sensitive to *P. infestans*, the first two were reported to have a modest tolerance to this pathogen.

The genotypes UdenarStGua53 (CIP 300046.22), UdenarStGua61 (CIP 391011.17), UdenarStGua68 (CIP 392557.171), UdenarStGua73 (CIP 393079.4), UdenarStGua75 (CIP393280.82), UdenarStGua77 (CIP 393371.164), and UdenarStGua78 (CIP 393371.58) from the CIP, Peru, were reported to be tolerant to attack by *P. infestans* (CIP 2019) and also showed good performance with respect to yield and late blight tolerance at Columbia (Lagos-Regalado et al. 2021).

According to Mihailovich et al. (2014), accessions LOP-868, HUA-332, and OCH-7643—new sources of PLRV resistance—had remarkably high GCA for this trait after being validated through mass selection in *S. tuberosum* (Andigenum Group) accessions kept at CIP, Peru. Supporting trait research has been built on further characterization of these sources, with the goal of increasing PLRV resistance frequencies and levels in various potato populations.

Rladg allele for potato leaf roll virus resistance was traced to potato chromosome V using a population of dihaploid individuals derived from LOP-868. Its numerous resistance mechanisms and a duplex allelic conformation of a main dominant gene in the donor accession were shown to be responsible for its high heritability (Velásquez et al. 2007). To aid in the introgression and tracking of this important gene in advanced breeding populations, a PCR-based marker (RGASC850) situated 2 cM from the gene was created (Mihovilovich et al. 2014). *Ryadg* information from Kasai et al. (2000) has been used to add this marker into a multiplex PCR method for marker assisted selection of PLRV and PVY resistance.

Suganthy and Udayasoorian (2016) studied the effect of enhanced ozone levels on ten Indian potato genotypes (Kufri Himalini, Kufri Himsona, Kufri Surya, Kufri Muthu, Kufri Jyothi, Kufri Jawahar, Kufri Giridhari, Kufri Giriraj, Kufri Swarna, Kufri Chipsona) and found that Kufri Surya was moderately resistant to ozone elevation levels (100, 150, and 200 ppb) showing appreciable yield of 650g/plant of fresh tuber and least visible injury along with favorable related parameters, while Kufri Himalini was the most susceptible bearing no tuber at all and showing maximum visible injury besides other traits.

Govindakrishnan et al. (1999) reported Kufri Asoka and MS-78-1 to be low nitrogen requiring potato genotypes and hence eco-friendly.

Gelmesa et al. (2024) carried out screening of 23 diverse potato genotypes at two locations of Ethiopia against heat stress, drought stress, and heat + drought combined stress. The genotypes obtained from CIP, viz., Cl-34(CIP-397079.6), Cl-15(CIP-304394.56), Cl-24(CIP-392781.1), Cl-28(CIP-395436.8), and

Table 2 Potato genotypes from CIP, Peru, having abiotic and biotic stress tolerance

Stress	Tolerant genotype
Heat	CIP-112 (380606.6) CIP-127 (392820.1) CIP-139 (396311.1) CIP-139 (396311.1) CIP-202 (302476.108) CIP-203 (302498.70) CIP-220 (304366.46) CIP-229 (304387.17) CIP-232 (304394.56) CIP-235 (304405.47)
Salinity	CIP-112 (380606.6) CIP-139 (396311.1) CIP-205 (302499.3) CIP-221 (304368.46)
Late blight	CIP-221 (304368.46) LB-7 (393371.58)

Source: Modified from Gelmesa et al. (2024)

Cl-14(CIP-304387.39) were identified as the best drought tolerant; Cl-34(CIP-397079.6), Cl-15(CIP-304394.56), Cl-14(CIP-304387.39), Cl-32(CIP-397036.7), and Cl-10(CIP-304369.22) were noted to be the best heat tolerant; Cl-12(CIP-304371.67), Cl-13(CIP-304383.80), Cl-2(CIP-302499.30), Cl-34(CIP-397079.6), Cl-15(CIP-304394.56), and Cl-14(CIP-304387.39), were reported to be most tolerant under combined drought-heat stress tolerant genotypes (Table 2).

From the 1980s through the late 1990s, the International Potato Centre (CIP), Peru, worked on strengthening its lowland tropical potato population with multiple viral disease resistance aiming at the lowland tropics virus resistant (LTVR) population. An open-recurrent selection approach was utilized in the development of the broad-based LTVR population to transfer the genes from foreign elite and advanced potato clones into the lowland tropical breeding population. Day lengths of 10–14 h, lowest night-time temperatures of 18–20 °C, mean maximum temperatures above 25 °C , and mean yearly soil temperatures of 22 °C or higher at 50 cm depth are the typical characteristics of lowland tropical ecosystems. The aim of developing these LTVR population was to bring viral disease (PVX and PVY) resistance, source of which was the native Andigenum and long-day adapted Andigenum group (Neotuberosum) accessions (Muñoz et al. 1975; Gálvez et al. 1992), heat tolerance, and early bulking ability (source being *S. tuberosum* germplasm bred under the summer conditions of the northern Hemisphere) (Cubillos and Plaisted 1976) on the other (Plaisted 1987; Mendoza 1990).

In every trial where LTVR families were assessed, strict screening for PVX and PVY in early generation was regularly carried out and susceptible progenies were eliminated. The CIP experimental stations in Peru served as testing grounds; these are hot, humid lowland environments that are rather dry and humid. Hybrids from *S. tuberosum* × Neotuberosum crosses outperformed those from

S. tuberosum × *S. tuberosum* or *S. tuberosum* × Andigenum Group in experiments comparing the yield stability of various clones throughout different locations. The superior performance of crossings between *S. tuberosum* and Neotuberosum clones may have been explained by the higher level of heterozygosity in conjunction with earlier selection of the divergent parental materials for adaptation (Mendoza and Estrada 1979).

Additionally, these researchers observed that *S. tuberosum* × Neotuberosum hybrids could yield tubers in 60–70 days after planting, even if they are grown under short days and were adapted to hot and humid conditions, and might have been the reason for the adoption of potato in warmer locations. Many multiplex (genetic makeup of a polyploid with several copies of an allele, RRrr or RRRr in this case, guaranteeing that approximately 100% of its cross-progeny will inherit the resistance imparted by a significant gene) PVY and/or PVX extreme resistant progenitors were selected during the ensuing genotypic recurrent selection cycles, and of these, some also were good combiners for yield and dry matter content (Mendoza et al. 1996).

During each recurrent selection cycle, clones with higher yield coupled with superior tuber quality were chosen in tandem with population growth and parental line development. Using an open-recurrent selection technique, a second breeding stage sought to include foreign elite and advanced clones. The genetic factors in a North Carolina Design II assessment yielded a heritability of 0.54–0.69 for resistance to PLRV infection. When sprouted tubers were used for controlled inoculations with greenhouse-reared infective aphids prior to tuber families being planted in the field, higher estimates were obtained, suggesting that under controlled inoculation escape was not possible and errors were minimized due to higher load and uniformity in distribution of inoculum.

Resistance variability was found to be mostly influenced by additive genetic effects (74%), followed by dominance effects (10%) and additive × environment (location) effects (5%), suggesting that future cycles of recurrent selection may bring about greater breeding advancements (Salas 2002).

7.4 Modern Breeding Techniques

The majority of breeding techniques place little emphasis on genotypic selection at the molecular level and mostly rely on phenotypic selection to improve germplasm. Conventional breeding is a hard and time-consuming process that attempts to gradually integrate desired features from superior individuals. In comparison, marker-assisted breeding can reduce the time of potato genotypic improvement and may also be less input-demanding and hence cost-effective (Slater et al. 2013).

Few potato breeding programs have used marker-assisted selection (MAS), especially those that have markers specific to disease resistance alleles (Schultz et al. 2012; Ortega and Lopez-Vizcon 2012). MAS has been more efficient over traditional phenotypic selection alone. Under this system, selection can be made for other more complicated features such as yield, cooking characteristics, and most physical traits of plants and tubers (van Eck et al. 2007; Ramakrishnan et al. 2015).

MAS in potato can enable selection of major genes and identification quantitative trait loci (QTL) with significant impacts at an early stage (Sharma et al. 2014; Gebhardt 2013). QTLs for important traits such as yield and starch content of tubers and specific markers for some important traits have been developed in recent years (Schönhals et al. 2016), and also for viral disease resistances (del Rosario et al. 2018; Mihovilovich et al. 2014), and other essential traits (Ramakrishnan et al. 2015). Marker-assisted selection can be economical if used multiplexed and at an early stage of clonal production (Slater et al. 2013).

In order to shorten the breeding cycle and increase section intensity, estimated breeding values for traits that can be deduced from pedigree data have also been proposed (Slater et al. 2014a, 2014b). According to Slater et al. (2016), genomic selection is also suggested as a way to enhance the combination of unknown QTL early in the breeding phase. Several researchers (Hamilton et al. 2011; Uitdewilligen et al. 2013; Vos et al. 2015) have developed high-density and high-throughput polymorphic marker systems in potatoes. Genome Wide Assisted Studies (GWAS) for allele/gene identification have been limited due to the very small number of meiosis that separate modern cultivars, as corroborated by recent estimates of linkage disequilibrium in populations of potato cultivars (Vos et al. 2017; Sharma et al. 2018).

For over 40 years, somatic hybridization has been utilized in the potato to overcome crossing barriers in cultivated and wild species and successful fusion products containing various characteristics have been obtained from 23 different species of *Solanum* (Tiwari et al. 2018). Somatic hybrids have been used in many researches to isolate candidate genes and better understand the genetic architecture of critical traits. However, it appears that no variety has been created via somatic hybrids with cultivated potatoes. This is probably because tetraploid potato breeding is limited in its ability to effectively eliminate unwanted alleles.

Compared to normal breeding methods, direct introgression of genes through transgenics is a very new approach to potato improvement. However, it has proven extremely successful for a few qualities where exploitation of natural genetic diversity was not feasible. In the late 1980s, the first transgenic potato was developed, and the combined resistance to PVX and PVY in a commercial cultivar was first documented by Lawson et al. (1990). Soon after, new commercial transgenic cultivars were created by Perlak et al. (1993) as a result of the engineering for insect resistance.

Between 1995 and 2001, the first generation of transgenic potatoes were sold under the brand name NewLeaf™ in the United States and Canada. However, the production of the NewLeaf potatoes soon declined due to their high production cost without much benefits (Thornton 2003), and by 2004 their commercial sale stopped. There were no new transgenic potatoes released for 10 years. Recently, a number of transgenic potatoes have been developed, featuring lower levels of reducing sugars, low acrylamide potential, reduced black spot, reduced bruising and browning, and resistance to late blight.

There are many options for creating engineered novel potato with features through which both the grower and the consumer may benefit (Halterman et al. 2016), but the acceptance of transgenic among the consumers is still erratic and unstable, and many potato-producing nations currently lack science-based regulatory frameworks. Therefore the scope of using genetic engineering for developing genotypes targeting specific traits becomes limited.

The most recent genetic engineering method that targets specific DNA sequences is genome editing (Yin et al. 2017; Scheben et al. 2017). This technique aims at targeted mutagenesis of specific genes through knock-out or deletion or allelic alterations. Several researchers have worked on genome editing of potatoes and created novel and useful products (Wang et al. 2015; Nicolia et al. 2015; Butler et al. 2015, 2016; Clasen et al. 2016; Andersson et al. 2017; Ma et al. 2017).

Two genome editing methods, TALEN (transcription activator-like effector nucleases) and CRISPR/Cas9, have been used for genome editing in potato, which can overcome the problem of linkage drag of traditional breeding. The desired mutation was produced via transient expression using PEG-mediated protoplast transfection or *Agrobacterium*-mediated leaf infiltration; however, these techniques can be labor-intensive due to the lack of a selective agent and somatic variation in plants regenerated from protoplasts. Virus vectors might facilitate the delivery of the editing reagent, but there are extra challenges associated with their removal and propagation. Thus, although still in its early stages of development, genome editing technology has significant prospects for the potato industry.

A novel approach to potato breeding called true hybrid potato is becoming more widely popular since it can remove many of the problems associated with traditional tetraploid potato breeding. A self-incompatibility inhibitor gene was identified in the wild species *Solanum chacoense* two decades ago (Hosaka and Hanneman 1998a, 1998b). The S locus inhibitor gene (*Sli*), after being introgressed into a diploid cultivated potato, made the genotype self-compatible (Phumichai et al. 2005), and introgression of this *Sli* gene into diploid potato lines became one of the breeding goals of potato breeders. This resulted in the production of S_3 diploid lines, which had good yield and agronomic performance with 80% homozygosity (Lindhout et al. 2011). To create recombinant inbred line populations, an inbred line of *S. chacoense* (M_6) was created concurrently (Jansky et al. 2014). The utilization of the wild *S. chacoense*, however, should be avoided by using new sources of the self-compatibility system. Self-compatible landraces, are rare, but occur within the diploid cultivated potato germplasm and may be utilized to create inbred lines from other gene pools, including the Phureja and Stenotomum groups (Haynes and Guedes 2018). Using the CRISPR–Cas9 gene editing technique, the self-incompatibility gene S-RNase is knocked off and this technique is more efficient (Ye et al. 2018).

Hybrid varieties are propagated by true seeds and may have rapid genetic gain through the fixation of main genes for critical qualities such as disease resistance and the exploitation of heterosis through hybridization. Because of their low weight, low pathogen content, good storability, feasibility for beneficial coating, and rapid rate of multiplication, botanical seeds have long been recognized as an incredibly intriguing substitute for tuber seeds.

Several potato varieties have been developed through potato hybridization methods, but adoption of these varieties is dependent on supply of the seed tubers at reasonable prices (Almekinders et al. 2009). Therefore, various organizations working on potato research aimed to develop and popularize true potato seeds (TPS), which is the botanical seed of potato as a complement for the traditional propagule, i.e., tubers. But till today the adoption rate of TPS among farmers is meagre. Oliver F_1, a TPS variety recently produced by a Dutch breeding company Bejo Zaden B.V. (http://www.bejo.com/magazine/bejo-introduces-its-first-true-potato-seed-variant), is being used currently in several African nations where it is difficult to find high-quality seed tuber.

The benefits of true hybrid potato seeds are enormous, but development of superior potato inbred parental lines having disease resistance and other desirable traits, suited to wide ecology, for the purpose will require large amount of time. Whether the technology of true hybrids in potatoes will be accepted by the farmers, particularly small and marginal, of poor and developing nations, is to be seen.

8 Conclusion

There are ways for improving conventional potato breeding. However, new techniques and instruments are needed to overcome the narrow genetic base of cultivated potatoes and the various other hindrances of breeding of polyploid crop, besides the low coefficient of seed tuber propagation, and rapid inbreeding depression of potato. The new breeding techniques especially with biotechnological interventions will complement the traditional potato breeding and result in genotypes for farmers that are widely adapted, with disease and pest resistance having high quality and having consumer preference.

References

Almekinders CJM, Chujoy E, Thiele G (2009) The use of true potato seed as pro-poor technology: the efforts of an international agricultural research institute to innovating potato production. Potato Res 52:275–293

Andersson M, Turesson H, Nicolia A, Fält AS, Samuelsson M, Hofvander P (2017) Efficient targeted multiallelic mutagenesis in tetraploid potato (Solanum tuberosum) by transient CRISPRCas9 expression in protoplasts. Plant Cell Rep 36(1):117–128. https://doi.org/10.1007/s00299-016-2062-3

Barquero M, Gómez L, Brenes A (2005) Resistencia al tizón tardío (Phytophthora infestans)en clones promisorios de papa en Costa Rica. Agron Costarricense 29(3):31–45. https://www.mag.go.cr/rev_agr/v29n03_031.pdf

Bonierbale MW, Plaisted RL, Tanksley SD (1993) A test of the maximum heterozygosity hypothesis using molecular markers in tetraploid potatoes. Theor Appl Genet. 86(4):481–91. https://doi.org/10.1007/BF00838564. PMID: 24193596

Bonierbale MW, Amoros WR, Salas E, de Jong W (2020) Potato breeding. In: Campos H, Ortiz O (eds) The potato crop. https://doi.org/10.1007/978-3-030-28683-5_6

Bradshaw JE (2009) Potato breeding at the Scottish Plant Breeding Station and the Scottish Crop Research Institute: 1920–2008. Potato Res 52:141–172. https://doi.org/10.1007/s11540-009-9126-5

Bradshaw JE (2017) Review and analysis of limitations in ways to improve conventional potato breeding. Potato Res 60:171–193. https://doi.org/10.1007/s11540-017-9346-z

Bradshaw JE, Ramsay G (2005) Utilisation of the commonwealth potato collection in potato breeding. Euphytica 146(1–2):9–19. https://doi.org/10.1007/s10681-005-3881-4

Burton WG (1989) The potato, 3rd edn. Longman Scientific and Technical, New York/Harlow. 742 p

Butler NM, Atkins PA, Voytas DF, Douches DS (2015) Generation and inheritance of targeted mutations in potato (Solanum tuberosum L.) using the CRISPR/Cas system. PLoS One 10(12):e0144591. https://doi.org/10.1371/journal.pone.0144591

Butler NM, Baltes NJ, Voytas DF, Douches DS (2016) Geminivirus-mediated genome editing in potato (Solanum tuberosum L.) using sequence-specific nucleases. Front Plant Sci 7:1045. https://doi.org/10.3389/fpls.2016.01045

Chase SS (1963) Analytic breeding in Solanum tuberosum L.—a scheme utilizing parthenotes and other diploid stocks. Can J Genet Cytol 5:359–363

CIP (2019) Catálogo de clones avanzados del CIP. En: Centro Internacional de la Papa. 1 p. https://research.cip.cgiar.org/cipcatlg_ac/Catalogue.php?cipnumber=CIP391002.6. Retrieved Nov 2019

Clasen BM, Stoddard TJ, Luo S, Demorest ZL, Li J, Cedrone F, Coffman A (2016) Improving cold storage and processing traits in potato through targeted gene knockout. Plant Biotechnol J 14(1):169–176. https://doi.org/10.1111/pbi.12370

Comai L (2005) The advantages and disadvantages of being polyploid. Nat Rev Genet 6:836–846

Cubillos AG, Plaisted RL (1976) Heterosis for yield in hybrids between S.tuberosum ssp. tuberosum and tuberosum ssp. andigena . American Potato Journal 53, 143–150. https://doi.org/10.1007/BF02854117

De Nettancourt D (1977) Incompatibility in angiosperms. Springer, Berlin

del Rosario HM, Vidalon LJ, Montenegro JD, Riccio C, Guzman F, Bartolini I, Ghislain M (2018) Molecular and genetic characterization of the Ryadg locus on chromosome XI from Andigena potatoes conferring extreme resistance to potato virus Y. Theor Appl Genet 131(9):1925–1938. https://doi.org/10.1007/s00122-018-3123-5

FAOSTAT (2020) Statistical Database. Food and Agriculture Organization of the United Nations, Rome.

Finkers-Tomczak A, Bakker E, de Boer J, van der Vossen E, Achenbach U, Golas T, Suryaningrat S, Smant G, Bakker J, Goverse A (2011) Comparative sequence analysis of the potato cyst nematode resistance locus H1 reveals a major lack of co-linearity between three haplotypes in potato (Solanum tuberosum ssp.). Theor Appl Genet 122(3):595–608. https://doi.org/10.1007/s00122-010-1472-9

Frankel R, Galun E (1977) Pollination mechanisms, reproduction, and plant breeding. Springer, New York

Gabriel J, Carrasco E, García W, Equise H, Navia O, Torres R, Ortuño N, Franco J, Thiele G, yEstrada N. (2001) Experiencias y logros sobre mejoramiento convencional y selecciónparticipativa de papa en Bolivia. Revista Latinoamericana de la Papa 12:169–192

Gallais A (2004) Quantitative genetics and selection theory in autopolyploid plants. INRA, Paris

Gálvez R, Mendoza HA, Fernández-Northcote EN (1992) Herencia de la inmunidad al virus Y de la papa (PVY) en clones derivados de Solanum tuberosum spp. Andigena. Fitopatología 27:8–15

Gavrilenko T, Antonova O, Shuvalova A, Krylova E, Alpatyeva N, Spooner DM, Novikova L (2013) Genetic diversity and origin of cultivated potatoes based on plastid microsatellite polymorphism. Genet Resour Crop Evol 60(7):1997–2015. https://doi.org/10.1007/s10722-013-9968-1

Gebhardt C (2013) Bridging the gap between genome analysis and precision breeding in potato. Trends Genet 29:248–256. https://doi.org/10.1016/j.tig.2012.11.006

Gelmesa D, Dechassa N, Mohammed W, Etissa E, Gebre E, Monneveux P (2024) Index-based selection of potato (*Solanum tuberosum* L.) genotypes for tolerance to soil moisture and ambient heat stress. Crop Sci 64:1266–1283. https://doi.org/10.1002/csc2.21061

Govindakrishnan PM, Sharma RC, Khan IA, Sharma HC (1999) A technique for selecting potato *(Solanum tuberosum)* genotypes with low nitrogen Requirement. Indian J Agric Sci 69(5):350–354

Grüneberg WJ, Mwanga R, Andrade M, Espinoza J (2009) Chapter 5: selection methods Part 5: breeding clonally propagated crops. In: Ceccarelli S, Guimarães EP, Weltzien E (eds) Plant breeding and farmer participation. FAO, Rome, pp 275–322

Halterman D, Guenthner J, Collinge S, Butler N, Douches D (2016) Biotech potatoes in the 21st century: 20 years since the first biotech potato. Am J Potato Res 93(1):1–20. https://doi.org/10.1007/s12230-015-9485-1

Hamilton JP, Hansey CN, Whitty BR, Stoffel K, Massa AN, Deynze AV, De Jong WS, Douches DS, Buell CR (2011) Single nucleotide polymorphism discovery in elite North American potato germplasm. BMC Genomics 12:302

Hanneman RE Jr (1994) The testing and release of transgenic potatoes in the North American Center of diversity. In: Krattiger AF, Rosemarin A (eds) Biosafety of sustainable agriculture: sharing biotechnology regulatory experiences of the Western Hemisphere. ISAAA/SEI, Stockholm/Ithaca, pp 47–67

Hardigan MA, Laimbeer FPE, Newton L, Crisovan E, Hamilton JP, Vaillancourt B, Wiegert-Riningera K, Wooda JC, Douches DS, Farréa EM, Veilleux RE, Buell CR (2017) Genome diversity of tuber-bearing Solanum uncovers complex evolutionary history and targets of domestication in the cultivated potato. Proc Natl Acad Sci 114(46):E9999–E10008. https://doi.org/10.1073/pnas.1714380114

Harlan JR, de Wet JMJ (1971) Toward a rational classification of cultivated plants. Taxon 20(4):509–517. https://doi.org/10.2307/1218252

Haverkort AJ, Struik PC, Visser RGF, Jacobsen E (2009) Applied biotechnology to combat late blight in potato caused by Phytophthora infestans. Potato Res 52(3):249–264

Hawkes JG (1990) The potato: evolution, biodiversity and genetic resources. Belhaven Press, London, p 259

Hayder A, Ahmed MB, Hannan MM, Razvy MA, Mandal MA, Salahin M, Karim R, Hossain KM (2007) Analysis of genetic diversity in some potato varieties grown in Bangladesh. Middle-East J Sci Res 2(3–4):143–145

Haynes KG, Guedes ML (2018) Self-compatibility in a diploid hybrid population of Solanum phureja–S. stenotomum. Am J Potato Res 95(6):729–734. https://doi.org/10.1007/s12230-018-9680-y

Hosaka K, Hanneman RE (1998a) Genetics of self compatibility in a self-incompatible wild diploid potato species Solanum chacoense. I. Detection of an S locus inhibitor (Sli) gene. Euphytica 99:191–197

Hosaka K, Hanneman RE (1998b) Genetics of self-compatibility in a selfcompatible wild diploid potato species Solanum chacoense. 2. Localization of an S-locus inhibitor (Sli) gene on the potato genome using DNA markers. Euphytica 103:265–271. http://www.bejo.com/magazine/bejo-introduces-its-first-true-potato-seed-variant https://cipotato.org/genebankcip/potato-cultivated/. Accessed 13 Aug 2024

Irikura YB (1989) Collected data of plant genetic resources 2. Matsuo T (ed). Kodansha Scientific, Tokyo, pp 498–518

Jansky SH, Spooner DM (2017) The evolution of potato breeding. Plant Breed Rev 41:169–211

Jansky SH, Chungand Y, Kittipadukal P (2014) M6: a diploid potato inbred line for use in breeding and genetics research. J Plant Regist 8:195–199. https://doi.org/10.3198/jpr2013.05.0024crg

Kasai K, Morikawa Y, Sorri VA, Valkonen JP, Gebhardt C, Watanabe KN (2000). Development of SCAR markers to the PVY resistance gene Ryadg based on a common feature of plant disease resistance genes. Genome. 43(1):1–8. PMID: 10701106

Kumari M, Kumar M, Solankey SS (2018) Breeding potato for quality improvement. In: Potato—from incas to all over the world. Intech Open, London, pp 37–59. https://doi.org/10.5772/intechopen.71482

Lagos Regalado JJ, Lagos Burbano TC, Duarte Alvarado DE, Lagos Santander LK (2021) Selection of potato genotypes Solanum tuberosum group Andigena by their tolerance to Phytophthora infestans (Mont.) of Bary. Revista Facultad Nacional de Agronomía Medellín, 74(1), 9361–9372. https://doi.org/10.15446/rfnam.v74n1.87566

Lawson C, Kaniewski W, Haley L, Rozman R, Newell C, Sanders P, Tumer NE (1990) Engineering resistance to mixed virus infection in a commercial potato cultivar: resistance to potato virus X and potato virus Y in transgenic Russet Burbank. Nat Biotechnol 8(2):127

Levin DA (1983) Polyploidy and novelty in flowering plants. Am Nat 122:1–25

Lindhout P, Meijer D, Schotte T, Hutten RC, Visser RG, van Eck HJ (2011) Towards F 1 hybrid seed potato breeding. Potato Res 54:301–312

Lutaladio N, Ortiz O, Haverkort A, Caldiz D (2009) Sustainable potato production; guidelines for developing countries. FAO, Rome

Ma J, Xiang H, Donnelly DJ, Meng FR, Xu H, Durnford D, Li XQ (2017) Genome editing in potato plants by agrobacterium-mediated transient expression of transcription activator-like effector nucleases. Plant Biotechnol Rep 11(5):249–258. https://doi.org/10.1007/s11816-017-0448-5

Mendoza HA (1990) Mejoramiento poblacional. Una estrategia para la utilizaci6n del germoplasma de papa. En: Mejoramiento de papa en el Cono Sur, pp 47–62

Mendoza HA, Estrada NR (1979) Breeding potatoes for tolerance to stress: heat and frost. In: Mussel H, Staples RR (eds) Stress physiology in crop plants. Wiley, New York, pp 227–262

Mendoza HA, Mihovilovich EJ, Saguma F (1996) Identification of triplex (YYY) potato virus Y (PVY) immune progenitor derived from Solanum tuberosum spp. Andigena. Am Potato J 73(1):13–19. ISSN 0003-0589

Mihovilovich E, Aponte M, Lindqvist-Kreuze H, Bonierbale M (2014) An RGA-derived SCAR marker linked to PLRV resistance from Solanum tuberosum ssp. andigena. Plant Mol Biol Report 32(1):117–128

Muñoz FJ, Plaisted RL, Thurston HD (1975) Resistance to potato virus Y in S. tuberosum spp. Andigena. Am Portao J 52:107–115

Nicolia A, Proux-Wéra E, Åhman I, Onkokesung N, Andersson M, Andreasson E, Zhu LH (2015) Targeted gene mutation in tetraploid potato through transient TALEN expression in protoplasts. J Biotechnol 204:17–24. https://doi.org/10.1016/j.jbiotec.2015.03.021

Ning Y, Liu W, Wang GL (2017) Balancing immunity and yield in crop plants. Trends Plant Sci 22(12):1069–1079. https://doi.org/10.1016/j.tplants.2017.09.010

Ochoa CM (1990) The potatoes of South America: Bolivia. Cambridge University Press, Cambridge

Ortega F, Lopez-Vizcon C (2012) Application of molecular marker-assisted selection (MAS) for disease resistance in a practical potato breeding programme. Potato Res 55:1–13. https://doi.org/10.1007/s11540-011-9202-5

Ortiz R (1998) Potato breeding via ploidy manipulations. Plant Breed Rev 16:15–85

Ovchinnikova A, Krylova E, Gavrilenko T, Smekalova T, Zhuk M, Knapp S, Spooner DM (2011) Taxonomy of cultivated potatoes (Solanum section Petota: Solanaceae). Bot J Linn Soc 165:107–155

Perlak FJ, Stone TB, Muskopf YM, Petersen LJ, Parker GB, McPherson SA, Fischhoff DA (1993) Genetically improved potatoes: protection from damage by Colorado potato beetles. Plant Mol Biol 22(2):313–321

Phumichai C, Mori M, Kobayashi A et al (2005) Toward the development of highly homozygous diploid potato lines using the self-compatibility controlling Sli gene. Genome 48:977–984

Plaisted RL (1987) Advances and limitations in the utilization of Neotuberosum in potato breeding. In: Jellis GJ, Richardson DE (eds) The production of new potato varieties. Cambridge University Press, Cambridge, pp 186–196

Plaisted RL, Hoopes RW (1989) The past record and future prospects for the use of exotic potato germplasm. Am Potato J 66:603–627

Pushkarnath P (1942) Studies on sterility in potatoes. 1. The genetics of self- and cross-incompatibilities. Indian J Genet Pl Br 2:11–36

Ram HH (1998) Vegetable breeding. Principles and practices. Kalyani Publisher, Ludhiana

Ramakrishnan AP, Ritland CE, Sevillano RHB, Riseman A (2015) Review of potato molecular markers to enhance trait selection. Am J Potato Res 92(4):455–472. https://doi.org/10.1007/s12230-015-9455-7

Rauf S, Silva JAT, Khan AA, Naveed A (2010) Consequences of plant breeding on genetic diversity. Int J Plant Breed 4(1):1–21

Rodríguez L, Ñustez C, Estrada N (2009) Criolla Latina, Criolla Paisa y Criolla Colombia,nuevos cultivares de papa criolla para el departamento de Antioquia (Colombia). Agron Colombiana 27(3):289–303

Rodríguez F, Ghislain M, Claused AM, Jansky SH, Spooner DM (2010) Hybrid origins of cultivated potatoes. Theor Appl Genet 121:1187–1198

Salas E (2002) Estudio de la variabilidad genética de la resistencia al Virus del Enrollamiento de la papa (PLRV) en una población tetraploide. Tesis para optar el Grado de Magister Scientiae. Universidad Nacional Agraria—La Molina, Lima

Sanford JC, Hanneman RE (1982) Intermating of potato haploids and spontaneous sexual polyploidization—effects on heterozygosity. Am J Potato Res 59:407–414

Scheben A, Wolter F, Batley J, Puchta H, Edwards D (2017) Towards CRISPR/Cas crops–bringing together genomics and genome editing. New Phytol 216(3):682–698. https://doi.org/10.1111/nph.14702

Schönhals EM, Ortega F, Barandalla L, Aragones A, De Galarreta JR, Liao JC, Sanetomo R, Walkemeier B, Tacke E, Ritter E, Gebhardt C (2016) Identification and reproducibility of diagnostic DNA markers for tuber starch and yield optimization in a novel association mapping population of potato (Solanum tuberosum L.). Theor Appl Genet 129(4):767–785. https://doi.org/10.1007/s00122-016-2665-7

Schultz L, Cogan NOI, McLean K, Dale MFB, Bryan GJ, Forster JW et al (2012) Evaluation and implementation of a potential diagnostic molecular marker for H1-conferred potato cyst nematode resistance in potato (*Solanum tuberosum* L.). Plant Breed 131:315–321. https://doi.org/10.1111/j.1439-0523.2012.01949.x

Scott GJ, Suarez V (2011) Growth rates for potato in India and their implications for industry. Potato J 38(2):100–112

Sharma R, Bhardwaj V, Dalamu D, Kaushik SK, Singh BP, Sharma SK, Umamaheshwari R, Baswaraj R, Kumar V, Gebhardt C (2014) Identification of elite potato genotypes possessing multiple disease resistance genes through molecular approaches. Sci Hortic 179:204–211. https://doi.org/10.1016/j.scienta.2014.09.018

Sharma SK, MacKenzie K, McLean K, Dale F, Daniels S, Bryan GJ (2018) Linkage disequilibrium and evaluation of genome-wide association mapping models in tetraploid potato. G3 (Bethesda) 8(10):3185–3202. https://doi.org/10.1534/g3.118.200377

Slater AT, Cogan NOI, Forster JW (2013) Cost analysis of the application of marker-assisted selection in potato breeding. Mol Breed 32(2):299–310. https://doi.org/10.1007/s11032-013-9871-7

Slater AT, Wilson GM, Cogan NO, Forster JW, Hayes BJ (2014a) Improving the analysis of low heritability complex traits for enhanced genetic gain in potato. Theor Appl Genet 127(4):809–820. https://doi.org/10.1007/s00122-013-2258-7

Slater AT, Cogan NOI, Hayes BJ, Schultz L, Dale MFB, Bryan GJ et al (2014b) Improving breeding efficiency in potato using molecular and quantitative genetics. Theor Appl Genet 127:2279–2292. https://doi.org/10.1007/s00122-014-2386-8

Slater AT, Cogan NOI, Forster JW, Hayes BJ, Daetwyler HD (2016) Improving genetic gain with genomic selection in autotetraploid potato. Plant Genome 9(3):1–15. https://doi.org/10.3835/plantgenome2016.02.0021

Solano J, Acuña I, Esnault F, Brabant P (2014) Resistance to Phytophthora infestans in Solanum tubersosum landraces in Southern Chile. Trop Plant Pathol 39(4):307–315. https://doi.org/10.1590/S1982-56762014000400005

Sood S, Bhardwaj V, Pandey SK, Chakrabarti SK (2017) History of potato breeding: improvement diversification and diversity. In: Chakrabarti SK, Xie C, Tiwari JK (eds) The potato genome. Springer, Cham, pp 31–72

Spooner DM, Núñez J, Trujillo G, Del Rosario Herrera M, Guzmán F, Ghislain M (2007) Extensive simple sequence repeat genotyping of potato landraces supports a major reevaluation of their gene pool structure and classification. Proc Natl Acad Sci USA 104:19398–19403

Spooner DM, Gavrilenko T, Jansky SH, Ovchinnikova A, Krylova E, Knapp S, Simon R (2010) Ecogeography of ploidy variation in cultivated potato (Solanum sect. Petota). Am J Bot 97:2049–2060

Spooner DM, Ghislain M, Simon R, Jansky SH, Gavrilenko T (2014) Systematics, diversity, genetics, and evolution of wild and cultivated potatoes. Bot Rev 80:283–383. https://doi.org/10.1007/s12229-014-9146

Sprague GF, Tatum A (1941) General vs. specific combining ability in single crosses of corn. J Am Soc Agron 34:923–932

Suganthy VS, Udayasoorian C (2016) Screening and selection of ten potato genotypes (Solanum tuberosum L.) for their relative tolerance and susceptibility to the fumigation of the elevated O3 levels. Asian J Environ Sci 11(1):1–6. https://doi.org/10.15740/HAS/AJES/11.1/1-6

Thiele G, Theisen K, Bonierbale M, Walker T (2010) Targeting the poor and hungry with potato science. Potato J37:75–86

Thornton M (2003) The rise and fall of Newleaf potatoes. Biotechnology: science and society at a crossroads. Nat Agric Biotech Council Rep 15:235–243

Tiwari JK, Devi S, Ali N, Luthra SK, Kumar V, Bhardwaj V, Chakrabarti SK (2018) Progress in somatic hybridization research in potato during the past 40 years. Plant Cell Tiss Org 132:225–238. https://doi.org/10.1007/s11240-017-1327-z

Ugent DL, Pozorski S, Pozorski T (1982) Archaeological potato tuber remains from the Casma Valley of Peru. Econ Bot 36:182–192

Ugent DL, Dillehay T, Ramirez C (1987) Potato remains from a Late Pleistocene settlement in southcentral Chile. Econ Bot 41:17–27

Uitdewilligen JG, Wolters AML, D'hoop BB, Borm TJA, Visser RGF, van Eck HJ (2013) A next generation sequencing method for genotyping-by-sequencing of highly heterozygous autotetraploid potato. PLoS One 8(5):e62355. https://doi.org/10.1371/journalpone0062355

Van Eck J, Conlin B, Garvin DF, Mason H, Navarre DA, Brown CR (2007) Enhancing beta-carotene content in potato by RNAi-mediated silencing of the beta-carotene hydroxylase gene. Am J Potato Res 84(4):331

Velásquez AC, Mihovilovich E, Bonierbale M (2007) Genetic characterization and mapping of major gene resistance to potato leafroll virus in Solanum tuberosum ssp. Andigena. Theor Appl Genet 114(6):1051–1058

Vos PG, Uitdewilligen JGAML, Voorrips RE, Visser RGF, van Eck HJ (2015) Development and analysis of a 20K SNP array for potato (Solanum tuberosum): an insight into the breeding history. Theor Appl Genet 128:2387–2401. https://doi.org/10.1007/s00122-015-2593-y

Vos PG, Paulo MJ, Voorrips RE, Visser RGF, van Eck HJ, van Eeuwijk FA (2017) Evaluation of LD decay and various LD-decay estimators in simulated and SNP-array data of tetraploid potato. Theor Appl Genet 130(1):123–135. https://doi.org/10.1007/s00122-016-2798-8

Walker T, Thiele G, Suarez V, Crissman C (2011) Hindsight and foresight about potato production and consumption. Social sciences working paper 2011-5. In: International Potato Centre (CIP), Lima

Wang S, Zhang S, Wang W, Xiong X, Meng F, Cui X (2015) Efficient targeted mutagenesis in potato by the CRISPR/Cas9 system. Plant Cell Rep 34(9):1473–1476. https://doi.org/10.1007/s00299-015-1816-7

Watanabe K, Arbizu C, Schmiediche PE (1992) Potato germplasm enhancement with disomic tetraploid Solanum acaule I. Efficiency of introgression. Genome 35(1):53–57. https://doi.org/10.1139/g92-009

Wricke G, Weber WE (1986) Quantitative genetics and selection in plant breeding. de Gruyter, Berlin

Ye M, Peng Z, Tang D, Yang Z, Li D, Xu Y, Huang S (2018) Generation of self-compatible diploid potato by knockout of S-RNase. Nat Plants 4:651–654. https://doi.org/10.1038/s41477-018-0218-6

Yin K, Gao C, Qiu JL (2017) Progress and prospects in plant genome editing. Nat Plants 3:17107. https://doi.org/10.1038/nplants.2017.107

Impact of Climate Change on Potatoes and Mitigation Strategies

Menka Pathak, Kartik Pramanik, Satyaprakash Barik, Mahendra Prasad Das, Himansu Sekhar Gouda, and Debashis Dash

Abstract Crop production is a constantly evolving concept and faces significant constraints due to various factors. Climate change, along with biotic and abiotic stresses, poses potential threats to crop growth and development. This phenomenon particularly challenges the production of cool season vegetables compared to warm season vegetables. Potatoes, for example, prefer cooler weather, making their tuber development phase especially sensitive to extreme climate changes. These changes pose global threats to potato production and food security. Potatoes are sensitive to temperature fluctuations, water stress, soil conditions, as well as biotic stresses such as disease evolution and insect attacks. While warmer temperatures can promote plant growth, they also increase the risk of heat stress during tuber formation. On the other hand, colder temperatures or frost events can damage potato crops affecting both yield and quality. To address these impacts on potato cultivation resulting from climate change, farmers may need to adopt various adaptation strategies. These include switching to more heat-tolerant potato varieties, introducing water-efficient irrigation techniques, crop rotation, bettering soil management practices that enhance resilience against extreme weather events, and integrating pest and disease monitoring and management plans into overall crop management strategies.

Keywords Climate change · Heat-tolerant potato varieties · Crop rotation · Frost

---

M. Pathak (✉) · M. P. Das
Department of Vegetable Science, Odisha University of Agriculture and Technology, Bhubaneswar, Odisha, India

K. Pramanik
M. S. Swaminathan School of Agriculture, Centurion University of Technology and Management, Paralakhemundi, Odisha, India

S. Barik · H. S. Gouda · D. Dash
Department of Agriculture and Allied Sciences, CV Raman Global University, Bhubaneswar, Odisha, India

S. S. Solankey (ed.), *Advances in Research on Potato Production*, Advances in Olericulture, https://doi.org/10.1007/978-3-031-82710-5_3

1 Introduction

The potato (*Solanum tuberosum* L.) is christened as "king of vegetables" and is the most crucial vegetable crop across the globe, third important after rice and wheat in terms of food value, crop production, and consumption (FAO 2009). Potatoes are dubbed as "perfect" for human food and nutrition because they are a good source of carbohydrates, as well as numerous vitamins and minerals like Ca, P, Fe, and vitamins B_1, B_2, B_6, and C (Tesfaye 2016). Potato tubers are utilized for a wide range of table, canned, dehydrated, processed, frozen, livestock feed, and industrial applications, and they offer an incredibly high yield per acre (Feustel 1987; Talburt 1987). For the Sustainable Development Goal 2 (SDG 2-Zero Hunger), potatoes support "access, availability, stability, and utilization," the four pillars of food security. Starch makes up about 75% of their dry weight, while the exact proportion varies according to the cultivar. Despite having a low protein level, potatoes are more biologically valuable and utilized because of metabolites and important amino acids like lysine (Gorska-Warsewicz et al. 2021). In addition to ensuring food security, potatoes generate revenue and employment (Devaux et al. 2021).

However, potato farming is confronted with substantial obstacles due to climate change. This chapter discusses several mitigation techniques to guarantee the resilience of potato farming systems while examining the complex effects of climate change on potatoes. Growing, yielding, and quality of potatoes are impacted by temperature rise, changed rainfall patterns, and an aggravate in adverse weather events. Varying degrees of climate change resilience are shown by different potato types. When creating adaptive techniques, it is essential to comprehend how vulnerable current types are. It has been demonstrated through research that in changing climatic conditions, some cultivars may be more resistant to pests and illnesses or more tolerant to heat stress. Temperature and humidity variations have an impact on the potato crop cycle, which in turn affects the crop's productivity directly and indirectly. A staple non-grain crop that is consumed across the world, potatoes are sensitive to environmental conditions. By controlling transpiration, photosynthesis, and respiration processes and influencing plant growth and development throughout their physiological cycles, climatic conditions have a direct impact on crop potential productivity. This suggests that there may be conflicting effects on potato yield, which calls for additional study in a variety of temporal and spatial contexts across the surface of the planet. As a result, understanding the climatic requirements of potatoes and their physiological reactions to their environment is crucial for assisting growers in producing large yields of high-quality tubers in a variety of meteorological circumstances.

2 Impact of Climate Change on Potatoes Production

2.1 Temperature

Temperature fluctuations play a pivotal role in determining the plant growth and development, tuber yield, and quality of potato. Potato plant undergoes various growth phases, i.e., vegetative growth phase, transition phase, and reproductive phase (Fig. 1). The transition phase including tuber initiation, tuber bulking, tuber maturation, and dormancy of tuber is most vulnerable to extreme temperature and water stresses resulting in less tuber production and affecting the storability of tuber by impacting the dormancy period of tuber. Understanding the temperature requirements and potential impacts on potato physiology is essential for optimizing cultivation practices and mitigating the adverse effects of temperature stress on potato production. The physiological processes of both the potato plant and the tuber stop when the base temperature drops below 2–4 °C. As photosynthesis cannot produce glucose due to zero enzyme activity, respiration can almost cease, preventing growth. Temperature quotient (Q10) = 2.5 indicates that all physiological processes accelerate above the base temperature. Accordingly, biochemical and physiological processes in plants occur 2.5 times more quickly for every 10 °C increase in temperature. Development and growth can occur at low temperatures (4–18 °C) because

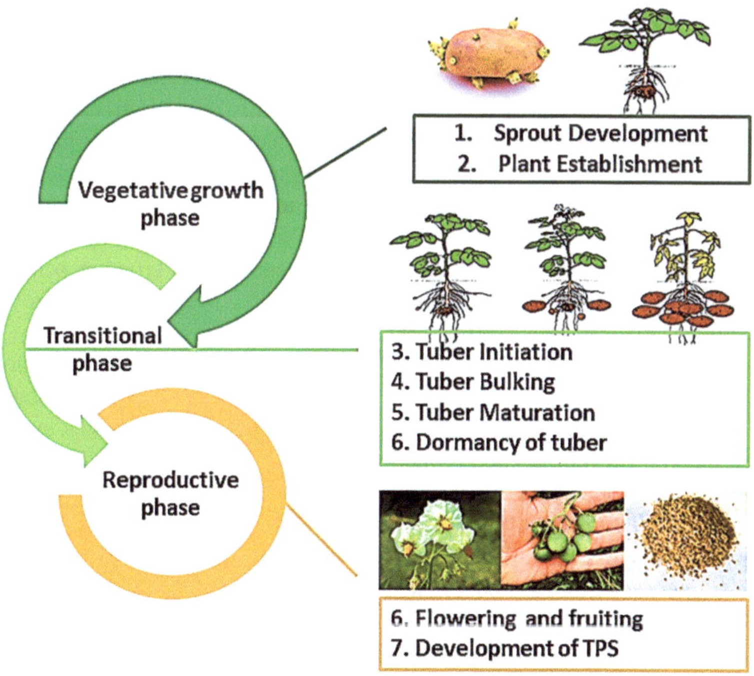

Fig. 1 Growth phases of potato

of the increased respiratory efficiency. Temperature increases most likely have a negative impact on potato yield due to disruption in plant growth, tuber initiation, and development.

The optimum temperature for its good plant growth is 15–20 °C. The ideal temperature range for potato sprouting is 22–34 °C. Temperatures below 21 °C are ideal for tuberization; however, temperatures around 21 °C have a negative impact and 29 °C and above are not conducive to tuberization. In India, potato is cultivated when the highest daytime temperature is less than 35 °C and the lowest nigh temperature is less than 20 °C. In bright, sunny days with chilly nights, good crop growth is noted. When the nighttime temperature exceeds 23 °C, most potato varieties do not tuberize, and these places are not favorable for potato cultivation. In the plains, the ideal temperature range for planting is between 30 and 32 °C and 10 and 20 °C at night. When the highs and lows temperature are between 20 and 22 °C, planting takes place in the hills. Closing of stomata to save moisture, sluggish tuber development and physical tuber damage occurs in potato when temeprature exceeds 30 °C during potato growth. Temperature ranges between 5 and 30 °C can significantly hinder tuber initiation, enlargement and yield. By 2100, the Intergovernmental Panel on Climate Change projects a 1.1–6.4 °C rise in global temperatures. The impact of rising temperatures on the production of potatoes in particular regions will differ based in part on the local temperature at the time. Levy et al. (2017) reported that temperatures above 30 °C can negatively impact potatoes in several ways, such as by reducing tuber initiation and growth, less starch being partitioned to physiological damage of the tubers (such as brown spots), shortening or eliminating tuber dormancy, and causing tubers to sprout prematurely. These impacts may lower crop yields as well as the quantity and mass of tubers. Thus, future potato crop yields are expected to be significantly reduced in regions (such as much of Sub-Saharan Africa) (Haverkort and Verhagen 2008) where present temperatures are close to the temperature range suitable for potato (Hijmans 2003). Potatoes are susceptible to frost damage at low temperatures, which can severely harm tubers and hinder growth (Haverkort and Verhagen 2008). Increase in temperatures may probably help potato plant in extending the growing periods and expanding the amount of potential potato-growing land in regions where frost damage risks are observed (high altitudes and high latitudes).

According to Levy and Veilleux (2007), heat stress results in malformation and necrosis of tubers, slows tuber initiation and bulking, and produces imbalances in source-sink interactions. Adaption to climate change needs heat-stress tolerance for the continued growth of potato production in areas that are currently unsuitable for it, such as subtropical India (Gaur and Pandey 2000), the semi-arid Middle East (Levy et al. 2017), and the tropics (Minhas et al. 2011), which includes Sub-Saharan Africa (Asfaw et al. 2015). Because of the inhibition of the tuberization initiation at higher temperatures, enlargement is sensitive to temperature changes (Ewing 1981; Hancock et al. 2014). According to Wolf et al. (1991), the temperature of the potato plant affects the transport of carbon to its sink organs. At higher temperatures, starch absorbs less carbon that has been digested in tuber. Since there is less sucrose transported to the tuber as a result, sink strength is lowered and sucrose accumulates at the phloem unloading sites. According to Reynolds et al. (1990), heat-sensitive *Solanum* species that bear tubers show an extreme suppression of CO_2 fixation and chlorophyll loss as compared to heat-tolerant species when temperatures are too high.

Various tuber disorders are caused due to high temperatures, such as secondary tuber development (which is frequently linked to excessive stolon extension and branching), chain tuberization, and uneven tuber shape. High temperatures also have detrimental effects on the periderm formation of potatoes (Montetberg 2017). As per Ginzberg et al. (2017), the tubers' relative health value may be diminished due to the observed reduced accumulation of anthocyanins. High temperatures can also cause pre-harvest sprouting and influence tuber dormancy throughout maturation. The whole tuber yield and quality may therefore be impacted by a wide range of intricate heat stress reactions. Screening potato germplasm has revealed a wide range in heat tolerance (Levy and Veilleux 2007). However, heat-tolerant ideotypes should retain yield along with other desirable traits like processing and quality as well as nutritional and health-promoting properties. In a controlled environment, Thornton et al. (1996) reported that the growth rate of tubers was more impacted by high temperatures than the growth of the entire plant. When compared to low temperatures, all clones showed a decrease in the yield of tuber dry matter; however, Russet Burbank showed a significant loss.

2.2 *Carbon Dioxide (CO_2)*

It is projected that higher atmospheric carbon dioxide concentrations will be beneficial to potato plants and crop harvests. For potatoes, high carbon dioxide concentration is the main advantage, that leads to increase in photosynthetic rates, which ultimately speeds up growth and development in potato. Since potatoes divert more starch to tubers for enlargement under higher CO_2 levels, potato yields are likewise expected to benefit (Haverkort et al. 2008). This is expected to lead to an increase in the water use efficiency in potato plants, or the quantity of carbon assimilated per unit of water lost (Haverkort et al. 2008). A higher microclimate relative humidity and larger biomass due to elevated CO_2 may result in C_3 plant canopy density and size. The slower rate of decomposition resulting from the increased biomass may help in tolerance of pathogens when these conditions are met by warmer winter temperatures. Plant diseases including rusts, powdery mildews, leaf spots, and blights may be encouraged by this (Das et al. 2016; Singh et al. 2013; Londhe 2016; Schoville et al. 2017).

Because of its extensive carbon sinks below ground and effective phloem loading mechanism, potatoes are projected to respond significantly to rising atmospheric CO_2. In potatoes, there is a positive correlation between CO_2 content and absorption. One of the rare negative impacts of high CO_2 concentration is a decrease in leaf chlorophyll content, especially in the later growth season following tuber start. Senescence later in the season is also anticipated to cause a progressive decrease in leaf photosynthesis in the increasing CO_2 environment. The idea that plants cultivated in high CO_2 environments experience an acceleration of leaf senescence is further supported by the observation that leaf N concentrations in those leaves deplete more quickly. Potato cultivar productivity in India will not be affected by temperature change and increased CO_2 in 2020 compared to the baseline scenario, but it may slightly decline in 2055 in Punjab, according to simulation studies. On the other hand, in West Bengal and Bihar, the combined effects of temperature change and increased CO_2 indicate that potato productivity will drop to a high level.

Russet Burbank, Denali, and Norland cultivars of potatoes were cultivated under conditions of 350 or 1000 μmol mol^{-1} CO_2, 400 or 800 μE $m^{-2}s^{-1}$ photosynthetic photon flux (PPF), and 12 or 24 h of sunshine each day. More CO_2 resulted in higher tuber yields at low PPF but lower tuber yields under high PPF. While the yield for Russet Burbank declined, the tuber yield for Denali improved with an increase in PPF. Denali responded to higher CO_2 enrichment with the largest gains in tuber and overall weight (21 and 18%, respectively). Norland exhibited the lowest response (9 and 9%, respectively), whereas Russet Burbank displayed an intermediate response, with increases almost equal to those of Denali under a 12-h photoperiod (18%) but lower than those of Denali under a 24-h photoperiod. With reduced PPF and a short photoperiod, CO_2 enrichment showed a trend of increased potato plant growth. Enhancements in potato yield could be achieved by prolonging the photoperiod or raising the net daily photosynthetically active radiation (PAR) (Stuttle et al. 1996). In potato crop models, temperature is a more significant source of uncertainty than CO_2 and water (Fleisher et al. 2016). Growing crops of potatoes requires CO_2 because it promotes the growth of the potato. According to the International Potato Centre (CIP 2013) and Daccache et al. (2011), the photosynthetic rates increase which enrich potatoes with more CO_2 to supply photosynthate to bulk faster.

2.3 Water Availability

Water is vital element needed for plant growth and development. The total amount of water needed varies from 350 to 550 mm based on the kind of soil, climate, and length of potato crop. Irrigation prior to planting is beneficial for consistent germination. About a week later, a second irrigation is administered, and more are added as needed. Light and regular irrigations are preferable. Stolon formation, tuber initiation, and tuber development stages are considered as critical periods for irrigation in potato. About ten days before crop harvest, irrigation is halted to allow the skin of the tuber to firm.

Changes in precipitation patterns, including erratic rainfall and prolonged droughts, pose a threat to potato farming. Insufficient water availability during critical growth stages can result in water stress and reduced tuber quality. A deficiency or severe fluctuation in available moisture during the time of tuber formation may result in growth cracks or mishappen tubers, along with initiation of knobby secondary growth. High yields of potatoes are obtained when the available soil moisture is maintained between 50% and 65% throughout the irrigation season. The crop should be protected from moisture stress during stolon initiation and tuberization; otherwise, it will affect production and delay the maturity of the crop. Changes in precipitation patterns may also lead to waterlogging or drought, further compromising yields. Tuber yield and size are adversely affected by water stress. For good growth and development, potatoes have a high-water requirement. The plants are shallow-rooted and develop a relatively poor root system. A few roots penetrate deeply, but they are not numerous enough to penetrate the soil thoroughly at greater depths. Elevated temperatures, coupled with water scarcity, can lead to decreased potato yields.

The accessibility of water resources is anticipated to alter due to climate change in many parts of the world, according to IPCC estimates (Intergovernmental Panel on Climate Change 2007). Water resources are predicted to decline in many places, especially in semi-arid regions (Haverkort et al. 2008). Even in regions where a decline in the average annual rainfall is anticipated, there is a forecast for an increase in adverse environment events, such as severe flooding (Haverkort et al. 2008). According to Adavi et al. (2018) and Banu et al. (2017), the length, timing, and intensity of water shortage during potato development impact the consequences of drought on the potato plant. For instance, it is anticipated that there will be a minimum 75% reduction in the quantity of arable land in the UK that can be used to produce rainfed potatoes (Daccache et al. 2012). Modifying seasonal rainfall patterns presents issues for potato crops in addition to changes in overall rainfall. For instance, Bolivia has had a shorter growing season for potatoes due to the reduction in duration of the rainy days in recent decades (Climate change- can potato tolerate the heat 2012).

Potatoes are susceptible to drought mostly because of their shallow root systems, which make it difficult for them to absorb water, and because their photosynthetic system cannot regenerate after being stressed by heat and water (Iwama and Yamaguchi 2006). By itself, drought inhibits the growth of plants (Deblonde and Ledent 2001), shortens the growth cycle (Kumar et al. 2007), and reduces the size and quantity of tubers (Eiasu et al. 2007; Schafleitner et al. 2011). According to Schafleitner et al. (2011) and Simelton et al. (2012), climate change is expected to increase the frequency of drought occurrences in many places, harming crops like potatoes that are prone to drought. In addition to potentially affecting the quality of tuber by decreasing dry matter and raising sugar content, this is probably going to have an impact on potato yields (Haverkort and Verhagen 2008).

The morphological structure of the potato root system is altered by soil drying (Wishart et al. 2013). This includes higher root thickness, lower root elongation, and higher proliferation of lateral roots, as well as the emergence of root hairs that preserve root-soil interaction in soil that contract from roots during drying (Haling et al. 2013). Furthermore, smaller xylem vessels and higher root hydraulic resistance system boost the ability of plant to withstand drought by slowing the pace at which water is drawn from the soil without drastically lowering the total amount of water available. Permeability of cell membrane and aquaporin pronouncement can regulate hydraulic conductivity, which alters along the root depending on tissue age and root types (Doussan et al. 1998). According to Maurel et al. (2010), the process of soil drying causes an initial rise in root hydraulic conductivity, which is necessary for maintaining water intake. However, as water is extracted further, hydraulic conductivity in root lowers soil loss. These variations are attributed to the activity and regulation of aquaporin. Enhancement in suberization of the root exodermis/endodermis over time further reduces the hydraulic conductivity of the root (Vandeleur et al. 2009). Xylem embolism, a phenomenon where air plunges into the xylem vessels and interrupts the flow of sap, can further impair the hydraulic conductivity of the roots in a plant following extended droughts (Cruiziat et al. 2002).

Larger root architecture in potato genotypes, especially those connected to stolons, improve resistance to drought and enable yield maintenance in drought-prone environments (Puertolas et al. 2014; Wishart et al. 2013, 2014). It's interesting to note that genotypes with larger root systems also saw faster canopy closure. This means that the quantity of water dispersed from the soil through evaporation other than transpiration by the plant will be reduced more quickly and the time of maximal photosynthetic potential would be extended. In dry soil, plants have reduced stomatal density, which leads to decreased transpiration rates (Yan et al. 2012; Sun et al. 2014). The plants resistant to drought have five times lesser stomata in the upper leaf surface as compared to plant sensitive to drought and larger root architecture (Boguszewska-Mankowska et al. 2020). There is heat tolerance, and reduced transpirational cooling is noticed in drought-resistant plant varieties which maintain chlorophyll fluorescence under water stresses and counterfeits the production of reactive oxygen species (ROS) and harmful metabolites.

2.4 Photoperiod

In potato, tuberization takes place under short days and low temperatures. Long days promote haulm growth without formation of tubers. The highest tuber yield of potato is obtained under short-day (12-h photoperiod) condition at 20 °C. In phenomenon, the early maturing potato varieties prefer long days for growth and development while late maturing varieties performed better under short days for tuber initiation and development. In plant, endogenous hormones influence the photoperiodic activities. In potato, higher gibberellic acid concentration negatively affects the tuber development whereas, higher abscisic acid leads to enhanced tuber growth in potato. Increased gibberellic acid is attributed with long photoperiods in some potato varieties that reduce tuber growth. The physiological process is detrimental for temperate zones having long photoperiods during main cropping. In contrast, some American potato genotypes are attributed with marginal gibberellic-photoperiod response due to neutral photoperiodism in nature (Dwelle 1985). Potato growth is affected by lack of sunshine.

Stuttle et al. (1996) highlighted on the influence of photoperiod on the net carbon assimilation rate (NCAR) and accumulation of starch in freshly developed canopy potato leaves under varied PPF conditions. NCAR reduction of 16.1% was observed if the photoperiod was extended from 12 to 18 h, with the largest declines occurring under high photoperiod levels. When high PPF treatments were used instead of low light treatments, the maximal starch concentrations were achieved at a shorter photoperiod. NCAR does not appear to be correlated with the amount of starch in individual leaves, but it does appear to be regulated by a feedback mechanism under conditions of high PPF, high CO_2, and extended photoperiod. This shows that under extended photoperiod and high PPF conditions for Norland, maximum NCAR cannot be preserved with increased CO_2 enrichments.

Dam et al. (1996) determined that early tuber growth was best suited for average minimum temperatures (15–19 °C) and short day length (<12 h). These facilitate higher tuber growth and partitioning of energy to dry matter production, coinciding with early bulking and growth onsets. Further increases have a significant effect on the start of tuber growth and absolute growth rates; slight temperature increases also dramatically lowered partitioning rates. Treatment differences varied in the quantity of tubers started. Using potato cuttings, Snyder and Ewing (1989) revealed that genotypes have different responses to varied temperatures and day length on tuber growth and yield. Midmore and Prange (1992) revealed low temperature and low irradiation increase relative growth rate. Low irradiance had the reverse impact at high temperatures, resulting in the lowest relative growth rates and net assimilation rates. The high temperature significantly decreased the number and weight of tubers. Almost no tubers were generated when high temperatures and low irradiance were combined. Lower irradiance can exacerbate these facts, which are in line with earlier findings showing decreased potato growth at high temperatures. Gawronska and Dwelle (1989) noticed that potato plants in high light produced auxiliary branches and with low light levels witnessed small, asymmetrical low-weight tubers, less accumulation of total biomass, and increased stem elongation.

2.5 *Soil Factors*

Alluvial, black, red, and laterite soils with pH values between 5.5 and 8.0 are the soil types in which potatoes are cultivated. Potato cultivation is not recommended in heavy clay soil. The best soils are the deep alluvial soils found in the Indo-Gangetic plains, which have an almost neutral or alkaline soil reaction. Alluvial soils have the most area covered by potatoes, followed by red and black soils. Soils that are saline, alkaline, or sodic are not conducive to the growth of potatoes. The ideal soil is rich in organic matter, loose, and free of hard layers that obstruct root penetration. Water drainage is further limited by compacted layers. Soils are rich in organic matter and perfect for growing potatoes that hold moisture, good drainage, and aeration for the development of roots, stolons initiation, and tubers development. In this direction, Kufri Safed is a hardy variety and suitable for poor soils with meagre facilities of irrigation.

2.5.1 Soil Temperature

The soil temperature affects the sprouting of tubers planted in soil. For most of the potato cultivars, 6 °C produces very little sprout elongation and 9 °C is the slowest temperature for elongation while 18 °C is the maximum temperature. Sale (1979) demonstrated linear relationship of emergence to the average soil temperature and relative independence from daily variations, with an optimum temperature of 22–24 °C. According to the Western Potato Council (2003), the ideal soil temperature for starting tubers is between 16 and 19 °C. Reynolds and Ewing (1989) revealed that the induction of tuberization by the leaves and shoot growth were not adversely affected by heating the soil to cool air temperatures. There was no subsurface tuber formation and significant stolonization compromise under high soil temperatures.

Soil temperature had less impact on the induction of tuberization by the leaves than air temperature; nonetheless, high soil temperatures may inhibit the signal to tuberize comparable to that of exogenously supplied gibberellin. As soil temperatures climb beyond 20 °C, tuber development decreases, and at soil temperatures above 30 °C, tuber growth almost completely ceases (Mares et al. 1985). The influence of soil temperature on potato tuber size and quality during growth and development has not been well studied. Soil temperature and weather condition from mid-June to mid-July are regarded as critical phase for potato quality enhancement along with reduced soil temperature using sprinkler irrigation increases sugar in tubers (Kincaid et al. 1993). Yamaguchi et al. (1964) observed that tubers of Russet Burbank and 'White Rose' had a higher yield, specific gravity, and starch content and a lower sugar content when cultivated at soil temperatures between 15 and 24 °C. Management practices, such as plant density, mulching, and irrigation, may manipulate the soil temperature substantially within the root zone that influence stolon formation, tuber initiation and enlargement, and tuber maturation in particular regions where availability of solar irradiance might be a non-limiting factor for potato growing.

Mulch boosted tuber yield by 20% as reported by Midmore et al. (1986). Manrique and Meyer (1984) reported mulching had no influence on tuber yields during the winter season, while mulching improved soil moisture and yield increases by 58%. Mulch decreases the daily maximum soil temperature at a 15 cm depth by 1.5–4.5 °C (Mahmood and Lingrani 2002). Mulching led to earlier crop growth, faster emergence, and higher tuber yields. The advantages of mulch in potato production systems are highlighted by many workers (Jaiswal 1995; Ruiz et al. 1999) as an effective substitute to reduce heat and water stressors and increase crop output.

2.5.2 Salinity

Salinity negatively impacts soil structure, aeration, lowering water infiltration, and availability of oxygen for the roots. It also increases the soil osmotic pressure of water and decreases root water uptake. Potato genotypes that are tolerant to salinity have to deal with ion toxicity, which manifests as high amounts of boron and sodium as well as overall nutritional imbalances (Koyro et al. 2012). Osmoprotectants are produced, such as proline, glycine, and betaine, which improve osmotic adjustment and function as chaperones to proteins to stop their denaturation, which can mitigate salinity effects (Rontein et al. 2002; Koyro et al. 2012). Reduced CO_2 absorption, which increases the formation of ROS that are particularly harmful to nucleic acids, lipids, and proteins, is another effect of salinity and drought together (Huchzemeyer and Koyro 2005). Plants produce chemicals and specialized enzymes like glutathione, superoxide dismutase, and catalase that act as antioxidants that control reactive oxygen species (ROS) (Blokhina et al. 2003). Salinity stress initiates intense maturation and imbalance of nutritional components, reducing tuber growth and yield in potato. Salinity stress causes the metabolism to alter during the reproductive and vegetative phases, lowering the water potential, toxicity, ion deficiency, stomata closure, and reduced CO_2 uptake (Chourasia et al. 2021). According to Blom-Zandstra and Verhagen (2015), potato plants are very sensitive to saline water and are severely effected by irrigation with saline water, especially at the initiation of tuber formation.

2.5.3 Frost

Frost can affect at any stage of the development in potato. Frost produces burns, withering leaves on foliage, and tissue collapse in potato tubers near the soil's surface, resulting in a mushy, blackened portion (Chang et al. 2014; Condori et al. 2014). Frost can affect potatoes. The varieties Kufri Sheetman and Kufri Dewa are resistant to frost.

2.5.4 Atmospheric Humidity and Wind

In potato storage, RH has a significant role in tuber weight loss as well as the frequency and severity of diseases and pests. Wheeler et al. (1989) found no appreciable variations in the overall dry weight of the plants across the atmospheric humidity treatments; plants grown at 85% relative humidity yielded larger tuber yields. When air is drier and more humid, leaf areas become larger, and the color of the leaves tended to be darker green. Increased humidity seemed to cause a shift in photosynthetic allocation patterns, favoring tubers over leaves and stems. Potato encourages growth and more effective use of irrigation water in coastal Peru under cool and humid weather (Bowen 2003). When compared to summertime, water use efficiency is improved during the winter months due to reduced soil water evaporation brought on by a smaller vapor pressure deficit (VPD). As a result of a lower VPD, Sinclair (1984) also have shown that generally more humid settings offer improved water use efficiency.

Potatoes are significantly affected by wind. According to Pavlista (2002), leaves are damaged by rubbing the leaves against one another causing bronzed patches on the leaves, which are brown with a shining surface due to wind speeds. The bronzed sections split under pressure, creating a rip with a sharp edge that tears through the damaged tissue. Winds can also have an impact on stems and causes a slight weakness in the tissues. Strong winds can twist vines, causing a break or weakness in the stems that resembles a hinge. The vine may twist completely, and the stem may collapse if it is exposed to severe winds for several hours. This would stop the passage of nutrients through the phloem between the vine and the tubers. Wind has an impact on crop output and photosynthetic activity by influencing transpiration rates. Reduced potato yield can result from higher stomatal resistance in locations where winds are consistently strong throughout the year (Pavlista 2002; Sun and Dickinson 1997).

Windbreaks had a modest impact on leaf length and width, but they did increase the height and number of leaves on potato plants. Sturrock (1981) observed that windbreaks boosted yield by 35%, although windbreaks enhanced production by as much as 7.7%. Wright and Brooks (2002) observed windbreaks reduced wind damage to the leaves of shielded potato plants and enhanced tuber yield for the sheltered portion of the field by 4.8–9.3%. Windbreaks enhanced yield at distances from the windbreak of three to eight times of windbreak length in seasons when wind speeds were higher than usual. According to Cleugh (2012), yields from potato crops were much greater in the sheltered zone between 2 and 18 times the height of the windbreak than in the unprotected portions.

2.5.5 Pests and Disease Pressure

More than a hundred diseases that are brought on by bacteria, fungus, viruses, and mycoplasmas can affect potatoes. However, degeneration and deterioration of the potato crop, along with yield losses to the crop, have been observed to be caused by *Phytophthora infestans*, *Alternaria solani*, *Rhizoctonia solani*, *Sclerotium rolfsii*, *Streptomyces* spp., *Ralstonia solanacearum*, and several virus pathogens (Shekhawat et al. 1993) presented in Table 1. Climate change-related variations in temperature and precipitation patterns have the potential to impact not only the physiology and resilience of the host plant, but also the growth stage, development pace, and pathogenicity of infectious pathogens (Chakarborty and Datta 2003). A longer growing season can occasionally be caused by an increase in pest pathogen pressure and vector activity with greater rates of multiplication brought on by temperature and atmospheric CO_2 increases (CIP International Potato Center 2013).

The positive effects of high atmospheric CO_2 include increased predation by predators, the effect of foliar Bt application, carbon-based plant defense, and decreased insect development rates, as reviewed by Das et al. (2011). High atmospheric CO_2 increases food consumption by caterpillars and aphid reproduction and decreases the nitrogen-based plant defense. Virus incidence has become a major constraint for potato production especially in the subtropics and tropics. Rising temperatures accelerate the rate of multiplication of most potato viruses. Viral attacks do not cause as much harm in temperate regions because of this. However, the whole potato virus situation in the temperate world could be altered by global warming. Although hilly areas are thought to be ideal for producing potato seeds,

Table 1 Climatic requirement of different diseases of potato

Name	Casual organism	Optimum climatic requirement	References
Fungal diseases of potato			
Late blight	*Phytophthora infestans*	Temperature range:10–27 °C/10–16 °C (D/N), relative humidity >90%, light rain, fog, or heavy dew	Luck et al. (2012) and Kirk et al. (2013)
Early blight	*Alternaria solani*	Temperature range: 10–35 °C/13.5–24 °C (D/N), relative humidity >90%, rainy weather	Johnson et al. (2015)
Wart disease	*Synchytrium endobioticum*	Temperature: 14–24 °C, soil moisture content: 100%	Singh (2000)
Black scurf	*Rhizoctonia solani*	Temperature: 18–28 °C, soil temperature: 16–23 °C	Frohning (2013)
Charcoal rot	*Macrophomina phaseolina*	Ambient temperature: 30 °C or above	Somani (2007)
Bacterial diseases of potato			
Bacteria wilt/brown rot	*Ralstonia solanacearum*	Temperature: 15–35 °C, acidic soil	Hayward (1991)
Common scab	*Streptomyces scabies*	Soil temperature: 19–24 °C with low moisture level	Healy and Lambart (1991)
Wet rot	*Sclerotium rolfsii*	Temperature 4–37 °C Cool to moderate maximum daily temperatures (29 °C) and moisture from rain, fog, dew, or high relative humidity.	Duarte et al. (2004)
Mosaic disease	Virus	Temperatures of 16–20 °C with cloudy weather	Sutic et al. (1999)

this may change after two to three decades. The bulk of potatoes are grown on subtropical plains, which offer a distinct environment. While the direct impact of global warming on potato viruses may be remote, the changing biology of insect vectors could have major consequences. The temperature rise will increase the number of vectors, which will raise the need for insecticide sprays to control the vector population.

3 Adapting Potato Production to Climate Change

Knowledge regarding the climate change, especially weather forecasting can assist in sustainable crop production and in enabling the cultivation of potatoes in regions where existing commercial cultivars are not performing well due to climatic constraints. This can be achieved by adapting potato farming techniques and varieties. Relocating producing locations, enhancing water use, and cultivating novel resistant potato types are some ways to adapt potatoes to climate change.

3.1 Creating New Potato Cultivars

To produce new potato varieties, two primary strategies are used: genetic alteration and traditional plant breeding techniques. These methods could be crucial in developing novel cultivars (Table 2) that can sustain yields in the face of climate change-related challenges.

The following traits, especially the capacity to sustain tuber growth and initiation in high temperatures, may be useful in mitigating the detrimental effects of climate change on potato production: heat stress tolerance. In nations where potato production areas are close to the maximum temperature limitations of present cultivars (such as Sub-Saharan Africa and India), developing cultivars with increased resistance to heat stress is essential for sustaining yields (2012, World Potato Congress).

3.2 Drought Tolerance

This includes potatoes that can withstand brief droughts and recover to offer respectable yields, as well as improved water use efficiency (amount of food produced per amount of water used). Since most commercial potato cultivars require frequent irrigation due to their shallow roots, deeper root systems may potentially be advantageous (Water and potato resources 2012).

Table 2 List of varieties released in India indicating their main climate change-related attributes and preferences to Indian farmers

Variety name	Yield potential (q/ha)	Crop maturity	Heat tolerance	Drought tolerance	Late blight resistance
Popular varieties					
Kufri Pukhraj	35–40	Early	Sensitive	High	Medium
Kufri Jyoti	25–30	Medium	Sensitive	Medium	Medium
Kufri Bahar	30–35	Medium	Sensitive	Medium	Sensitive
Kufri Aloo	24	Late	Low	Sensitive	Sensitive
Kufri Chipsona	30–35	Medium	Sensitive	Medium	High
Kufri Sindhuri	30–35	Late	High	Medium	Sensitive
Kufri Chandramukhi	20–25	Early	Sensitive	High	Sensitive
Kufri Khyati	25–30	Early	Sensitive	High	High
Promising varieties					
Lady Rosetta	30	Early	High	High	Sensitive
Kufri Megha	25–30	Medium	Sensitive	Medium	High
Kufri Kanchan	25–30	Medium	Sensitive	Medium	Medium
Kufri Arun	30–35	Medium	Sensitive	High	High
Kufri Surya	25–30	Early	High	Medium	Sensitive

3.3 Early Maturation and Rapid Growth

In addition to reducing the number of life cycles of pests like potato tuber moth, that can complete in a single growing season, faster growing potatoes may assist some regions adapt to shorter growing seasons (Climate change: can withstand the heat 2012).

3.4 Disease Resistance

Resistances in potatoes to diseases and pests observed in the area could be beneficial, particularly in terms of adjusting to diseases that spread to other regions.

Certain potato cultivars, hybrids, and related species possess characteristics that could potentially mitigate the adverse effects of climate change, both directly and indirectly, on potato production. Zahov is a polyploid hybrid that is heat-tolerant and a cross between numerous wild potato species (Levy et al. 2017). In comparison to *Solanum tuberosum*, certain wild potato species have been discovered to be more heat tolerant. *Solanum verrucosum* is a kind of wild potato that possesses genes that provide strong resistance to late blight (Liu and Halterman 2006). Sullu is an Andean potato variety that may withstand droughts better (American Society of Plant Biologists 2012). With the cry1Ia1 gene from the potato tuber moth, SpuntaG2, a genetically altered variation of the Spunta potato cultivar is resistant to potato tuber moth. Species closely related to potatoes (*Solanum tuberosum*) are an important source of genetic material for developing new potato cultivars that can be crossed with potato. Additionally, there are numerous natives to the Andes, the original home of the commercial potato, types of *Solanum tuberosum* that include genetic information that may be valuable (Genebank, CIP 2009).

3.5 Apical Rooted Cutting

Apical rooted cutting is a new seed production system in potato adopted in non-conventional area of potato seed production. It is a low-cost and rapid method of potato seed production technology in which a single node terminal cutting is obtained from hardened tissue culture plants having juvenile round leaves. The terminal cutting is then planted in soilless media to form a seedling after successful rooting. Rooted apical stem cuttings are the easiest and cheapest means of propagating potato (Struik and Wiersema 1999). The traditional method for propagation of commercial potato cultivars is through tubers. However, the multiplication ratio at 1: 3 to 1:15 (1 tuber yields 3–15 seed tubers) is very low (Bryan et al. 1981; Rashid 1994; Marissa et al. 2004) requiring number of multiplication cycles to get marketable quantity seed tubers (Hossain and Vecchio 2000). Whereas the use ARC technology increases the multiplication ratios of 1: 40 to 1: several thousand cuttings per year (Bryan et al. 1981; Hossain et al. 1999; Tsoka et al. 2012).

3.6 *Shifting Growing Areas*

According to Haverkort et al. (2008) and Hijmans (2003), changes in water and temperature regimes would primarily cause potato yields to rise in some regions such as northern Russia while decreasing in others, such as Sub-Saharan Africa. It is also anticipated that potato production would be able to expand into high latitude and altitude regions where it has previously been restricted by frost damage. There will likely be modifications to the locations where potato crops can be produced profitably because of these crop yield adjustments (Hijmans 2003). Potato-producing locations could be shifted to minimize yield declines in certain nations due to rising temperatures and reduced water availability (Hijmans 2003). The competition for land between potato crops and other crops and land uses could be a challenge in changing potato production.

3.7 *Improving Water Capture, Use, and Irrigation*

To sustain production, potatoes need regular irrigation while the tubers are developing. Subsurface drip irrigation (SDI) is one irrigation method that has been adopted on potatoes with the goal of using less water (Potato Success Story 2012). To lessen the effects of shifting precipitation patterns, better water management techniques are crucial. These include rainwater collection, effective irrigation systems, and soil moisture monitoring. Techniques for farming that use less water can contribute to the preservation of this essential resource.

3.8 *Precision Agriculture Techniques*

Using data analytics and sensor technology in precision agriculture can maximize resource utilization and boost output. This method minimizes the negative effects of potato cultivation on the environment by ensuring effective water and nutrient management.

3.9 *Crop Rotation and Diversification*

Rotating crops and diversifying farming systems help break pest and disease cycles while promoting soil health. Intercropping potatoes with other compatible crops can also enhance biodiversity and improve overall resilience.

3.10 Integrated Pest Management (IPM)

Adopting IPM practices, including biological control, crop monitoring, and judicious use of pesticides, can mitigate the impact of pests and diseases. This approach minimizes environmental harm while maintaining crop health.

3.11 Government Subsidy on the Fertilizers

It is observed that even with government subsidies for fertilizers, there is rarely any availability, possibly due to limited supply. Therefore, growers are forced to purchase fertilizers from independent merchants, raising production costs and making the purchase unfeasible from an economic standpoint.

3.12 High Cost of Production

A small number of farmers are considering switching from potatoes to wheat due to the crop's susceptibility and high production costs, while a larger number, particularly those in their youth, intend to leave the agricultural industry altogether.

3.13 Weather-Based Insurance

To assist farmers in lowering the risk of crop failure because of these occurrences, efficient crop insurance programs should be developed. Programs for private and public insurance, both official and informal, must be established to lessen income losses brought on by the effects of climate change. But to create laws that promote profitable insurance options, data are required.

3.14 Mulching Raises Organic Matter

The amount of organic matter in the soil can be increased via mulching. Consequently, this raises the soil's capacity to hold water, which lowers runoff and erosion and improves the amount of water available to plants.

3.15 *Policy and Education*

Government Policies: Governments should enact policies that incentivize sustainable and climate-resilient farming practices. Financial support, research funding, and education programs can encourage farmers to adopt mitigation strategies.

Educational Outreach: Educating farmers about the impact of climate change and disseminating knowledge about effective mitigation strategies is crucial. Extension services and farmer training programs can play a pivotal role in building awareness and promoting sustainable practices.

4 Conclusion

Climate change poses the most urgent challenge to humanity at present. Its impact on agricultural production has a detrimental effect on farmer incomes. Farmers are aware of this issue and are striving to adjust their practices to sustain agricultural output and product quality while mitigating the negative consequences of climate change. However, they encounter difficulties with implementing low-cost adaptation measures, obtaining subsidized supplies, and receiving government support. In addition, farmers' income is either fluctuating or decreasing annually, leading some potato growers to consider switching crops or leaving the industry altogether. Despite government awareness and efforts to address the severity of the problem, current actions seem insufficient. It is suggested that careful consideration be given by the government for streamlining agricultural assistance programs to ensure both people's welfare and agricultural self-sufficiency. Many regions observe heightened vulnerability in potato production due to climate change-induced variations in yield quantity and quality linked with increased damage such as crop decay; higher replanting costs; elevated pest, disease, and production expenses; as well as amplified soil erosion from extreme weather events like rising temperatures, droughts, and excessive rainfall causing floods. Therefore, introducing more resilient, sustainable, and productive techniques becomes imperative for agriculture to adapt successfully. Understanding how climate change affects potato farming is crucial as it continues transforming our agricultural landscapes. By integrating scientific advancements with sustainable farming methods aided by policy support, we can safeguard the flexibility and resilience of potato agricultural systems thus securing this vital food supply for future generation.

5 Future Line of Work

To better understand the effects of climate change and identify new areas for potato cultivation, additional research is needed. This includes quantifying regional vulnerability and impact assessments, developing early warning systems for diseases, breeding heat- and short-duration cultivars, breeding salinity- and drought-tolerant cultivars, developing improved agronomic management techniques for efficient use of water and fertilizer, developing virus- and late blight-resistant varieties, rescheduling chemical sprays based on newly emerging pathogen populations, and developing integrated pest management strategies. A study of the ideal fertility level, the adoption of resilient types resistant to climate change, and the timing of the start, end, and seasonal dry spells of the rains are also recommended.

References

Adavi Z, Moradi R, Saeidnejad AH, Tadayon MR, Mansouri H (2018) Assessment of potato response to climate change and adaptation strategies. Sci Hortic 228:91–102

Asfaw A, Bonierbale M, Khan MA (2015) Integrative breeding strategy for making climate-smart potato varieties for sub-Saharan Africa. In: Potato and sweetpotato in Africa: transforming the value chains for food and nutrition security. CABI, Wallingford, pp 134–142

Bellon-Maurel V, Fernandez-Ahumada E, Palagos B, Roger JM, McBratney A (2010) Critical review of chemometric indicators commonly used for assessing the quality of the prediction of soil attributes by NIR spectroscopy. TrAC Trends Anal Chem 29(9):1073–1081

Blokhina O, Virolainen E, Fagerstedt KV (2003) Antioxidants, oxidative damage and oxygen deprivation stress: a review. Ann Bot 91(2):179–194

Blom-Zandstra M, Verhagen A (2015) Potato production systems in different agro ecological regions and their relation with climate change (No. 614). Wageningen UR

Boguszewska-Mańkowska D, Zarzyńska K, Nosalewicz A (2020) Drought differentially affects root system size and architecture of potato cultivars with differing drought tolerance. Am J Potato Res 97:54–62

Bowen WT (2003) Water productivity and potato cultivation. In: Water productivity in agriculture: limits and opportunities for improvement. CABI Publishing, Wallingford, pp 229–238

Bryan JE, Jackson MT, Melendez GN (1981) Rapid multiplication techniques for potatoes. Training and Communication Department, International Potato Center, Lima, p 22

Chakraborty S, Datta S (2003) How will plant pathogens adapt to host plant resistance at elevated CO_2 under a changing climate? New Phytol 159(3):733–742

Chang YK, Chi L, Etnier JL, Wang CC, Chu CH, Zhou C (2014) Effect of acute aerobic exercise on cognitive performance: role of cardiovascular fitness. Psychol Sport Exerc 15(5):464–470

Chourasia KN, Lal MK, Tiwari RK, Dev D, Kardile HB, Patil VU et al (2021) Salinity stress in potato: understanding physiological, biochemical and molecular responses. Life 11(6):545

CIP (2009) Colombia. International Potato Center: World Potato Atlas "Climate change—can potato stand the heat? 2012. WREN Media. http://www.new ag.info/en/focus/focusItem.php?a=532. Retrieved 7 Nov 2012

CIP International Potato Center (2013) Potato faces up to climate change challenges. International Potato Center. https://cipotato.org/blog/potato-faces-up-to-climatechangechallenges

Cleugh H (2012) Trees for shelter windbreaks for Australian farms. Rural Industries Research and Development Corporation Thind, T.S. Fungicides in crop health security. Indian Phytopathol 65:109–115

Climate Change 2007: Synthesis Report. Intergovernmental Panel On Climate Change. Retrieved 7 Nov 2012

Condori B, Hijmans RJ, Ledent JF, Quiroz R (2014) Managing potato biodiversity to cope with frost risk in the high Andes: a modeling perspective. PLoS One 9(1):e81510

Cruiziat P, Cochard H, Améglio T (2002) Hydraulic architecture of trees: main concepts and results. Ann For Sci 59(7):723–752

Daccache A, Weatherhead EK, Stalham MA, Knox JW (2011) Impacts of climate change on irrigated potato production in a humid climate. Agric For Meteorol 151(12):1641–1653
Daccache A, Keay C, Jones RJ, Weatherhead EK, Stalham MA, Knox JW (2012) Climate change and land suitability for potato production in England and Wales: impacts and adaptation. J Agric Sci 150(2):161–177
Das L, Annan JD, Hargreaves JC, Emori S (2011) Centennial scale warming over Japan: are the rural stations really rural? Atmos Sci Lett 12(4):362–367
Das T, Majumdar MHD, Devi RT, Rajesh T (2016) Climate change impacts on plant diseases. SAARC J Agric 14(2):200–209
Deblonde PMK, Ledent JF (2001) Effects of moderate drought conditions on green leaf number, stem height, leaf length and tuber yield of potato cultivars. Eur J Agron 14(1):31–41
Devaux A, Goffart JP, Kromann P, Andrade-Piedra J, Polar V, Hareau G (2021) The potato of the future: opportunities and challenges in sustainable agri-food systems. Potato Res 64(4):681–720
Doussan C, Vercambre G, Pagè L (1998) Modelling of the hydraulic architecture of root systems: an integrated approach to water absorption—Distribution of axial and radial conductances in maize. Ann Bot 81(2):225–232
Duarte V, De Boer SH, Ward LJ, de Oliveira MC (2004) Characterization of atypical Erwinia carotovora strains causing blackleg ofpotato in Brazil. J Appl Microbiol 96:535–545
Dwelle RB (1985) Photosynthesis and photoassimilate partitioning. Potato Physiol:35–58
Eiasu BK, Soundy P, Hammes PS (2007) Response of potato (Solarium tuberosum) tuber yield components to gel-polymer soil amendments and irrigation regimes. N Z J Crop Hortic Sci 35(1):25–31
Ewing EE (1981) Heat stress and the tuberization stimulus. Am Potato J 58:31–49
FAO (2009) International year of the potato 2008: new light on a hidden treasure. http://www.fao.org/potato-2008/en/events/book.html
Farisa Banu S, Rubini D, Rakshitaa S, Chandrasekar K, Murugan R, Wilson A et al (2017) Antivirulent properties of underexplored Cinnamomum tamala essential oil and its synergistic effects with DNase against Pseudomonas aeruginosa Biofilms—an in vitro study. Front Microbiol 8:1144
Feustel IC (1987) Miscellaneous products from potatoes. In: Talburt WF, Smith O (eds) Potato processing, 4th edn. Van Nostrand, New York, pp 727–746
Fleisher DH, Timlin DJ, Reddy VR (2016) Climate change and potato: responses to carbon dioxide, temperature, and drought. In: Improving modeling tools to assess climate change effects on crop response, vol 7, pp 69–90
Frohning J (2013) Evaluation of Rhizoctonia solani-Heterodera glycines interactions on soybean. Doctoral dissertation, University of Illinois at Urbana-Champaign
Gaur PC, Pandey SK (2000) Potato improvement in sub-tropics. In: Potato, global research & development. Proceedings of the Global Conference on Potato, New Delhi, India, 6–11 December, 1999: Volume 1. Indian Potato Association, pp 52–63
Gawronska H, Dwelle RB (1989) Partitioning of photoassimilates by potato plants (*Solanum tuberosum* L.) as influenced by irradiance I. Partitioning patterns in cultivar Russet Burbank grown under high and low irradiance. Am Potato J 66(4):201–213
Genebank. CIP. http://cipotato.org/genebank. Retrieved 7 Nov 2012
Ginzberg I, Oren-Shamir M, Hirschberg J, Mandolino G, Parisi B, Faigenboim A, Fogelman E, Ovadia R (2017) The effect of hot climate on anthocyanins, carotenoids and glycoalkaloids biosynthesis. In: Proceedings of the 20th triennial EAPR conference 9th–14th July 2017, Versaille
Gorska-Warsewicz H, Rejman K, Kaczorowska J, Laskowski W (2021) Vegetables, potatoes and their products as sources of energy and nutrients to the average diet in Poland. Int J Environ Res Public Health 18(6):3217
Haling RE, Brown LK, Bengough AG, Young IM, Hallett PD, White PJ, George TS (2013) Root hairs improve root penetration, root–soil contact, and phosphorus acquisition in soils of different strength. J Exp Bot 64(12):3711–3721
Hancock RD, Morris WL, Ducreux LJ, Morris JA, Usman M, Verrall SR et al (2014) Physiological, biochemical and molecular responses of the potato (*Solanum tuberosum* L.) plant to moderately elevated temperature. Plant Cell Environ 37(2):439–450

Haverkort AJ, Verhagen A (2008) Climate change and its repercussions for the potato supply chain. Potato Res 51:223–237

Haverkort AJ, Boonekamp PM, Hutten R, Jacobsen E, Lotz LAP, Kessel GJT, Visser RGF, van der Vossen EAG (2008) Societal costs of lateblight in potato and prospects of durable resistance through cisgenic modification. Potato Res 51:47–57

Hayward AC (1991) Biology and epidemiology of bacterial wilt caused by Pseudomonas solanacearum. Annu Rev Phytopathol 29(1):65–87

Healy FG, Lambert DH (1991) Relationships among Streptomyces spp. causing potato scab. Int J Syst Evol Microbiol 41(4):479–482

Hijmans RJ (2003) The effect of climate change on global potato production. Am J Potato Res 80:271–279

Hossain MJ, Vecchio V (2000) Potential of sprout cutting on the growth, yield and multiplication rate of potato. Ital J Agron 3(1):1–6

Hossain MJ, Nahar MS, Ahmad AU (1999) Sprout and top-shoot cutting for rapid multiplication rate potato in Bangladesh. J Agric Sci 132:437–443

Huchzermeyer B, Koyro H-W (2005) Salt and drought stress effects on photosynthesis. Enzyme cohesion and high turn over metabolite shuttling, essential for functioning of pathways, is impaired by changes in cytosolic water potential. In: Pessarakli M (ed) Handbook of photosynthesis, 2nd edn. CRC Press/Taylor and Francis, Boca Raton, pp 751–777

Hunting for drought tolerance genes in ancient Andean landraces. American of Society Plant Biologists. http://www.aspb.org/PressReleases/potatoes.cfm. Retrieved 7 Nov 2012

Information highlights from World Potato Congress, Kunming, China, April 2004. World Potato Congress. http://www.peracto.com.au/publications/world-potato-congress.pdf. Retrieved 7 Nov 2012

Intergovernmental Panel on Climate Change (IPCC) (2007) Climate change, a synthesis report. Contribution of working Group 1,11 and 111 to the Fourth Assessment Report of the Intergovernmental Panel on Climate Change.Core—writing team Pachauri RK, ResingerA (eds) IPCC, Geniva Switzerland

Iwama K, Yamaguchi J (2006) Abiotic stresses. In: Handbook of potato production, improvement, and postharvest management. CRC Press, pp 231–278

Jaiswal VP (1995) Response of potato (Solanum tuberosum) cultivars to date of planting and mulching under warm-temperature condition

Johnson E, Grabowski M, Orshinsky A (2015) Early blight of tomato: commercial fruit and vegetable production. University of Minnesota Extension, pp 612–624

Kincaid DC, Westermann DT, Trout TJ (1993) Irrigation and soil temperature effects on Russet Burbank quality. Am Potato J 70:711–723

Kirk H, Dorn S, Mazzi D (2013) Molecular genetics and genomics generate new insights into invertebrate pest invasions. Evol Appl 6(5):842–856

Koyro H-W, Ahmad P, Geissler N (2012) Abiotic stress responses in plants: an overview. In: Ahmad P, Prasad MNV (eds) Environmental adaptations and stress tolerance of plants in the era of climate change. Springer Science and Business Media, pp 1–28

Kumar P, Pandey SK, Singh BP, Singh SV, Kumar D (2007) Effect of nitrogen rate on growth, yield, economics and crisps quality of Indian potato processing cultivars. Potato Res 50:143–155

Levy D, Veilleux RE (2007) Adaptation of potato to high temperatures and salinity-a review. Am J Potato Res 84:487–506

Levy JG, Mendoza A, Miller JC, Tamborindeguy C, Pierson EA (2017) Global gene expression in two potato cultivars in response to 'Candidatus Liberibacter solanacearum'infection. BMC Genomics 18(1):1–16

Liu Z, Halterman D (2006) Identification and characterization of RB-orthologous genes from the late blight resistant wild potato species Solanum verrucosum. Physiol Mol Plant Pathol 69(4–6):230–239

Londhe S (ed) (2016) Sustainable potato production and the impact of climate change. IGI Global

Luck J, Asaduzzaman M, Banerjee S, Bhattacharya I, Coughlan K, Debnath GC, De Boer D, Dutta S, Forbes G, Griffiths W, Hossain D, Huda S, Jagannathan R, Khan S, O'Leary G, Miah, G, Saha A, Spooner-Hart R (2012) The effects of climate change on pests and diseases of major

food crops in the Asia Pacific Region. Final report for APN (Asia-Pacific Network for Global Change Research) project, P-73
Mahmood GI, Ligrani PM (2002) Heat transfer in a dimpled channel: combined influences of aspect ratio, temperature ratio, Reynolds number, and flow structure. Int J Heat Mass Transf 45(10):2011–2020
Manrique LA, Meyer R (1984) Effects of soil mulches on soil temperature, plant growth and potato yields in an aridic isothermic environment in Peru
Mares MA, Willig MR, Lacher TE Jr (1985) The Brazilian Caatinga in South American zoogeography: tropical mammals in a dry region. J Biogeogr:57–69
Marissa IA, Maura LS, Gabriel LS, Malab BS, Legaspi M (2004) Rapid multiplication of disease-free raniag lowland potato seed tubers. Mariano Marcos State University, Laoog City
Midmore DJ, Prange RK (1992) Growth responses of two Solanum species to contrasting temperatures and irradiance levels: relations to photosynthesis, dark respiration and chlorophyll fluorescence. Ann Bot 69(1):13–20
Midmore DJ, Berrios D, Roca J (1986) Potato (Solanum spp.) in the hot tropics II. Soil temperature and moisture modification by mulch in contrasting environments. Field Crop Res 15(2):97–108
Minhas JS, Rawat S, Govindakrishnan PM, Kumar D (2011) Possibilities of enhancing potato production in non-traditional areas. Potato J 38(1)
Monteberg EL (2017) Influence of soil characteristics on skin finish of Norwegian potatoes. In: Proceedings of the 20th triennial EAPR conference 9th–14th July 2017, Versaille
Pavlista AD (2002) Skin set evaluation by skin shear measurements. Am J Potato Res 79:301–307
Potato and water resources. FAO. http://www.potato2008.org/en/potato/water.html. Retrieved 7 Nov 2012
Potato success story, China. Netafirm. http://www.netafim.com.au/article/subsurface-drip-irrigation%2D%2Dpotato. Retrieved 7 Nov 2012
Puértolas J, Ballester C, Elphinstone ED, Dodd IC (2014) Two potato (Solanum tuberosum) varieties differ in drought tolerance due to differences in root growth at depth. Funct Plant Biol 41(11):1107–1118
Rashid MM (1994) Transferable technologies in tuber crops. Tuber Crops Research Center, Bangladesh Agricultural Research Institute, Joydebpur, pp 1–14
Reynolds MP, Ewing EE (1989) Effects of high air and soil temperature stress on growth and tuberization in Solanum tuberosum. Ann Bot 64(3):241–247
Reynolds MP, Ewing EE, Owens TG (1990) Photosynthesis at high temperature in tuber-bearing Solanum species: a comparison between accessions of contrasting heat tolerance. Plant Physiol 93(2):791–797
Rontein D, Basset G, Hanson AD (2002) Metabolic engineering of osmoprotectant accumulation in plants. Metab Eng 4(1):49–56
Ruiz JM, Hernandez J, Castilla N, Romero L (1999) Potato performance in response to different mulches. 1. Nitrogen metabolism and yield. J Agric Food Chem 47(7):2660–2665
Sale PJM (1979) Aust J Agric Res 30:667
Schafleitner R, Ramirez J, Jarvis A, Evers D, Gutierrez R, Scurrah M (2011) Adaptation of the potato crop to changing climates. In: Crop adaptation to climate change, pp 287–297
Schoville SD, Chen YH, Andersson MN, Benoit JB, Bowsher JH, Brevik K, Cappelle K, Chen MM, Anna K, Childers C (2017) A model species for agricultural pest genomics: the genome of the Colorado potato beetle, *Leptinotarsa decemlineata* (Coleoptera: Chrysomelidae). Sci Rep 8:1–18
Shekhawat NS, Rathore TS, Singh RP, Deora NS, Rao SR (1993) Factors affecting in vitro clonal propagation of Prosopis cineraria. Plant Growth Regul 12:273–280
Simelton E, Fraser ED, Termansen M, Benton TG, Gosling SN, South A et al (2012) The socioeconomics of food crop production and climate change vulnerability: a global scale quantitative analysis of how grain crops are sensitive to drought. Food Secur 4:163–179
Sinclair J (1984) Naturalness in language. In: Corpus linguistics. Brill, pp 203–210

Singh PH (2000) Wart disease of potato. In: Khurana SMP (ed) Diseases and pests of potato- a manual. Central Potato Research Institute, Shimla

Singh AK, Sharma V, Pal AK, Acharya V, Ahuja PS (2013) Genome-wide organization and expression profiling of the NAC transcription factor family in potato (Solanum tuberosum L.). DNA Res 20(4):403–423

Snyder RG, Ewing EE (1989) Interactive effects of temperature, photoperiod, and cultivar on tuberization of potato cuttings. HortScience 24(2):336–338

Somani AK (2007) Saving potato tubers from charcoal rot. Indian Hortic 52(2):26–27

Struik PC, Wiersema SG (1999) Production of pre-basic seed. In: Seed potato technology. Wageningen Academic, Wageningen, pp 173–216

Sturrock RR (1981) Microglia in the prenatal mouse neostriatum and spinal cord. J Anat 133(Pt 4):499

Stuttle GW, Yorio NC, Wheeler RM (1996) Interacting effects of photoperiod and photosynthetic photon flux on net carbon assimilation and starch accumulation in potato leaves. J Am Soc Hortic Sci 121:264–268

Sun D, Dickinson GR (1997) Early growth of six native Australian tree species in windbreaks and their effect on potato growth in tropical northern Australia. For Ecol Manag 95(1):21–34

Sun Y, Yan F, Cui X, Liu F (2014) Plasticity in stomatal size and density of potato leaves under different irrigation and phosphorus regimes. J Plant Physiol 171(14):1248–1255

Sutic DD, Ford RE, Tosic MT (1999) Handbook of plant virus diseases. CRC Press

Talburt WF (1987) History of potato processing. In: Talburt WF, Smith O (eds) Potato processing, 4th edn. Van Nostrand, New York, pp 1–10

Tesfaye HT (2016) A review on potato (Solanum tuberosum L.) production situations in Ethiopia. Food Sci Qual Manage 57:109–112

Thornton MK, Malik NJ, Dwelle RB (1996) Relationship between leaf gas exchange characteristics and productivity of potato clones grown at different temperatures. Am Potato J 73(2):63–77

Tsoka O, Demo P, Nyende AB, Ngamau K (2012) Potato seed tuber production from *in vitro* and apical stem cutting under aeroponic system. Afr J Biotechnol 11(63):12612–12618

Van Dam J, Kooman PL, Struik PC (1996) Effects of temperature and photoperiod on early growth and final number of tubers in potato (*Solanum tuberosum* L.). Potato Res 39:51–62

Vandeleur RK, Mayo G, Shelden MC, Gilliham M, Kaiser BN, Tyerman SD (2009) The role of plasma membrane intrinsic protein aquaporins in water transport through roots: diurnal and drought stress responses reveal different strategies between isohydric and anisohydric cultivars of grapevine. Plant Physiol 149(1):445–460

Western Potato Council (2003) Guide to commercial potato production on the Canadian prairies. Western Potato Council, Portage la Prairie

Wheeler RM, Tibbitts TW, Fitzpatrick AH (1989) Potato growth in response to relative humidity. HortScience 24(3):482–484

Wishart J, George TS, Brown LK, Ramsay G, Bradshaw JE, White PJ, Gregory PJ (2013) Measuring variation in potato roots in both field and glasshouse: the search for useful yield predictors and a simple screen for root traits. Plant Soil 368:231–249

Wishart J, George TS, Brown LK, White PJ, Ramsay G, Jones H, Gregory PJ (2014) Field phenotyping of potato to assess root and shoot characteristics associated with drought tolerance. Plant Soil 378:351–363

Wolf S, Marani A, Rudich J (1991) Effect of temperature on carbohydrate metabolism in potato plants. J Exp Bot 42(5):619–625

Wright AJ, Brooks SJ (2002) Effect of windbreaks on potato production for the Atherton Tablelands of North Queensland. Aust J Exp Agric 42(6):797–807

Yamaguchi M, Timm H, Spurr AR (1964) Effects of soil temperature on growth and nutrition of potato plants and tuberization, composition, and periderm structure of tubers

Yan F, Sun Y, Song F, Liu F (2012) Differential responses of stomatal morphology to partial root-zone drying and deficit irrigation in potato leaves under varied nitrogen rates. Sci Hortic 145:76–83

Advances in Breeding for the Development of New Potato Varieties

Pushpam Patel, Anuradha Sinha, Shirin Akhtar, and Randhir Kumar

Abstract Potato (*Solanum tuberosum* L.) is a vital food crop globally that is facing challenges such as climate change, evolving pest and disease pressures, and increasing demand and breeding efforts that play a crucial role in enhancing its agronomical traits. These things explore recent advancements in potato breeding techniques, including traditional methods, transgenic approaches, marker-assisted selection (MAS), genomic selection (GS), and genome editing (GE). Traditional breeding methods involve controlled cross-pollination to produce offspring with desired traits, while MAS accelerates the process by identifying genetic markers linked to desirable traits. GS leverages genomic data to predict plant performance, particularly for complex traits. Transgenic approaches introduce foreign genes to confer desirable traits, albeit with some controversy. Genome editing technologies like CRISPR-Cas9 offer precise alteration to the potato genome, hastening variety development. These advancements have significantly contributed to the creation of new potato types with superior features, such as disease resistance, yield, and nutritional content, ensuring global food security and sustainability. Continued research and investment in breeding technologies are crucial for further enhancing potato varieties to meet future challenges.

P. Patel (✉)
Krishi Vigyan Kendra, Barh, Patna, Bihar Agricultural University, Sabour, Bhagalpur, Bihar, India

A. Sinha
Krishi Vigyan Kendra, Aurangabad, Bihar Agricultural University, Sabour, Bhagalpur, Bihar, India

S. Akhtar
Department of Horticulture (Vegetable Science), Bihar Agricultural College, Bihar Agricultural University, Sabour, Bhagalpur, Bihar, India

R. Kumar
Department of Horticulture (Vegetable Science), Nalanda College of Horticulture, Noorsarai, Nalanda, Bihar Agricultural University, Sabour, Bhagalpur, Bihar, India

S. S. Solankey (ed.), *Advances in Research on Potato Production*,
Advances in Olericulture, https://doi.org/10.1007/978-3-031-82710-5_4

Keywords Potato · Climate change · Traits · Traditional methods · Transgenic · MAS · Genome editing

1 Introduction

The shifting climate makes it difficult for farmers to grow crops, which ultimately puts a great deal of pressure on the capacity to feed the world's growing population. Climate change is causing unpredictable weather patterns and unstable air temperatures. Numerous foreign researchers have cautioned us against this, claiming that if we do not adopt sustainable farming methods, climate change may result in a reduction in the world's food supply, rendering it unfeasible for humans to produce food of a good standard. Rich countries do, however, contribute a lot more to greenhouse gas emissions than developing ones do. According to projections from the UN Food and Agriculture Organization (FAO), to supply food to an additional 2.3 billion people by 2050, there will need to be a 70% increase in world food production.

The aggregation of greenhouse gases in the surroundings, such as methane (CH_4), carbon dioxide (CO_2), nitrous oxide (N_2O), and ozone (O_3), is the cause of this problem. These gases are created by people as a result of long-term, rigorous industrial enlargement and high-consumption lifestyles in developed countries. India needs a national policy that addresses climate change first, enhances the ecological sustainability of the country's growth trajectory, and collaborates with other nations to jointly and collectively tackle this issue. As the world's population rises, there will be a corresponding need for a corresponding growth in the food supply. On the other hand, several factors like urbanization, industrialization, and climate change have put too much strain on the food supply and agricultural areas that are currently available (Badami and Ramankutty 2015). Various biotic and abiotic stress factors that consistently affect crops worldwide are other reasons contributing to the fall in food production. A number of crops with refined nourishing profiles have been generated using the recent gene pool thanks to technology developments and cooperative public-private alliances (Ricroch and Henard-Damave 2016; Ma et al. 2017b).

After maize, rice, and wheat, potatoes (*Solanum tuberosum* L.), a solanaceous family food crop, come in fourth place among the main staple crops. When it comes to worldwide consumption, potatoes are the most significant non-grain food crop (Devaux et al. 2020). It was initially domesticated approximately 10,000 years ago in Peru. Given that food security is becoming more of a concern and that there will be 9.7 billion people in the sphere by 2050, potatoes will likely become a major crop and staple food in the future (Hameed et al. 2018; FAO 2019). Because potatoes contain a number of important phytochemicals that have been shown to have major health advantages, they have high nutritional values (Rasheed et al. 2022). The main components of potatoes are a variety of phytonutrients, including phenolics, carotenoids, anthocyanins, and flavonoids; mineral elements, including sodium, potassium, copper, magnesium, iron, and zinc; and vitamins, including vitamin B_6, B_1,

B_9, C, and E (White et al. 2012; Ngobese et al. 2017). These phytonutrients are vital for preserving human health because of their antioxidant qualities (Rasheed et al. 2022).

Compared to other possible food crops, it can yield higher amounts of carbs, vitamins, proteins, and minerals per unit area and time (Zaheer and Akhtar 2016). Potato is widely employed in the industry not only as a marketable raw material but also for the manufacture of processed foods, alcohol, and bio-fuel (Liang and McDonald 2014). Potato tubers typically include fewer than 3% proteins, vitamins, dietary fiber, minerals, and other substances; 77% of water; and 20% of carbs (Zaheer and Akhtar 2016). Since potatoes originated in South America, they have expanded over many different geographical borders due to their short crop length and extensive environmental tolerance. Presently, over three thousand potato cultivars are extensively dispersed throughout over 125 nations, mostly in temperate, subtropical, and tropical climates, accounting for a significant portion of the worldwide agricultural market's economic share (Birch et al. 2012). Over the past 20 years, emerging nations like Bangladesh, India, and China have also seen notable growth in the cultivation and use of potatoes (Zaheer and Akhtar 2016). Burlingame et al. (2009) provide in-depth information about the tuber composition of various potato varieties.

Potato may be utilized as economical foods that sustainably furnish abundant calories (93 kcal/100 g tuber) to support regular survival, and they can substitute other expensive foods in limited-income food shortage countries (Burlingame et al. 2009). The significance of potato to the world is undeniable, and in 2008, the United Nations declared it to be the "International Year of the Potato" to promote their ability to combat poverty, food shortages, and, most importantly, malnutrition. Enhancing the crop processing tuber quality while increasing production is the primary goal. Proper genotypes and growth conditions impact potato yield, which is strongly connected with sprout frequency, plant height, shoot diameter, and leaf length (Tiwari et al. 2020). Table potato market prices are determined by the visual characteristics of cultivars, such as eye depth, flesh color, tuber form, and cooking appropriateness. The chemical makeup of the tuber determines how well potato products cook and look. For instance, the texture of the cooked product is influenced by the starch content, and the color of the fried product is directly influenced by the sugar level (Andrivon 2017). Several molecular and breeding methods have been applied to improve potato characteristics. Traditional breeding techniques aim to increase potato output, processing, and storage quality (Halterman et al. 2016). Potato breeders were able to include resistance against early and late blight disease by mating hybrid lines with wild species (*S. brevidens* and *S. bulbocastanum*) that acquired resistance versus fungal infections (Naess et al. 2000; Tek et al. 2004).

The process of identifying the best individuals in future generations has slowed down traditional breeding, even though it's useful for improving specific traits with little variation within the species. Additionally, it can be hard to find useful genetic diversity in wild relatives and to introduce that diversity into cultivated plants. The potato's highly diverse genome and complex structure (The Potato Genome Sequencing Consortium 2011) create challenges for breeding efforts to improve it

(Lindhout et al. 2011). Intra-species incompatibility and inbreeding depression are two further factors that make it difficult for traditional breeding techniques to introduce characteristics into polyploid crops. Within this framework, novel breeding technologies (NBTs) present a cutting edge for improving agricultural plant traits and give a solid foundation for accurate and reliable plant genome editing. New breeding technologies (NBTs), including modern genome editing methods such as transcription activator-like effectors nucleases, zinc-finger nucleases, and CRISPR/Cas9, represent significant advancements in the field (Schaart et al. 2016; Weeks et al. 2016).

The CRISPR system, though relatively new, has already been competently applied to enhance the traits of various important crops such as rice, wheat, maize, soybean, cassava, cotton, and potatoes (Puchta 2017). Traits introduced through this method include resistance to bacterial, fungal, and viral infections; herbicide tolerance; drought resistance; and longer shelf life—contributing to improved overall quality and yield. The action mechanisms and potential of these non-biological terminators in genetic engineering have been profusely researched (Zaidi et al. 2017a, b, 2018; Weeks 2017; Puchta 2017; Schiml and Puchta 2016; Mahfouz et al. 2016; Bortesi and Fischer 2015). The current study offers a thorough analysis of the many genetic techniques consisting of new breeding techniques (NBTs) that have been effectively used to improve the nutritional content of potatoes. Additionally, we list the key issues raised by the regulatory clearances of transgenic potatoes and a summary of the data on potatoes that have been sold thus far.

2 Breeding History

Modern potato varieties have been developed through extensive breeding efforts that involve both wild species and various cultivar groups. These varieties have been grown outside of South America since the late 1500s. During the evolution and domestication of potatoes, polyploidization was influenced by factors such as self-incompatibility, unaltered gametes, and reproductive barriers between species. Potatoes naturally reproduce both sexually and asexually. Sexual reproduction occurs through insect pollination of the flowers, resulting in true potato seeds (TPS) in berries, while asexual reproduction takes place through tubers, which can sprout into new potato plants from genetically unique seedlings derived from TPS.

That is a great summary of the shift in potato breeding practices! The move from open pollination to artificial hybridization was crucial for improving potato varieties. By creating controlled crosses, breeders could combine specific traits and generate new genetic variations, which were essential for developing better cultivars. The establishment of gene banks, like the International Potato Centre (CIP) in 1971, has been instrumental in preserving genetic diversity and providing resources for breeding programs worldwide. This work has played a significant role in enhancing potato yield, disease resistance, and adaptability.

3 Early Years of Potato Breeding (1901–1937)

- Redcliffe N. Salaman, a key figure in England, played a significant role in early potato breeding and genetics. Salaman's work included studying genetic factors related to immunity, resistance to diseases, skin color, flesh color, tuber shape, and other traits. Salaman's work was constrained by the limited understanding of genetic principles and meiosis in potatoes at the time. The lack of comprehensive genetic knowledge and insights into meiosis restricted his ability to predict and control genetic outcomes in potato breeding. His foundational efforts paved the way, but it was only with later advances in genetics and a clearer understanding of meiosis that breeders could fully utilize artificial hybridization to enhance potato cultivars.
- Salaman's research on resistance to late blight included breeding efforts with *Solanum edinense*, a natural hybrid of *S. demissum* and *S. tuberosum*. This hybrid provided valuable genetic material for developing potato varieties with enhanced resistance to the disease.
- William Black in Scotland continued the work on late blight resistance, discovering new factors and establishing a breeding program.

4 Assessment of Yield and Quantitative Traits

- Potato breeders, including Salaman, recognized that yield was influenced by both genetic and environmental factors.
- Yield trials were conducted, and early experiments on the manurial response provided insights into variety and treatment interactions.

5 Challenges in Genetic Understanding

- Before 1937, potato breeders lacked knowledge of the tetraploid nature of potatoes and tetrasomic inheritance.
- Even with limited genetic knowledge, breeders managed to evaluate and analyze yield and other quantitative traits across identical generations through randomized and replicated trials. This approach allowed them to assess the performance of different potato varieties and make informed decisions about which traits to select for further breeding.

6 Breeding Challenges

Diploid potatoes display out-crossing because of a gametophytic self-incompatibility system, which prevents inbreeding and encourages genetic diversity. This mechanism ensures that fertilization occurs between genetically distinct individuals, leading to increased variation and potentially improved traits in the offspring. However, tetraploid potatoes are self-compatible. Selfing results in inbreeding depression in most potatoes, making it challenging to eliminate unfavorable alleles.

7 Conventional Breeding Techniques

Conventional potato breeding focuses on improving output, processing qualities, and retention quality (Halterman et al. 2016). Traditional breeding of potato varieties involves the traditional methods of selecting and crossing plants to grow new varieties with desired attributes. Conventional breeding relies on natural processes and genetic variation within the existing gene pool of the crop. Potato breeders have employed conventional breeding methods to introduce resistance to early and late blight diseases. By crossing hybrid lines with wild species such as *S. brevidens* and *S. bulbocastanum*, which possess resistance to fungal diseases, potato breeders have successfully embedded immunity to early and late blight diseases (Naess et al. 2000; Tek et al. 2004).

- **Selection of Parental Lines:** Breeders start by selecting two or more potato plants with desirable traits in potato breeding, including disease resistance, high yield, good taste, or adaptation to specific environmental conditions. These plants serve as the parental lines for the breeding process.
- **Crossing:** The selected parental lines are cross-pollinated to produce a population of plants with a combination of their genetic traits. Potatoes are typically cross-pollinated through the process of moving pollen from the flowers of one plant to the stigma of a different plant. However, some varieties can also be self-pollinated.
- **Generation of Offspring (Seed):** After successful pollination, the plants produce seeds. Potato seeds are small, green berries that contain the true seeds of the plant. These seeds are collected and used to grow the next generation of potato plants.
- **Field Trials and Selection:** The resulting plants from the seeds are grown in field trials. Breeders evaluate the offspring for various traits, including yield, disease resistance, tuber quality, and adaptability to different environmental conditions. This step involves a rigorous selection process to identify the plants that exhibit the desired characteristics.
- **Repetition of the Breeding Cycle:** The selected plants become the new parental lines for the next round of breeding. This cyclic process is repeated over several generations, with each cycle aimed at refining and improving the traits of interest.
- **Development of New Potato Varieties:** After several cycles of selection and crossbreeding, breeders identify potato lines with the desired combination of traits. These lines are then multiplied and tested for stability and consistency in different environments.

- **Release and Commercialization:** Once a new potato variety has been thoroughly evaluated and proven to meet the necessary standards, it is officially released and made available to farmers for commercial cultivation.

8 Challenges in Conventional Breeding

Despite successful trait improvement through conventional breeding, there are challenges, including slow progress and limitations due to the need for phenotypic characterization. Searching for valuable genetic variability in wild relatives can be a labor-intensive process. The introgression of beneficial traits from wild relatives into cultivated varieties is considered a challenging task. Potato genome shows high heterozygosity and tetraploid nature (The Potato Genome Sequencing Consortium 2011), which are major challenges, and this process can be a disadvantage in breeding activities to enhance potato crops due to allelic suppression occurring at each breeding cross (Lindhout et al. 2011).

9 Advances in Potato Breeding Methods

9.1 Introduction of New Breeding Technologies (NBTs)

NBTs are highlighted as playing a crucial role in trait improvement in crops. New breeding technologies (NBTs) offer a foundation for accurate and reliable plant genome editing as well as a competent and leading hand for improving traits in agricultural plants. These NBTs encompass advanced genome editing techniques like zinc-finger nucleases (ZFNs), transcription activator-like effector nucleases (TALENs), and clustered regularly interspaced short palindromic repeats/CRISPR-associated protein 9 (CRISPR/Cas9) (Jinek et al. 2012; Schaart et al. 2016; Weeks et al. 2016).

9.2 Application of CRISPR System in Potato Improvement

The CRISPR system, despite its recent development, has been effectively applied to enhance the traits of countless economically important crops, such as maize, wheat, soybean, cassava, rice, cotton, and potato (Puchta 2017). The introduced traits encompass resistance to bacterial, fungal, and viral diseases; herbicide tolerance; drought resistance; and increased longevity, all of which improve overall quality and production. Technologies such as CRISPR-Cas9 have revolutionized genome editing, enabling precise modifications to specific genes in the potato genome. Genome editing allows for targeted improvements, such as enhancing disease resistance, nutritional content, or tolerance to environmental stress. This technique offers greater precision compared to traditional breeding and can expedite the development of new varieties. Numerous reviews have been written about the working methods and expected applications of these new breeding technologies (NBTs) in plant genetic engineering (Bortesi and Fischer 2015; Mahfouz et al. 2016; Schiml and Puchta 2016; Puchta 2017; Weeks 2017; Zaidi et al. 2017a, b, 2018).

1. **Marker-Assisted Selection (MAS):** MAS is a great apparatus for assisting plant breeding, but many agronomic traits are complex and controlled by several loci. An improved and new breeding technique approach has been developed, which is genomic selection (GS). GS helps estimate plants' genetic success based on all molecular markers' information simultaneously instead of a few ones like in MAS (Meuwissen et al. 2001). MAS involves the use of molecular markers linked to specific traits of interest. By analyzing the genetic markers, breeders can identify plants or crops carrying desired traits more quickly and accurately than traditional methods. This allows for the selection of promising potato variants in earlier generations, speeding up the breeding process. In MAS-related breeding, 99% genome can be recovered with only three backcross generations instead of six to seven generations required to recover the same percentage of the genome without the use of molecular markers. RFLP, RAPD, AFLP, and SSR were commonly used for MAS breeding.
2. **Genomic Selection:** Enhanced exactness of high-throughput genotyping procedures, decreased cost of sequencing techniques, and marker selection have reinforced the value of GS (Jonas and de Koning 2013). GS can predict the genetic value/breeding value. There are two populations: the training population that is genotyped and phenotyped, and the testing population that is genotyped but not phenotyped. Genomic selection uses advanced statistical models to predict the performance of a plant based on its genomic profile. By analyzing the entire genome, breeders can estimate the potential of a plant for specific traits, allowing for more informed decisions in the selection process. Various statistical models are used to predict the traits of interest, such as Kinship GAUSS, PR-BLUP, Bayesian LASSO, Bayes B, and Bayes Cp (Crossa et al. 2017).
3. **Cisgenesis and Intragenesis:** Cisgenesis involves transferring genes between closely related species, while intragenesis involves the introduction of genes from the same species. These techniques aim to incorporate beneficial traits into potato varieties more precisely and with minimal disturbance to the natural genetic makeup.
4. **Phenomics:** Phenomics involves the high-throughput measurement of plant traits, allowing for the rapid and detailed analysis of plant characteristics. Advanced imaging and sensor technologies are used to collect data on traits such as growth, morphology, and stress responses. This information can aid in the selection of plants with desired traits more efficiently.
5. **Bioinformatics and Data Analysis:** The use of bioinformatics tools and data analysis techniques has become integral in modern breeding programs. These tools help breeders analyze large datasets, including genomic information, to identify patterns, associations, and potential candidate genes for specific traits.

6. **Doubled Haploid Technology:** Doubled haploids are plants with a doubled set of chromosomes derived from a single, haploid cell. This technology allows for the rapid generation of genetically uniform lines, enabling quicker trait evaluation and variety development.
7. **Transgenic Approaches**: Transgenic approaches involve the introduction of genes from other organisms into the genome of a target organism, such as a potato plant. Transgenic approaches have been used in potato breeding to bestow specific traits that may not be easily accomplishable through conventional breeding (Hameed et al. 2017). Here are some transgenic approaches applied in potato breeding:

   - **Disease Resistance:** One of the significant applications of transgenic technology in potato breeding is the introduction of genes that confer resistance to specific diseases. For example, the introduction of the Bt (*Bacillus thuringiensis*) gene in potatoes helps in producing a protein toxic to certain pests, insects that provide built-in pest resistance.

     – **Late Blight Resistance:** The oomycete pathogen *Phytophthora infestans* is the primary cause of late blight, which poses a serious risk to potato harvests. Transgenic potatoes have been developed with genes from wild potato relatives or other sources that confer resistance to late blight. This helps reduce the reliance on chemical fungicides.
   - **Virus Resistance:** Transgenic potatoes have been engineered to resist viral infections. Genes from viruses or other sources that interfere with the replication or spread of common potato viruses, such as Potato Virus Y (Hameed et al. 2017), have been introduced to enhance resistance.
   - **Improved Tuber Quality:** Transgenic approaches have been used to modify starch composition in potatoes. For example, genes responsible for altering starch structure or content have been introduced to enhance the standard of potato tubers for specific industrial or culinary purposes.
   - **Enhanced Nutritional Content:** Genes responsible for the biosynthesis of specific nutrients, such as vitamins or proteins, can be introduced to enhance the nutritional content of potatoes. This can resolve nutritional deficiencies in certain populations.
   - **Tolerance to Abiotic Stress:** Transgenic potatoes with genes conferring tolerance to environmental stresses, such as drought or high salinity, have been developed. These varieties are designed to withstand challenging growing conditions, contributing to increased resilience.
   - **Reduced Acrylamide Formation:** Acrylamide is a possible carcinogen that can develop in some foods during high-temperature cooking methods. Transgenic potatoes have been developed with reduced levels of the precursors that lead to acrylamide formation, addressing food safety concerns. The development and commercialization of transgenic potatoes are subject to rigorous regulatory processes to ensure that are safe for human consumption and do not pose risks to the environment. Many countries have established regulatory frameworks to assess the potential risks and benefits associated with

genetically modified organisms (GMOs). While transgenic approaches offer valuable tools for addressing specific challenges in potato cultivation, public perception, regulatory hurdles, and concerns about the potential environmental impact of GMOs have led some researchers and breeders to explore alternative breeding methods, such as genome editing, which allows for precise modifications without introducing foreign genes. The selection of a breeding method is influenced by the objectives, regulatory environment, and public acceptance.

9.3 Somatic Hybridization

A small portion of *Solanum* diversity contains untamed *Solanum* species used in the breeding of potato. Many useful genes contain these species, but they are not successfully transferred to cultivated potato through traditional breeding methods because of sexual barriers. However, it is currently possible to conquer these barriers using somatic hybridization. Somatic hybridization is targeted to reinforce the gene pool of potato by integrating genes from wild species. ICAR-CPRI, Shimla, produced interspecific potato somatic hybrids from wild species via protoplast fusion between *Solanum tuberosum* 'C-13' and wild *Solanum cardiophyllum; Solanum tuberosum* 'C-13' and *Solanum pinnatisectum* for late blight resistance. Wild *Solanum cardiophyllum and Solanum pinnatisectum* are resistant to late blight with broad genetic bases containing W/γ, W/α, and T/β type's diverse cytoplasm types. Somatic hybrids possess ≥24% dry matter content, outstanding keeping quality and late blight resistance along with adaptability for sub-tropical conditions. ICAR-CPRI, Shimla, produced another interspecific potato somatic hybrid between *Solanum tuberosum* 'C-13' + *Solanum etuberosum.* The resultant *somatic hybrid is resistance to* potato virus Y and showed different cytoplasm types (W/γ, W/α, and T/β) (Tiwari et al. 2021).

10 Constraints to Potato Prolificacy, Productivity and Quality

The challenges to potato prolificacy, productivity, and quality mentioned were broadly categorized into biotic and abiotic stresses. Abiotic and biotic stressors that consistently harm crops are among the other factors contributing to the fall in food supply around the world. With advancements in technology and increased public-private collaboration, numerous crops with improved nutrient profiles have progressed using operational gene pools or germplasm collections (Ricroch and Henard-Damave 2016; Ma et al. 2017b). Addressing these challenges is crucial for sustaining potato production. Some of the major constraints are widely discussed below:

10.1 Biotic Stresses

Viruses Potatoes are susceptible to various viral infections, such as Potato Virus Y (PVY) Potato Virus X (PVX), and others. These viruses can significantly reduce crop yields and affect tuber quality (Fletcher 2012; Hameed et al. 2014; Steinger et al. 2014).

Bacteria and Fungi Pathogenic bacteria like *Ralstonia solanacearum* can lead to diseases such as bacterial wilt, resulting in significant crop losses. Additionally, bacterial diseases including soft rot from *Dickeya solani* and common scab from *Streptomyces scabies* (Buttimer et al. 2017) are impacting potato quality. Fungal diseases, such as late blight caused by *Phytophthora infestans* and powdery scab from *Spongospora subterranea* (Arora et al. 2014; Balendres et al. 2016), are also contributing to the global decline in potato quality.

Insect Pests Insects such as Colorado potato beetle (Casagrande 2014) and aphids (Bass et al. 2014) can transmit diseases and directly damage potato plants, impacting both yield and quality.

10.2 Abiotic Stresses

Drought Water scarcity can negatively impact potato growth and yield. Potatoes require consistent moisture throughout their growth cycle, and drought conditions can lead to stunted growth and reduced tuber formation.

Salinity High soil salinity can affect potato plants by disrupting water uptake and nutrient absorption, leading to reduced productivity.

Temperature Extremes Potatoes are sensitive to temperature fluctuations. High temperatures can affect tuber development, while frost can damage plants and tubers, reducing yields.

Post-harvest Problems The accumulation of reducing sugars during cold storage can lead to issues like the development of sweet potatoes and undesirable changes in color and taste, impacting the quality of the stored potatoes.

Research, innovation, and collaboration among farmers, scientists, and policy-makers are essential for developing sustainable solutions to these constraints and ensuring the long-term productivity and quality of potato crops. Addressing these constraints requires an integrated and sustainable approach, including:

10.3 Agronomic Attributes Affecting Potato

- **Selecting Resilient and Resistant Varieties:** Developing and adopting potato varieties that are resistant to diseases and pests can mitigate biotic stresses. Selection of varieties with desirable traits such as high yield potential, good taste, and texture.

- **Crop Management Practices:** Implementing proper crop rotation, using disease-resistant seed potatoes, and employing good agricultural practices can help manage both biotic and abiotic stresses. Rotation of potato with other non-related crops to improve soil health and nutrient balance. Implement a crop rotation system to break the life cycles of pests and diseases, reducing the risk of build-up in the soil.
- **Irrigation Management:** Efficient water management practices can help mitigate the impact of drought and salinity stress.
- **Climate-Smart Agriculture:** Enforcing climate-smart agricultural approaches can help adapt to temperature extremes and changing climatic conditions. Improving productivity and quality in potato cultivation involves addressing a range of factors, including the management of biotic and abiotic stresses, adopting best agricultural practices, and leveraging technological innovations. Here are some strategies to enhance productivity and quality in potato production.

10.4 Post-harvest Technologies Affecting Potato

Implementing improved storage technologies and practices to minimize sugar accumulation during cold storage is essential for maintaining potato quality. Many infections, including black dots, soft rot, and Fusarium dry rot, are linked to long-term storage and can severely affect tuber quality, necessitating its discarding for additional refining (Usall et al. 2016). *Colletotrichum coccodes* causes black dot disease, which is transmitted through soil and results in brown necrotic lesions on the tuber surface, accelerating their break down (Brierley et al. 2015). By integrating these strategies, farmers can work toward achieving sustainable potato production with increased yields and improved quality, contributing to food security and economic stability. Temperature and period of storage are two critical elements that tuber's sensitivity to black dot disease can be influenced by certain factors and may be handled to obstruct losses in potato quality (Peters et al. 2016). According to Makani et al. (2015), tuber quality during long-term preservation is also influenced by the tuber harvesting date.

10.5 Challenges with Seed Tubers

Propagating seeds from true seeds is a process that requires considerable time and labor, and the resulting plants can exhibit wide variations. Tubers are usually preferred for propagation due to their ability to produce genetically identical plants (clones). Unlike early harvested, immature tubers, which experienced a notable decrease in ascorbic acid concentrations and dry matter, full-maturity tubers preserved their excellence throughout storage (Makani et al. 2015). Tubers, however, can carry pests and diseases, affecting tuber quality and harvest yields. Challenges such as high transportation and storage costs and a low multiplication rate are associated with using seed tubers. Till date, ICAR-CPRI, Shimla, has developed many high-yielding, disease resistance, nutritionally enriched as well as processing varieties (Table 1).

11 Constrains to Potato Productivity and Quality

Both biotic (such as viruses, bacteria, fungi, and insect pests) and abiotic (such as drought, salt, temperature, frost, and postharvest issues, such as the buildup of reducing sugars during cold storage) stressors pose a threat to the sustainable production of potatoes.

Diseases and Insect Pests Affecting Potato Potato virus Y, Potato leaf roll virus, and Potato virus X are among the major viruses affecting potatoes globally. Most potato diseases stem from the widespread presence of phytopathogens, with viruses being particularly significant (Salazar 1996). Cultivated potatoes are prone to about 40 viral species and 2 viroid species (Salazar 1996). Symptoms of viral infections, such as necrotic lesions on leaves and tubers, mosaic patterns, and overall plant growth retardation, result in reduced yield and diminished tuber quality. In addition, various fungal diseases, such as late blight and powdery scab (Arora et al. 2014; Balendres et al. 2016), as well as bacterial diseases like soft rot and common scab, also severely affect potato quality around the world.

Potato affected by late blight display water-soaked areas on leaves with necrotic lesions and tubers with irregularly colored tissue, resulting in dry, hard tubers that are more susceptible to additional microbial infections. To address viral infections, potatoes have been genetically modified using various methods, varying from traditional breeding to advanced genetic engineering. Due to potatoes' polyploid nature, which complicates the spread of resistance genes, transgenic approaches are often more effective for enhancing viral resistance compared to conventional breeding. For instance, genetic modifications have targeted single or multiple RNA viruses, such as PVY resistance (Missiou et al. 2004), resistance to PVY and PLRV (Chung et al. 2013), and resistance to PVX, PVY, and Potato virus S (Hameed et al. 2017). RNA interference-mediated resistance targeting viral coat proteins and genetically modified potatoes with viral resistance are currently available in many countries (Mathur et al. 2017). Additionally, potato pests, such as the Colorado potato beetle (Casagrande 2014), peach-potato aphid (Bass et al. 2014), and potato tuber moth (Liu et al. 2018), either directly damage the plants by causing necrosis or deformities or indirectly contribute to pathogen spread, particularly viruses. The extensive use of chemical pesticides has led to concerns about the development of insecticide-resistant pests.

Genetic engineering has introduced several effective methods to reduce pest occurrence in plants, including the incorporation of insecticidal proteins/toxins (Palma et al. 2014), CRISPR-Cas9-based crop protection, and RNA

Table 1 List of potato varieties developed by ICAR-CPRI, Shimla

S. No.	Variety	Year of release	Selection number	Parentage
1.	Kufri Kisan	1958	PS 454	Up-to-date × Sd. 16
2.	Kufri Kumar	1958	S 1758	Lumbri × Katahdin
3.	Kufri Kuber	1958	ON 2236	(*S. curtilobum* × *S. tuberosum*) x *S. andigenum*
4.	Kufri Red	1958	–	Clonal selection from Darjeeling Red Round
5.	Kufri Kundan	1958	Hybrid-9	Ekishirazu x Katahdin
6.	Kufri Neela	1963	A 1528	Katahdin × Shamrock
7.	Kufri Sindhuri	1967	C 140	Kufri Red × Kufri Kundan
8.	Kufri Alankar	1968	A 3649	Kennebec × ON 2090
9.	Kufri Jeevan	1968	SLB/E 427	M 109-3 × Seedling 698-D
10.	Kufri Naveen	1968	SLB/E-402	3070d (4) × Seedling 692-D
11.	Kufri Chamatkar	1968	ON 1202	Ekishirazu × Phulwa
12.	Kufri Chandramukhi	1968	A 2708	Seedling 4485 × Kufri Kuber
13.	Kufri Jyoti	1968	SLB/Z--389(b)	3069d (4) × 2814a (1)
14.	Kufri Neelamani	1968	A 7871	Kufri Kundan × 134-D
15.	Kufri Khasigaro	1968	SLB/A-67	Taborky × Seedling 698-D
16.	Kufri Sheetman	1968	C 3745	Craigs Defiance × Phulwa
17.	Kufri Muthu	1971	SLB/Z-785	3046 (1) × M109-3
18.	Kufri Lauvkar	1972	A 7416	Serkov × Adina
19.	Kufri Dewa	1973	C 3804	Craigs Defiance × Phulwa
20.	Kufri Badshah	1979	JF 4870	Kufri Jyoti × Kufri Alankar
21.	Kufri Bahar	1980	E 3797	Kufri Red × Gineke
22.	Kufri Lalima	1982	BS/C-1753	Kufri Red × AG 14 (Wis. X 37)
23.	Kufri Sherpa	1983	F 5242	Ultimus × Adina
24.	Kufri Swarna	1985	PCN/76-110	Kufri Jyoti × (VTn)262.33.3PCN/76
25.	Kufri Megha	1989	SS/C-562	SLB/K-37× SLB/Z-73
26.	Kufri Ashoka	1996	–	–
27.	Kufri Jawahar	1996	–	–
28.	Kufri Chipsona-1	1998	MP/90-83	CP2416 × MS/78-79
29.	Kufri Chipsona-2	1998	MP/91-G	CP2346 (F-6 from Peru) × QB/B 92-4
30.	Kufri Giriraj	1998	–	–
31.	Kufri Pukhraj	1998	–	Craigs Defiance x Kufri Jyoti
32.	Kufri Anand	1999	–	MS/92-1090 X CP1704
33.	Kufri Kanchan	1999	–	–
34.	Kufri Arun	2005	–	Kufri Lalima and MS/82-797

(continued)

Table 1 (continued)

S. No.	Variety	Year of release	Selection number	Parentage
35.	Kufri Pushkar	2005	–	–
36.	Kufri Shailja	2005	–	–
37.	Kufri Chipsona-3	2006	MP/97-583	MP/91-86 × Kufri Chipsona-2
38.	Kufri Himalini	2006	–	–
39.	Kufri Surya	2006	–	–
40.	Kufri Girdhari	2008	–	Kufri Megha x Bulk pollen (CP 2132, CP 2175, CP 2183, CP 2366, CP 3245, CP 3246, CP 3247, CP 3268, CP 3269 & CP 3789)
41.	Kufri Himsona	2008	MP/97-644	MP/92-35 × Kufri Chipsona-2
42.	Kufri Khyati	2008	–	–
43.	Kufri Sadabahar	2008	–	MS/81-145 x PH/F-1545
44.	Kufri Frysona	2009	MP/98-71	MP/90-30 × MP/90-94
45.	Kufri Neelima	2010	–	–
46.	Kufri Garima	2012	–	PH/F-1045 and MS/82-638
47.	Kufri Gaurav	2012	–	–
48.	Kufri Lalit	2013	2001- P-55	85-P-670 x CP3192
49.	Kufri Mohan	2016		MS/92-1090 × CP1704
50.	Kufri Ganga	2018	–	MS/82-638 × Kufri Gaurav
51.	Kufri Lima	2018	–	–
52.	Kufri Neelkanth	2018	MP/97-644	MS/78-62 X Kufri Neela
53.	Kufri Chipsona-4	2019	MP/01-916	Atlantic × MP/92-35
54.	Kufri FryoM	2019	MP/04-578	Kufri Chipsona-1 × MP/92-35
55.	Kufri Karan	2019	SM/00-42	Cruza 27 (CP 2376) and HB/83-39
56.	Kufri Manik	2019	–	–
57.	Kufri Sahyadri	2019	–	–
58.	Kufri Sangam	2019	–	–
59.	Kufri Thar-1	2019	–	–
60.	Kufri Thar-2	2019	–	–
61.	Kufri Thar-3	2019	–	–
62.	Kufri Daksh	2023	WS/07-113	CP1748 × LT-1
63.	Kufri Chipsona-5	2024	–	–
64.	Kufri Jamunia	2024	–	–
65.	Kufri Bhaskar	2024	–	–

Source: Modified from Tiwari et al. (2021)

interference-mediated insect resistance (Zhang et al. 2017; Douglas 2017). For instance, RNA interference technology has been used to develop resistance in potatoes against the Colorado potato beetle. According to Zhang et al. (2015), RNA interference reagents, delivered through sap as long double-stranded RNAs in chloroplasts, triggered a lethal RNA response that significantly affected the larvae of the beetle and reduced the expression of more than 80% of targeted genes, including the β-actin gene.

Agronomic Attributes Affecting Potato Crop rotation, tillage, and other agronomic practices have a major impact on the production and quality of potato tubers in farming systems. Potato requires a good amount of nitrogen inputs to sustain their tuber vigor and productivity because of their shallow root system (Alva et al. 2011). Different farming techniques have been found to have an impact on metrics related to tuber quality, such as increased nutrient/mineral concentration, tuber dry mass accumulation, and output improvements (Brazinskiene et al. 2014; Tein et al. 2014; Nyiraneza et al. 2015). According to Qin et al. (2017), using a potato-legume crop rotation instead of continuous potato cultivation resulted in beneficial effects on soil microbiota and increased tuber production by up to 19%. Tein et al. (2014) found that incorporating organic and mineral fertilizers, including nitrogen, phosphorus, and potassium, into crop rotations notably altered the nutrient content of potato tubers compared to non-fertilized controls, affecting levels of N, nitrate, magnesium (Mg), and phosphorus (P). Leonel et al. (2017) studied five potato cultivars to determine how varying P levels affected tuber chemical composition under consistent cultivation practices. They observed that higher P fertilization led to an increase in dry matter, protein, and starch content in the tubers, along with a reduction in total sugar content. Understanding the chemical composition of potato tubers is crucial for evaluating their nutritional value and processing quality. In line with the importance of organic produce, Lombardo et al. (2017) compared the nutritional content of conventionally and organically grown potatoes. Their field tests revealed that yellow-fleshed potatoes grown organically had higher phenolic content, lower nitrate levels, and a more attractive tuber flesh color.

Climatic and Soil Factors Affecting Potato Various abiotic factors, such as soil composition and climate, significantly influence potato yield and quality during both the growing season and after harvest. As reported by Kim et al. (2017), potato cultivation thrives in cooler climates (around 19 °C) but is highly sensitive to heat. Low temperatures are essential for initiating the first tuber set, but rapid temperature spikes during early tuberization can greatly affect both yield and size (Zhou et al. 2017). Global studies show a 3–11% decrease in potato production for every 1 °C increase in temperature (Fleisher et al. 2017; Kim et al. 2017). On the downside, frost and freezing during the early growth stage can severely damage young plants, leading to lower yields and poorer tuber quality (Chang et al. 2014).

Drought and salinity are two other major abiotic challenges. Drought leads to varied physiological and morphological changes in potato plants, often resulting in lower yields due to reduced internal water pressure during tuber bulking and maturation (Stark et al. 2013). While some potato varieties exhibit better drought tolerance, the effects of water stress are still a considerable limitation. On the other hand, salt stress inhibits plant growth and development by increasing oxidative stress and reducing photosynthesis, which lowers tuber yield. Salinity also poses long-term risks to soil health, making it a persistent challenge for sustainable cultivation. However, interspecific breeding has led to improved resistance to oxidative stress and salinity in potatoes, offering a potential solution (Jbir-Koubaa et al. 2015). The downside to these breeding efforts is that they can be time-consuming and expensive and may not always yield consistent results across different environmental conditions.

Post-harvest Factors Affecting Potato The storage conditions of potato tubers after harvest significantly impact their final processing quality. Long-term storage often leads to various diseases, such as black dots, soft rot, and Fusarium dry rot, which severely degrade tuber quality and make them unsuitable for further processing (Usall et al. 2016). Black dot disease, caused by *Colletotrichum coccodes*, results in brown necrotic lesions on the tuber skin, accelerating skin deterioration (Brierley et al. 2015). Key factors influencing a tuber's susceptibility to black dot include storage temperature and duration, both of which can be managed to prevent quality loss (Peters et al. 2016).

Harvest timing also affects tuber quality during storage. Research by Makani et al. (2015) showed that tubers harvested at full maturity retained their quality throughout storage, unlike early harvested, less mature tubers, which experienced significant reductions in dry matter and ascorbic acid levels.

Dormancy is another factor that affects tuber quality during storage. Dormant tubers tend to sprout, which can lead to quality issues. Phytohormones like ethylene can be applied to control dormancy by inhibiting sprouting (Sonnewald and Sonnewald 2014).

Typically, potatoes are stored in cold conditions (4–8 °C) to ensure a year-round supply. However, this cold storage can lead to increased levels of reduced sugars in the tubers, a phenomenon known as "cold-induced sweetening" (CIS) (Bhaskar et al. 2010). During cold-induced sweetening (CIS), starch in the tubers is converted to sugars by hydrolytic enzymes (Sowokinos 2001). The increased sucrose is then converted to glucose and fructose by the enzyme vacuolar acid invertase (VInv) (Sowokinos 2001; Bhaskar et al. 2010). When CIS-affected tubers are processed at high temperatures, acrylamide—a dark brown, bitter compound—can accumulate, raising serious concerns for food safety (Vinci et al. 2012). Studies have shown that fried potato products, such as fries and chips, can contain dangerously high levels of acrylamide (up to 70%) (Pedreschi et al. 2014; McCombie et al. 2016; Esposito et al. 2017).

Increased Protein Content In regions where diets are commonly deficient in protein, there is an increased risk of protein deficiencies (Chakraborty et al. 2010). Unfortunately, cultivated potatoes have a low protein content (0.85–4.2%) and lack essential amino acids like lysine and tyrosine (Burlingame et al. 2009). To address this limitation, researchers have genetically modified potatoes to express the tuber-specific gene *Amaranthus hypochondriacus1* (*AmA1*), which enhances protein content. The *AmA1* gene encodes for albumin, a seed protein that is non-allergenic and rich in essential amino acids. In transgenic potatoes, this protein is localized in the cytoplasm or vacuole, resulting in up to 60% more protein in the tubers compared to non-modified controls (Chakraborty et al. 2010). Additionally, these transgenic potatoes showed an increased rate of photosynthesis, leading to greater biomass and plant production.

Recently, the transgenic potato cultivar Desirée (cv.) exhibited a significant increase in methionine levels, an important amino acid involved in various metabolic processes (Kumar and Jander 2017). By employing RNA interference (RNAi) technology to suppress the host gene *S. tuberosum* methionine γ-lyase (*StMGL*) and overexpressing the exogenous gene *Arabidopsis thaliana* cystathionine γ-synthase (*AtCGS*), researchers achieved nearly double the amount of free methionine in the transgenic tubers compared to control tubers (Kumar and Jander 2017). Experimental investigations indicated no morphological or yield differences between the altered plants and control plants. Despite these advancements, other attempts to increase potato protein content have yielded inconsistent results and sometimes negative impacts on yield (Zeh et al. 2001; Dancs et al. 2008; Rinder et al. 2008; Galili and Amir 2013).

Increased Vitamin and Carotenoid Contents Numerous studies have aimed to increase the vitamin content in potatoes. For instance, research by Bulley et al. (2012) demonstrated that introducing the *A. thaliana* L-galactose phosphorylase (GDP) gene into potatoes led to a threefold increase in ascorbate (vitamin C) levels. Carotenoids, which are phytonutrients with antioxidant properties and are precursors to vitamins, are found in various fruits and vegetables (Dellapenna and Pogson 2006). Li et al. (2012) found that incorporating the cauliflower Orange (*Or*) gene resulted in a significant increase in carotenoid content (pro-vitamin A) in cold-stored tubers. Potatoes typically have lower β-carotene levels compared to other carotenoids such as lutein and zeaxanthin (Ezekiel et al. 2013). By silencing the β-carotene hydroxylase (bch) gene using RNA interference (RNAi), researchers achieved a notable increase in lutein and β-carotene levels in the tubers (Van Eck et al. 2007). Additionally, expressing three bacterial genes involved in carotenoid biosynthesis boosted carotenoid content in tubers by a factor of 20 (Diretto et al. 2007).

Regarding calcium content, potatoes are not typically a rich source of calcium (Weaver et al. 1999). To address this, Park et al. (2005) employed a transgenic approach by introducing the Arabidopsis H^+/Ca^{2+} transporter (sCAX1) gene into the Russet Norkotah cultivar. This resulted in a significant increase in calcium content in the tubers, up to three times higher than in controls. Field tests over three crop generations confirmed the stable integration of this trait without negatively impacting growth or yield. Such calcium-enriched potatoes could offer nutritional benefits, particularly in regions where potatoes are a dietary staple (Park et al. 2005).

Increased Phenolic Contents Caffeoylquinic acids (CQAs) account for nearly 80% of the phenolic compounds found in potatoes (Brown 2005). Research by Li et al. (2016) focused on increasing CQAs in potato tubers. Their study achieved a more than threefold increase in CQAs and total flavonoid content through the tuber-specific constitutive expression of an exogenous gene, a flavonol-specific transcriptional activator (AtMYB12) derived from *A. thaliana*. Notably, they employed a selectable marker-free approach to streamline regulatory approvals (Daniell 2002). Higher phenolic content in plants not only offers health benefits but also exhibits antibacterial properties, particularly in reducing fungal infections (Li et al. 2016).

Increased Starch Contents In nations where food is scarce, potato tubers may supply a substantial amount of calories due to their high starch content. Two structural elements that makeup starch are amylose and amylopectin, which are produced by the combined efforts of many enzymes. Many components of the starch production process have been engineered in a number of experiments to increase the amount of digestible amylose in potatoes (Schwall et al. 2000; Hofvander et al. 2004). Potatoes with increased amylose content were produced as a result of tuber-specific RNA silencing of two host genes (*SBEI*, *SBEII*), implicated in the starch branching pathway (Andersson et al. 2006). Through marker-assisted crossover, a recessive allele (gene marker: *IAm*) from wild potato (*S. sandemanii*) was recently introduced into cultivated potato (*S. tuberosum*), resulting in a considerable (28–59%) increase in amylose levels in non-genetically modified potatoes (Krunic et al. 2018). Soon, plant-based oils show promise as a possible source of renewable energy feedstock. The current focus of biofuel research is mostly on genetically manipulating the lipids/triacylglycerol (TAG) production pathways in crops to create crops with increased oil contents (Vigeolas et al. 2007; Vanhercke et al. 2014; Zale et al. 2016). Potato tubers have a very low proportion of oil (lipids), in contrast to their high starch content. According to some recent research, the modified potato tubers have a higher TAG content (Hofvander et al. 2016; Liu et al. 2017). The TAG content of the tuber increased 100-fold when three genes (WRI1, DGAT1, and OLEOSIN) were constitutively expressed in tissue as opposed to controls (Liu et al. 2017). However, there was also a depletive nutritional impact associated with this TAG rise, as seen by much lower amounts of accumulated sugar (sucrose) and starch (amylose). Improved knowledge of the detrimental effects of TAG buildup on the amylose and phosphate contents of tubers, as well as the necessity of optimizing genetic engineering for certain features, were obtained by further investigation of this process (Mitchell et al. 2017).

Reduction of Anti-nutrient Contents in Potato Reducing the anti-nutrient components of food is another way to raise its nutritional value. Potato consumption has not been linked to any fatal toxicity (Zaheer and Akhtar 2016). Nevertheless, during crop maturation, certain anti-nutrient elements such as steroidal glycoalkaloids (SGAs) (0.071–175 mg/100 g), mainly α-solanine and α-choconine, accumulate in tubers (Burlingame et al. 2009). Higher concentrations of these SGAs in food may result in neurotoxic effects and/or issues with nutrition absorption (Itkin et al. 2013). Glycosyltransferase and uridine 5′-diphosphate, two essential enzymes that biochemically react with sugars, cholesterol, and other nitrogenous substances to build up glycoalkaloid molecules, are expressed concurrently during the biosynthesis of SGAs (Itkin et al. 2011). The amount of SGAs in potato leaves and tubers was significantly reduced (up to 74-fold) when the host gene, glycoalkaloid metabolism 4 (*GAME4*), was silenced by RNA interference (Itkin et al. 2013). Importantly, many wild potato species naturally generate large amounts of SGAs (Gregory et al. 1981). For this reason, when using wild germplasm as a genetic resource, breeders need to be cautious in mapping the SGAs-gene(s) connection with the desired characteristics. Before being consumed, raw potatoes are often cooked (fried, baked, mashed, microwaved) into a variety of food items such as chips, fries, snaps, etc. (Zaheer and

Akhtar 2016). Processing potatoes can occasionally produce dangerous substances that lead to obesity, cardiovascular illness, and/or neurotoxicity. Food regulatory agencies and the processing industry debated the first report of acrylamide, a possible neurotoxic and carcinogenic ingredient, being present in potato fried goods (Tareke et al. 2002). Since then, a number of investigations have been carried out to investigate the formation of acrylamide during the "Maillard Reaction," which is a reaction between the asparagine contents of tubers, reducing sugars (mainly glucose and fructose), and free α-radicals present in cooking oil during high-temperature potato processing (Stadler et al. 2002; Friedman 2003; Vinci et al. 2012). Certain diploid species of wild potatoes are inherently resistant to CIS, in contrast to the farmed tetraploid potato (*S. tuberosum*). Using additional genetic techniques and quantitative trait loci (QTL) mapping, researchers have discovered a few recessive genes linked to CIS resistance in wild animals. With little success, potato breeders attempted to introduce these genes into farmed potatoes to lessen the CIS impact. Thus, via indirectly controlling the CIS process at the cellular level, several molecular techniques have been used to minimize the synthesis of acrylamide (Bhaskar et al. 2010; Li et al. 2013; Zhu et al. 2016; Hameed et al. 2018). When the *VInv* gene was expressed constitutively in tuber-specific antisense binary constructs, the amount of sugar in cold-stored tubers was significantly reduced. Food items made from these transgenic lines had an eightfold reduction in acrylamide concentration when processed at high temperatures compared to controls (Ye et al. 2010).

New Breeding Technologies Used to Improve the Nutritional Quality of Potato Conventional breeding approaches for gene pyramiding in polyploid crops present significant challenges due to their complexity, time requirements, and labor intensity (Weeks 2017). Attempts to enhance specific traits in potatoes using germplasm from wild species have yielded mixed results (Carputo and Barone 2005). The tetraploid nature of cultivated potatoes (*S. tuberosum*), which possess four sets of chromosomes ($2n = 4x = 48$), complicates precise genetic modifications using traditional breeding methods (The Potato Genome Sequencing Consortium 2011). Consequently, newer breeding technologies (NBTs), such as CRISPR/Cas9, TALENs, and ZFNs, offer promising alternatives by providing more accurate and efficient genome editing with reduced time investment (Mahfouz et al. 2016; Petolino et al. 2016; Schiml and Puchta 2016).

In potato research, precise genome editing has been successfully achieved using CRISPR/Cas9 (Butler et al. 2015; Wang et al. 2015) and TALENs (Sawai et al. 2014; Nicolia et al. 2015; Clasen et al. 2016). The first successful application of TALENs for modifying the potato genome was in 2014, setting a foundation for future advancements. Sawai et al. (2014) employed TALENs to target and silence the sterol side chain reductase 2 (*StSSR2*) gene, which plays a role in cholesterol production and the formation of toxic steroidal glycoalkaloids (SGAs). This led to site-specific mutations and a significant reduction in SGA content without affecting plant growth (Sawai et al. 2014).

Further research by Nicolia et al. (2015) demonstrated a notable mutation frequency (7–8%) in the acetolactate synthase (*ALS*) gene in transgenic potato cv. Desirée, with no observable phenotypic differences from control plants, confirmed targeted gene disruption (Nicolia et al. 2015). Wang et al. (2015) utilized the CRISPR/Cas9 system to induce targeted mutations in the auxin/indole-3-acetic acid (*StIAA2*) gene, leading to altered protein expression. In a subsequent study, Butler et al. (2015) applied CRISPR/Cas9 to the *ALS1* gene in both diploid and tetraploid potato varieties, achieving site-specific mutations ranging from 3% to 60%, which were stably inherited in subsequent generations (Butler et al. 2015).

Additionally, TALENs technology has been employed to address cold-induced sweetening (CIS) by targeting and disrupting *VInv* activity to reduce reducing sugar accumulation in stored potatoes, enhancing their nutritional quality (Clasen et al. 2016). Potato CV was transformed with the use of protoplasts. Plants transformed by Ranger Russet with TALENs constructs had *VInv* alleles knocked out (Clasen et al. 2016). It is interesting to note that 5 out of 18 transformed lines had almost negligible CIS activity and almost total *VInv* gene silence. Moreover, transgenic generated tubers processed at high temperatures (fried chips) produced light brown products with a markedly reduced quantity of dietary acrylamide. Crucially, a small percentage of transgenic lines demonstrated a total lack of TALEN sequences in downstream characterization, providing a transgene-free method. Other attempts to add nutritional qualities to potatoes include CRISPR/Cas9 (Andersson et al. 2017) and TALENs (Kusano et al. 2016) for starch modification. Granule-bound starch synthase (GBSS), a host gene in potatoes, was the subject of a site-specific mutation by Kusano et al. (2016) through the development of a new delivery mechanism known as the "Emerald–Gateway TALEN system." During starch granulation, GBSS is primarily involved in the production of amylose. According to Zeeman et al. (2010), its disruption may lower the amylose content and amylose/amylopectin ratio, which might have an impact on the starch quality of potato tubers. When potato cells were transformed using TALEN constructs by *Agrobacterium*, three stable mutation types were seen in the regenerated lines, with a deletion mutation (63 nucleotide loss) predominating (Kusano et al. 2016).

In potato cv. Kuras, the GBSS has recently been effectively silenced by the use of CRISPR/Cas9 technology (Andersson et al. 2017). In 2% of transformed lines, site-specific mutations in all four alleles of the GBSS gene were obtained by protoplast transformation using CRISPR/Cas9 constructs. When compared to controls, transgenic tubers with full knock-out of the targeted genomic locations had changed the starch quality and total loss of GBSS activity (Andersson et al. 2017). In addition to helping to better understand the molecular underpinnings of potato starch-related features, genome-wide investigations in conjunction with transcriptomics, proteomics, and metabolomics may also help to speed up or facilitate starch modifications utilizing NBTs. In order to fulfill the demands of the food and industrial sectors, research on the manufacture of high-quality starch from potatoes may be of current interest. Butler et al. (2016) used a geminivirus replicon (GVR) vector in a different investigation to target the potato herbicide tolerance gene ALS1 using sequence-specific nucleases (SSNs). Herbicide susceptibility was demonstrated for regenerated transformants that had an ALS1 point mutation. Forsyth et al. (2016) utilized the TALEN method along with a molecular marker, specifically the mutant acetolactate synthase (*ALS*) gene, to achieve targeted integration of a transgene into a transcriptionally active region of the potato genome. After validating the system in *Nicotiana benthamiana*, they targeted the potato *Ubi7* gene with TALENs. Agrobacterium-mediated transformation was then employed to create transgenic potatoes of the *S. tuberosum* cv. Ranger Russet. The study found that most transgenic lines had a single copy of the transgene, which simplifies the process of creating multiple lines for characterization. This demonstrated TALEN's effectiveness for precise genome editing in potatoes. However, regulatory concerns may arise due to the use of TALEN reagents containing bacterial genes (TALE DNA binding domain from *Xanthomonas*). In another study, TALENs were used to introduce targeted mutations in two potato cultivars, Russet Burbank and Shepody, through non-viral Agrobacterium-mediated transformation (Ma et al. 2017a). The TALEN constructs aimed to induce mutations in the Vacuolar invertase 2 gene (*StvacINV2*) and the 1,4-alpha-glucan branching enzyme (*SBE1*). Illumina deep sequencing confirmed targeted chromosomal mutations, predominantly deletions, in the regenerated lines. This approach may efficiently enhance desirable traits in superior potato cultivars. Zinc-finger nucleases (ZFNs), another early genome editing technology, create double-strand breaks (DSBs) by fusing synthetic zinc finger proteins with the FokI restriction endonuclease's cleavage domains (Kim and Kim 2014). While ZFNs have been used in crops like tobacco, Arabidopsis, and soybean (Cai et al. 2009; Zhang et al. 2010; Curtin et al. 2011), they have not yet been applied to genetic engineering in potatoes or other horticultural crops (Gaur et al. 2018). The poor success rate (about 24%), low or variable mutation rate (approximately 10%), strong off-target effects, and technically challenging feasible design might be the cause of the few examples of ZFNs-mediated genome alteration in plants. (Xiong et al. 2015; Zhang et al. 2017). The range of ZFNs technology that can be used by the scientific community has been significantly reduced by these difficulties. The papers included in this part offer detailed information about the suitability of certain NBTs for altering the potato genome. While these technologies have countless uses in the field of genetic engineering, choosing appropriate genomic targets and an effective editing tool is a necessary first step in achieving the intended results. For instance, desirable features might be included in the TALEN/CRISPR system through temporary expression, even in the absence of stable transgenic integration. Modified crops with undetectable foreign DNA or RNA may be easier to get regulatory or public approval for and appear more like their natural counterparts. When it comes to designing complicated genetic features requiring several biochemical pathways, such as the creation of proteins, vitamins, carbohydrates, etc., NBTs may provide a viable alternative to laborious multistep engineering utilizing traditional approaches.

12 The Commercialization of GM Potatoes Thus Far: Risk Assessments and Regulations

The last 20 years have seen a significant increase in biotech crops, which has demonstrated the power of genetic engineering to modify a wide range of agronomic, environmental, and mostly health-related plant features. Biotech crops are already grown on more than two billion hectares of agricultural land, demonstrating their significance and capacity to adapt to upcoming difficulties by producing beneficial phenotypes in plants (Parisi et al. 2016). According to Ricroch and Henard-Damave (2016), the majority of GM potato cultivars that have been commercialized so far have traits like resistance to viruses and other phytopathogens incorporated into them. In order to provide resistance to the Colorado beetle (*Leptinotarsa decemlineata*), Monsanto R's "NewLeaf" genetically modified potato was the first to hit the market in 1995 (Kilman 2001). In March 2010, a different genetically modified potato variety surfaced. Named "Amflora," this genetically modified potato variety was created by BASF Plant Science with the goal of increasing its amylopectin content, or waxy tuberous starch, for the processing industry. The European Commission authorized it (Lucht 2015; Zaheer and Akhtar 2016). Twenty-three genetically modified potato lines, ten at the precommercial, eleven at the regulatory, and two at the advance development stages are awaiting regulatory clearance. These lines are developed for a variety of agronomical and quality-related features (Parisi et al. 2016). The production of three genetically modified potatoes (GM) designed to fend off fungal (late blight) infections and the production of acrylamide was authorized by the Food and Drug Administration (FDA) and the Environmental Protection Agency (EPA) in 2017. The potatoes were developed by *Simplot Corp. R*. It's significant that these GM potatoes were approved by regulators and created with a variety of contemporary breeding techniques. Numerous potato varieties that have been modified for different agronomic qualities are being developed or are undergoing field testing and biosafety assessments in several nations (Ricroch and Henard-Damave 2016). Due to several technological, ethical, and social/public constraints, the research and marketing of genetically modified crops presents a significant challenge for scientists and regulatory authorities (Podevin et al. 2012). Further impeding producer and consumer adaptations of these GM goods is the economic value of using them. However, according to Smyth et al. (2017), the expenses associated with GM development and approval typically surpass $35 million, making them inappropriate for a large number of low-income public-private organizations. This reduces many poor nations' interest in funding GM technologies (Perez-Massot et al. 2013). The public's acceptance of GM-labeled items raises further concerns for the government, legislative branch, and/or environmental authority. As technology advances, NBTs may be utilized to provide a highly targeted genetic change that is identical to natural variations or mutants, greatly allaying GM fears (Wolt et al. 2016). For instance, a transgene-free potato with enhanced processing features was created in 2014 by the worldwide biotech company Cellectis R (Wolt et al. 2016). By protoplast transformation of external genetic material from a plant pest (*Phytophthora infestans*), they introduced a base loss in the potato genome using TALEN technology. The regenerated crops' phenotypic and molecular validation revealed no foreign material in the segregating generation; as a result, they were deemed transgene-free and were not subject to regulatory action by APHIS, a USDA regulatory jurisdiction under the Plant Protection Act: CFR 7 (Wolt et al. 2016). Many desirable nutritional traits in potato have been enhanced using the transgenic technology, as presented in Table 2.

Table 2 Transgenic potato enhanced for nutritional traits

Nutrient	Cultivar(s)	Molecular function	Targeted gene(s)	Technology used	Results	References
Protein	*S. tuberosum* var-A16 and some other local cultivars	Encodes seed-specific albumin protein	*Amaranthus hypochondriacus*1 (*AmA1*)	Transgenesis	Up to 60% increase in tuber protein	Chakraborty et al. 2000, 2010
Desirée, Breeding lines (MGL6 and MGL34)	*AtCGS*: methionine biosynthesis; *StMGL*: prevent methionine degradation	*A. thaliana* cystathionine g-synthase (*AtCGS*); *S. tuberosum* methionine g-lyase (*StMGL*)	Transgenesis and endogenous gene silencing	Up to two-fold increase in tuber methionine	Kumar and Jander 2017	Carotenoid
	Desirée, Breeding line (91E22), Yema de Huevo	Involved in carotenoid biosynthesis pathways	Beta-carotene hydroxylase (*bch*)	RNAi silencing	Significant increase in tuber b-carotene and lutein	Van Eck et al. 2007
Desirée	Involved in b-carotene biosynthesis pathways	Three bacterial (Erwinia) genes (*CrtB*, *CrtI*, and *CrtY*)	Transgenesis	Up to 20-fold increase in tuber carotenoids	Diretto et al. 2007	Vitamins
	Ranger Russet	Involved in ascorbate biosynthesis pathways	*A. thaliana* L-galactose phosphorylase (GDP)	Transgenesis	Up to three-fold increase in tuber vitamin C	Bulley et al. 2012

Desirée	Regulate carotenoid accumulation	Cauliflower Orange (*Or*)	Transgenesis	Up to ten-fold increase in tuber pro-vitamin A	Li et al. 2012	Taedong Valley
	Involved in vitamin C biosynthesis pathways	Rat-cells L-gulono-g-lactone oxidase (*GLOase*)	Transgenesis	Enhanced (141%) L-Ascorbic acid in tubers	Upadhyaya et al. 2010	Starch
	Kuras	Involved in starch branching pathways	Starch branching enzyme family (*SBEI* and *SBEII*)	RNAi silencing	Significant increase in tuber amylose	Andersson et al. 2006
Calcium	Russet Norkotah	Involved in H+/ Ca^{2+} transport	*A. thaliana* H+/Ca^{2+} transporter (*sCAX1*)	Transgenesis	Up to a three-fold increase in tuber Ca	Park et al. 2005
Caffeoyl quinic acids (CQAs)	Desirée	Activates the CQA biosynthetic pathway	*A. thaliana* flavonol- specific transcriptional activator (*AtMYB12*)	Transgenesis	More than three-fold increase in tuber CQAs	Li et al. 2016
ANTI-NUTRIENT	Acrylamide	Katahdin, Russet Burbank	Involved in the accumulation of reducing sugars inside the vacuole and acrylamide precursors	Vacuolar acid invertase (*VInv*); asparagine synthetase genes (*StAS1* and *StAS2*)	RNAi silencing	Significant reduction in acrylamide content
Bhaskar et al. 2010; Zhu et al. 2016						
Steroidal glycoalkaloids (SGAs)	Desirée	Involved in SGAs biosynthesis pathways	Glycoalkaloid metabolism 4 (*GAME4*)	RNAi silencing	Up to 74-fold decrease in tuber SGAs	Itkin et al. 2013

Transgenesis: Introducing an exogenous gene "transgene" into a living organism so that the organism will stably exhibit a new property and transmit that property to the next generation.

However, application of mutatagenesis and some new breeding techniques have also been found useful for targeted trait improvement in potato (Table 3).

Table 3 Applications of some new breeding technologies for potato trait improvement

Objective	Target gene(s)	Molecular function	Edit tool	Results	References
Targeted mutagenesis	IAA /Aux gene family (*StIAA2*)	Involved in indole-3-acetic acid/auxin and protein synthesis	CRISPR/Cas9	Altered Aux/IAA protein expression	Wang et al. 2015
Reduced anti-nutrient element	Vacuolar invertase gene (*VInv*)	Accumulation of reducing sugars inside tubers	TALENs	Reduced acrylamide and resistance against CIS	Clasen et al. 2016
Targeted mutagenesis	Acetolactate synthase1 gene (*StALS1*)	Biosynthesis of branched-chain amino acids	CRISPR/Cas9	Site-specific mutation	Butler et al. 2015
Reduced antinutrient element	Sterol side chain reductase 2 (*SSR2*)	Involved in cholesterol biosynthesis	TALENs	Reduced level of toxic steroidal glycoalkaloids	Sawai et al. 2014
Altered starch quality	Granule-bound starch synthase (GBBS)	Involved in the starch synthesis pathway	CRISPR/Cas9	High mutation frequencies at targeted sites (up to 67%)	Andersson et al. 2017
Targeted mutagenesis	Acetolactate synthase gene (*ALS*)	Biosynthesis of branched-chain amino acids	TALENs	Site-specific mutation with frequency rate (7–8%)	Nicolia et al. 2015
Targeted mutation	Acetolactate synthase1 gene (*ALS1*)	Biosynthesis of branched-chain amino acids	Geminivirus replicon (GVR)-mediated TALENs and CRISPR/Cas9 delivery	Site-specific mutation and herbicide susceptibility	Butler et al. 2016
Altered starch composition	Granule-bound starch synthase (*GBBS*)	Involved in the starch synthesis pathway	TALENs	Targeted mutation	Kusano et al. 2016

13 Conclusion and Future Prospects

Research on food security and safety can offer significant solutions to the growing demand for food, particularly in nations with food shortages. Plant genetic engineering is developing quickly, and this has given rise to new and innovative technologies for producing crops with improved nutritional qualities and production. Given this, the potato crop has a great deal of potential to support food security by producing high-energy, low-cost food on a resilient basis. Nutritious characteristics like higher protein content, β-carotene level, vitamin C content, triacylglycerol, tuber methionine, and amylose content are incorporated into potatoes. Reducing anti-nutritional chemicals in potatoes, such as acrylamide, steroidal glycoalkaloids and other dietary toxins, is the focus of further research goals. The introduction of significant agronomical features in a variety of agricultural plants has been made possible by the recent appearance of NBTs like TALENs, ZFNs, CRISPR/Cas9, etc. These technologies offer robust, precise, and site-specific genome editing capabilities. NBTs also provide superior features including increased yield, abiotic and biotic stress tolerance, and improved nutrition.

References

Alva A, Fan M, Qing C, Rosen C, Ren H (2011) Improving nutrient-use efficiency in Chinese potato production: experiences from the United States. J Crop Improv 25:46–85. https://doi.org/10.1080/15427528.2011.538465

Andersson M, Melander M, Pojmark P, Larsson H, Bulow L, Hofvander P (2006) Targeted gene suppression by RNA interference: an efficient method for production of high-amylose potato lines. J Biotechnol 123:137–148. https://doi.org/10.1016/j.jbiotec.2005.11.001

Andersson M, Turesson H, Nicolia A, Falt AS, Samuelsson M, Hofvander P (2017) Efficient targeted multiallelic mutagenesis in tetraploid potato (Solanum tuberosum) by transient CRISPR-Cas9 expression in protoplasts. Plant Cell Rep 36:117–128. https://doi.org/10.1007/s00299-016-2062-3

Andrivon D (2017) Potato facing global challenges: how, how much, how well? Potato Res 60:389–400

Arora R, Sharma S, Singh B (2014) Late blight disease of potato and its management. Potato J 41:61–40

Badami MG, Ramankutty N (2015) Urban agriculture and food security: a critique based on an assessment of urban land constraints. Glob Food Secur 4:8–15. https://doi.org/10.1016/j.gfs.2014.10.003

Balendres M, Tegg R, Wilson C (2016) Key events in pathogenesis of spongospora diseases in potato: a review. Aus Plant Pathol 45:229–240. https://doi.org/10.1007/s13313-016-0398-3

Bass C, Puinean AM, Zimmer CT, Denholm I, Field LM, Foster SP et al (2014) The evolution of insecticide resistance in the peach potato aphid, *Myzus persicae*. Insect Biochem Mol Biol 51:41–51. https://doi.org/10.1016/j.ibmb.2014.05.003

Bhaskar PB, Wu L, Busse JS, Whitty BR, Hamernik AJ, Jansky SH et al (2010) Suppression of the vacuolar invertase gene prevents cold-induced sweetening in potato. Plant Physiol 154:939–948. https://doi.org/10.1104/pp.110.162545

Birch PR, Bryan G, Fenton B, Gilroy EM, Hein I, Jones JT et al (2012) Crops that feed the world 8: potato: are the trends of increased global production sustainable? Food Secur 4:477–508. https://doi.org/10.1007/s12571-012-0220-1

Bortesi L, Fischer R (2015) The CRISPR/Cas9 system for plant genome editing and beyond. Biotechnol Adv 33:41–52. https://doi.org/10.1016/j.biotechadv.2014.12.006

Brazinskiene V, Asakaviciute R, Miezeliene A, Alencikiene G, Ivanauskas L, Jakstas V et al (2014) Effect of farming systems on the yield, quality parameters and sensory properties of conventionally and organically grown potato (Solanum tuberosum L.) tubers. Food Chem 145:903–909. https://doi.org/10.1016/j.foodchem.2013.09.011

Brierley J, Hilton A, Wale S, Peters J, Gladders P, Bradshaw N et al (2015) Factors affecting the development and control of black dot on potato tubers. Plant Pathol 64:167–177. https://doi.org/10.1111/ppa.12238

Brown C (2005) Antioxidants in potato. Am J Potato Res 82:163–172. https://doi.org/10.1007/BF02853654

Bulley S, Wright M, Rommens C, Yan H, Rassam M, Lin-Wang K et al (2012) Enhancing ascorbate in fruits and tubers through over-expression of the l-galactose pathway gene GDP-l-galactose phosphorylase. Plant Biotechnol J 10:390–397. https://doi.org/10.1111/j.1467-7652.2011.00668.x

Burlingame B, Mouillé B, Charrondière R (2009) Nutrients, bioactive nonnutrients and antinutrients in potatoes. J Food Comp Anal 22:494–502. https://doi.org/10.1016/j.jfca.2009.09.001

Butler NM, Atkins PA, Voytas DF, Douches DS (2015) Generation and inheritance of targeted mutations in potato (Solanum tuberosum L.) using the CRISPR/Cas system. PLoS One 10:e0144591. https://doi.org/10.1371/journal.pone.0144591

Butler NM, Baltes NJ, Voytas DF, Douches DS (2016) Gemini virus mediated genome editing in potato (*Solanum tuberosum* L.) using sequence specific nucleases. Front Plant Sci 7:1045. https://doi.org/10.3389/fpls.2016.0104

Buttimer C, McAuliffe O, Ross RP, Hill C, O'Mahony J, Coffey A (2017) Bacteriophages and bacterial plant diseases. Front Microbiol 8:34. https://doi.org/10.3389/fmicb.2017.00034

Cai CQ, Doyon Y, Ainley WM, Miller JC, DeKelver RC, Moehle EA et al (2009) Targeted transgene integration in plant cells using designed zinc finger nucleases. Plant Mol Biol 69:699–709. https://doi.org/10.1007/s11103-008-9449-7

Carputo D, Barone A (2005) Ploidy level manipulations in potato through sexual hybridisation. Ann Appl Biol 146:71–79. https://doi.org/10.1111/j.1744-7348.2005.04070.x

Casagrande R (2014) The Colorado potato beetle: 125 years of mismanagement. Bull Entomol Soc Am 33:142–150. https://doi.org/10.1093/besa/33.3.142

Chakraborty S, Chakraborty N, Datta A (2000) Increased nutritive value of transgenic potato by expressing a nonallergenic seed albumin gene from Amaranthus hypochondriacus. Proc Natl Acad Sci USA 97:3724–3729. https://doi.org/10.1073/pnas.97.7.3724

Chakraborty S, Chakraborty N, Agrawal L, Ghosh S, Narula K, Shekhar S et al (2010) Next-generation protein-rich potato expressing the seed protein gene AmA1 is a result of proteome rebalancing in transgenic tuber. Proc Natl Acad Sci USA 107:17533–17538. https://doi.org/10.1073/pnas.1006265107

Chang DC, Sohn HB, Cho JH, Im JS, Jin YI, Do GR et al (2014) Freezing and frost damage of potato plants: a case study on growth recovery, yield response, and quality changes. Potato Res 57:99–110. https://doi.org/10.1007/s11540-014-9253-5

Chung BN, Yoon JY, Palukaitis P (2013) Engineered resistance in potato against Potato leafroll virus, Potato virus A and Potato virus Y. Virus Genes 47:86–92. https://doi.org/10.1007/s11262-013-0904-4

Clasen BM, Stoddard TJ, Luo S, Demorest ZL, Li J, Cedrone F et al (2016) Improving cold storage and processing traits in potato through targeted gene knockout. Plant Biotechnol J 14:169–176. https://doi.org/10.1111/pbi.12370

Crossa J, Pérez-Rodríguez P, Cuevas J, Montesinos-López O, Jarquín D, De Los Campos G, Burgueño J, GonzálezCamacho JM, Pérez-Elizalde S, Beyene Y (2017) Genomic selection in plant breeding: methods, models, and perspectives. Trends Plant Sci 22:961–975

Curtin SJ, Zhang F, Sander JD, Haun WJ, Starker C, Baltes NJ et al (2011) Targeted mutagenesis of duplicated genes in soybean with zinc-finger nucleases. Plant Physiol 156:466–473. https://doi.org/10.1104/pp.111.172981

Dancs G, Kondrak M, Banfalvi Z (2008) The effects of enhanced methionine synthesis on amino acid and anthocyanin content of potato tubers. BMC Plant Biol 8:65. https://doi.org/10.1186/1471-2229-8-65

Daniell H (2002) Molecular strategies for gene containment in transgenic crops. Nat Biotechnol 20:581–586. https://doi.org/10.1038/nbt0602-581

Dellapenna D, Pogson BJ (2006) Vitamin synthesis in plants: tocopherols and carotenoids. Annu Rev Plant Biol 57:711–738. https://doi.org/10.1146/annurev.arplant.56.032604.144301

Devaux A, Goffart J-P, Petsakos A, Kromann P, Gatto M, Okello J, Suarez V, Hareau G (2020) Global food security, contributions from sustainable potato agri-food systems. In: Campos H, Ortiz O (eds) The potato crop: its agricultural, nutritional and social contribution to humankind. Springer International Publishing, Cham, pp 3–35

Diretto G, Al-Babili S, Tavazza R, Papacchioli V, Beyer P, Giuliano G (2007) Metabolic engineering of potato carotenoid content through tuber-specific overexpression of a bacterial mini-pathway. PLoS One 2:e350. https://doi.org/10.1371/journal.pone.0000350

Douglas AE (2017) Strategies for enhanced crop resistance to insect pests. Annu Rev Plant Biol 69:637–660. https://doi.org/10.1146/annurev-arplant-042817-040248

Esposito F, Nardone A, Fasano E, Triassi M, Cirillo T (2017) Determination of acrylamide levels in potato crisps and other snacks and exposure risk assessment through a Margin of Exposure approach. Food Chem Toxicol 108:249–256. https://doi.org/10.1016/j.fct.2017.08.006

Ezekiel R, Singh N, Sharma S, Kaur A (2013) Beneficial phytochemicals in potato—a review. Food Res Int 50:487–496. https://doi.org/10.1016/j.foodres.2011.04.025

FAO (2019) FAOSTAT. Food and Agriculture Organization of the United Nations, Rome

Fleisher DH, Condori B, Quiroz R, Alva A, Asseng S, Barreda C et al (2017) A potato model intercomparison across varying climates and productivity levels. Glob Chang Biol 23:1258–1281. https://doi.org/10.1111/gcb.13411

Fletcher J (2012) A virus survey of New Zealand fresh, process and seed potato crops during 2010–11. N Z Plant Prot 65:197–203

Forsyth A, Weeks T, Richael C, Duan H (2016) Transcription activatorlike effector nucleases (TALEN)-mediated targeted DNA insertion in potato plants. Front Plant Sci 7:1572. https://doi.org/10.3389/fpls.2016.01572

Friedman M (2003) Chemistry, biochemistry, and safety of acrylamide. A review. J Agric Food Chem 51:4504–4526. https://doi.org/10.1021/jf030204+

Galili G, Amir R (2013) Fortifying plants with the essential amino acids lysine and methionine to improve nutritional quality. Plant Biotechnol J 11:211–222. https://doi.org/10.1111/pbi.12025

Gaur RK, Verma RK, Khurana SM (2018) Genetic engineering of horticultural crops: present and future. In: Gregory P, Sinden SL, Osman SF, Tingey WM, Chessin DA (eds) Genetic engineering of horticultural crops. Elsevier, London, pp 23–46

Gregory P, Sinden SL, Osman SF, Tingey WM, Chessin DA (1981) Glycoalkaloids of wild, tuber-bearing Solanum species. J Agric Food Chem 29:1212–1215

Halterman D, Guenthner J, Collinge S, Butler N, Douches D (2016) Biotech potatoes in the 21st century: 20 years since the first biotech potato. Am J Potato Res 93:1–20. https://doi.org/10.1007/s12230-015-9485-1

Hameed A, Iqbal Z, Asad S, Mansoor S (2014) Detection of multiple potato viruses in the field suggests synergistic interactions among potato viruses in Pakistan. Plant Pathol J 30:407–412. https://doi.org/10.5423/PPJ.OA.05.2014.0039

Hameed A, Tahir MN, Asad S, Bilal R, Van Eck J, Jander G et al (2017) RNAi-mediated simultaneous resistance against three RNA viruses in potato. Mol Biotechnol 59:73–83. https://doi.org/10.1007/s12033-017-9995-9

Hameed A, Zaidi SSEA, Shakir S, Mansoor S (2018) Applications of new breeding technologies for potato improvement. Front Plant Sci 9:925

Hofvander P, Andersson M, Larsson CT, Larsson H (2004) Field performance and starch characteristics of high-amylose potatoes obtained by antisense gene targeting of two branching enzymes. Plant Biotechnol J 2:311–320. https://doi.org/10.1111/j.1467-7652.2004.00073.x

Hofvander P, Ischebeck T, Turesson H, Kushwaha SK, Feussner I, Carlsson AS et al (2016) Potato tuber expression of Arabidopsis WRINKLED1 increase triacylglycerol and membrane lipids while affecting central carbohydrate metabolism. Plant Biotechnol J 14:1883–1898. https://doi.org/10.1111/pbi.12550

Itkin M, Rogachev I, Alkan N, Rosenberg T, Malitsky S, Masini L et al (2011) GLYCOALKALOID METABOLISM1 is required for steroidal alkaloid glycosylation and prevention of phytotoxicity in tomato. Plant Cell 23:4507–4525. https://doi.org/10.1105/tpc.111.088732

Itkin M, Heinig U, Tzfadia O, Bhide A, Shinde B, Cardenas P et al (2013) Biosynthesis of antinutritional alkaloids in solanaceous crops is mediated by clustered genes. Science 341:175–179. https://doi.org/10.1126/science.1240230

Jbir-Koubaa R, Charfeddine S, Ellouz W, Saidi MN, Drira N, GargouriBouzid R et al (2015) Investigation of the response to salinity and to oxidative stress of interspecific potato somatic hybrids grown in a greenhouse. Plant Cell Tissue Organ Cult 120:933–947. https://doi.org/10.1007/s11240-014-0648-4

Jinek M, Chylinski K, Fonfara I, Hauer M, Doudna JA, Charpentier E (2012) A programmable dual-RNA-guided DNA endonuclease in adaptive bacterial immunity. Science 337:816–821. https://doi.org/10.1126/science.1225829

Jonas E, de Koning D-J (2013) Does genomic selection have a future in plant breeding? Trends Biotechnol 31:497–504. [CrossRef] [PubMed]

Kilman S (2001) Monsanto's genetically modified potatoes find slim market, despite repelling bugs. Wall St. J. 22. Available online at: http://online.wsj.com/article/SB985128671233949916.html

Kim H, Kim JS (2014) A guide to genome engineering with programmable nucleases. Nat Rev Genet 15:321–334. https://doi.org/10.1038/nrg3686

Kim YU, Seo BS, Choi DH, Ban HY, Lee BW (2017) Impact of high temperatures on the marketable tuber yield and related traits of potato. Eur J Agron 89:46–52. https://doi.org/10.1016/j.eja.2017.06.005

Krunic SL, Skryhan K, Mikkelsen L, Ruzanski C, Shaik SS, Kirk HG et al (2018) Non-GMO potato lines with an altered starch biosynthesis pathway confer increased-amylose and resistant starch properties. Starch 70:1600310. https://doi.org/10.1002/star.201600310

Kumar P, Jander G (2017) Concurrent overexpression of Arabidopsis thaliana cystathionine gamma-synthase and silencing of endogenous methionine gamma-lyase enhance tuber methionine content in Solanum tuberosum. J Agric Food Chem 65:2737–2742. https://doi.org/10.1021/acs.jafc.7b00272

Kusano H, Onodera H, Kihira M, Aoki H, Matsuzaki H, Shimada H (2016) A simple Gateway-assisted construction system of TALEN genes for plant genome editing. Sci Rep 6:30234. https://doi.org/10.1038/srep30234

Leonel M, do Carmo EL, Fernandes AM, Soratto RP, Eburneo JAM, Garcia EL et al (2017) Chemical composition of potato tubers: the effect of cultivars and growth conditions. J Food Sci Technol 54:2372–2378. https://doi.org/10.1007/s13197-017-2677-6

Li L, Yang Y, Xu Q, Owsiany K, Welsch R, Chitchumroonchokchai C et al (2012) The Or gene enhances carotenoid accumulation and stability during post-harvest storage of potato tubers. Mol Plant 5:339–352. https://doi.org/10.1093/mp/ssr099

Li M, Song B, Zhang Q, Liu X, Lin Y, Ou Y et al (2013) A synthetic tuber-specific and cold-induced promoter is applicable in controlling potato cold-induced sweetening. Plant Physiol Biochem 67:41–47. https://doi.org/10.1016/j.plaphy.2013.02.020

Li Y, Tang W, Chen J, Jia R, Ma L, Wang S et al (2016) Development of marker-free transgenic potato tubers enriched in caffeoylquinic acids and flavonols. J Agric Food Chem 64:2932–2940. https://doi.org/10.1021/acs.jafc.6b00270

Liang S, McDonald AG (2014) Chemical and thermal characterization of potato peel waste and its fermentation residue as potential resources for biofuel and bioproducts production. J Agri Food Chem 62:8421–8429. https://doi.org/10.1021/jf5019406

Lindhout P, Meijer D, Schotte T, Hutten RC, Visser RG, Van Eck J (2011) Towards F1 hybrid seed potato breeding. Potato Res 54:301–312. https://doi.org/10.1007/s11540-011-9196-z

Liu Q, Guo Q, Akbar S, Zhi Y, El Tahchy A, Mitchell M et al (2017) Genetic enhancement of oil content in potato tuber (Solanum tuberosum L.) through an integrated metabolic engineering strategy. Plant Biotechnol J 15:56–67. https://doi.org/10.1111/pbi.12590

Liu Y, Hu CH, Wang CY, Xiong Y, Li ZK, Xiao C (2018) Occurrence of parthenogenesis in potato tuber moth. J Insect Sci 18:14. https://doi.org/10.1093/jisesa/iey003

Lombardo S, Pandino G, Mauromicale G (2017) The effect on tuber quality of an organic versus a conventional cultivation system in the early crop potato. J Food Compost Anal 62:189–196. https://doi.org/10.1016/j.jfca.2017.05.014

Lucht JM (2015) Public acceptance of plant biotechnology and GM crops. Viruses 7:4254–4281. https://doi.org/10.3390/v7082819

Ma J, Xiang H, Donnelly DJ, Meng F-R, Xu H, Durnford D et al (2017a) Genome editing in potato plants by agrobacterium-mediated transient expression of transcription activator-like effector nucleases. Plant Biotechnol Rep 11:249–258. https://doi.org/10.1007/s11816-017-0448-5

Ma X, Mau M, Sharbel TF (2017b) Genome editing for global food security. Trends Biotechnol 36:123–127. https://doi.org/10.1016/j.tibtech.2017.08.004

Mahfouz MM, Cardi T, Neal Stewart C Jr (2016) Next-generation precision genome engineering and plant biotechnology. Plant Cell Rep 35:1397–1399. https://doi.org/10.1007/s00299-016-2009-8

Makani MN, Sargent SA, Zotarelli L, Huber DJ, Sims CA (2015) Irrigation method and harvest time affect storage quality of two early-season, table stock potato (Solanum tuberosum L.) cultivars. Sci Horti 197:428–433. https://doi.org/10.1016/j.scienta.2015.09.055

Mathur V, Javid L, Kulshrestha S, Mandal A, Reddy AA (2017) World cultivation of genetically modified crops: opportunities and risks. In: Eric L (ed) Sustainable agriculture reviews. Springer, Cham, pp 45–87

McCombie G, Biedermann M, Biedermann-Brem S, Suter G, Eicher A, Pfefferle A (2016) Acrylamide in a fried potato dish (rosti) from restaurants in Zurich, Switzerland. Food Addit Contam Part B Surveill 9:21–26. https://doi.org/10.1080/19393210.2015.1102974

Meuwissen THE, Hayes BJ, Goddard ME (2001) Prediction of total genetic value using genome-wide dense marker maps. Genetics 157:1819–1829

Missiou A, Kalantidis K, Boutla A, Tzortzakaki S, Tabler M, Tsagris M (2004) Generation of transgenic potato plants highly resistant to potato virus Y (PVY) through RNA silencing. Mol Breed 14:185–197. https://doi.org/10.1023/B:MOLB.0000038006.32812.52

Mitchell M, Pritchard J, Okada S, Larroque O, Yulia D, Pettolino F et al (2017) Oil accumulation in transgenic potato tubers alters starch quality and nutritional profile. Front Plant Sci 8:554. https://doi.org/10.3389/fpls.2017.00554

Naess S, Bradeen J, Wielgus S, Haberlach G, McGrath JM, Helgeson J (2000) Resistance to late blight in Solanum bulbocastanum is mapped to chromosome 8. Theor Appl Genet 101:697–704. https://doi.org/10.1007/s001220051533

Ngobese NZ, Workneh TS, Alimi BA, Tesfay S (2017) Nutrient composition and starch characteristics of eight European potato cultivars cultivated in South Africa. J Food Compos Anal 55:1–11

Nicolia A, Proux-Wéra E, Åhman I, Onkokesung N, Andersson M, Andreasson E et al (2015) Targeted gene mutation in tetraploid potato through transient TALEN expression in protoplasts. J Biotechnol 204:17–24. https://doi.org/10.1016/j.jbiotec.2015.03.021

Nyiraneza J, Peters RD, Rodd VA, Grimmett MG, Jiang Y (2015) Improving productivity of managed potato cropping systems in Eastern Canada: crop rotation and nitrogen source effects. Agron J 107:1447–1457. https://doi.org/10.2134/agronj14.0430

Palma L, Muñoz D, Berry C, Murillo J, Caballero P (2014) Bacillus thuringiensis toxins: an overview of their biocidal activity. Toxins (Basel) 6:3296–3325. https://doi.org/10.3390/toxins6123296

Parisi C, Tillie P, Rodriguez-Cerezo E (2016) The global pipeline of GM crops out to 2020. Nat Biotechnol 34:31–36. https://doi.org/10.1038/nbt.3449

Park S, Kang TS, Kim CK, Han JS, Kim S, Smith RH et al (2005) Genetic manipulation for enhancing calcium content in potato tuber. J Agric Food Chem 53:5598–5603. https://doi.org/10.1021/jf050531c

Pedreschi F, Mariotti MS, Granby K (2014) Current issues in dietary acrylamide: formation, mitigation and risk assessment. J Sci Food Agric 94:9–20. https://doi.org/10.1002/jsfa.6349

Perez-Massot E, Banakar R, Gómez-Galera S, Zorrilla-López U, Sanahuja G, Arjó G et al (2013) The contribution of transgenic plants to better health through improved nutrition: opportunities and constraints. Genes Nutr 8:29–41. https://doi.org/10.1007/s12263-012-0315-5

Peters J, Harper G, Brierley J, Lees A, Wale S, Hilton A et al (2016) The effect of post-harvest storage conditions on the development of black dot (Colletotrichum coccodes) on potato in crops grown for different durations. Plant Pathol 65:1484–1491. https://doi.org/10.1111/ppa.12535

Petolino JF, Srivastava V, Daniell H (2016) Editing plant genomes: a new era of crop improvement. Plant Biotechnol J 14:435–436. https://doi.org/10.1111/pbi.12542

Podevin N, Devos Y, Davies HV, Nielsen KM (2012) Transgenic or not? No simple answer! EMBO Rep 13:1057–1061. https://doi.org/10.1038/embor.2012.168

Puchta H (2017) Applying CRISPR/Cas for genome engineering in plants: the best is yet to come. Curr Opin Plant Biol 36:1–8. https://doi.org/10.1016/j.pbi.2016.11.011

Qin S, Yeboah S, Cao L, Zhang J, Shi S, Liu Y (2017) Breaking continuous potato cropping with legumes improves soil microbial communities, enzyme activities and tuber yield. PLoS One 12:e0175934. https://doi.org/10.1371/journal.pone.0175934

Rasheed H, Ahmad D, Bao J (2022) Genetic diversity and health properties of polyphenols in potato. Antioxidants 11:603

Ricroch AE, Henard-Damave MC (2016) Next biotech plants: new traits, crops, developers and technologies for addressing global challenges. Crit Rev Biotechnol 36:675–690. https://doi.org/10.3109/07388551.2015.1004521

Rinder J, Casazza AP, Hoefgen R, Hesse H (2008) Regulation of aspartate-derived amino acid homeostasis in potato plants (Solanum tuberosum L.) by expression of E. Coli homoserine kinase. Amino Acids 34:213–222. https://doi.org/10.1007/s00726-007-0504-5

Salazar LF (1996) Potato viruses and their control. International Potato Center, Lima

Sawai S, Ohyama K, Yasumoto S, Seki H, Sakuma T, Yamamoto T et al (2014) Sterol side chain reductase 2 is a key enzyme in the biosynthesis of cholesterol, the common precursor of toxic steroidal glycoalkaloids in potato. Plant Cell 26:3763–3774. https://doi.org/10.1105/tpc.114.130096

Schaart JG, van de Wiel CC, Lotz LA, Smulders MJ (2016) Opportunities for products of new plant breeding techniques. Trends Plant Sci 21:438–449. https://doi.org/10.1016/j.tplants.2015.11.006

Schiml S, Puchta H (2016) Revolutionizing plant biology: multiple ways of genome engineering by CRISPR/Cas. Plant Methods 12:8. https://doi.org/10.1186/s13007-016-0103-0

Schwall GP, Safford R, Westcott RJ, Jeffcoat R, Tayal A, Shi YC et al (2000) Production of very-high-amylose potato starch by inhibition of SBE A and B. Nat Biotechnol 18:551–554. https://doi.org/10.1038/75427

Smyth SJ, Kerr WA, Phillips PW (2017) Domestic regulatory approval costs. In: Biotechnology regulation and trade. Springer, Cham, pp 33–53

Sonnewald S, Sonnewald U (2014) Regulation of potato tuber sprouting. Planta 239:27–38. https://doi.org/10.1007/s00425-013-1968-z

Sowokinos JR (2001) Biochemical and molecular control of coldinduced sweetening in potatoes. Am J Potato Res 78:221–236. https://doi.org/10.1007/BF02883548

Stadler RH, Blank I, Varga N, Robert F, Hau J, Guy PA et al (2002) Food chemistry: acrylamide from Maillard reaction products. Nature 419:449–450. https://doi.org/10.1038/419449a

Stark J, Love S, King B, Marshall J, Bohl W, Salaiz T (2013) Potato cultivar response to seasonal drought patterns. Am J Potato Res 90:207–216. https://doi.org/10.1007/s12230-012-9285-9

Steinger T, Gilliand H, Hebeisen T (2014) Epidemiological analysis of risk factors for the spread of potato viruses in Switzerland. Ann Appl Biol 164:200–207. https://doi.org/10.1111/aab.12096

Tareke E, Rydberg P, Karlsson P, Eriksson S, Tornqvist M (2002) Analysis of acrylamide, a carcinogen formed in heated foodstuffs. J Agric Food Chem 50:4998–5006. https://doi.org/10.1021/jf020302f

Tein B, Kauer K, Eremeev V, Luik A, Selge A, Loit E (2014) Farming systems affect potato (Solanum tuberosum L.) tuber and soil quality. Field Crop Res 156:1–11. https://doi.org/10.1016/j.fcr.2013.10.012

Tek AL, Stevenson WR, Helgeson JP, Jiang J (2004) Transfer of tuber soft rot and early blight resistances from Solanum brevidens into cultivated potato. Theor Appl Genet 109:249–254. https://doi.org/10.1007/s00122-004-1638-4

The Potato Genome Sequencing Consortium (2011) Genome sequence and analysis of the tuber crop potato. Nature 475:189–195. https://doi.org/10.1038/nature10158

Tiwari A, Mishra A, Saxena V, Saxena S, Raghuvanshi AP (2020) Association of characters and selection parameters for high tuber yield in potato (*Solanum tuberosum* L.). Trop Plant Res 7:205–208

Tiwari JK, Buckseth T, Dalamu, Luthra SK, Singh RK, Bhardwaj V, Vanishree G, Kardile HB, Kumar V, Kumar M (2021) Plant tissue culture and somatic hybridization in potato, eTechnical Bulletin No. 1, ICAR - Central Potato Research Institute, Shimla (HP)

Upadhyaya CP, Akula N, Young KE, Chun SC, Kim DH, Park SW (2010) Enhanced ascorbic acid accumulation in transgenic potato confers tolerance to various abiotic stresses. Biotechnol Lett 32:321–330. https://doi.org/10.1007/s10529-009-0140-0

Usall J, Ippolito A, Sisquella M, Neri F (2016) Physical treatments to control postharvest diseases of fresh fruits and vegetables. Postharvest Biol Technol 122:30–40. https://doi.org/10.1016/j.postharvbio.2016.05.002

Van Eck J, Conlin B, Garvin D, Mason H, Navarre D, Brown C (2007) Enhancing beta-carotene content in potato by RNAi-mediated silencing of the beta-carotene hydroxylase gene. Am J Potato Res 84:331–342. https://doi.org/10.1007/BF02986245

Vanhercke T, El Tahchy A, Liu Q, Zhou XR, Shrestha P, Divi UK et al (2014) Metabolic engineering of biomass for high energy density: oilseedlike triacylglycerol yields from plant leaves. Plant Biotechnol J 12:231–239. https://doi.org/10.1111/pbi.12131

Vigeolas H, Waldeck P, Zank T, Geigenberger P (2007) Increasing seed oil content in oil-seed rape (Brassica napus L.) by overexpression of a yeast glycerol-3-phosphate dehydrogenase under the control of a seed-specific promoter. Plant Biotechnol J 5:431–441. https://doi.org/10.1111/j.1467-7652.2007.00252.x

Vinci RM, Mestdagh F, De Meulenaer B (2012) Acrylamide formation in fried potato products–present and future, a critical review on mitigation strategies. Food Chem 133:1138–1154. https://doi.org/10.1016/j.foodchem.2011.08.001

Wang S, Zhang S, Wang W, Xiong X, Meng F, Cui X (2015) Efficient targeted mutagenesis in potato by the CRISPR/Cas9 system. Plant Cell Rep 34:1473–1476. https://doi.org/10.1007/s00299-015-1816-7

Weaver CM, Proulx WR, Heaney R (1999) Choices for achieving adequate dietary calcium with a vegetarian diet. Am J Clin Nutr 70:543S–548S. https://doi.org/10.1093/ajcn/70.3.543s

Weeks DP (2017) Gene editing in polyploid crops: wheat, camelina, canola, potato, cotton, peanut, sugar cane, and citrus. Prog Mol Biol Transl Sci 149:65–80. https://doi.org/10.1016/bs.pmbts.2017.05.002

Weeks DP, Spalding MH, Yang B (2016) Use of designer nucleases for targeted gene and genome editing in plants. Plant Biotechnol J 14:483–495. https://doi.org/10.1111/pbi.12448

White PJ, Broadley MR, Hammond JP, Ramsay G, Subramanian NK, Thompson J, Wright G (2012) Bio-fortification of potato tubers using foliar zinc-fertiliser. J Hortic Sci Biotechnol 87:123–129

Wolt JD, Wang K, Yang B (2016) The regulatory status of genome-edited crops. Plant Biotechnol J 14:510–518. https://doi.org/10.1111/pbi.12444

Xiong J-S, Ding J, Li Y (2015) Genome-editing technologies and their potential application in horticultural crop breeding. Hortic Res 2:15019. https://doi.org/10.1038/hortres.2015.19

Ye J, Shakya R, Shrestha P, Rommens CM (2010) Tuber-specific silencing of the acid invertase gene substantially lowers the acrylamideforming potential of potato. J Agric Food Chem 58:12162–12167. https://doi.org/10.1021/jf1032262

Zaheer K, Akhtar MH (2016) Potato production, usage, and nutrition–a review. Crit Rev Food Sci Nutr 56:711–721. https://doi.org/10.1080/10408398.2012.724479

Zaidi SS, Mahfouz MM, Mansoor S (2017a) CRISPR-Cpf1: a new tool for plant genome editing. Trends Plant Sci 22:550–553. https://doi.org/10.1016/j.tplants.2017.05.001

Zaidi SS, Tashkandi M, Mahfouz MM (2017b) Engineering molecular immunity against plant viruses. Prog Mol Biol Transl Sci 149:167–186. https://doi.org/10.1016/bs.pmbts.2017.03.009

Zaidi SS, Mukhtar MS, Mansoor S (2018) Genome editing: targeting susceptibility genes for plant disease resistance. Trends Biotechnol. https://doi.org/10.1016/j.tibtech.2018.04.005. [Epub ahead of print]

Zale J, Jung JH, Kim JY, Pathak B, Karan R, Liu H et al (2016) Metabolic engineering of sugarcane to accumulate energy-dense triacylglycerols in vegetative biomass. Plant Biotechnol J 14:661–669. https://doi.org/10.1111/pbi.12411

Zeeman SC, Kossmann J, Smith AM (2010) Starch: its metabolism, evolution, and biotechnological modification in plants. Annu Rev Plant Biol 61:209–234. https://doi.org/10.1146/annurev-arplant-042809-112301

Zeh M, Casazza AP, Kreft O, Roessner U, Bieberich K, Willmitzer L et al (2001) Antisense inhibition of threonine synthase leads to high methionine content in transgenic potato plants. Plant Physiol 127:792–802. https://doi.org/10.1104/pp.010438

Zhang F, Maeder ML, Unger-Wallace E, Hoshaw JP, Reyon D, Christian M et al (2010) High frequency targeted mutagenesis in Arabidopsis thaliana using zinc finger nucleases. Proc Natl Acad Sci USA 107:12028–12033. https://doi.org/10.1073/pnas.0914991107

Zhang J, Khan SA, Hasse C, Ruf S, Heckel DG, Bock R (2015) Pest control. Full crop protection from an insect pest by expression of long double stranded RNAs in plastids. Science 347:991–994. https://doi.org/10.1126/science.1261680

Zhang H, Zhang J, Lang Z, Botella JR, Zhu J-K (2017) Genome editing—principles and applications for functional genomics research and crop improvement. Crit Rev Plant Sci 36:291–309. https://doi.org/10.1080/07352689.2017.1402989

Zhou Z, Plauborg F, Kristensen K, Andersen MN (2017) Dry matter production, radiation interception and radiation use efficiency of potato in response to temperature and nitrogen application regimes. Agric For Meteorol 232:595–605. https://doi.org/10.1016/j.agrformet.2016.10.017

Zhu X, Gong H, He Q, Zeng Z, Busse JS, Jin W et al (2016) Silencing of vacuolar invertase and asparagine synthetase genes and its impact on acrylamide formation of fried potato products. Plant Biotechnol J 14:709–718. https://doi.org/10.1111/pbi.12421

Advances in Molecular Breeding, Genome Editing, and Genetic Transformation of Potatoes

Jyoti Prakash Sahoo, Sunanya Das, Nandini Sahu, Madhuri Pradhan, Menka Pathak, Barsa Tripathy, and Satyaprakash Barik

Abstract Potato stands as one of the world's most essential crops, serving as a staple food for a significant portion of the global population. The continuous growth in demand for potatoes, coupled with challenges such as evolving pathogens, climate change, and the need for increased crop resilience, necessitates innovative approaches in breeding and genetic enhancement. In recent years, molecular breeding, genome editing, and genetic transformation technologies have emerged as powerful tools in addressing these challenges and elevating the potential of potato cultivation. The integration of molecular techniques, such as marker-assisted selection, has revolutionized traditional breeding methods, expediting the development of improved potato varieties with heightened resilience, disease resistance, and nutritional content. Notably, the advent of genome editing technologies, including *CRISPR-Cas9*, has assisted in a new era of precision in genetic modifications, allowing for targeted enhancements in potato traits. These advancements hold immense promise for addressing challenges in potato production, such as increasing yield and adapting to changing environmental conditions. This chapter highlights recent advances in molecular breeding, genome editing, and genetic transformation technologies aimed at enhancing the genetic potential of potatoes. This chapter also emphasizes the need for consideration of ethical, environmental, and regulatory aspects associated with the genetic transformation of potatoes. However, challenges

J. P. Sahoo · N. Sahu · M. Pradhan · S. Barik (✉)
Faculty of Agriculture and Allied Sciences, C.V. Raman Global University, Bhubaneswar, Odisha, India

S. Das
Department of Botany, School of Applied Sciences, Centurion University of Technology and Management, Bhubaneswar, Odisha, India

M. Pathak
Department of Vegetable Science, Odisha University of Agriculture and Technology, Bhubaneswar, Odisha, India

B. Tripathy
Institute of Agricultural Sciences, Siksha 'O' Anusandhan Deemed to be University, Bhubaneswar, Odisha, India

S. S. Solankey (ed.), *Advances in Research on Potato Production*,
Advances in Olericulture, https://doi.org/10.1007/978-3-031-82710-5_5

related to public acceptance, regulatory frameworks, and ethical concerns must be addressed to fully realize the potential of these technologies in shaping the future of potato cultivation.

Keywords Molecular breeding · Genome editing · Genetic transformation · Crop improvement · Food security

1 Introduction

The potato (*Solanum tuberosum* L.) stands as a globally significant crop, producing nutrient dense tubers and holding promise in addressing global poverty and hunger challenges (Aksoy et al. 2021). Despite its potential, the potato faces various biotic and abiotic stresses, leading to substantial production losses (Nadakuduti et al. 2018). To enhance its traits, both conventional breeding and molecular breeding are considered, with genetic engineering offering the advantage of manipulating genes without altering successful commercial cultivars (Hameed et al. 2018). While conventional breeding is time-consuming, molecular breeding and genetic engineering, including plastid transformation, present diverse methods for potato improvement (Ahmad et al. 2022). However, molecular breeding allows for precise selection of desirable traits related to productivity, nutritional quality, and disease resistance (del Mar Martínez-Prada et al. 2021; Das Dangol et al. 2019). Despite considerable efforts to optimize protocols, the success of potato transformation remains highly dependent on the genotype (Kaur et al. 2024). The predominant approach is *Agrobacterium*-mediated transformation, although alternative methods like particle bombardment and protoplast transfection are effective (Solomon-Blackburn and Barker 2001; Singh et al. 2024). Genome editing emerges as a promising tool, allowing precise modifications of targeted sequences (Permyakova and Deineko 2024).

Genetic modification and gene editing offer solutions, with current varieties showcasing resistance to pests and diseases, reduced acrylamide, and modified starch content (Gupta et al. 2024). However, advanced gene editing technologies like base and prime editing further enhance the potential for precise edits across the genome (Vats et al. 2024). CRISPR/Cas-mediated genome editing, particularly with a streamlined regulatory process, provides new possibilities for developing commercially viable potato varieties (Şuţan and Popescu 2024; Marhadour and Prodhomme 2023). In this era of rapid technological advancement, the convergence of molecular breeding, genome editing, and genetic transformation promises to reshape the future of potato cultivation (Angmo et al. 2023). These innovative approaches offer the potential to overcome longstanding challenges, unlocking new opportunities for developing potato varieties that are resilient, nutritious, and capable of meeting the demands of an ever-growing global population (Altaf et al. 2023; Rashid et al. 2024). This chapter aims to provide insights into the exciting developments at the intersection of science and agriculture, paving the way for a more sustainable and productive potato farming landscape.

2 Constraints in Potato Production and Breeding Objectives

The main challenges to the commercial production of potatoes are diseases and insect pests. During storage, the Colorado potato beetle, leaf miner flies, cyst nematodes, and potato tuber moths are the main pests that infest potatoes (Kikuchi et al. 2015). Blackleg during storage, bacterial wilt or brown rot, potato virus Y, late blight, early blight, and potato leaf roll virus are among the major potato diseases (Nelson and Thoreson 1981). Another significant problem with potato production is weeds, which can lower yields by directly competing with the crop for nutrients, moisture, and light or by harboring pests and diseases that damage the crop.

Moreover, when researchers compared potatoes cultivated in weed-free environments, weeds can negatively affect tuber yield and ultimately affect potato cultivation (Nelson and Thoreson 1981).

Weeds present during harvest have the potential to reduce yield by slowing down the harvesting process and increasing mechanical damage to the tubers. Potatoes are not very resilient to abiotic stress (George et al. 2017). Potatoes are particularly vulnerable to drought stress because of their shallow root system (Aksoy et al. 2015). Potatoes are susceptible to salt stress in addition to drought stress. Frost is one of the lowest temperatures that harms crop the most from an agricultural perspective (Kikuchi et al. 2015). The primary goal of developing this crop better is to increase yield while enhancing the quality of the processed tubers. Potato yield is highly connected with the number of sprouts, plant height, shoot diameter, and length of leaves, and is influenced by appropriate genotypes and growth conditions (Tiwari et al. 2020a). For instance, the texture of the cooked product is influenced by the starch content, and the color of the fried product is directly influenced by the sugar content (Andrivon 2017). The nutritious richness and high yield potential of potatoes make them an essential crop.

The breeding objectives for potatoes aim to develop varieties with high tuber yields, early bulking for intensive cropping systems, and photoperiod-insensitivity for wider adaptability. These varieties should respond well to fertilizers and produce high-quality tubers that are medium-sized, oblong, yellow-fleshed, with shallow eyes, uniform dry matter distribution, high vitamin C and protein contents, high specific gravity, low sugar content, minimal black spot bruising and glycoalkaloids, long shelf life, and good cooking quality. Additionally, the objectives include creating transgenic potatoes with enhanced nutritional traits, high starch content, and resistance to biotic and abiotic stresses. There is also a focus on developing varieties and agronomic practices for organic baby potato production. For processing purposes, the aim is to develop varieties with high specific gravity suitable for French fries, chips, and dehydrated products that have low sugar content to prevent browning, as well as "cold chippers" that do not accumulate reducing sugars at low temperatures. Improving storage quality by ensuring resistance or tolerance to shrinkage, rotting, sugar accumulation (especially reducing sugars), and maintaining reasonable dormancy is another key objective. The breeding program also seeks to develop varieties with field resistance or tolerance to major biotic stresses such as late blight, early blight, charcoal rot, wart, common scab, bacterial wilt, soft rot, viral diseases (potato virus X and Y), cyst nematode, root knot nematode, aphids, and tuber moth. Additionally, the focus extends to varieties resistant or tolerant to abiotic stresses like heat, drought, frost, and soil salinity (Permyakova and Deineko 2024). Ultimately, these breeding objectives strive to create potato varieties that enhance productivity, reduce environmental impact, and improve food security.

3 Concept of Molecular Breeding, Genome Editing, and Genetic Transformation of Potato

Conventional crop improvement methods are often labor-intensive and time-consuming, with the production of new commercial varieties hindered by the rarity and randomness of significant mutations that yield desirable traits (Permyakova and Deineko 2024). Combining genetic mapping with transgenic techniques provides an effective approach to acquiring desired features while aiding in the identification of appropriate markers for molecular breeding (Ahmad et al. 2022). In the context of a growing global population, advanced plant breeding tools and molecular techniques, such as TALENS, CRISPR-Cas9, RNAi, and cisgenesis, have proven effective in enhancing the nutritional content and production of potatoes (Ahmad et al. 2022). Molecular markers for marker-assisted selection made available by QTL mapping will also help improve potato output and quality. Using genome-wide markers, genomic selection accelerates the identification of potato breeding lines with the highest yield and best quality performance while improving selection accuracy (McCue et al. 2018).

Potato genome editing is a promising option due to its readily accessible genome sequence data and well-established techniques for genetic transformation and regeneration. According to Andersson et al. (2017), these methods can enhance crop quality and productivity without affecting the optimal allele combinations in existing varieties. An example is the Sterol side chain reductase 2 (StSSR2), which is involved in synthesizing the anti-nutritional sterol glycoalkaloid (SGA) and has four alleles. This was the first successful application of TALENs in a tetraploid potato cultivar (Sawai et al. 2014). The significant breakthrough in 2015 was the effective use of both CRISPR/Cas9 and TALENs gene-editing techniques for potato genome modification (Weeks 2017). Recent advancements in gene editing have shown that most, if not all, chromosome sets in polyploid plants can be quickly and effectively modified (Weeks 2017).

Genetic transformation is a valuable technique for meeting the growing global demand for crops with higher agricultural productivity and nutritional value. The discovery of *Agrobacterium*-mediated gene delivery, enabling the production of transgenic potato plants in the 1980s, marked the beginning of the plant transformation era (Horsch et al. 1984). Today, numerous techniques exist for genetically modifying crops to improve their quality. The two commonly employed techniques are particle bombardment and *Agrobacterium*-mediated transformation, with the latter offering significantly higher transformation frequency and efficiency. For transforming multiple genes, particle bombardment is suggested as an alternative (Zhao et al. 2007). DNA can also be transferred directly into protoplasts via electroporation, polyethylene glycol (PEG) treatment, and microinjection (Chakravarty et al. 2007). However, these methods can result in complex integration patterns, high copy numbers, transgenic rearrangements, and gene silencing (Nahimak et al. 2022).

4 Molecular Breeding Approaches in Potato Improvement

4.1 Combating Fungal and Viral Diseases

Marker-assisted selection (MAS) is becoming increasingly economically viable due to the development of convenient PCR-based marker assays. Potato breeding stands to benefit significantly from emerging technologies and research focused on providing reliable, high-throughput markers at reasonable costs. Late blight, the most severe disease in potatoes, is notorious for its association with the Irish Potato Famine and causes annual global losses of approximately $6 billion due to control expenses and product damage (Haverkort et al. 2009). Historically, resistance breeding against late blight relied on 11 dominant R genes (R1 to R11) from *S. demissum*, with molecular markers linked to these genes developed for MAS. Recent advancements, such as the work by Kim et al. (2012), demonstrate broad-spectrum late blight resistance in the potato differential set plant MaR9. This resistance is conferred by stacking seven R genes, including Rpi-abpt, R1, R3a, R3b, R4, R8, and R9. The presence of at least four (R3a, R3b, R4, and R8) and seven (R1, Rpi-abpt1, R3a, R3b, R4, R8, and R9) R genes in MaR8 and MaR9, respectively, was confirmed.

Sokolova et al. (2011) developed three SCAR markers—R1-1205, R3-1380, and Sdms-523—associated with the race-specific genes R1 and R3. These markers were validated for MAS application by screening 209 accessions of 21 wild *Solanum* species. Sedlák et al. (2005) validated the R1 gene in Czech potato genetic resources using previously developed markers SPUD237 and GP21 (De Jong et al. 1997) and R1F/R (76-2sf2/76-2SR) (Ballvora et al. 2002) through MAS. Several DNA markers, including BA47f2 and CosA (Gebhardt et al. 2004), GP179 (Meksem et al. 1995), and GP76 (Oberhagemann et al. 1999), have proven useful in MAS for late blight resistance conferred by the R1 gene (Gebhardt et al. 2004).

Currently, the Rpi-blb genes (Rpi-blb1/RB, Rpi-blb2, and Rpi-blb3) are the most crucial R genes in potatoes, developed from *S. bulbocastanum*, to recognize a broad spectrum of late blight resistance. Successful applications of MAS have been observed with the RB gene in potato breeding, with SCAR markers RB-629/638, Sblb-509 (Sokolova et al. 2011), RB-1223, and RB-629 (Pankin et al. 2011) linked to the RB gene (Rpi-blb1) of *S. bulbocastanum* validated for MAS. Advancements in the genetic analysis of resistance sources against potato cyst nematodes (*Globodera pallida* and *G. rostochiensis*) have resulted in the mapping of genes and

Table 1 Genes/QTLs associated with disease in major biotic stresses in cultivated potato

Pathogen	Gene	Source	Chromosome location	References
P. infestans	*R8 (Rpi-Smira2)*	*S. demissum*	9	Blatnik et al. 2022
	Rpi-vnt1.1	*S. venturii*	9	Ghislain et al. 2019
	Rpi-sto1	*S. stoloniferum*	8	Haverkort et al. 2016
	Rpi-blb1	*S. bulbocastanum*	8	Chen et al. 2017
	Rpi-blb2	*S. bulbocastanum*	6	Ghislain et al. 2019;
	Rpi-chc1	*S. chacoense*	10	Haverkort et al. 2016;
	Rpi-blb3	*S. bulbocastanum*	4	Rakosy-Tican et al. 2020
	Rpi-amr1	*S. americanum*	11	Witek et al. 2021
	ELR	*S. microdontum*	12	Du et al. 2017
	StDMR6-1	*S. tuberosum*		Kieu et al. 2021
	StCHL1	*S. tuberosum*		Kieu et al. 2021
PVY	Ry_{adg}	*S. tuberosum* ssp. *andigena*	11	Valkonen 2015
	Ry_{chc}	*S. chacoense*	9	Valkonen 2015
	Ry_{sto} ($Ry\text{-}f_{sto}$)	*S. stoloniferum*	12	Grech-Baran et al. 2020
	$Ry(o)_{phu}$	*S. tuberosum* Group Phureja	9	Torrance et al. 2020
PVX	*Rx1 (Rx)*	*S. tuberosum* ssp. *andigena*	12	Bendahmane et al. 1999
	Rx2	*S. acaule*	5	Ritter et al. 1991
PLRV	Rlr_{etb}	*Solanum etuberosum*	4	Kelley et al. 2009
	Rl_{adg}	*S. tuberosum* ssp. *andigena*	5	Carneiro et al. 2017

QTLs (Table 1). In certain instances, functional alleles have been successfully identified. This breakthrough has facilitated the creation of molecular markers closely associated with resistance phenotypes, which can be utilized in potato breeding programs. A diverse range of viruses and viroids poses a threat to potatoes, with approximately 40 different viruses and 2 viroid species known to infect them (Salazar 1992). Among these, three stand out as particularly detrimental to global potato production: potato virus Y (PVY; genus Potyvirus), potato leafroll virus (PLRV; genus Polerovirus), and potato virus X (PVX; genus Potexvirus) (Fletcher 2012; Hameed et al. 2014). The impact of these viruses is profound, leading to various degradation types such as mosaic, leaf curl, necrosis, bundle top, plant dwarfing, and leaf yellowing (Kolychikhina et al. 2021). However, an up-to-date list of genes/QTLs associated with virus resistance is given in Table 2.

Table 2 List of described genes and QTLs involved in potato cyst nematode resistance

Species of origin	Gene/ QTLs	Chromosome	Pathotype	References
Genes				
S. spegazzinii	*Gro1–4*	7	$Ro1^{r}$	Paal et al. 2004
S. tuberosum ssp. *andigena*	*Gpa2*	12	$Pa2/3^{p}$	Van der Vossen et al. 2000
S. pimpinellifolium	*Hero*	4	$Ro1^{r}$, $Pa2/3^{p}$	Ernst et al. 2002
QTLs				
S. multidissectum	*H2*	5	Pa 1^{p}, $Pa2/3^{p}$	Strachan et al. 2019
S. sparsipilum	$Gpa^{V}spl$	5	$Pa2/3^{p}$	Caromel et al. 2005
	Gpa^{XI} *spl*	11	$Pa2/3^{p}$	
S. spegazzinii	*Gpa*	5	$Pa2/3^{p}$	Caromel et al. 2005
	GpaM1	5	$Pa2/3^{p}$	Caromel et al. 2003
	GpaM2	6	$Pa2/3^{p}$	
	GpaM3	12	$Pa2/3^{p}$	
	Gro 1.2	10	Ro 1^{r}	Park et al. 2019
S. tarijense	Gpa^{XII} *tar*	11	Pa 3^{p}	Tan et al. 2009
S. tuberosum ssp. *andigena*	*H1*	5	Ro 1^{r}, Ro 4^{r}	Finkers-Tomczak et al. 2011
	H3	4	$Pa2/3^{p}$	Bryan et al. 2004
S. vernei	*GpaV*	5	$Pa2/3^{p}$	Van Eck et al. 2017
	GpaVI	9	$Pa2/3^{p}$	Bryan et al. 2002
S. tuberosum ssp. *andigena*, *S. vernei*, *S. oplocense*, *S. tuberosum*	*Grp1*	5	Ro 5^{r}, $Pa2/3^{p}$	Finkers-Tomczak et al. 2009
S. vernei, *S. tuberosum* ssp. *andigena*	*Ro2_A*	5	Ro 2^{r}	Park et al. 2019
	Ro2_B	5	Ro 2^{r}	
	Pa 2/3_A	5	$Pa2/3^{p}$	
	Pa 2/3_B	10	$Pa2/3^{p}$	

r: *G. rostochiensis*; p: *G. pallida*

4.2 Abiotic Stresses Tolerance

Modern cultivars have low resistance to abiotic stress since traditional crop breeding has concentrated on enhancing crop quality, production, and resistance to disease (Sanwal et al. 2022). Drought can be caused globally by insufficient rainfall or a lack of irrigation water, which is extremely dangerous for plant productivity. Due to its effects on tuber output and quality, drought is a particularly damaging stress during tuber development (Aliche et al. 2018). Transgenic breeding or biotechnology procedures are crucial to find candidate genes that might be implicated in drought stress resistance (Ashraf 2010). Several molecular techniques, such as QTL mapping, genomic selection (GS), and marker-assisted selection (MAS), are used to identify candidate genes. A list of strategies used for abiotic stress tolerance in potatoes is indexed in Table 3.

Table 3 List of strategies used for abiotic stress tolerance in potato

Name of genes involved	Compounds/ solutes/osmolytes involved	Source	Tolerance	References
Betaine aldehyde dehydrogenase (BADH) transgene	Glycine betaine	Spinach	Drought, high salinity, and cold	Kikuchi et al. 2015
codA gene		*Arthrobacter globiformis* (rhizobacterium)	Drought	Kumari et al. 2023
TPS1 gene	Trehalose	Yeast	Drought	Stiller et al. 2008
GG-phosphate phosphatase/ synthase (*ggpPS* gene)	Glucosylglycerol	Moderate halo-tolerant cyanobacteria	Drought and salinity	Sievers et al. 2013
Sucrose: sucrose 1-fructosyltransferase (1-SST) and fructan: fructan 1-fructosyltransferase (1-FFT) genes	Inulin	Globe artichoke	Drought	Moon et al. 2019
YUCCA7		Rice	Drought	Cheol Park et al. 2013
Mannitol-1- phosphate dehydrogenase (mtlD) gene	Mannitol	*E. coli*	Salinity	Chourasia et al. 2021
Pyrroline-5-carboxylate synthetase (P5CS) gene from Arabidopsis	Proline	Arabidopsis	Salinity	Topfer et al. 2020
Glyceraldehyde-3-phosphate dehydrogenase (GPD) gene		Oyster mushrooms	Salinity	Cho et al. 2014
HvNHX2) gene		Barley	Salinity	Bayat et al. 2010
AtNHX1 gene		Arabidopsis	Salinity	Wang et al. 2013
Δ12-Desaturase gene		Cyanobacteria	Salinity	Maali-Amiri et al. 2007
ω9-Desaturase		*Solanum commersonii*	Drought, high salinity, and cold	Kikuchi et al. 2015
Yeast invertase		Yeast	Salinity	Khan et al. 2015

5 Genome Editing Tools in Potato Improvement

Gene targeting involves inserting a DNA molecule into a cell to replace a corresponding chromosomal segment via homologous recombination, offering precise genome manipulation (Kim et al. 1996). The introduction of Clustered Regularly Interspaced Short Palindromic Repeats (CRISPR) technology has been a pivotal moment in genetic engineering. CRISPR is a precise and adaptable gene-editing technique that uses guide RNA to target specific DNA sequences and a CRISPR-associated protein (Cas) to induce alterations. It mimics the bacterial defense mechanism against viruses by using short RNA sequences to recognize and cut invading viral DNA, rendering them harmless (Gasiunas et al. 2012; Jinek et al. 2012, 2013; Cong et al. 2013). Scientists have adapted CRISPR for targeted genome editing across a wide range of organisms. CRISPR-Cas9, the most widely used variant, involves guide RNA designed to match the target DNA sequence and Cas9 as the molecular scissors, precisely cutting DNA at the specified location.

This precision significantly reduces off-target effects, increasing the safety and efficacy of gene editing procedures (Hsu et al. 2014). Its versatility allows modifications across diverse organisms, from bacteria to plants and animals, enabling varied applications in biological research and therapeutics (Shan et al. 2013). While Zinc Finger Nucleases (ZFNs) have made significant strides in genome editing, they have limitations, including the potential for unintended off-target effects, which can result in genetic changes outside the targeted site (Gupta and Musunuru 2014). Additionally, designing and constructing ZFNs requires high expertise and effort, limiting their effectiveness in precise genetic modifications (Cathomen and Joung 2008). Similar to ZFNs, TALENs use the nonspecific FokI domain for DNA cleavage and function as dimers.

However, their distinctive aspect lies in the DNA binding domains, derived from a series of sequential repeats similar to TALEs found in plant pathogenic bacteria within the genus *Xanthomonas* (Bogdanove and Voytas 2011). Each repeat, spanning 33–35 amino acids, identifies a single nucleotide, with the final repeat, termed a "half-repeat," typically encompassing only 20 amino acids (Muñoz Bodnar et al. 2013). TALENs demonstrate notably reduced off-target effects and cytotoxicities compared to ZFNs, making them a highly efficient tool for genome editing (Ding et al. 2013; Mussolino et al. 2011). A DNA-free genome editing method is implemented by delivering CRISPR-Cas9 ribonucleoproteins (RNPs) to potato protoplasts, targeting the gene encoding a granule-bound starch synthase (Andersson et al. 2018). Additionally, geminivirus-mediated genome editing in potatoes is reported, using sequence-specific nucleases to target the ACETOLACTATE SYNTHASE1 (ALS1) gene to incorporate herbicide-inhibiting point mutations within the ALS1 locus (Butler et al. 2016). A specific gRNA has been designed to successfully knock out a newly discovered starch synthase gene (SS6) in potatoes, constructing a Single-Nucleotide Polymorphism (SNP) map of the *Solanum tuberosum* L. cv. Desiree genome (Sevestre et al. 2020). CRISPR/Cas9 reagents targeting the potato PHYTOENE DESATURASE (*StPDS*) gene have been successfully expressed in hairy root clones and regenerated, broadening the genotypes amenable to *Agrobacterium*-mediated transformation while reducing chimerism in primary events and accelerating the generation of edited materials (Butler et al. 2020).

In potatoes, traits such as improved resistance to cold-induced sweetening, processing efficiency, herbicide tolerance, modified starch quality, and self-incompatibility have been targeted using CRISPR/Cas9 and TALEN reagents in diploid and tetraploid clones (Nadakuduti et al. 2018). Multiplex CRISPR-Cas9 gene editing can deliver potato cultivars with reduced browning and acrylamide (Ly et al. 2023). A recent study focuses on mitochondrial DNA editing in potatoes using mitoTALEN and mitoTALECD to edit the mitochondrial genome of potatoes, a vegetatively propagated crop, by knocking out the mitochondrial target sequence (orf125) (Nicolia et al. 2024). Furthermore, CRISPR/Cas9-guided editing of a novel susceptibility gene in potatoes has improved Phytophthora resistance without a growth penalty (Bi et al. 2024).

6 RNAi Technique in Potato

Double-stranded RNA-mediated interference (RNAi) provides a straightforward and swift approach for suppressing gene expression (Agrawal et al. 2003). RNA interference (RNAi) is a biological process where the degradation of mRNA is initiated by double-stranded small interfering RNAs (siRNA) with complementary sequences, leading to the suppression of target gene expression (Xu et al. 2019). The Colorado potato beetle (CPB) (*Leptinotarsa decemlineata*) is the primary insect pest for potato plants (Radcliffe and Lagnaoui 2007; Weber 2003). Various target genes have been identified for managing CPB through RNA interference technology. These include VATPase, A, B, E (Baum et al. 2007; Zhu et al. 2011), Sec23, COPβ, Actin (Zhu et al. 2011), Prohibitin (Ochoa-Campuzano et al. 2013), SAHase (Zhou et al. 2013), FTZ-F1 (Liu et al. 2014), Shd (Kong et al. 2014), NAT1 (Fu et al. 2015), JHEH (Lü et al. 2015), Alt, p5cdh (Wan et al. 2015), UAP (Shi et al. 2016a), ChS (Shi et al. 2016b), TPS and TREs (Shi et al. 2016c), E75 (Guo et al. 2016), JHAMT, ILP2 (Fu et al. 2016), HR4 (Xu et al. 2018), CncC (Kong et al. 2014), EcR (Hussain et al. 2019), and Mesh (Petek et al. 2020).

7 Cisgenic, Intragenic, and Transgenic Potatoes

Gene editing is a ground-breaking technology that allows scientists to make precise changes to the DNA of organisms, including plants, animals, and humans. The primary tool used for gene editing is the CRISPR/Cas9 system, but other technologies such as TALENs and ZFNs are also utilized. To study gene function, improve crop traits, extend shelf life, or enhance resistance to pests and diseases, gene silencing techniques are employed to suppress the expression of specific genes. These genes are typically downregulated using various biotechnological tools like antisense RNA, RNAi, small-interfering RNAs (siRNAs), or microRNAs (miRNAs), which act as natural cellular processes that regulate gene expression (Germundsson and Valkonen 2006). For instance, to produce amylose-free potatoes, the GBSSI gene, a key component of starch synthesis, was silenced by introducing either antisense or sense copies of GBSSI cDNA or genomic DNA (Kuipers et al. 1997).

RNAi also provides an antiviral defense mechanism by triggering post-transcriptional gene silencing (PTGS) through the invasion of nucleic acids, which is a sequence-specific regulatory mechanism initiated by dsRNA (Vaucheret et al. 2001). A specific viral HC-Pro gene (potyviral helper-component protease) siRNA was introduced into potato seedlings (cv. Agria) to improve resistance to aphid-mediated potyvirus infection (Petrov et al. 2015). Potato Virus A (PVA), another member of the Potyviridae group, was controlled by suppressing the VPg protein (Viral Genome Linked Protein) associated with virulence through transgene-mediated RNA silencing (Osmani et al. 2020). Additionally, Bao et al. (2019) suppressed the expression of the StTCP23 transcript, which encodes a potato transcription factor, using a siRNA derived from PSTVd to combat Potato Spindle Tuber Viroid disease via an RNAi-based strategy (Germundsson and Valkonen 2006).

Virus-Induced Gene Silencing (VIGS) is another biotechnological tool that uses modified plant viruses to carry genetic material matching the target gene's sequence. Recent developments of VIGS vectors in transgenic potato plants aimed to suppress silencing; VIGS triggers an RNA-mediated defense mechanism targeting the invading viral genome (Zhao et al. 2020). Faivre-Rampant et al. (2004) constructed a PVX-based VIGS vector (PVX.PDSAS) that silenced PVX virulence in both wild and tetraploid cultivars of *Solanum* species through several generations. CRISPR/Cas9, although often associated with gene editing, can also induce gene silencing by introducing mutations or deletions in the coding region of a target gene. Zhan et al. (2019) used the Cas13a protein to confer resistance to three PVY strains (RNA virus) by targeting the p3.CI, Nib, and CP viral genes, and *Phytophthora infestans*. Additionally, knocking out StDMR6-1 and StCHL-1 improved resistance against *P. infestans* (Kieu et al. 2021). Beyond disease and pest resistance, nitrogen use efficiency was enhanced by manipulating nitrogen metabolism factors via CRISPR/Cas9 (Tiwari et al. 2020b).

Zheng et al. (2021) eliminated the anti-nutritional factor steroidal glycoalkaloid (SDA) by silencing the sterol side chain of StSSR2 in cultivated potatoes using CRISPR/Cas9. Mutation of the StPPO2 gene through CRISPR/Cas9 reduced enzymatic browning in tetraploid potatoes by silencing polyphenol oxidase activity (González et al. 2020). Self-incompatibility is another important constraint in cultivated tuber breeding. Enciso-Rodriguez et al. (2019) generated a self-compatible diploid potato line by using a dual single-guide RNA to create knockout mutants, conserving the regions of the S-RNase gene. The qualitative and quantitative improvement of potatoes through transgenics is summarized in Table 4.

Table 4 Application of CRISPR/Cas genome editing for biotic and abiotic stress resistance/tolerance and tuber quality in potato

Trait	Target gene	Delivery/transformation system	Reference
Biotic stress resistance			
PVY, PVS, and PVA resistance	P3, CI, NIb, or CP (RNA virus genes)	*Agrobacterium*-mediated transformation method (ATMT)	Zhan et al. 2019
Late blight resistance	Caffeoyl-CoAO methyl transferase (StCCoAOMT)	ATMT	Hegde et al. 2021
	StDND1, StCHL1 and StDMR6-1 (S-genes: Susceptibility genes)	ATMT	Kieu et al. 2021
Abiotic stress tolerance			
Phosphate transport (roots)	StMYB44 (MYB transcription factor)	ATMT	Zhou et al. 2017
Tuber quality traits			
Enzymatic browning	Polyphenol oxidases (PPOs) gene (StPPO2)	Protoplast culture	González et al. 2020
Steroidal glycoalkaloids (SGAs)	Sterol side chain reductase 2 (StSSR2)	ATMT	Zheng et al. 2021
Glycoalkaloids	St16DOX	Electroporation method	Nakayasu et al. 2018
Carotenoid biosynthesis	PDS and coilin	In vitro study without delivery	Khromov et al. 2018
	StPDS	ATMT	Butler et al. 2020
	PHYTOENE desaturase (PDS)	ATMT	Bánfalvi et al. 2020
Starch quality	GBBS	Protoplast culture	Andersson et al. 2017; Johansen et al. 2019
	SBE1, SBE2	Protoplast culture	Zhao et al. 2021
	Starch-branching enzymes (SBEs) genes SBE1, SBE2	ATMT	Tuncel et al. 2019
	GBSS I	ATMT	Kusano et al. 2018; Veillet et al. 2019
Starch biosynthesis	Starch synthase gene (StSS6)	ATMT	Sevestre et al. 2020

8 Challenges in Molecular Breeding, Genome Editing, and Genetic Transformation of Potato

Various genetic transformation techniques, such as the biolistic method, *Agrobacterium*-mediated transformation, floral dip, and protoplast methods, have been employed to enhance the qualitative and quantitative traits of potatoes (Sandhya et al. 2020). However, deleting the Cas9 complex from the potato plant genome and delivering ribonucleoprotein complexes through selfing or backcrossing is challenging in vegetatively propagated tetraploid potatoes (Koltun et al. 2018). True potato seed, a by-product of meiosis, is genetically distinct from other seeds, making the maintenance of clonal identity crucial. Polyethylene glycol (PEG)-mediated protoplast transformation has been reported as an effective method for delivering Cas9/gRNA-RNPs in potatoes (Andersson et al. 2017). DNA-free preassembled Cas9/gRNA-RNPs can be directly delivered into plant cells to induce mutations, as demonstrated in lipofection-mediated DNA-free delivery (Liu et al. 2020). However, this approach has limitations, such as establishing suspension cultures, isolating protoplasts, and developing whole plants from the protoplast system (Sandhya et al. 2020). A viable delivery system for CRISPR/Cas9 is Virus-Induced Gene Editing (VIGE). VIGE uses a vector derived from a plant virus to deliver sgRNAs to potatoes quickly and effectively, bypassing the more complex and time-consuming processes of plant transformation and regeneration (Butler et al. 2015; Butler et al. 2016). Gene knockout is currently the preferred method in plants. For instance, in the potato StGBSS gene, Cas9 was able to modify all four alleles (Andersson et al. 2017). One limitation of SpCas9 is its PAM requirement (NGG), highlighting the need for a more diverse CRISPR/Cas toolbox (Veillet et al. 2020).

9 Regulations on Governing Genetically Modified Potato

Regulations governing genetically modified (GM) potatoes vary across different countries and regions, reflecting diverse perspectives on biotechnology, environmental concerns, and food safety (van Hove and Gillund 2023). In the United States, the cultivation and release of GM crops, including potatoes, are overseen by the US Department of Agriculture (USDA), particularly its Animal and Plant Health Inspection Service (APHIS) (Ghouri et al. 2023). Developers must obtain regulatory approval through a rigorous process that assesses potential environmental impacts and ensures compliance with the Plant Protection Act (Forbes et al. 2023). The Environmental Protection Agency (EPA) may also be involved if the GM potato incorporates pesticidal traits (Amirhusin 2023). In the European Union, GM crops are regulated by the European Food Safety Authority (EFSA), and their cultivation is subject to a comprehensive authorization process that includes risk assessments, public consultations, and approval by member states (Sivakamavalli and Pusam 2023). Many countries have established labeling requirements to inform consumers about the presence of GMOs in food products, contributing to transparency and consumer choice (Babar and Xu 2023). It is crucial for developers to navigate and comply with these regulatory frameworks to bring genetically modified potatoes to market legally (Ghag 2024).

10 Conclusion

Recent advances in molecular breeding, genome editing, and genetic transformation of potatoes have significantly propelled the field of potato crop improvement. The utilization of molecular techniques, such as marker-assisted selection, has expedited the breeding process, enabling the development of improved varieties with enhanced traits such as disease resistance, increased yield, and improved nutritional content. Genome editing technologies, such as CRISPR-Cas9, have revolutionized the precision with which specific genes can be modified, offering unprecedented control over the potato's genetic makeup. These advancements not only accelerate the breeding timeline but also provide opportunities to address complex traits and challenges, contributing to sustainable and resilient potato production. However, the adoption and deployment of genetically modified potatoes also raise important ethical, environmental, and regulatory considerations that require careful attention. As researchers continue to refine these technologies, a balanced approach must be maintained, considering both the potential benefits and the broader implications associated with the genetic modification of this important global food crop.

References

Agrawal N, Dasaradhi PV, Mohmmed A, Malhotra P, Bhatnagar RK, Mukherjee SK (2003) RNA interference: biology, mechanism, and applications. Microbiol Mol Biol Rev 67(4):657–685

Ahmad D, Zhang Z, Rasheed H, Xu X, Bao J (2022) Recent advances in molecular improvement for potato tuber traits. Int J Mol Sci 23(17):9982

Aksoy E, Demirel U, Ozturk ZN, Caliskan S, Caliskan ME (2015) Recent advances in potato genomics, transcriptomics, and transgenics under drought and heat stresses: a review. Turk J Bot 39(6):920

Aksoy E, Demirel U, Bakhsh A, Zia MAB, Naeem M, Saeed F, Çalışkan ME (2021) Recent advances in potato (Solanum tuberosum L.) breeding. Advances in plant breeding strategies: vegetable crops: volume 8: bulbs, roots and tubers 409–487

Aliche EB, Oortwijn M, Theeuwen TPJM, Bachem CWB, Visser RGF, Linden CGVD (2018) Drought response in field grown potatoes and the interactions between canopy growth and yield. Agric Water Manag 206:20–30

Altaf MA, Mandal S, Behera B, Mangal V, Naz S, Kumar R, Aftab T (2023) Salinity stress tolerance in Solanaceous crops: current understanding and its prospects in genome editing. J Plant Growth Regul 42(7):4020–4036

Amirhusin B (2023) Genetically modified organism status, regulation, approval, labeling, and consumer perception in ASEAN. In: GMOs and political stance. Academic, pp 129–150

Andersson M, Turesson H, Nicolia A, Fält AS, Samuelsson M, Hofvander P (2017) Efficient targeted multiallelic mutagenesis in tetraploid potato (*Solanum tuberosum*) by transient CRISPR-Cas9 expression in protoplasts. Plant Cell Rep 36:117–128

Andersson M, Turesson H, Olsson N, Fält AS, Ohlsson P, Gonzalez MN, Hofvander P (2018) Genome editing in potato via CRISPR-Cas9 ribonucleoprotein delivery. Physiol Plant 164(4):378–384

Andrivon D (2017) Potato facing global challenges: how, how much, how well? Potato Res 60:389–400

Angmo D, Sharma SP, Kalia A (2023) Breeding strategies for late blight resistance in potato crop: recent developments. Mol Biol Rep 50(9):7879–7891

Ashraf M (2010) Inducing drought tolerance in plants: recent advances. Biotechnol Adv 28(1):169–183

Babar U, Xu R (2023) Agricultural genetically modified organisms (GMOs) regulation in China. In: GMOs and political stance. Academic, pp 53–74

Ballvora AMR, Ercolano J, Weiß K, Meksem C, Bormann A, Oberhagemann P, Salamini F, Gebhardt C (2002) The *R1* gene for potato resistance to late blight (*Phytophthora infestans*) belongs to the leucine zipper/NBS/LRR class of plant resistance genes. Plant J 30:361–371

Bánfalvi Z, Csákvári E, Villányi V, Kondrák M (2020) Generation of transgene-free PDS mutants in potato by *Agrobacterium*-mediated transformation. BMC Biotechnol 20(1):25

Bao S, Owens RA, Sun Q, Song H, Liu Y, Eamens AL, Feng H, Tian H, Wang MB, Zhang R (2019) Silencing of transcription factor encoding gene StTCP23 by small RNAsderived from the virulence modulating region of potato spindle tuber viroid is associated with symptom development in potato. PLoS Pathog 15:e1008110

Baum JA, Bogaert T, Clinton W, Heck GR, Feldmann P, Ilagan O, Johnson S, Plaetinck G, Munyikwa T, Pleau M (2007) Control of coleopteran insect pests through RNA interference. Natl Biotechnol 25:1322–1326

Bayat F, Shiran B, Belyaev DV, Yur'Eva NO, Sobol'Kova GI, Alizadeh H, Babakov AV (2010) Potato plants bearing a vacuolar Na+/H+ antiporter HvNHX2 from barley are characterized by improved salt tolerance. Russ J Plant Physiol 157:696–706

Bendahmane A, Kanyuka K, Baulcombe DC (1999) The Rx gene from potato controls separate virus resistance and cell death responses. Plant Cell 11(5):781–791

Bi W, Liu J, Li Y, He Z, Chen Y, Zhao T, Liang X, Wang X, Meng X, Dou D, Xu G (2024) CRISPR/Cas9-guided editing of a novel susceptibility gene in potato improves *Phytophthora* resistance without growth penalty. Plant Biotechnol J 22(1):4–6

Blatnik E, Horvat M, Berne S, Humar M, Dolničar P, Meglič V (2022) Late blight resistance conferred by Rpi-Smira2/R8 in potato genotypes in vitro depends on the genetic background. Plant 11(10):1319

Bogdanove AJ, Voytas DF (2011) TAL effectors: customizable proteins for DNA targeting. Science 333(6051):1843–1846

Bryan GJ, McLean K, Bradshaw JE, De Jong WS, Phillips M, Castelli L (2002) Mapping QTLs for resistance to the cyst nematode *Globodera pallida* derived from the wild potato species *Solanum vernei*. Theor Appl Genet 105:68–77

Bryan GJ, McLean K, Pande B, Purvis A, Hackett CA, Bradshaw JE (2004) Genetical dissection of H3-mediated polygenic PCN resistance in a heterozygous autotetraploid potato population. Mol Breed 14:105–116

Butler NM, Atkins PA, Voytas DF, Douches DS (2015) Generation and inheritance of targeted mutations in potato (*Solanum tuberosum* L.) using the CRISPR/Cas system. PLoS One 10(12):e0144591

Butler NM, Baltes NJ, Voytas DF, Douches DS (2016) Geminivirus-mediated genome editing in potato (*Solanum tuberosum* L.) using sequence-specific nucleases. Front Plant Sci 7:1045

Butler NM, Jansky SH, Jiang J (2020) First-generation genome editing in potato using hairy root transformation. Plant Biotechnol J 18(11):2201–2209

Carneiro OLG, Ribeiro SRR de P, Moreira CM, Guedes ML, Lyra DH, Pinto CABP (2017) Introgression of the Rladg allele of resistance to potato leafroll virus in *Solanum tuberosum* L. Crop Breed Appl Biotechnol 17(3):242–249

Caromel B, Mugniery D, Lefebvre V, Andrzejewski S, Ellisseche D, Kerlan MC (2003) Mapping QTLs for resistance against *Globodera pallida* (Stone) Pa2/3 in a diploid potato progeny originating from *Solanum spegazzinii*. Theor Appl Genet 106:1517–1523

Caromel B, Mugniery D, Kerlan MC, Andrzejewski S, Palloix A, Ellisseche D (2005) Resistance quantitative trait loci originating from *Solanum sparsipilum* act independently on the sex ratio of *Globodera pallida* and together for developing a necrotic reaction. Mol Plant-Microbe Interact 18:1186–1194

Cathomen T, Joung JK (2008) Zinc-finger nucleases: the next generation emerges. Mol Ther 16(7):1200–1207

Chakravarty B, Wang-Pruski G, Flinn B, Gustafson V, Regan S (2007) Genetic transformation in potato: approaches and strategies. Am J Potato Res 84:301–311

Chen S, Borza T, Byun B (2017) DNA markers for selection of late blight resistant potato breeding lines. Am J Plant Sci 8(6):1197–1209

Cheol Park H, Cha JY, Yun DJ (2013) Roles of YUCCAs in auxin biosynthesis and drought stress responses in plants. Plant Signal Behav 8(6):337–349

Cho JI, Lim HM, Siddiqui ZS, Park SH, Kim AR, Kwon TR, Lee GS (2014) Over-expression of PsGPD, a mushroom glyceraldehyde-3-phosphate dehydrogenase gene, enhances salt tolerance in rice plants. Biotechnol Lett 36:1641–1648

Chourasia KN, Lal MK, Tiwari RK, Dev D, Kardile HB, Patil VU, Pramanik D (2021) Salinity stress in potato: understanding physiological, biochemical and molecular responses. Life 11(6):545

Cong L, Ran FA, Cox D, Lin S, Barretto R, Habib N, Zhang F (2013) Multiplex genome engineering using CRISPR/Cas systems. Science 339(6121):819–823

Das Dangol S, Barakate A, Stephens J, Çalıskan ME, Bakhsh A (2019) Genome editing of potato using CRISPR technologies: current development and future prospective. Plant Cell Tissue Organ Cult (PCTOC) 139:403–416

De Jong W, Forsyth A, Leister D, Gebhardt C, Baulcombe DC (1997) A potato hypersensitive resistance gene against potato virus X maps to a resistance gene cluster on chromosome 5. Theor Appl Genet 95:246–252

del Mar Martínez-Prada M, Curtin SJ, Gutiérrez-González JJ (2021) Potato improvement through genetic engineering. GM Crops Food 12(1):479–496

Ding Q, Lee YK, Schaefer EA, Peters DT, Veres A, Kim K, Cowan CA (2013) A TALEN genome-editing system for generating human stem cell-based disease models. Cell Stem Cell 12(2):238–251

Du J, Verzaux E, Chaparro-Garcia A (2017) Elicitin recognition confers enhanced resistance to *Phytophthora infestans* in potato. Nat Plants 1(4):15034

Enciso-Rodriguez F, Manrique-Carpintero NC, Nadakuduti SS, Buell CR, Zarka D, Douches D (2019) Overcoming self-incompatibility in diploid potato using CRISPR-Cas9. Front Plant Sci 10:376

Ernst K, Kumar A, Kriseleit D, Kloos DU, Phillips MS, Ganal MW (2002) The broad-spectrum potato cyst nematode resistance gene (Hero) from tomato is the only member of a large gene family of NBS-LRR genes with an unusual amino acid repeat in the LRR region. Plant J 31:127–136

Faivre-Rampant O, Gilroy EM, Hrubikova K, Hein I, Millam S, Loake GJ, Birch P, Taylor M, Lacomme C (2004) Potato virus X-induced gene silencing in leaves and tubers of potato. Plant Physiol 134:1308–1316

Finkers-Tomczak A, Danan S, van Dijk T, Beyene A, Bouwman L, Overmars H (2009) A high-resolution map of the Grp1 locus on chromosome V of potato harbouring broad-spectrum resistance to the cyst nematode species *Globodera pallida* and *Globoderarostochiensis*. Theor Appl Genet 119:165

Finkers-Tomczak A, Bakker E, de Boer J, van der Vossen E, Achenbach U, Golas T (2011) Comparative sequence analysis of the potato cyst nematode resistance locus H1 reveals a major lack of co-linearity between three haplotypes in potato (*Solanum tuberosum* ssp.). Theor Appl Genet 122:595–608

Fletcher JD (2012) A virus survey of New Zealand fresh process and seed potato crops during 2010-11. NZ Plant Protec 65:197–203

Forbes E, Wulff-Vester AK, Hvoslef-Eide TA (2023) Will genetically modified late blight resistant potatoes be the first GM crops to be approved for commercial growing in Norway. Front Plant Sci 14:1137598

Fu KY, Guo WC, Ahmat T, Li GQ (2015) Knockdown of a nutrient amino acid transporter gene LdNAT1 reduces free neutral amino acid contents and impairs *Leptinotarsa decemlineata* pupation. Sci Rep 5:18124

Fu KY, Zhu TT, Guo WC, Ahmat T, Li GQ (2016) Knockdown of a putative insulin-like peptide gene LdILP2 in *Leptinotarsa decemlineata* by RNA interference impairs pupation and adult emergence. Gene 581:170–177

Gasiunas G, Barrangou R, Horvath P, Siksnys V (2012) Cas9–crRNA ribonucleoprotein complex mediates specific DNA cleavage for adaptive immunity in bacteria. Proc Natl Acad Sci 109(39):2579–2586

Gebhardt CA, Ballvora B, Walkemeier P, Oberhagemann SK (2004) Assessing genetic potential in germplasm collections of crop plants by marker-trait association: a case study for potatoes with quantitative variation of resistance to late blight and maturity type. Mol Breed 13:93–102

George TS, Taylor MA, Dodd IC (2017) Climate change and consequences for potato production: a review of tolerance to emerging abiotic stress. Potato Res 60:239–268

Germundsson A, Valkonen J (2006) P1- and VPg-transgenic plants show similar resistance to Potato virus A and may compromise long distance movement of the virus in plant sections expressing RNA silencing-based resistance. Virus Res 116:208–213

Ghag SB (2024) Genetically modified organisms and their regulatory frameworks. In: Global regulatory outlook for CRISPRized plants. Academic, pp 147–166

Ghislain M, Byarugaba AA, Magembe E (2019) Stacking three late blight resistance genes from wild species directly into African highland potato varieties confers complete field resistance to local blight races. Plant Biotech J 17(6):1119–1129

Ghouri MZ, Munawar N, Aftab SO, Ahmad A (2023) Regulation of CRISPR edited food and feed: legislation and future. In: GMOs and political stance. Academic, pp 261–287

González MN, Massa GA, Andersson M, Turesson H, Olsson N, Fält AS (2020) Reduced enzymatic browning in potato tubers by specific editing of a polyphenol oxidase gene via ribonucleoprotein complexes delivery of the CRISPR/Cas9 system. Front Plant Sci 10:1649

Grech-Baran M, Witek K, Szajko K (2020) Extreme resistance to Potato virus Y in potato carrying the Rysto gene is mediated by a TIR-NLR immune receptor. Plant Biotechnol J 18(3):655–667

Guo WC, Liu XP, Fu KY, Shi JF, Lü FG, Li GQ (2016) Nuclear receptor ecdysone-induced protein 75 is required for larval–pupal metamorphosis in the Colorado potato beetle *Leptinotarsa decemlineata* (Say). Insect Mol Bio 25:44–57

Gupta RM, Musunuru K (2014) Expanding the genetic editing tool kit: ZFNs, TALENs, and CRISPR-Cas9. J Clinic Invest 124(10): 4154-4161

Gupta P, Karkute SG, Dash PK, Singh AK (2024) Genome editing for improvement of nutrition and quality in vegetable crops. Appl Gen Eng Plants:222–241

Hameed A, Iqbal Z, Asad S, Mansoor S (2014) Detection of multiple potato viruses in the field suggests synergistic interactions among potato viruses in Pakistan. Plant Pathol J 30:407

Hameed A, Zaidi SSEA, Shakir S, Mansoor S (2018) Applications of new breeding technologies for potato improvement. Front Plant Sci 9:925

Haverkort AJ, Struik PC, Visser RGF, Jacobsen E (2009) Applied biotechnology to combat late blight in potato caused by *Phytophthora infestans*. Potato Res 52:249–264

Haverkort AJ, Boonekamp PM, Hutten R (2016) Durable Late blight resistance in potato through dynamic varieties obtained by cisgenesis: scientific and societal advances in the DuRPh project. Potato Res 59(1):35–66

Hegde N, Joshi S, Soni N, Kushalappa AC (2021) The caffeoyl-CoA O-methyltransferase gene SNP replacement in russet Burbank potato variety enhances late blight resistance through cell wall reinforcement. Plant Cell Rep 40(1):237–254

Horsch RB, Fraley RT, Rogers SG, Sanders PR, Lloyd A, Hoffman N (1984) Inheritance of functional foreign genes in plants. Science 223:496–498

Hsu PD, Lander ES, Zhang F (2014) Development and applications of CRISPR-Cas9 for genome engineering. Cell 157(6):1262–1278

Hussain T, Aksoy E, Çalıskan ME, Bakhsh A (2019) Transgenic potato lines expressing hairpin construct of molting-associated EcR gene exhibit enhanced resistance against Colorado potato beetle (*Leptinotarsa decemlineata*, Say). Transgenic Res 28:151–164

Jinek M, Chylinski K, Fonfara I, Hauer M, Doudna JA, Charpentier E (2012) A programmable dual-RNA–guided DNA endonuclease in adaptive bacterial immunity. Science 337(6096):816–821

Jinek M, East A, Cheng A, Lin S, Ma E, Doudna J (2013) RNA-programmed genome editing in human cells. Elife 2:e00471

Johansen IE, Liu Y, Jørgensen B, Bennett EP, Andreasson E, Nielsen KL (2019) High efficacy full allelic CRISPR/Cas9 gene editing in tetraploid potato. Sci Rep 9:17715

Kaur G, Jain S, Bhushan S, Das N, Sharma M, Sharma D (2024) Role of microRNAs and their putative mechanism in regulating potato (*Solanum tuberosum* L.) life cycle and response to various environmental stresses. Plant Physiol Biochem 108334

Kelley KB, Whitworth JL, Novy RG (2009) Mapping of the potato leafroll virus resistance gene, Rlretb, from *Solanum etuberosum* identifies interchromosomal translocations among its E-genome chromosomes 4 and 9 relative to the A-genome of Solanum L. sect. Petota. Mol Breed 23(3):489–500

Khan MS, Ahmad D, Khan MA (2015) Utilization of genes encoding osmoprotectants in transgenic plants for enhanced abiotic stress tolerance. Electron J Biotechnol 18(4):257–266

Khromov AV, Gushchin VA, Timberbaev VI, Kalinina NO, Taliansky ME, Makarov VV (2018) Guide RNA design for CRISPR/Cas9-mediated potato genome editing. Dokl Biochem Biophys 479(1):90–94

Kieu NP, Lenman M, Wang ES, Petersen BL, Andreasson E (2021) Mutations introduced in susceptibility genes through CRISPR/Cas9 genome editing confer increased late blight resistance in potatoes. Sci Rep 11(1):4487

Kikuchi A, Huynh HD, Endo T, Watanabe K (2015) Review of recent transgenic studies on abiotic stress tolerance and future molecular breeding in potato. Breed Sci 65:85–102

Kim YG, Cha J, Chandrasegaran S (1996) Hybrid restriction enzymes: zinc finger fusions to Fok I cleavage domain. Proc Natl Acad Sci 93(3):1156–1160

Kim HJ, Lee HR, Jo KR, Mortazavian SM, Huigen DJ, Evenhuis B, Kessel G, Visser RG, Jacobsen E, Vossen JH (2012) Broad spectrum late blight resistance in potato diferential set plants MaR8 and MaR9 is conferred by multiple stacked R genes. Theor Appl Genet 124:923–935

Koltun A, Corte ED, Mertz-Henning LM, Gonçalves LS (2018) Genetic improvement of horticultural crops mediated by CRISPR/Cas: a new horizon of possibilities. Horti Bras 36:290–298

Kolychikhina MS, Beloshapkina OO, Phiri C (2021) Change in potato productivity under the impact of viral diseases. In IOP Conference Series: Earth and Environ Sci 663(1): 012035)

Kong Y, Liu XP, Wan PJ, Shi XQ, Guo WC, Li GQ (2014) The P450 enzyme Shade mediates the hydroxylation of ecdysone to 20-hydroxyecdysone in the Colorado potato beetle, *Leptinotarsa decemlineata*. Insect Mol Bio 23:632–643

Kuipers AGJ, Jacobsen E, Visser RGF (1997) Applications of antisense technology in plants. In: Lichtenstein C, Nellen W (eds) Antisense technology: a practical approach. Oxford University Press, Oxford, pp 191–219

Kumari K, Patra A, Guha S, Goyal T, Kumar M, Mehta S (2023) The biotechnological story of microbial genes from soil to transgenic plants. In: Singh N, Chattopadhyay A, Lichtfouse E (eds) Sustainable agriculture reviews 60. Springer, Cham

Kusano H, Ohnuma M, Mutsuro-Aoki H, Asahi T, Ichinosawa D, Onodera H (2018) Establishment of a modified CRISPR/Cas9 system with increased mutagenesis frequency using the translational enhancer dMac3 and multiple guide RNAs in potato. Sci Rep 8:13753. https://doi.org/10.1038/s41598-018-32049-2

Liu XP, Fu KY, Lü FG, Meng QW, Guo WC, Li GQ (2014) Involvement of FTZ-F1 in the regulation of pupation in Leptinotarsa decemlineata (Say). Insect Biocem Mol Biol 55:51–60

Liu W, Rudis MR, Cheplick MH, Millwood RJ, Yang JP, Ondzighi-Assoume CA, Stewart CN (2020) Lipofection-mediated genome editing using DNA-free delivery of the Cas9/gRNA ribonucleoprotein into plant cells. Plant Cell Rep 39:245–257

Lü FG, Fu KY, Guo WC, Li GQ (2015) Characterization of two juvenile hormone epoxide hydrolases by RNA interference in the Colorado potato beetle. Gene 570:264–271

Ly DNP, Iqbal S, Fosu-Nyarko J, Milroy S, Jones MG (2023) Multiplex CRISPR-Cas9 gene-editing can deliver potato cultivars with reduced browning and acrylamide. Plan Theory 12(2):379. https://doi.org/10.3390/plants12020379

Maali-Amiri R, Goldenkova-Pavlova IV, Yur'Eva NO, Pchelkin VP, Tsydendambaev VD, Vereshchagin AG, Nosov AM (2007) Lipid fatty acid composition of potato plants transformed with the Δ12-desaturase gene from cyanobacterium. Russ J Plant Physiol 54:600–606. https://doi.org/10.1134/S1021443707050056

Marhadour S, Prodhomme C (2023) Recent trends in genetics studies and molecular breeding of potato. In: Potato production worldwide. Academic, pp 273–301

McCue KF, Breksa A, Vilches A, Belknap WR (2018) Modification of potato steroidal glycoalkaloids with silencing RNA constructs. Am J Potato Res 95:317

Meksem K, Leister D, Peleman J, Zabeau M, Salamini M, Gebhardt C (1995) A high-resolution map of the vicinity of the *R1* locus on chromosome V of potato based on RFLP and AFLP markers. Mol Gen Genet 249:74–81

Moon KB, Ko H, Park JS (2019) Expression of Jerusalem artichoke (*Helianthus tuberosus* L.) fructosyltransferases, and high fructan accumulation in potato tubers. Appl Biol Chem 62:74. https://doi.org/10.1186/s13765-019-0481-x

Muñoz Bodnar A, Bernal A, Szurek B, López CE (2013) Tell me a tale of TALEs. Mol Biotechnol 53:228–235

Mussolino C, Morbitzer R, Lütge F, Dannemann N, Lahaye T, Cathomen T (2011) A novel TALE nuclease scaffold enables high genome editing activity in combination with low toxicity. Nucleic Acids Res 39(21):9283–9293

Nadakuduti SS, Buell CR, Voytas DF, Starker CG, Douches DS (2018) Genome editing for crop improvement–applications in clonally propagated polyploids with a focus on potato (*Solanum tuberosum* L.). Front. Plant Sci 9:1607

Nahimak V, Almasia NI, Gonzalez MN, Massa GA, Decima Oneto CA, Feingold SE, Hopp HE, Vazquez Rovere C (2022) State of the art of genetic engineering in potato: from the first report to its future potential. Front Plant Sci 12:768233. https://doi.org/10.3389/fpls.2021.768233

Nakayasu M, Akiyama R, Lee HJ, Osakabe K, Osakabe Y, Watanabe B (2018) Generation of α-solanine-free hairy roots of potato by CRISPR/Cas9 mediated genome editing of the St16DOX gene. Plant Physiol Biochem 131:70–77. https://doi.org/10.1016/j.plaphy.2018.04.026

Nelson DC, Thoreson MC (1981) Competition between potatoes (*Solanum tuberosum*) and weeds. Weed Sci 29(6):672–677. https://doi.org/10.1017/S004317450004025X

Nicolia A, Scotti N, D'Agostino N, Festa G, Sannino L, Aufiero G, Cardi T (2024) Mitochondrial DNA editing in potato through mitoTALEN and mitoTALECD: molecular characterization and stability of editing events. Plant Methods 20:4

Oberhagemann P, Chatot-Balandras C, Schäfer-Pregl R, Wegener D, Palomino C, Salamini F, Bonnel E, Gebhardt C (1999) A genetic analysis of quantitative resistance to late blight in potato: towards marker-assisted selection. Mol Breed 5:399–415

Ochoa-Campuzano C, Martínez-Ramírez AC, Contreras E, Rausell C, Real MD (2013) Prohibitin, an essential protein for Colorado potato beetle larval viability, is relevant to *Bacillus thuringiensis* Cry3Aa toxicity. Pesticide Biochem Physiol 107:299–308

Osmani Z, Sabet MS, Nakahara KS, Mokhtassi BA, Vahabid K, Moienia A, Shams BM (2020) Identification of a defense response gene involved in signaling pathways against PVA and PVY in potato. GM Crops Food 12:86–105

Paal J, Henselewski H, Muth J, Meksem K, Menendez CM, Salamini F (2004) Molecular cloning of the potato Gro1-4 gene conferring resistance to pathotype Ro1 of the root cyst nematode *Globoderarostochiensis*, based on a candidate gene approach. Plant J 38:285–297

Pankin A, Sokolova E, Rogozina E, Kuznetsova M, Deahl K, Jones R, Khavkin E (2011) Allele mining in the gene pool of wild *Solanum* species for homologues of late blight resistance gene *RB/Rpi-blb1*. Plant Genet Resour 9:305–308

Park J, Hackett C, Dandurand L, Wang X, Jong WDD (2019) QTL for resistance to *Globodera rostochiensis* pathotype Ro2 and *G. pallida* pathotype Pa2/3 in autotetraploid potato. Am J Potato Res 96:552–563

Permyakova NV, Deineko EV (2024) Crop improvement: comparison of transgenesis and gene editing. Horticulturae 10(1):57

Petek M, Coll A, Razinger J, Gruden K (2020) Validating the potential of double-stranded RNA targeting Colorado potato beetle mesh gene in laboratory and field trials. Front Plant Sci 11:1250

Petrov N, Stoyanova M, Andonova TA (2015) Induction of resistance to potato virus Y strain NTN in potato plants through RNAi. Biotechnol Biotec Eq 29:21–26

Radcliffe EB, Lagnaoui A (2007) Pests and diseases. In: Vreugdenhil D, Bradshaw J, Gebhardt C, Govers F, Taylor MA, MacKerron DK, Ross HA (eds) Potato biology and biotechnology: advances and perspectives, 1st edn. Elsevier, Oxford, pp 545–554

Rakosy-Tican E, Thieme R, König J (2020) Introgression of two broad-spectrum late blight resistance genes, Rpi-Blb1 and Rpi-Blb3, from *Solanum bulbocastanum* Dun plus race-specific R genes into potato pre-breeding lines. Front Plant Sci 11:699

Rashid MAR, Pan Z, Wang Y, Shaheen T, Ahmed HGMD (2024) Biofortification of potatoes to reduce malnutrition. In: Biofortification of grain and vegetable crops. Academic, pp 223–237

Ritter E, Debener T, Barone A, Salamini F, Gebhardt C (1991) RFLP mapping on potato chromosomes of two genes controlling extreme resistance to potato virus X (PVX). Mol Gen Genet 227(1):81–85

Salazar LF (1996) Potato Viruses and Their Control. International Potato Center; Lima, Peru

Sandhya D, Jogam P, Allini VR, Abbagani S, Alok A (2020) The present and potential future methods for delivering CRISPR/Cas9 components in plants. J Gen Eng Biotech 18:1–11

Sanwal SK, Kumar P, Kesh H, Gupta VK, Kumar A, Kumar A, Meena BL, Colla G, Cardarelli M, Kumar P (2022, 1842) Salinity stress tolerance in potato cultivars: evidence from physiological and biochemical traits. Plan Theory 11(14)

Sawai S, Ohyama K, Yasumoto S, Seki H, Sakuma T, Yamamoto T (2014) Sterol side chain reductase 2 is a key enzyme in the biosynthesis of cholesterol, the common precursor of toxic steroidal glycoalkaloids in potato. Plant Cell 26:3763–3774

Sedlák P, Vej P, Melounová M, Křenek P, Domkářová J, Zoufalá J (2005) Characterisation of resistance genes resources against late blight available for Czech potato breeding by means of selected DNA markers. Plant Soil Environ 51:82–86

Sevestre F, Facon M, Wattebled F, Szydlowski N (2020) Facilitating gene editing in potato: a Single-Nucleotide Polymorphism (SNP) map of the *Solanum tuberosum* L. cv. Desiree genome. Sci Rep 10(1):2045. https://doi.org/10.1038/s41598-020-58985-6

Shan Q, Wang Y, Li J, Zhang Y, Chen K, Liang Z, Gao C (2013) Targeted genome modification of crop plants using a CRISPR-Cas system. Nat Biotechnol 31(8):686–688

Shi JF, Mu LL, Chen X, Guo WC, Li GQ (2016a b) RNA interference of chitin synthase genes inhibits chitin biosynthesis and affects larval performance in *Leptinotarsa decemlineata* (Say). Inter J Biol Sci 12:1319

Shi JF, Xu QY, Sun QK, Meng QW, Mu LL, Guo WC, Li GQ (2016b c) Physiological roles of trehalose in Leptinotarsa larvae revealed by RNA interference of trehalose-6-phosphate synthase and trehalase genes. Insect Biochem Physiol 77:52–68

Shi JF, Fu J, Mu LL, Guo WC, Li GQ (2016c) Two Leptinotarsa uridine diphosphate N-acetylglucosamine pyrophosphorylases are specialized for chitin synthesis in larval epidermal cuticle and midgut peritrophic matrix. Insect Biochem Mol Biol 68:1–12

Sievers N, Muders K, Henneberg M, Klähn S, Effmert M, Junghans H, Hagemann M (2013) Establishing glucosylglycerol synthesis in potato (*Solanum tuberosum* L. cv. Albatros) by expression of the ggpPS gene from *Azotobacter vinelandii*. J Plant Sci Mol Breed 2(1):10–7243. https://doi.org/10.7243/2050-2389-2-1

Singh B, Lokya V, Kaundal P, Tiwari S (2024) Genetic engineering for potato improvement: current challenges and future opportunities. In: Genetic engineering of crop plants for food and health security: volume 1. Singapore, Springer Nature Singapore, pp 213–232

Sivakamavalli J, Pusam Y (2023) GMO regulations in India. In: GMOs and political stance. Academic Press, pp 75–91

Sokolova EA, Pankin M, Beketova M, Kuznetsova S, Spiglazova E, Rogozina I, Yashina KE (2011) SCAR markers of the *R*-genes and germplasm of wild *Solanum* species for breeding late blight-resistant potato cultivars. Plant Genet Resour 9:309–312

Solomon-Blackburn RM, Barker H (2001) Breeding virus resistant potatoes (*Solanum tuberosum*): a review of traditional and molecular approaches. Heredity 86(1):17–35

Stiller I, Dulai S, Kondrák M, Tarnai R, Szabo L, Toldi O, Banfalvi Z (2008) Effects of drought on water content and photosynthetic parameters in potato plants expressing the trehalose-6-phosphate synthase gene of *Saccharomyces cerevisiae*. Planta 227:299–308

Strachan SM, Armstrong MR, Kaur A, Wright KM, Lim TY, Baker K (2019) Mapping the H2 resistance effective against *Globodera pallida* pathotype Pa1 in tetraploid potato. Theor Appl Genet 132:1283–1294. https://doi.org/10.1007/s00122-019-03278-4

Şuţan NA, Popescu A (2024) Genome editing by different site-specific nucleases and their applications in improving horticultural crops. In: Genome editing and global food security. Routledge, pp 42–75

Tan AMY, Park TH, Alles R, Hutten RC, Visser RG, van Eck HJ (2009) GpaXIl tar originating from *Solanum tarijense* is a major resistance locus to *Globoderapallida* and is localised on chromosome 11 of potato. Theor Appl Genet 119:1477–1487

Tiwari A, Mishra A, Saxena V, Saxena S, Raghuvanshi AP (2020a) Association of characters and selection parameters for high tuber yield in Potato (*Solanum tuberosum* L.). Trop Plant Res 7:205–208. https://doi.org/10.22271/tpr.2020.v7.i1.026

Tiwari JK, Buckseth T, Singh RK, Kumar M, Kant S (2020b) Prospects of improving nitrogen use efficiency in potato: lessons from transgenics to genome editing strategies in plants. Front Plant Sci 11:597481

Topfer N, Braam T, Shameer S, Ratcliffe RG, Sweetlove LJ (2020) Alternative crassulacean acid metabolism modes provide environment-specific water-saving benefits in a leaf metabolic model. Plant Cell 32(12):3689–3705

Torrance L, Cowan GH, McLean K (2020) Natural resistance to Potato virus Y in *Solanum tuberosum* Group Phureja. Theor Appl Genet 133(3):967–980

Tuncel A, Corbin KR, Ahn-Jarvis J, Harris S, Hawkins E, Smedley MA et al (2019) Cas9-mediated mutagenesis of potato starch-branching enzymes generates a range of tuber starch phenotypes. Plant Biotechnol J 17(12):2259–2271

Valkonen JPT (2015) Elucidation of virus-host interactions to enhance resistance breeding for control of virus diseases in potato. Breed Sci 65(1):69–76

Van der Vossen EA, van der Voort JN, Kanyuka K, Bendahmane A, Sandbrink H, Baulcombe DC (2000) Homologues of a single resistance-gene cluster in potato confer resistance to distinct pathogens: a virus and a nematode. Plant J 23:567–576

Van Eck HJ, Vos PG, Valkonen JPT, Uitdewilligen JGAML, Lensing H, de Vetten N (2017) Graphical genotyping as a method to map Nyonsto and Gpa5 using a reference panel of tetraploid potato cultivars. Theor Appl Genet 130:515–528

van Hove L, Gillund F (2023) Is it only the regulatory status? Broadening the debate on cisgenic plants. In: Cisgenic crops: safety, legal and social issues. Springer International Publishing, Cham, pp 269–288

Vats S, Kaur S, Chauhan A, Biswas DK, Deshmukh R (2024) Advancement in delivery systems and vector selection for CRISPR/Cas-mediated genome editing in plants. Appl Gen Engin Plants 52–77

Vaucheret H, Béclin C, Fagard M (2001) Post-transcriptional gene silencing in plants. J Cell Sci 114:3083–3091

Veillet F, Chauvin L, Kermarrec MP, Sevestre F, Merrer M, Terret Z (2019) The Solanum tuberosum GBSSI gene: a target for assessing gene and Base editing in tetraploid potato. Plant Cel Rep 38(9):1065–1080

Veillet F, Perrot L, Guyon-Debast A, Kermarrec MP, Chauvin L, Chauvin JE, Nogué F (2020) Expanding the CRISPR toolbox in P. patens using SpCas9-NG variant and application for gene and base editing in solanaceae crops. Int J Mol Sci 21(3):1024

Wan PJ, Fu KY, Lü FG, Guo WC, Li GQ (2015) Knockdown of a putative alanine aminotransferase gene affects amino acid content and flight capacity in the Colorado potato beetle *Leptinotarsa decemlineata*. Amino Acids 47:1445–1454

Wang L, Zhang J, Wang D, Zhang J, Cui Y, Liu Y, Binyu (2013) Assessment of salt tolerance in transgenic potato carrying AtNHX1 gene. Crop Sci 53(6): 2643–2651

Weber D (2003) Colorado beetle: Pest on the move. Pesticide Outlook 14:256–259

Weeks DP (2017) Gene editing in polyploid crops: wheat, camelina, canola, potato, cotton, peanut, sugar cane, and citrus. Gene Edit Plants 65-80. https://doi.org/10.1016/bs.pmbts.2017.05.002

Witek K, Lin X, Karki HS (2021) A complex resistance locus in *Solanum americanum* recognizes a conserved Phytophthora effector. Nat Plants 7(2):198–208

Xu L, Zhang Y, Zhang S, Deng J, Lu M, Zhang L, Zhang J (2018) Comparative analysis of the immune system of an invasive bark beetle, *Dendroctonusvalens*, infected by an entomopathogenic fungus. Dev Comp Immunol 88:65–69

Xu W, Jiang X, Huang L (2019) RNA interference technology. Compre Biotech:560–575

Zhan X, Zhang F, Zhong Z, Chen R, Wang Y, Chang L (2019) Generation of virus-resistant potato plants by RNA genome targeting. Plant Biotechnol J 17:1814–1822

Zhao Y, Qian Q, Wang HZ (2007) Co-transformation of gene expression cassettes via particle bombardment to generate safe transgenic plant without any unwanted DNA. In Vitro Cell Dev Biol Plant 43:328–334

Zhao J, Rios CG, Song J (2020) Potato virus X-based microRNA silencing (VBMS) in potato. Jove J Vis Exp 159:e61067

Zhao X, Jayarathna S, Turesson H, Fält AS, Nestor G, González MN (2021) Amylose starch with no detectable branching developed through DNA-free CRISPR-Cas9 mediated mutagenesis of two starch branching enzymes in potato. Sci Rep 11(1):4311

Zheng Z, Ye G, Zhou Y, Pu X, Su W, Wang J (2021) Editing sterol side chain reductase 2 gene (StSSR2) via CRISPR/Cas9 reduces the total steroidal glycoalkaloids in potato. All Life 14:401–413

Zhou LT, Jia S, Wan PJ, Kong Y, Guo WC, Ahmat T, Li GQ (2013) RNA interference of a putative S-adenosyl-L-homocysteine hydrolase gene affects larval performance in *Leptinotarsa decemlineata* (Say). J Insect Physio 59:1049–1056

Zhou X, Zha M, Huang J, Li L, Imran M, Zhang C (2017) StMYB44 negatively regulates phosphate transport by suppressing expression of PHOSPHATE1 in potato. J Exp Bot 68:1265–1281

Zhu F, Xu J, Palli R, Ferguson J, Palli SR (2011) Ingested RNA interference for managing the populations of the Colorado potato beetle, *Leptinotarsa Decemlineata*. Pest Management Sci 67:175–182

Advances in Research on Potato Seed Production

Meenakshi Kumari, Anuradha Sinha, Shashank Shekhar Solankey, and Swapnil

Abstract Potato ranks third among food crops in terms of importance for human consumption. India ranks second globally in potato production. This is called the friend of a poor man. It is an inexpensive crop to grow and the principal source of glucose, frequently used in place of grains. Together with amino acids (tryptophan, isoleucine, and leucine) it also contains vitamins (thiamin, riboflavin, pyridoxine, and ascorbic acid), protein, and minerals (calcium, phosphorus, and iron). The most crucial component in the production of potatoes is their seed. Higher rate of seed (2.5–3.0 tons/ha), minimum multiplication rate, increasing viral deterioration, transportation, and storage are the main challenges facing the production of potato seed in the country. Also, a significant portion of the expense of producing potatoes comes from potato seed alone, i.e., it alone accounts for 40–50% of the total production of potato cost, which puts back enormous amounts of potentially edible food in the soil. Associated problems include the poor multiplication rate of new improved potato varieties and the frequent need to replace seeds owing to degeneration. Potato researchers are always working to create novel technologies that would enable the rapid replication of the original breeder's seed of the released varieties in enough numbers to fulfill our nation's needs. To increase seed quality and speed up healthy disease-free seed multiplication, major global potato growers have switched from traditional to high-tech seed production processes. Many of the issues surrounding

M. Kumari (✉)
Department of Vegetable Science, Faculty of Agricultural Sciences, SGT University, Gurugram, Haryana, India

A. Sinha
Krishi Vigyan Kendra, Aurangabad, Bihar Agricultural University, Sabour, Bhagalpur, Bihar, India

S. S. Solankey
Department of Horticulture (Vegetable Science), Nalanda College of Horticulture, Noorsarai, Nalanda, Bihar Agricultural University, Sabour, Bhagalpur, Bihar, India

Swapnil
Department of Genetics and Plant Breeding, Centurion University of Technology and Management, Bhubaneswar, Odisha, India

S. S. Solankey (ed.), *Advances in Research on Potato Production*, Advances in Olericulture, https://doi.org/10.1007/978-3-031-82710-5_6

the production of potato seeds can be solved by new inventions, especially in tropical and subtropical nations.

Keywords Advanced technique · Potato · Seed production · High-tech · Disease free

1 Introduction

The potato (*Solanum tuberosum* L.), which makes up 11.3% of all potato acreage and 12.5% of all potato production worldwide, is quickly rising to prominence in India as one of the main food crops. In 2021–2022, India produced 53,387,000 metric tons in an area of 2,202,000 ha (Agriculture Statistics at a Glance 2022). The production of potatoes in India is still restricted by a shortage of high-quality seed, a dearth of cultivars with a range of economic attributes that offer growers better alternatives, and a lack of suitable storage facilities. The lack of sufficient high-quality planting material is the most significant of them since it directly affects the production of crops (Singh and Sharma 2018). Bulky, seed potatoes have a high seed rate and a slow rate of multiplication (five to six times), which presents a special set of difficulties for seed potato production and its supply chain (Young 1990). As the potato is vegetatively multiplied, tuber serves as a seed, and it degenerates quickly and needs a new seed every year (Singh et al. 2018). The limited number of seed-producing regions (aphid-free) in the nation exacerbates the problems associated with seeds even further. It has been shown that using high-quality seeds can boost output by 15–20% (Shaheb et al. 2016). There is a big gap between the supply and demand for certified seed potatoes in India (NAAS 2021). Advanced seed potato production including true potato seed, production of micro-tubers from micro-plants, tissue culture, aeroponics, micropropagation, apical root cutting, etc. is a substitute to address the lack of premium seed potatoes in our country due to the advantage of producing seed from basic material that has been tested from pathogens and can be produced under controlled conditions in any region.

2 Traditional Seed Potato Production

In India, in the 1930s, the technique of producing good quality seeds of potato started in the hills. For production, hilly areas were selected because of their reduced aphid infestation and slower rate of seed degradation. However, the formalized process of producing potato seeds started in the 1960s. It is predicated on the field replication, clonal selection, and indexing of tuber. The conventional approach suggested multiplying seed tubers every seven generations because of the poor rate of seed multiplication caused by the vegetative mode of propagation under a stringent quality control system to produce an adequate number of certified seeds.

2.1 Stepwise Production of Traditional Potato Seed

1. Always use quality seeds that are free from all fungal, bacterial, viral, tuber-borne, and nematode diseases.
2. Adopt region-wise sowing date.
3. Apply timely weedicides, manure and fertilizers, earthing-up, and water as per the dose recommended.
4. Follow proper plant protection schedule against various diseases like late blight and also manage insect vectors.
5. In a crop season seed crops should be inspected at least three times and also rogued out off type regularly.
6. An important operation dehaulming should be done before aphids reach a critical limit and after dehaulming regular visits to the seed crops are a must to control the regrowth of seed crops.
7. After 15 days of dehaulming seed crop should be harvested.
8. After harvesting, for 15 days seed crop is kept in a shady place for curing purposes; after that remove diseased, cut, and cracked seed crops; and in the last, the crop should be graded according to size.
9. Treat the seed crop by spraying boric acid 3% solution and dry in air.
10. Packaging is also important, so treated seed crops must be packed in gunny bags, labeled with all information, and should be stored at 4 °C.

2.2 Problems in Traditional Seed Potato Cultivation

Cultivation of seed potatoes mostly depends on the accessibility of disease-free seed material, temperature, and plant protection measures. Potato plants are extremely sensitive to biological factors such as rainfall, temperature, and photoperiod (Singh 2002). As potato is vegetatively propagated, its viral, fungal, and bacterial diseases are easily spread through the tubers. These diseases are responsible for the decrease in vigor and productivity. Moreover, Indian farmers are using traditional methods for the cultivation of potatoes and face many difficulties (Fig. 1).

Therefore, to overcome the above limitation and increase the production of healthy potatoes using recently developed innovative technology for raising seed quality, lowering field exposures, and a strong certification and quality control system guaranteeing, the seed's quality supplied by independent seed producers is the only option to get over the aforementioned limitations. This will increase the quality, decrease field exposure, and generate planting material free of viruses. This chapter focuses on the most current changes in India's seed potato industry.

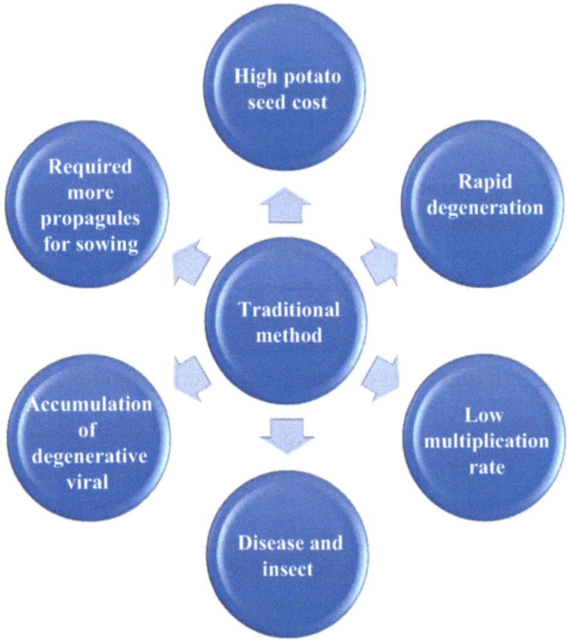

Fig. 1 Problem faced during traditional methods for the cultivation of potato

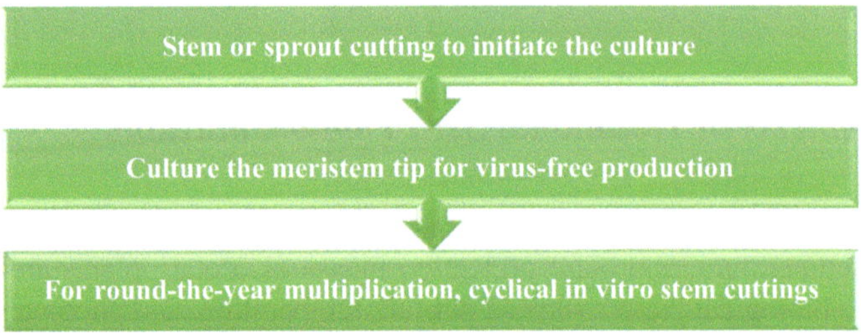

Fig. 2 Steps of the tissue culture-based potato seed production

3 System Based on Tissue Culture

Tissue culture is the science of growing cells, organs, or tissues on artificial media which is in liquid, solid, or semi-solid state. It is a promising technology that unravels issues related to potato seed production. In 1951, an attempt was made to create tissue culture from potato tubers for the first time (Steward and Caplin 1951). A variety of tissues from different plant parts, including leaves, internode segments, petioles, ovaries, stems, anthers, roots, and shoot tips, have been successfully demonstrated. The important steps of the tissue culture-based potato seed production system are displayed in Fig. 2.

However, the micro-plants produced from tissue culture-based systems are generally not directly used for commercial seed production but are used for either mini-tuber production under aeroponic systems or greenhouse or micro-tuber production in vitro. The mini-/micro-tubers obtained from tissue culture are multiplied for three to four succeeding generations before being used for commercial seed production.

3.1 Some of the Advantages of Tissue Culture Technology

- It helps to regenerate whole plants from any part of the plant such as the stem, roots, tuber, leaves, or from a single cell.
- A fast and large number of productions of potato plants make it possible to fill the gap of increasing demand for good quality potato seed.
- It produces identical, uniform, disease-free seed material in a short duration of time.
- Moreover, it is not limited by the time or weather. Healthy potato seeds can be grown all year round in the laboratory.
- It helps to produce haploid plants.
- It offers novel potential for breeding programs for potato.

4 Using Meristem Tip Culture to Eradicate Viruses

It is one of the significant tissue culture applications for the eradication of viruses (Badoni and Chauhan 2010). Using this method, a tiny growth point or meristem about 0.1–0.3 mm in size is removed from the potato plant's stem or tuber sprout and put in a test tube or other container with culture material to create stock that is free of viruses under controlled conditions. The production of virus-free material is based on the belief that the viruses are not capable of infecting the apical meristems, and thus a virus-free plant is produced from a small piece of meristems (Kassanis 2008). Apical meristem has several unique characteristics that make it possible to eradicate viruses such as the following:

- In the meristematic region, the vascular system for virus transmission is not developed.
- Elevated auxin content and chromosome multiplication at the time of mitosis in meristem may hinder virus multiplication.
- The presence of a virus inactivation system with higher activity is more in the apical region as compared to other regions (Naik and Karihaloo 2007).

However, combining chemotherapy and/or thermotherapy with the meristem culture (Fig. 3) approach increases the possibility of growing plants free of viruses (Sarkar et al. 2011). Only a small number of virus-free mericlones are obtained, even after all safety measures for viral eradication have been taken. Thus, each mericlone must undergo an enzyme-linked immunosorbent assay (ELISA) test for

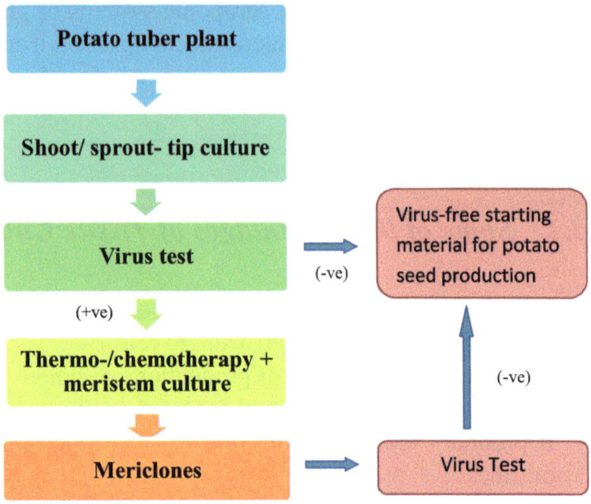

Fig. 3 Procedure of virus-free meristem tip culture

viruses, at intervals of 3 weeks. Thus, under controlled conditions, it can yield approximately 59,049 micro-plants in 7–8 months from a single culture tube (NAAS 2021).

5 Mini-tuber Production in the Soil

Mini-tubers are tiny seed potato tubers that are grown either from highly densely (Buckseth et al. 2020) or low-density planting (Struik and Lommen 1990) sown plantlets that were propagated in vitro in a glasshouse, aphid-proof net houses, in containers with diverse substrate mixtures (Jones 1988), in hydroponics system (Rolot and Seutin 1999), or through aeroponic system (Kang et al. 1996). This system links between in vitro and field multiplication. Two methods for producing mini-tubers include employing micro-plants as planting material or transplanting hardened virus-free micro-tubers inside a net house. The plants produced from plantlets or micro-tubers are subjected to destructive harvest or repeated harvest methods for the production of mini-tubers.

(a) Destructive harvest method: In this method, in vitro, plantlets are planted directly in greenhouses or screenhouses. The harvesting is done at the end of the crop cycle. To increase the tuber numbers, plantlets are sown at low density. This system produces a lesser number but larger size tubers per square meter (Sharma et al. 2013).
(b) Repeated harvest method: In this method, in vitro, plantlets are planted in a glasshouse under tuber-inducing environment and the tubers at regular intervals. About 3500 mini-tubers are produced per square meter with >5 mm size when 350 plants were harvested after 4, 7, and 10 weeks (Lommen and Struik 1992).

Before being given to farmers as seed, the mini-tubers so produced are usually increased in three consecutive generations. The primary benefits of micro-tubers over micro-plants are their increased hardiness, ease of handling and transportation, and less need for maintenance during and after planting. However, the creation of micro-tubers in the lab requires an extra 2–3 months of subsequent storage to break dormancy. Furthermore, using micro-plants directly to produce mini-tubers is preferable over using micro-tubers because micro-tubers give fewer yields than micro-plants.

6 Aeroponic System

In the potato sector, a more modern technology known as aeroponics has gained popularity during the last few decades. The purpose of this technology is to produce mini-tubers by using healthy in vitro plants (Pandey and Singh 2014; Buckseth et al. 2016). This technology makes it easier to produce and adopt phytosanitary standards all year round (Singh et al. 2016) by planting micro-plants on top of the chamber and using nutritional solution to fog the root zone that is hanging inside the chamber (Tierno et al. 2014; Buckseth et al. 2016). Standardizing specific growth conditions and nutrient solutions for different varieties in this system is crucial since different potato varieties have varying nutritional requirements and photoperiods (Millam and Sharma 2007; Otazu 2008). In one square meter, 771 mini tubers are possible to grow in soil-based nursery beds whereas, 2094 are in an aeroponic system (Fig. 4). Although the planning, operating costs, and uniformity of genotype-responsive nutrient solutions for this system are highly specialized, the resulting high-quality seeds have transformed the seed potato industry in India (Buckseth et al. 2022).

Fig. 4 Mini-tubers production through aeroponic system

7 Apical Rooted Cuttings (ARC) System

There are several methods for cutting and rooting the potato plants' rapidly growing young vegetative portion (Bryan et al. 1981). In the potato production belt, farmer producer organizations (FPOs), progressive farmers, and other small company owners can readily implement ARC technology. ARCs have several advantages over aeroponic technology, including lower capital costs and shorter production cycles because of their simplicity, effectiveness, ease of implementation, and small production cycle (Buckseth et al. 2019). On nursery beds (400 micro-plants m^2) in ARC, the robust buffer supply of micro-plants serves as the initial planting material (Fig. 5). After the mother stock has been planted for 15–20 days, the first cutting cycle begins. The growing micro plant apical section (1.5–2.0 cm) is clipped and then placed in pro-trays with soilless media such as cocopeat. After the mother plant's apical dominance is eliminated, lateral buds begin to form. These buds eventually develop into additional branches, which are then utilized to accelerate the rate at which potato plants multiply. Every 7–10 days, laterals from mother plants are cut sequentially and continue for more than 35–45 days, and it depends on the variety's fresh shoot growth. Just three to four cuttings are recommended in the north-west central plains of India to fit within the window for seed production and adhere to the seed plot approach, which involves growing potato seed crops during a period of aphid-free season. Rooted cuttings that are 7–10 days old are sown beneath an insect-proof net house to produce tubers.

8 Micropropagation

Many issues with the conventional seed production method can be resolved by using micropropagation (tissue culture) techniques in seed potato production (Singh et al. 2019). To create new micro-plants, virus-free potato mericlones single-node

Fig. 5 Micro-plants serving initial planting material

Fig. 6 Potato tissue culture plants

cuttings are cultivated in liquid or semisolid culture medium in an aseptic environment (Naik et al. 2000). Murashige and Skoog's (MS) medium with the addition of 30 g L^{-1} sucrose and 2.0 mg L^{-1} D calcium pantothenate for large scale in vitro micropropagation of potato (Buckseth et al. 2016). The cultures are maintained at 20 ± 20C with a 16-h photoperiod (60 μ mol m^{-2} s^{-1} PFD). In each 25 × 150 mm culture tube, three single-node cuttings with one or two leaves are typically placed. In 3 weeks, the axillary/apical buds of these cuttings develop into fully grown plants. These microscopic plants are further sub-cultured on fresh culture material up to 10 times at intervals of 3 weeks to get tissue culture plants (Fig. 6). Thus, under-regulated circumstances, it may yield around 59,049 micro-plants from a single culture tube in 7–8 months (NAAS 2021).

9 Production of Micro-tubers from Micro-plants

Micro-tubers are very small tubers that are created in vitro in circumstances that induce tuber formation (Naik and Karihaloo 2007). These micro-tubers are extremely easy to handle, store, distribute, and conserve germplasm (Gopal et al. 2004). From each tissue culture flask, 15–20 micro-tubers each with an average size of around 100–200 mg are typically produced. Micro-tubers are mostly employed in greenhouses to produce mini-tubers rather than to cultivate commercial crops (Naik and Sarkar 2000). Aphid-proof net homes with 50 micro-tubers each are placed over the nursery beds containing the micro-tubers. Farmers have not taken much interest in this technique because of the low crop emergence (50–60%) and low micro-tuber survival (60–65%) in nursery beds (Venkataslam et al. 2017).

10 Seed Plot Technique (SPT)

Viruses are transmitted by aphids, as a result, the potato seed started to grow in high hills which were nearly free from aphids throughout the potato cultivation period. Although the issue of rapid degeneration in plains is due to aphid population, the production of seed potatoes is fully dependent on the hills to satisfy the requirement of seed in our country. Therefore, the whole pressure was put on the hills for good quality potato seed production which offers a wide number of issues such as the following:

- Restricted areas encompass an elevation of 750 m or more which are appropriate for seed production.
- Seeds grown in hills cannot be appropriate in plains for growing in the month of October due to dormancy.
- Limited means of transport ensuing in delay in planting.

To overcome these issues, the SPT technique was identified in the 1970s to facilitate quality seed potato production under low aphid conditions in Indian plains. This system assisted by a biotechnological approach for micropropagation, virus elimination, and effective diagnosis of the virus has sustained the national potato seed production program by producing approximately 2600 tons of breeder's seed yearly. This seed is promoted to about 432,000 tons of certified seed. The transference of potato propagation from hills to plains due to the advancement of SPT in India permits the improvement of varieties suited for diverse agro-climatic regions. Also, the huge stretch of fertile land from Punjab to Assam enhanced the area by 12 folds (Pandey et al. 2018), and the SPT method gives 30–40% higher yields free from diseases (Singh and Rana 2014). It facilitates the accessibility of enough breeder seed for all over the country (Buckseth et al. 2023).

11 True Potato Seed

A botanical seed created by sexual reproduction is known as a true potato seed (TPS). Dr. S. Ramanujam came up with the concept of true potato seed as early as 1949. TPS can develop into a whole plant, just like all other botanical seeds, but because each plant differs genetically, the population of TPS is diverse (Gaur and Pandey 1993). The ICAR-CPRI worked to create homogeneous TPS populations to achieve disease resistance and also give high tuber yields. Compared to crops grown from seed tubers, TPS populations provide several benefits. Less rate of seed (about 150 g/ha as against 2.5–3.0 t/ha of seed tubers), ease of storage and transportation, and planting material free of pests and diseases are the main advantages. TPS populations were created by ICAR-CPRI in the 1980s, including HPS-1/13, TPS C-3, and 92-PT-27.

In India the Indian Seed Certification Standards provide well-defined quality control standards for TPS production. When it comes to the cropping system's dynamics for producing potato, TPS technology is still not considered a ready-to-use technology when compared economically to seed tubers (Tiwari et al. 2017).

The hybrid TPS has reportedly been listed as a certified class seed that can be planted for any use, except for seed production. Conversely, farmers typically use the tuberlets grown in nurseries from TPS as commercial seeds. The nature of late maturity, inconsistent germination, and non-uniformity are the barriers that have hindered the adaption and broad usage of TPS. However, after the introduction of diploid F1 hybrid breeding, interest in TPS technology has grown again.

12 Diploid Hybrid TPS

Potatoes are autotetraploid crops that are heterozygous; selfing causes a significant inbreeding depression. A self-compatibility gene can be incorporated into self once ploidy is reduced to a diploid state, which can lead to homozygosity much more quickly than at the tetraploid level. Lindhout et al. (2011) have proved the effectiveness of this strategy. Several governmental and corporate entities, such as ICAR-CPRI, are involved in the production of diploid inbred lines and the use of heterosis in diploid potatoes (ICAR-CPRI Annual Report, 2020). At present the high-tech/tissue culture technique enhances the multiplication rates along with disease-free potato plant production, while the traditional method has also its own benefits; so far both are important for the improvement of the potato crop, and a comparison between traditional and high-tech approach is presented in Table 1.

Table 1 Traditional methods are compared to a high-tech approach for producing seed potatoes

Parameters	Traditional method	High-tech approach
Multiplication type	Based on clonal selection	Based on tissue culture
Process	Multiplication of nucleus seed tubers	Multiplication of tissue culture material
Exposure to field	High exposure to the field	Less field exposure
Multiplication speed	Slow	Fast
Multiplication rate of seed	Low	High
Incidence of disease/stock degeneration	Increased generation-to-generation degradation of seed stocks as a result of the build-up of bacteria, fungi, viruses, and viroids	Stock degeneration is reduced as a result of reduced field exposure
Land requirement	Need more space for early seed multiplication and more acreage because they are bulkier	Require less space because early seed multiplication results in smaller, more homogenous planting material and less gap between seeds
Input cost	High cost of seed due to planting of whole large size tubers for good quality crop	The initial outlay is substantial, but it can be recouped in three to four crop seasons
Advantage	Currently, the conventional system accounts for 70% of seed production in India, meeting the nation's annual seed requirements	In a relatively short amount of time, clean seed tubers devoid of any illnesses were acquired

Source: Adopted from Buckseth et al. (2022)

13 Conclusion

Seed potato production is mostly based on clonal selection and its multiplication in succeeding stages in the open field which is called a traditional system. Potato is propagated vegetatively by progeny tubers in the field; hence, they are susceptible to infection and accumulation of viruses which hinder the production of good quality potato seed. Moreover, the rate of seed multiplication is low and takes more time duration to get breeder seed. To conquer these problems and for the production of good quality potato seed, innovative technology is most important. This technology helps in the production of huge quantities with half of the time and space, preferred seed size, and round-the-year production. Moreover, uniform seed tubers of relatively smaller size, disease-free, good quality, and with very low cost are additional benefits of innovative technology.

References

Agriculture Statistics at a Glance (2022) Ministry of Agriculture and Farmers Welfare, Government of India

Badoni A, Chauhan JS (2010) Conventional *vis-a-vis* biotechnological methods of propagation in potato: a review. Stem Cell 1(1):1–6

Bryan J, Jackson M, Melendez N (1981) Rapid multiplication techniques for potatoes. International Potato Center (CIP), Lima

Buckseth T, Sharma AK, Pandey KK, Singh BP, Muthuraj R (2016) Methods of pre-basic seed potato production with special reference to aeroponics—a review. Sci Hortic 204:79–87

Buckseth T, Vanishree G, Singh RK, Tiwari JK, Sharma AK, Patil VU, Bhardwaj V (2019) Aeroponic minituber production through TPS. CPRI Newsletter No. 76

Buckseth T, Singh RK, Tiwari JK, Sharma AK, Singh S, Chakrabarti SK (2020) A novel sustainable aeroponic system for healthy seed production in India-an update. Indian J Agric Sci 90(2):243–248

Buckseth T, Singh RK, Tiwari JK, Bhardwaj V (2022) Advances in innovative seed potato production systems in India. Front Agron. https://doi.org/10.3389/fagro.2022.956667

Buckseth T, Kumar V, Sharma AK, Choudhary AK, Singh B (2023) Chronological outlook on the advancement in seed potato production technologies. Indian J Agron 68(XXII Biennial National Symposium Special issue):S231–S240

Gaur PC, Pandey SK (1993) "True potato seed (TPS) technology," in Advances in horticulture, vol. 7. Eds. K. L. Chadha and J. S. Grewal (New Delhi, India: Malhotra Publishing House), 85–111

Gopal J, Chamail A, Sarkar D (2004) In vitro production of microtubers for conservation of potato germplasm: Effect of genotype, abscisic acid, and sucrose. In Vitro Cell. Dev. Biol. Plant 40:485–490. https://doi.org/10.1079/IVP2004540

ICAR-CPRI Annual Report (2020) Annual report. Indian Council of Agricultural Research-Central Potato Research Institute, Shimla, Himachal Pradesh, India. Available at: https://cpri.icar.gov.in//11e44298b-584d-4d5c89adc37862d3f9eb.pdf

Jones ED (1988) A current assessment of in vitro culture and other rapid multiplication methods in North America and Europe. Am Potato J 65:209–220

Kang JG, Kim SY, Kim HJ, Om YH, Kim JK (1996) Growth and tuberization of potato (*Solanum tuberosum* L.) cultivars in aeroponic, deep flow technique and nutrient film technique culture systems. J Korean Soc Hortic Sci 39:65–69

Kassanis B (2008) The use of tissue cultures to produce virus free clones from infected potato varieties. Int J Ann Appl Biol 45(3):422–427
Lindhout P, Meijer D, Schotte T, Hutten RCB, Visser RGF, VanEck HJ (2011) Towards F1 hybrid seed potato breeding. Potato Res 54:301–312
Lommen WJM, Struik PC (1992) Production of potato mini-tubers by repeated harvesting: effects of crop husbandry on yield parameters. Potato Res 35:419–432
Millam S, Sharma SK (2007) Soil-free techniques. In: Vreugdenhil D (ed) Potato biology and biotechnology advances and perspective. Elsevier, Amsterdam
Naik PS, Karihaloo JL (2007) Micropropagation for production of quality potato seed in Asia pacific. Asia-Pacific Consortium on Agricultural Biotechnology (APCoAB), New Delhi
Naik PS, Sarkar D (2000) "In vitro approaches to propagation and conservation of genetic resources in potato," in Biotechnology in horticultural and plantation crops. Eds. K. L. Chadha, P. N. Ravindran and L. Sahijram (New Delhi, India: Malhotra Publishing House), 369–406
Naik PS, Chakrabarti SK, Sarkar D, Birman RK (2000) Potato biotechnology: Indian perspective. In: Khurana SMP, Shekhawat GS, Singh BP, Pandey SK (eds) Proceedings of potato: global research and development. Indian Potato Association, New Delhi, pp 194–211
National Academy of Agricultural Sciences (2021) Innovations in potato seed production. Strategy paper no. 14. National Academy of Agricultural Sciences, New Delhi
Otazu V (2008) Quality seed potato production using aeroponics. A potato production manual. CIP-International Potato Center, Lima
Pandey KK, Singh BP (2014) Effect of bower system on mini-tuber yield of different varieties in aeroponic system. In: Abstract in national seminar on emerging problems of potato. Indian Potato Association), Shimla, pp 193–194
Pandey NK, Singh DK, Singh BP, Jeevalatha A, Raigond B, Singh B (2018) Advances in quality potato production and post-harvest management. Agrotech Publishing Academy, Udaipur, pp 20–21
Rolot JL, Seutin H (1999) Soilless production of potato mini-tubers using hydroponic technique. Potato Res 42:457–469
Sarkar D, Pandey SK, Sharma S, Chandel P (2011) Potato. In: Singh HP, Parthasarathy VA, Nirmal BK (eds) Advance of horticulture biotechnology-regeneration systems-vegetables, ornamentals and tuber crops (Volume II). Westville Publishing House, New Delhi, pp 319–345
Shaheb MRS, Begum MM, Ahmed KU, Nazrul MI, Wiersema SG (2016) Challenges of seed potato (Solanum tuberosum l.) production and supply system in Bangladesh–a review. Agriculturists 13(1):173–188
Sharma AK, Venkatasalam EP, Kumar V (2013) Potato mini-tuber production during main and off crop seasons in high hills of north-western Himalaya. Potato J 40(1):29–37
Singh IP (2002) Seed production of vegetable crops. Aman Publication House, Meerut
Singh BP and Rana RK (2014) History of potato and its emerging problems in India. In: Souvenir-national seminar on emerging problems of potato, pp 1–2
Singh BP, Sharma S (2018) Potato seed production systems: then and now. Potato J 45(1):1–16
Singh RK, Dubey SK, Rakesh K, Singh SN, Buckseth T, Singh AK (2016) Farmer participatory seed potato production through krishi vigyan kendra networking: an action research for enhanced availability of seed potatoes in India. Potato J 43(2):193–199
Singh AK, Jankiram T, Chakrabarti SK, Vinay B, Tiwari JK (2018) Indian potato varieties. ICAR-Central Potato Research Institute, Shimla, p 178p
Singh RK, Buckseth T, Tiwari JK, Sharma AK, Chakrabarti SK (2019) Recent advances in production of healthy planting material for disease management in potato. Biotech Today Int J @ Bio Sci 9(1):7–15. https://doi.org/10.5958/2322-0996.2019.00001.2
Steward FC, Caplin SM (1951) Tissue culture from potato tuber: the synergistic action of 2,4-d and of coconut milk. Science 111:518–520. https://doi.org/10.1126/science.113.2940.518
Struik PC and Lommen WJM (1990) Production, storageand use of micro and mini-tubers. In: Proc. 11th TriConf of the Euro Assoc Potato Res (EAPR), Edinburg, pp 122–33

Tierno R, Carrasco A, Ritter E, de Galarreta JIR (2014) Differential growth response and minituber production of three potato cultivars under aeroponics and greenhouse bed culture. Am J Potato Res 91:346–353

Tiwari JK, Luthra SK, Vinod K, Bhardwaj V, Singh RK, Sridhar J et al (2017) Genomics in true potato seed technology: engineering cloning through seed. In: Ckakrabrti SK, Conghua X, Tiwari JK (eds) The potato genome. Springer Cham, Switzerland, pp 297–305

Venkataslam EP, Singh RK, Sharma AK, Singh V, Buckseth T, Chakrabarti SK (2017) Seed potato production techniques: principles and applications. ICAR-Central Potato Research Institute, Shimla, p 101p

Young N (1990) Seed potato systems in developed countries: Canada, The Netherlands and Great Britain. International Potato Centre, Lima, p 128

Advances in Potato Mini-tubers Production through Aeroponics and ARC Techniques

Anuradha Sinha, Shashank Shekhar Solankey, Shirin Akhtar, and Pushpam Patel

Abstract Potato is grown by many farmers, especially by small and marginal farmers, because of its high yield potential. But due to the growing demand of the rising population and changing lifestyle, we have to produce more potatoes which is not possible by conventional cultivation of potatoes, and one of the major bottlenecks in potato production is the unavailability of high-quality and certified seed potatoes because potato is propagated through tuber which is highly susceptible to infestation of several diseases and the chance of infestation become more when tuber comes in contact to soil. Thus, there is a need or time to revolutionize or advance potato seed production which is possible using innovative technology, i.e., aeroponic and apical rooted cutting (ARC) techniques. These technologies help to produce high-quality seed, which is equivalent to certified seed. Additionally, these systems save time, land, fertilizer, and water in an environment-friendly manner and give higher and uniform yields as compared to the conventional method.

Keywords Aeroponics · Apical rooted cutting (ARC) · Conventional · High-quality seed potatoes

A. Sinha (✉)
Krishi Vigyan Kendra, Aurangabad, Bihar Agricultural University, Sabour, Bhagalpur, Bihar, India

S. S. Solankey
Department of Horticulture (Vegetable Science), Nalanda College of Horticulture, Noorsarai, Nalanda, Bihar Agricultural University, Sabour, Bhagalpur, Bihar, India

S. Akhtar
Department of Horticulture (Vegetable Science), Bihar Agricultural College, Bihar Agricultural University, Sabour, Bhagalpur, Bihar, India

P. Patel
Krishi Vigyan Kendra, Barh, Patna, Bihar Agricultural University, Sabour, Bhagalpur, Bihar, India

S. S. Solankey (ed.), *Advances in Research on Potato Production*, Advances in Olericulture, https://doi.org/10.1007/978-3-031-82710-5_7

1 Introductions

Due to its high yield per unit input and nutritive value potato is recognized as "Food for the Future" by Food and Agriculture Organization. It has already held third rank for the most important food crop in the world. India grasps second ranks globally with productivity of 23.77 t/ha from an area of 2.16 million ha (FAO 2022). However, it is predicted that our country needs to produce 122 million MT of potatoes by 2050, to meet the demand of the rising population (Anonymous 2019). Potatoes become more important crops where the cultivable lands are limited. That's why marginal farmers have taken up potato cultivation. However, the possible yield of potatoes is limited by the lack of suitable cultivars for particular agro-climatic regions, climate change, unavailability of healthy quality seed tuber, higher cost of seed tuber, and incidence of diseases and insect pests. Among the various factors, a major hitch that contributes to lower yield in India is the shortage of quality seed tubers (Naik and Karihaloo 2007; Singh and Sharma 2018). Moreover, India follows the informal tuber seed production systems in potatoes, in which in most cases farmers recycle their planting materials from the preceding crop harvest. Currently, approximately 97% of demands for seeds are met by informal seed production systems (Wasilewska-Nascimento et al. 2020).

Potato is mostly propagated through tubers, which are called seed potatoes. However, seed potatoes are prone to buildup and further spread of diseases in succeeding seasons. This leads to a significant loss in yield and quality of tuber (Tsoka et al. 2012). Therefore, the production and multiplication of healthy seed potatoes play a very decisive role in a sustainable potato production system because the healthy planting material is key to increasing the productivity of potato crops. The innovation in the in vitro multiplication method has reduced the time required to produce large quantities of pre-basic seed potatoes. The certified seed that is used for commercial production by potato growers is derived from pre-basic seed potatoes or mini tubers (Donnelly et al. 2003; Mbiyu et al. 2012). The production of a huge amount of healthy mini tubers, in a short period at low costs and an ecologically friendly method is very important for improving yield (Ritter et al. 2001) and lowering the cost of production of seed potatoes (Mateus-Rodriguez et al. 2013).

2 Aspect of Seed Potatoes Production in India

Many potato growers recycle their local seed potatoes and those who don't do; they change their seed potatoes after 3–8 seasons (Gildemacher et al. 2009). The continuing uses of tubers from field production contribute to the gathering and rapid multiplication of pathogens such as bacteria, fungi, viruses, and nematodes, leading to their deterioration. The deterioration negatively influences the quality of the tuber and leads to a yield diminution in successive cycles of vegetative replication (Thomas-Sharma et al. 2016). The Government of India is dedicated to providing certified seeds to the farmers growing different crops. However, for high-volume

Table 1 Difference between convention and hi-tech production system

Parameters	Conventional	Hi-tech
Uniformity	Less uniform	More uniform
Field exposure	High	Low
Rate of multiplication	Slow	Fast
Disease infestation	High	Low
Land requirement	High	Low

crops such as potatoes, the supply of high-quality seed potatoes faces a unique challenge as they are prone to several pathogens during their multiplication process and their seed multiplication ratio is very low that is 1:6 (Venkataslam et al. 2017; Naik and Buckseth 2018). Besides its low multiplication rate, the conventional system may suffer from many other limitations like

(i) Requirement of a high volume of healthy planting material in the initial stage
(ii) The slower process of producing and multiplying healthy seed stock
(iii) The buildup of viral diseases on tubers in each field exposure
(iv) No system for monitoring the quality of the seed
(v) Very fast degeneration of tuber seed
(vi) The necessity of replacement of the seed every year

The seed-related problems are further provoked because of the limited availability of aphid-free regions in India. It is observed that the good-quality seed can increase productivity by 15–20% (Shaheb et al. 2016). However, there is a huge gap between the demand and supply of good quality potato seed in India (National Academy of Agricultural Sciences 2021). It is mostly impracticable to fill the gap between demand and supply using traditional methods. The conventional method of seed potato production in India consists of generating a nucleus seed followed by breeders' seed, foundation seed I, foundation seed II, and certified seed (Venkataslam et al. 2017; Singh et al. 2019) which takes more time. In addition to this, it is a labor-intensive method that favors disease accumulation (Badoni and Chauhan 2010; El-Komy et al. 2010), and the seed potatoes produced are of inferior quality (Table 1). Such practices contribute to lower yields with the spread of diseases (Gildemacher 2012; Chindi and Kassa 2013; Abdurahman et al. 2017). Hence, it is vital to advance a seed potato production system, through innovative technology such as aeroponic and apical rooted cutting. These techniques also help to achieve healthy planting material in a short duration of time.

3 Aeroponic Production of Mini Tubers

Aeroponics is an innovative soil-less method and it works well in greenhouse conditions (Ritter et al. 2001). In this system, the foliage part of the potato plant develops like earthy crops whereas roots are suspended in hydro-atomized moisture or

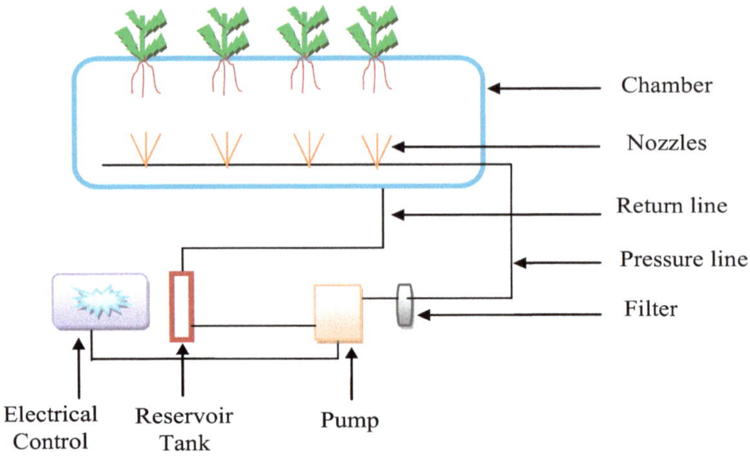

Fig. 1 Illustrative presentation of the aeroponic system

nutrient-rich environments in the dark chamber (Fig. 1). This system is different from the hydroponic system as the plant roots are suspended inside the growth chamber in which nutrients are supplied in fogged or misted form (Tierno et al. 2014) and also have conserved 70% water; however, in hydroponics system, crops are grown in an inert medium or nutrient solutions to provide mechanical support. If we compare it with soil-cultivated plants, these systems offer enhanced growth, yield, and quality metabolite production (Thakur et al. 2019; Partap et al. 2020; Kumar et al. 2023). In an aeroponics system, the plants access 100% carbon dioxide (450–780 ppm) for photosynthesis (Kumar et al. 2023), providing precision nutrient dosage, light synchronization, spray time, and pH levels. In addition to this, it delivers fresh with zero pesticides, the highest quality flavors, and maximum growth of the plant. Aeroponics is an advanced technology that is used for potato mini tubers production by utilizing in vitro plants (Singh et al. 2012). The size of mini tubers varies from 5 to 25 mm; however, larger size mini tubers are also common (Hassanpanah et al. 2009). In seed programs, it can be primarily used for virus-free production of seed potatoes (Ritter et al. 2001). This system is extensively accepted by many seed potato manufacturing organizations internationally (Otazu 2008).

3.1 Components of Aeroponics System

(a) Greenhouse: Greenhouse provides vertical space along with a favorable microclimate from the growth to the harvesting of healthy seed potatoes. That means the choosing of cladding material for the greenhouse is important to maintain the internal environment.

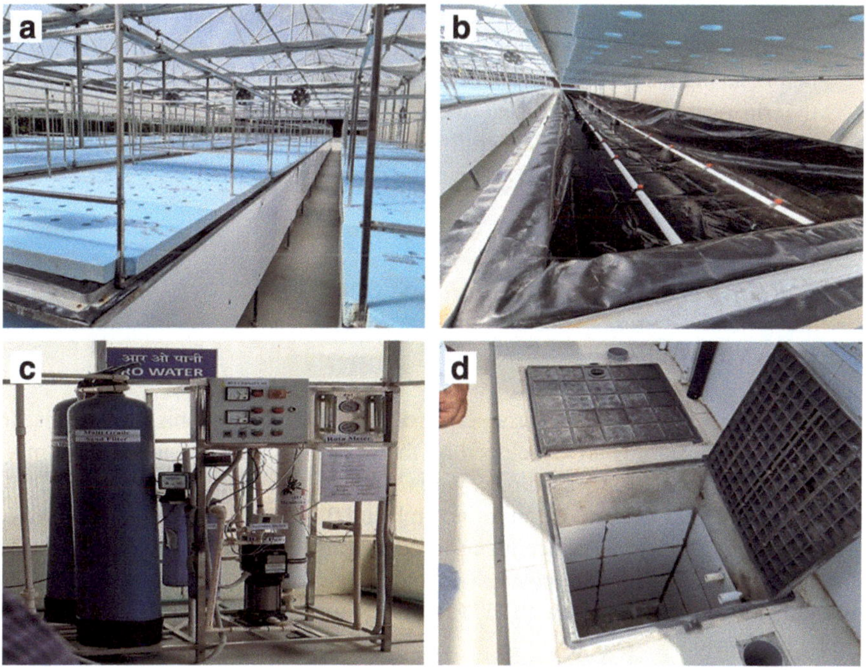

Fig. 2 Component of aeroponic system: (**a**) growth chamber, (**b**) spray nozzles/nebulizers, (**c**) high-pressure pump, (**d**) reservoir tank

(b) Growth chamber: It consists of a Styrofoam box with generally 6 m × 1.8 m × 1.5 m dimensions (Fig. 2a). The sides of the boxes were sliced to effortlessly harvest mini tubers.
(c) Spray misters: Misters are used to create very fine spray for a uniform supply of water and nutrient solution to plant rhizosphere (Fig. 2b). Misters require frequent maintenance as they clog from nutrients.
(d) Droplet size: The size of the droplet varies from 20 to 100 μm. The smaller droplets remain in the air as a fog inside the growth chamber and the larger droplets make contact with the roots.
(e) High-pressure pump: The pump can produce enough pressure to produce the ideal water droplet (Fig. 2c). These pumps are generally reverse osmosis, which creates a stable 80 PSI with necessary nutrient flow. Pumping water at high pressure through nozzles helps to achieve atomization. Nozzles are available in diverse spray patterns and orifices. Larger sprayer nozzles and orifices decrease the possibility of blockage but require pressure to function.
(f) Light and Temperature: Light and temperature are very essential for the growth and development of crops. In an aeroponic system, it can be maintained by fluorescent tubes. About 15,000–20,000 lux are required for vegetative growth whereas 35,000–40,000 lux are required for flowering and fruiting. The optimum temperature is 15–25 °C.

(g) Reservoir Tank: Reservoir tanks are required to hold water and nutrient solution for misting the root zone intermittently (Fig. 2d). The major benefit of intermittent is saving the running cost because the pump is operated for a short time, but the rhizosphere is still misted within the moisture, nutrient solution, and oxygen-rich environment. Generally, a 1–2 min of misting afterward 5 min off will ensure the moist conditions near the rhizosphere.

3.2 Requirements of Aeroponics System

3.2.1 In Vitro Culture

In vitro culture is used for the production of virus-free microplants of potatoes. In this system, the sprouts are developed on the basal medium containing a salt mixture (Murashige and Skoog 1962), vitamins (Linsmaier and Skoog 1965), 3% sucrose, 0.7% agar, 100 mg L^{-1} myo-inositol, and 0.5 mg L^{-1} 6-benzylaminopurine. The developing shoots are relocating on basal medium without 6-benzylaminopurine for the development of microplants. The microplants are subcultured every 30 days in a climate-controlled room with 21 °C temperature, light flux 90 μmol m^{-2} s^{-1}, and 16 h photoperiod.

3.2.2 Preparation of Microplants for Cultivation

Before transplanting, the microplants are cultivated in a mixture of sand and perlite in a ratio of 4:1 for acclimatization in the greenhouse for 20 days. During acclimatization, the plants are regularly irrigated with a nutrient solution.

3.2.3 Growing System

In the aeroponic system, the growing boxes are installed inside the greenhouse in which the microplants with 5 cm plant height, 2.5 cm root length, and 2–3 leaves are held firmly into the holes in an upright position and their roots are hung inside the box (Tierno et al. 2014). The lower leaf was cut off at regular an interval, which permits the plants to be pressed into the growing box to encourage stolon development. In these boxes, the root can get a favorable growing environment regarding oxygen and moisture. These environments permit excellent nutrition assimilation resulting in faster and healthier development of seed potato. The nutrient solution is changed every 15 days. The nutrient solution is applied for 20 s for every 5 min from 8:00 a.m. to 6:00 p.m. and 20 s for every 10 min from 6:00 p.m. to 8:00 a.m. (Broćić et al. 2022). This system allows a nonstop and round-the-year production cycle, water savings, and reduces labor. The top of the growing boxes is placed with a pivot and can unlock to pick the mini-tubers at regular intervals of time. The picking of mini tubers starts after 45–50 days of transplanting or when they reach a volume of 3–10 g. In 4–5 months around 10–12 pickings are taken during the crop cycle (Tiwari et al. 2019). Harvesting promotes tuber initiation and allows a size control of mini tubers (Ritter et al. 2001). The timing and number of pickings are essential aspects of mini tuber production. Depending upon the cultivar, 40–50 mini-tubers can be produced from an in vitro plant (Buckseth et al. 2020). The mini-tubers that are formed in the aeroponics system are generally planted in open fields to advance their generations (Fig. 3). The requirements of nutrient and photoperiods are different for different potato cultivars. Thus, it is essential to normalize the cultivar-specific, growing conditions, planning, and nutrient solutions (Otazu 2008), but it offers high-quality seed potatoes (Buckseth et al. 2022).

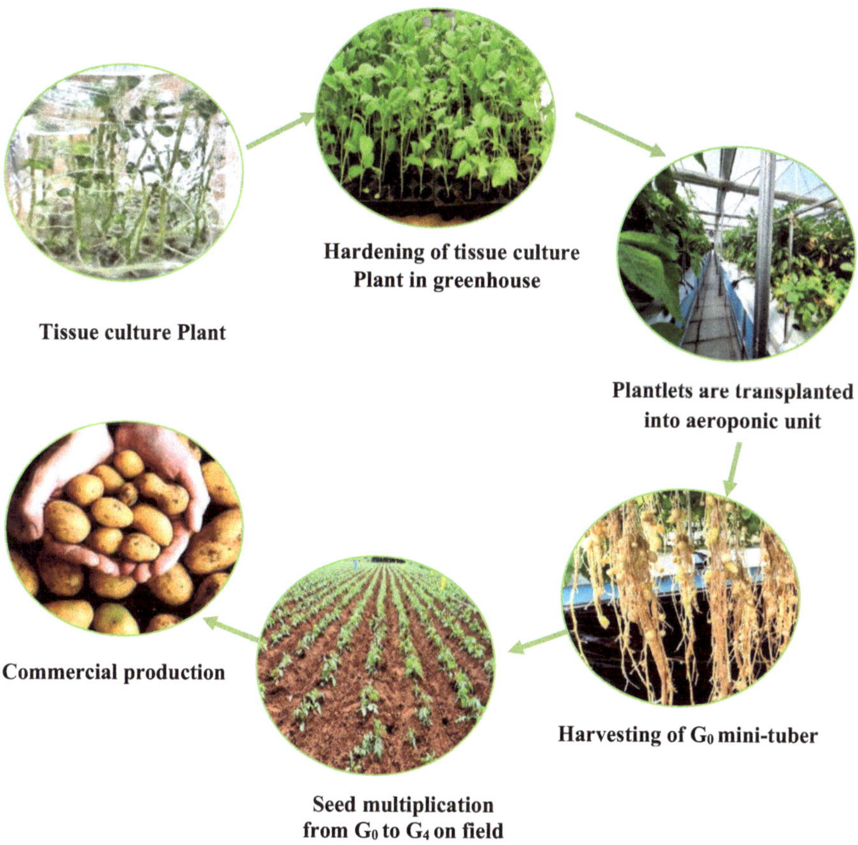

Fig. 3 Representation of pre-basic seed potato production using aeroponics system

3.3 Importance of Aeroponics

There are many advantages of the aeroponics system as compared to the soil-based system for mini-tuber production.

- The Aeroponics method gives higher mini-tubers as compared to conventional soil production. A single plant can give more than 100 mini-tubers (Otazu 2008) while conventional methods produce only 5–10 tubers per plant (Mbiyu et al. 2012). Also, the multiplication rates are more in aeroponic systems i.e. 1:45 as compared to conventional production, i.e., 1:6 (Mateus-Rodriguez et al. 2013).
- Multiple picking is possible in an aeroponic system, which offers an elevated number of mini tubers by 100–250% and yield compared to soil production (Oraby et al. 2015).
- Nutrient composition and pH are easily monitored which allows precise supplies of plant nutrients. Additionally, the closed-loop systems of aeroponic recycle the nutrient solution (Gopinath et al. 2017; Tessema et al. 2017), thus increasing nutrient use efficiency and increasing plant growth (Zebarth et al. 2006).
- Decreased the use of fertilizers (up to 60%) and plant protection chemicals (up to 100%), in the aeroponic system increases fertilizer use efficiency, reduces fertilizer residues, and protects the environment (Lakhiar et al. 2018).
- A separate ventilation structure is not required (Gwynn-Jones et al. 2018; Wang et al. 2018) as the roots are suspended in air, therefore there is not an issue of inadequate oxygen.
- Produce disease-free, true-to-type, and uniform-sized seed potatoes.
- An aeroponic system has high water use efficiency as it saves 95% more water than conventional farming.
- Aeroponics saves 98% more land than conventional methods by utilizing vertical as well as horizontal space.
- Growing potato plants in controlled conditions facilitates farmers to predict crop growing stage, harvesting time, and quality of plants.
- A closed environment eradicates probable contamination of pathogens, therefore there is no need to apply herbicides or pesticides, resulting in a more organic product.

3.4 Drawbacks of Aeroponics

- Slight failure of equipment can cause the loss of a crop.
- The misters required regular cleaning to keep them from clogging by the nutrient solution.
- Aeroponic systems rely on power to pump water and nutrients.
- Require high initial investment and technical knowledge for operations.
- It is labor-intensive due to manual harvesting.

4 Apical Rooted Cuttings (ARCs) Technology

Apical rooted cuttings are a substitute for mini tubers in contemporary seed potatoes. In this technology, the rooted cutting produced from tissue culture plantlets is transplanted in the screen house for the production of plantlets rather than the production of mini tubers. Further, the rooted cuttings from the screen house are transplanted into the main field for the production of seed tubers.

4.1 Principle of ARCs Technology

After cutting of apical portion, the concentration of IAA is lowered which permits to growth of lateral buds which further triggers the growth of new branches which are utilized to amplify the multiplication rate.

4.2 Working of ARC Technology

In this technology, the fast-growing vegetative part of tissue-cultured microplants can be slashed with three nodes and planted deep up to two nodes for rooting to set up a mother bed (Bryan et al. 1981) or nursery bed at the spacing of 5–15 cm × 5–15 cm filled with cocopeat and vermicompost mixture rather than allowing to produce mini tubers. The best time to cut the other plant is morning. From the mother bed, the apical or top portion of the 1.5–2.0 cm length of the micro plants is then cut at a 45-degree angle and replanted in portrays containing soilless media in the screen house to harden the cutting to be ready for transplanting (Fig. 4). These 15- to 20-days-old rooted cuttings with 2–3 true leaves are transplanted in open fields for

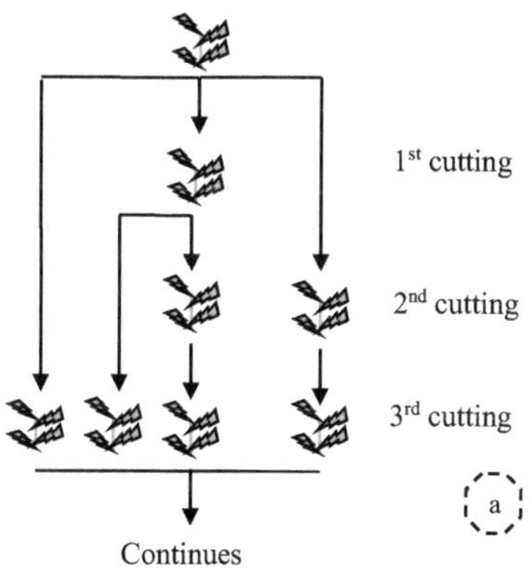

Days	Cumulative cutting
1	1
15	2
30	4
45	8
60	16
75	32
90	64
105	128
120	256

b

Fig. 4 Diagrammatic representation of production of cutting (**a**) work, (**b**) potential

seed production. Once the apical dominance is removed, the lateral buds are allowed to grow into new branches, which are used for chronological cuttings of lateral branches at an interval of 7–10 days from mother plants for multiplication, and it prolonged based on the growth of new branches. Thus, around 45-days-old micro-plants can generate 6–8 cuttings (Buckseth et al. 2022). Batch-wise cuttings are maintained to protect the cutting from the chance of infection.

4.3 Plant Nutrition

The cuttings are simultaneously irrigated with a nutrition solution after transplanting. Also, too much application of nitrogen delays the rooting of plants. Table 2 represents a basic nutrient composition for cuttings (Parker 2019).

4.4 Major Step for Apical Rooted Cuttings Production (Fig. 5)

1. Production of disease-free tissue culture plant
2. In vitro multiplication of tissue culture plant
3. Multiplication of mother plants in protected conditions
4. Production of rooted cuttings from mother plants in protected conditions
5. Transplanting of cutting in the field for one or two seasons for seed production
6. Commercial production of potato

4.5 Advantage

ARC technology is simple, cost-effective, and easy to execute. This technology has the potential to renovate potato seed production systems. Moreover, there is less than 1 or negligible incidence of viral diseases and off type plants. The fast and elevated rate of multiplication will extensively contribute to reducing the shortages of seed potatoes. Every cutting creates 7–10 tubers, and sometimes 15+, which are further replicated for one or two seasons and then the produce is sold as seed. Their

Table 2 Nutrient composition for the production of rooted cuttings

Sl. No.	Soluble fertilizer	500 liter tank (gram)
1	Calcium nitrate	128
2	Magnesium sulfate	50
3	Microsol B	6
4	Potassium nitrate	252
5	Potassium phosphate	68

Fig. 5 Pictorial depiction of procedure of apical rooted cutting production: (**a**) healthy tissue culture plant, (**b**) mother plants in protected conditions, (**c**) apical cutting, (**d**) planting of apical cutting in pro trays, (**e**) apical cutting ready for transplanting in the field

Table 3 The difference between ARC technology and conventional seed production system

Sl. No.	Particulars	ARC technology	Conventional seed production system
1	Production cost	Low	High
2	Seed system	Decentralized	Centralized
3	Risk	Low	High
4	Gestation period	Only two seasons are required	Six seasons are required
5	Technology	Simple	Complex
6	Scope for expansion	High	Low
7	Suitability	The small landholder can also become a seed producer	For only larger landholders

produce is equivalent to certified seed in the seed certification system. Also, this technology has several advantages over conventional seed production systems, which are presented in Table 3. This implies that the seed is of very high quality and will give a high per-unit yield.

4.6 Environmental Factor Affecting ARC Production

Environmental factors are very important for the maintenance of juvenile mother plants nursery beds. The photoperiod range of 11–13 h is very conducive for tuberization. Lengthening the day length to 16 h enhanced the production of apical cutting (VanderZaag and Escobar 1990). A light induction of 1–2 h extended day length which is helpful in apical cutting production. Warm air temperatures promote rapid plantlet growth and regrowth after apical cuttings harvesting. Ideal minimum air temperatures are above 16 °C at 800–1500 (Uyen and VanderZaag 1983). At lower altitudes, productivity may be reduced due to heat stress (Minh et al. 1990).

5 Conclusion

Even though the potato is not just a staple food, but also a cash crop, their yield is still limited, mostly due to scarcity of quality seed potatoes. To satisfy the rising demand, seed potatoes have still grown in soil for many generations. This enhances the threat of pathogen buildup and spread that trigger the seed deterioration. A realistic solution to this trouble is to use hi-tech techniques especially aeroponic and apical rooted cutting that allow the sustainable and proficient agro-practice to decline the number of generations in the field, which is essential to minimize the risk of soil-borne pathogens. These technologies may also work well in limited areas as well as in adverse ecological conditions, such as water scarcity or infertile soil.

References

Abdurahman A, Griffin D, Elphinstone J, Struik PC, Schulz S, Schulte-Geldermann E, Sharma K (2017) Molecular characterization of *Ralstonia solanacearum* strains from Ethiopia and tracing potential source of bacterial wilt disease outbreak in seed potatoes. Plant Pathol 6:826–834

Anonymous (2019) Advancement in potato production technology and its future prospects. ICAR-Central Potato Research Institute Regional Station, Modipuram-250110, Meerut

Badoni A, Chauhan JS (2010) Conventional vis-a vis biotechnological methods of propagation in potato: a review. Stem Cell 1:1–6

Broćić Z, Oljača J, Pantelić D, Rudić J, Momčilović I (2022) Potato aeroponics: effects of cultivar and plant origin on minituber production. Horticulturae 8:915. https://doi.org/10.3390/horticulturae8100915

Bryan J, Jackson M, Meléndez N (1981) Rapid multiplication techniques for potatoes. International Potato Center (CIP), Lima

Buckseth T, Singh RK, Tiwari JK, Sharma AK, Singh S, Chakrabarti SK (2020) A novel sustainable aeroponic system for healthy seed production in India-an update. Indian J Agric Sci 90(2):243–248

Buckseth T, Singh RK, Tiwari JK, Sharma S, Gautam A, Sharma AK et al (2022) Influence of microplant hardening methods on aeroponic potato minituber production. Potato Res 65:335–348. https://doi.org/10.1007/s11540-021-09530-z

Chindi A, Kassa B (2013) Seed tuber cycle and latent infection for the spread of potato bacterial Wilt *Ralstonia solanacearum* (Smith) a threat for seed production in Ethiopia. Asian J plant Pthol 7(2):74–83

Donnelly D, Coleman WK, Coleman SE (2003) Potato microtuber production and performance: a review. Am J Potato Res 80:103–115

El-Komy MH, Abou-Taleb EM, Aboshosha SM, El-Sherif EM (2010) Different expression of potato pathogenesis-related to proteins upon infection with late blight pathogene: a case study expression of potato osmotin-like protein. Int J Agric Biol 12(2):179–186

Food and Agriculture Organization (FAO) (2022) FAOSTAT database. Available at: http://www.fao.org/faostat/en/#data/QC

Gildemacher P (2012) Innovations in seed systems in East Africa. PhD thesis, Wageningen, The Netherlands

Gildemacher PR, Demo P, Barker I, Kaguongo W, Woldegiorgis G, Wagoire WW, Wakahiu M, Leeuwis C, Struik PC (2009) A description of seed potato systems in Kenya, Uganda and Ethiopia. Am J Potato Res 86:373–382

Gopinath P, Vethamoni PI, Gomathi M (2017) Aeroponics soilless cultivation system for vegetable crops. Chem Sci Rev Lett 6:838–849

Gwynn-Jones D, Dunne H, Donnison I, Robson P, Sanfratello GM, Schlarb-Ridley B, Hughes K, Convey P (2018) Can the optimization of pop-up agriculture in remote communities help feed the world? Glob Food Secur 18:35–43

Hassanpanah D, Hosienzadeh AA, Allahyari N (2009) Evaluation of planting date effects on yield and yield components of Savalan and Agria cultivars in Ardabil region. J Food Agric Environ 7(3 and 4):525–528

Kumar R, Rathore S, Kundlas K, Rattan S, Warghat AR (2023) Vertical farming and organic farming integration: a review. In: Organic farming (second edition), global perspectives and methods. Woodhead Publishing Series in Food Science, Technology and Nutrition, pp 291–315

Lakhiar IA, Gao J, Syed TN, Chandio FA, Buttar NA (2018) Modern plant cultivation technologies in agriculture under controlled environment: a review on aeroponics. J Plant Interact 13:338–352

Linsmaier EM, Skoog F (1965) Organic growth factor requirements of tobacco tissue cultures. Physiol Plant 18:100–127

Mateus-Rodriguez JR, de Haan S, Andrade-Piedra JL, Maldonado L, Hareau G, Barker I, Chuquillanqui C, Otazu V, Frisancho R, Bastos C et al (2013) Technical and economic analysis of aeroponic and other systems for potato mini-tuber production in Latin America. Am J Potato Res 90:357–368

Mbiyu MW, Muthoni J, Kabira J, Elmar G, Muchira C, Pwaipwai P, Ngaruiya J, Otieno S, Onditi J (2012) Use of aeroponics technique for potato (*Solanum tuberosum*) minitubers production in Kenya. J Hortic For 4:172–177

Minh TV, Uyen NV, VanderZaag P (1990) Rapid multiplication of potatoes: influence of environment and management on growth of juvenile apical cuttings. Am Potato J. 67:789–797

Murashige T, Skoog F (1962) A revised medium for rapid growth and bio assays with tobacco tissue cultures. Physiol Plant 15:473–497

Naik PS, Buckseth T (2018) Recent advances in virus elimination and tissue culture for quality potato seed production. In: Gosal SS (ed) Biotechnologies of crop improvement, vol 1. Springer International Publishing, Cham, pp 131–158

Naik PS, Karihaloo JL (2007) Micropropagation for production of quality potato seed in Asia-Pacific. Asia-Pacific Consortium on Agricultural Biotechnology, New Delhi, p 54

National Academy of Agricultural Sciences (2021) Innovations in potato seed production. Strategy, New Delhi, p 14

Oraby H, Lachance A, Desjardins Y (2015) A low nutrient solution temperature and the application of stress treatments increase potato mini-tubers production in an aeroponic system. Am J Potato Res 92:387–397

Otazu V (2008) Manual on quality seed potato production using aeroponics. Lima, International Potato Center (CIP)

Parker M (2019) Production of apical cuttings of potato. International Potato Center, Lima, p 9p
Partap M, Kumar P, Kumar A, Joshi R, Kumar D, Warghat AR (2020) Effect of elicitors on morpho-physiological performance and metabolites enrichment in *Valeriana jatamansi* cultivated under aeroponic conditions. Front Plant Sci 11:01263. https://doi.org/10.3389/fpls.2020.01263
Ritter E, Angulo B, Riga P, Herran C, Relloso J, San Jose M (2001) Comparison of hydroponic and aeroponic cultivation systems for the production of potato minitubers. Potato Res 44:127–135
Shaheb MRS, Begum MM, Ahmed KU, Nazrul MI, Wiersema SG (2016) Challenges of seed potato (*Solanum tuberosum* l.) production and supply system in Bangladesh–a review. Agriculturists 13(1):173–188
Singh BP, Sharma S (2018) Potato seed production systems: then and now. Potato J 45:1–16
Singh V, Pandey KK, Singh S, Dhruv K, Singh BP (2012) Aeroponics: an innovative method of potato seed production. In: Abstract in national consultation on potato research and development: way forward. OAU&T, Bhubaneswar, pp 45–49
Singh RK, Buckseth T, Tiwari JK, Sharma AK, Chakrabarti SK (2019) Recent advances in production of healthy planting material for disease management in potato. Biotech Today Int J Biol Sci 9(1):7–15. https://doi.org/10.5958/2322-0996.2019.00001.2
Tessema L, Chindi A, W/Giorgis G, Solomon A, Shunka E, Seid E (2017) Determination of nutrient solutions for potato (*Solanum tuberosum* L.) seed production under aeroponics production system. Open Agric 2:155–159
Thakur K, Partap M, Kumar D, Warghat AR (2019) Enhancement of picrosides content in *Picrorhiza kurroa* Royle ex Benth. mediated through nutrient feeding approach under aeroponic and hydroponic system. Ind Crop Prod 133:160–167. https://doi.org/10.1016/j.indcrop.2019.03.021
Thomas-Sharma S, Abdurahman A, Ali S, Andrade-Piedra L, Bao S, Charkowski AO, Crook D, Kadian M, Kromann P, Struik PC et al (2016) Seed degeneration in potato: the need for an integrated seed health strategy to mitigate the problem in developing countries. Plant Pathol 65:3–16
Tierno R, Carrasco A, Ritter E, Ignacio J, de Galarreta R (2014) Differential growth response and minituber production of three potato cultivars under aeroponics and greenhouse bed culture. Am J Potato Res 91:346–353. https://doi.org/10.1007/s12230-013-9354-8
Tiwari JK, Singh RK, Buckseth T, Chakrabarti SK (2019) Aeroponics: a novel precision system to study nutrient biology in potato. Indian Horti 64(5):33–34
Tsoka O, Demo P, Nyende AB, Ngamau K (2012) Potato seed tuber production from in vitro and apical stem cutting under aeroponic system. Afr J Biotechnol 11:12612–12618
Uyen NV, VanderZaag P (1983) Vietnamese farmers use tissue culture for commercial potato production. Am Potato J 60:873–879
VanderZaag P, Escobar V (1990) Rapid multiplication of potatoes in the warm tropics: rooting and establishments of cuttings. Potato Res 33:13–21
Venkataslam EP, Singh RK, Sharma AK, Singh V, Buckseth T, Chakrabarti SK (2017) Seed potato production techniques: principles and applications. ICAR-Central Potato Research Institute, p 101
Wang C, Wang X, Wang K, Hu J, Tang M, He W, Vander Zaag P (2018) Manipulating aeroponically grown potatoes with gibberellins and calcium nitrate. Am J Potato Res 95:351–361
Wasilewska-Nascimento B, Boguszewska-Mańkowska D, Zarzyńska K (2020) Challenges in the production of high-quality seed potatoes (*Solanum tuberosum* L.) in the tropics and subtropics. Agronomy 10(2):260. https://doi.org/10.3390/agronomy10020260
Zebarth BJ, Arsenault WJ, Sanderson JB (2006) Effect of seed piece spacing and nitrogen fertilization on tuber yield, yield components, and nitrogen use efficiency parameters of two potato cultivars. Am J Potato Res 83:289–296

Advances in Agronomical Management of Potatoes

Ayesha Fatima, P. Nagarjun, Shashank Shekhar Solankey, Rina Roy, Hena Parween, Bholanath Saha, and Swaraj Kumar Dutta

Abstract Potato (*Solanum tuberosum* L.) is an important tuber crop which is used as staple food in various developing countries and has always been the "poor man's friend". Potato plant belongs to the Solanaceae family. The origin of potato is believed to be South America, and it was introduced in India during the seventeenth century. It is fourth in ranking for food production worldwide and can provide food and nutritional security to the world. Keeping this in view, FAO has declared it as the "food for future". Potato tuber is a modified stem which is edible and nutritious. Potato is highly susceptible to several diseases which severely impact its overall productivity; hence, it is essential to use disease-free planting material for the propagation of potato. Micropropagation techniques combined with meristem culture have been successfully incorporated into potato seed production programme. Varieties suitable for different agro-ecologies have been developed which have tremendously contributed to increasing potato productivity. Location-specific integrated potato production technology has significantly impacted yield maximization without exerting much pressure on fertilizers and water which are already scarce resources. Prospects of potato diversification in newer areas might help in providing resilience to climate change and can serve as a major breakthrough in achieving self-sufficiency in potato production.

Keywords Disease · Management · Potato · Production · Tuber and varieties

A. Fatima (✉) · P. Nagarjun · R. Roy · H. Parween · B. Saha · S. K. Dutta
Department of Agronomy, Dr. Kalam Agricultural College, Kishanganj, Bihar Agricultural University, Sabour, Bhagalpur, Bihar, India

S. S. Solankey
Department of Horticulture (Vegetable Science), Nalanda College of Horticulture, Noorsarai, Nalanda, Bihar Agricultural University, Sabour,
Bhagalpur, Bihar, India

S. S. Solankey (ed.), *Advances in Research on Potato Production*,
Advances in Olericulture, https://doi.org/10.1007/978-3-031-82710-5_8

1 Introduction

Potato (*Solanum tuberosum* L.) is the fourth most important cash crop to improve food security and reduce poverty among marginal and small farmers in developing countries. In the early nineteenth century, India contributes only 1% of its total world production, but now it contributes 14% of the total world production and is the second largest producer in the world. Traditional breeding of potato faces many challenges due to its autotetraploid nature and its tendency towards inbreeding depression. Breeding programmes have focused on productivity, nutritional quality, and disease resistance. Gene editing (GE) or genetic modification (GM) can be used to improve potato crops and alleviate these issues. Some of the currently genetically engineered and gene-edited potato varieties have modified starch content, reduced acrylamide content, and resistance to late blight and Colorado potato beetle.

Potatoes are rich in starch, and they find a place in a multitude of culinary purposes. Once baked or boiled, they transform into versatile ingredients for dishes like potato pancakes, mashed potatoes, soup, and salad. In Europe, potatoes are heated to convert starch into sugar and consumed as alcoholic beverages. Dehydrated potato flakes enrich various potato-based products, while the flour can be used as a thickening agent in soups and gravies. Even potato peel and other by-products that are rich in starch can be liquefied and fermented into ethanol. Furthermore, potato starch finds applications in pharmaceuticals, woodworking, papermaking, and textiles, serving as an adhesive, binder, filler, and texture enhancer. Beyond its industrial uses, potato starch proves to have many health benefits, including the prevention of high blood pressure, cancer, inflammation, and scurvy, as well as skincare benefits. The potato's exceptional commercial value is highlighted by the fact that about 80% of its plant biomass is edible (Osaki et al. 1996). Till now, the yield and productivity of the crop have been far below the world average. So, this chapter discusses recent advances in the agronomical management of potatoes, focusing on breeding techniques, soil management, fertilization, pest and disease control, and sustainable farming practices for improving yield, quality, and sustainability of crop.

2 Varietal Development in Potato

In India, potato is grown under diverse agro-climatic conditions. Broadly the potato-growing zones in India could be classified into the northern hills, the northern plains, the eastern hills, the plateau region, and the southern hills. In India, the development of indigenous potato varieties was the most important step towards potato revolution. Introduced European varieties were repeated failures in the subtropical growing condition of the Indian plains, where 90% of potatoes are grown today. Therefore, an organized breeding programme was necessary to develop potato varieties which are suitable for cultivation under diverse agro-climatic conditions of India. Organized potato breeding programme started in the year 1935 when ICAR sanctioned a project to the Imperial Agricultural Research Institute, New Delhi, to initiate potato breeding work in northern India. Central Potato Research Institute (CPRI) was established in 1949 in Patna, but this was shifted to Shimla (Himachal Pradesh) in 1956 to maintain healthy seed stocks and to facilitate its hybridization and breeding research for increasing the genetic base of selection and for development of Indian potato varieties. There are different potato-growing zones established in India from 1956 to 1983 which are located at Kufri (Himachal Pradesh), Jalandhar (Punjab), Patna (Bihar), Shillong (Meghalaya), Modipuram (Uttar Pradesh), Gwalior (Madhya Pradesh), and Ootacamund (Tamil Nadu).

The invention of the "Seed Plot Technique" in 1963 was a significant advancement in the potato improvement programme, enabling the production of disease-free seed in the plains and establishing a national disease-free seed production in the hills and plains.

Since all of the nation's indigenous varieties were medium to long maturing (90–120 days), the main goal of varietal development was to develop early maturing varieties, particularly for warmer climates where the growing window for potato production is getting smaller. A significant number of varieties have been developed which are matured in 70–90 days (in plains) like Kufri Ashoka, Kufri Chandramukhi, Kufri Jawahar, Kufri Khyati, Kufri Pukhraj, Kufri Surya, etc. and 100–110 days (in hills).

A significant number of varieties have been developed with a primary emphasis on increasing tuber yield, wider adaptability, and disease resistance (Tessema et al. 2020). Systematic coordinated breeding efforts of potato varietal improvement programmes have resulted in the development of 61 improved potato varieties at ICAR-CPRI, Shimla, adaptable for cultivation under different agro-climatic zones of India. In addition to table varieties, processing varieties have also been developed. Numerous varieties have been developed that can be harvested within 60–70 days of planting, and also there are varieties which can be harvested whenever needed between 60 and 100 days of planting (Rawal et al. 2020). Some of the prominent varieties developed for cultivation under different agro-ecologies have been listed in Table 1. Potato production has been severely challenged by the preponderance of biotic stresses such as early blight, late blight, Colorado beetle infestation, etc. Many varieties have also been developed that are resistant to the above stresses (Table 1).

In order to optimize yields, reduce cost, and enhance the potato quality, the potato growers must implement best management techniques (BMPs) which are customized to their particular growing environment (Asci et al. 2015). This practice includes field preparation, irrigation scheduling, crop fertilization, weed management, harvest, and storage. Furthermore, precision agricultural technologies (PATs) are unique approach to farming that maximizes production by utilizing modern techniques and data-driven decision-making.

Table 1 Suitability of potato varieties in different agro-ecological zones of India

S. no.	Agro-ecology	Suitable varieties
1.	North Indian plains	Kufri Anand, Kufri Ashoka, Kufri Badshah, Kufri Bahar, Kufri Chandramukhi, Kufri Ganga, Kufri Gaurav, Kufri Khyati, Kufri Pushkar, Kufri Arun, Kufri Chipsona-1, Kufri Chipsona-2, Kufri Chipsona-3, Kufri Sangam, Kufri Neelkanth, Kufri Uday, Kufri Mohan, Kufri Daksh
2.	North-eastern plains	Kufri Red
3.	North-western plains	Kufri Sheetman
4.	North Indian hills	Kufri Himalini, Kufri Shailza, Kufri Kanchan, Kufri Himsona
5.	South Indian hills	Kufri Swarna, Kufri Neela, Kufri Neelamani
6.	North-eastern hills	Kufri Mehga, Kufri Naveen, Kufri Khasigaro

Source: Modified from Luthra et al. (2021)

3 Soil and Climatic Requirement

Potato generally requires relatively mild temperature throughout early growth and cool weather during tuber development. The ideal temperature range for potatoes is 25 °C during germination, 20 °C during vegetative growth, and 17–20 °C during tuberization and tuber development. Well-drained sandy loam and loam soils rich in organic matter with good drainage and aeration are most appropriate for cultivation of potato crops. Soils that are saline or alkaline are not good for growing potatoes. Since acidic soils (pH 5.0–6.5) tend to prevent scab disease, they are ideally suited to these conditions. Sunshine along with cooler nights limits the spread of diseases.

4 Field Preparation

A well-pulverized, soft and friable seedbed is required for optimum tuberization of potato crops. The field should be ploughed once with a soil-turning plough, 20–25 cm deep, shortly after the *kharif* crop is harvested. Subsoiler is an important primary tillage implement. It creates an excellent drainage condition by cracking the hardpan beneath the soil's surface. After that, four to five ploughings with local plough or two to three cross harrowings need to be done. To level and smoothen the soil surface, one or two plankings are also required. Optimum soil moisture at the time of planting is essential and 20–25 t/ha of well-decomposed FYM (Farmyard Manure) is mixed in the soil at the time of last ploughing. Potato is often planted on raised seedbeds/ridges that are constructed with the help of potato ridger. For uniform germination, pre-sowing irrigation must be completed prior to field preparation.

5 Seed Selection and Quality Seed Production

Production of quality seeds is highly crucial for sustainable potato production since seeds account for 30–50% of input cost of potato. Ideally, potato seeds must be produced in cooler areas to prevent infections and infestations by insect vectors of viruses (Ahmad et al. 2019). The type and duration of storage are also critical for ensuring good-quality potato seed. The storage area should be well ventilated in terms of light and air circulation. The age of potato seeds must be 3 months at the time of planting, and the seeds should be carefully chosen, in good condition, and free from bruises to prevent the incidence of storage diseases such as late blight, bacterial rot, and silver scurf. Healthy medium-sized or small-sized sprouting tubers with at least two to three "eyes" and a diameter of roughly 2.5 cm (25–40 g) should be chosen for planting. The seed rate of potato is 15–20 quintals per hectare. In the main crop, a 45–60 cm row-to-row and a 15–20 cm plant-to-plant spacing is recommended.

Many of the limitations of the conventional seed production system can be overcome by using micropropagation techniques in potato seed production (Singh et al. 2019). By utilizing rapid tissue culture method in virus-indexed and pathogen-free mericlones, significant quantities of pathogen-free cultures can be obtained in potatoes (Naik and Karihaloo 2007). Under micropropagation technique, three single-node cuttings are taken with one or two leaves and then planted in culture tubes of 25 × 150 mm dimensions, and the apical/axillary buds of these cuttings further develop into mature plants in about 3 weeks. Again, at 3-week intervals, these microplants are once again subcultured up to 10 times on fresh culture media. Thus, under controlled conditions, it is possible to generate approximately 59,049 microplants from a single culture tube in 7–8 months (NAAS 2021). Mini-tubers are small tubers used for seed potatoes that are grown from in vitro propagated plantlets sown at high density in glasshouses or aphid-proof net houses (Buckseth et al. 2020). Microplants or microtubers can be used as planting material to produce minitubers (Naik 2005). Soil-less aeroponics technology is yet another recent technology which has shown significant success in the production of quality seed potatoes. With this technology, the microplants are placed on the top of the growth chamber, and a nutrient solution is used to fog the root zone that develops inside the chamber (Buckseth et al. 2016). This technology can be employed for year-round production of quality seed potatoes. The concept of true potato seed (TPS) was first propounded by Dr. S. Ramanujam in 1949. It offers several advantages over potato seed tubers like planting material free from pests and diseases, convenience of transportation and storage, and a significantly lower seed rate (about 150 g/ha compared to 2.5–3.0 t/ha of seed tubers) (Buckseth et al. 2022).

6 Planting Methods of Potato

In India, there are three methods of planting potato:

1. *Planting potatoes on ridges*: Using a spade, ridges 45–60 cm apart are constructed once the field has been prepared. Potatoes are planted on the ridges with the aid of khurpi.

2. *Flat method:* Potatoes are planted in shallow furrows on a level surface. After germination, when plants reach a height of 10–12 cm, ridges are formed. Light soils are suited for this technique. To thicken the ridges later on, two or three earthing ups are done.
3. *Planting potatoes in ridges after a level surface*: This technique involves preparing a field and then opening shallow furrows on the level surface. Following the planting of tubers, small ridges are made. At a later stage, the ridges are thickened by earthing up from side soil.

7 Nutrient Management in Potato

In potato cultivation, variable rate technology (VRT) can be used to prepare the soil by mapping the changes and levels of nutrients in the soil, figuring out the best rate of fertilizer to utilize in each particular field region, and then applying the right amount of fertilizer. This helps to ensure adequate soil fertility for potato growth, minimize waste, and prevent overfertilizing some areas while underfertilizing others (Bohman et al. 2019). For the purpose of obtaining the highest possible proportion of process-grade tubers without compromising processing quality during harvest and storage at elevated temperatures (10–12 °C) with the application of sprout suppressant CIPC, the ideal requirements for nitrogen, phosphorus, and potassium are 270, 80, and 150 kg of N, P_2O_5, and K_2O per hectare. Applying nitrogen in two equal splits is recommended: half during planting and the remaining half right before earthing up with soil. On the other hand, basal applications of potassium and phosphorus are done at the time of tuber planting. Twenty-five to thirty per cent of nutrient requirement must be preferably supplied through organics since this aids in fulfilling the micronutrient needs and enhancing the soil moisture holding capacity. Balanced fertilization not only helps in lowering the reducing sugar content but also improves the colour of the produce. The problem of internal brown spot (IBS) in potatoes, especially in light soils, that can be reduced by the application of calcium at 200 kg/ha in two equal splits (i.e. at planting and earthing up) in the form of gypsum has been reported to be advantageous.

Potatoes need a consistent supply of nutrients. Deficiencies or variations in the supply of soluble nutrients (particularly N) lead to improper vine growth, increased susceptibility to insects and disease-causing pathogens, declined tuber yields, and reduced tuber quality (Stark et al. 2004). Potatoes have a high amount of fertilizer requirement because it is a heavy feeder, and also due to the presence of a shallow, ineffective root system (Munoz et al. 2005; Pack et al. 2006). The nutrients that are most frequently fertilized in the production of potatoes are nitrogen (N), phosphorus (P), and potassium (K) (Davenport et al. 2005).

7.1 Nitrogen (N)

Among all the important macronutrients, nitrogen (N) has the greatest influence on potato production (Bucher and Kossmann 2011; Silva et al. 2013). It has been demonstrated that altered N levels can lead to uneven tuber growth and enhance the risk of both external (deformed tuber) and internal (brown centre and hollow heart) tuber malformations (Stark et al. 2004; Westermann 2005). For plants to maximize the transportation of above-ground carbohydrates into tubers and improve "skin" production, N concentrations in the plant must decline as the plant matures. To optimize potato production, tuber quality, and N efficiency, it is advised to synchronize N availability with the demand of potato. During the first 4–5 weeks of growth, the potato crop has a relatively low need for nitrogen (N), and a high N supply can even inhibit or delay tuberization (Jackson 1999; Zebarth and Rosen 2007). Nitrogen promotes stem development throughout this stage of growth. As a result, tubers have less dry matter available. Nevertheless, the tubers may benefit from more aboveground biomass at a later stage of growth. Nitrogen has a significant effect on the number of newly appearing leaves, the rate of expansion of the leaves, and the plant's canopy development. Potatoes are known for their shallow root structure, which makes them highly prone to water shortages in addition to being inefficient users of nutrients like nitrogen (Liu et al. 2015). In addition to accounting for additional N sources, such as those provided by catch crops or intercrops like N-fixing leguminous plant species, an adequate N supply should also be based on estimates which correspond to the actual plant requirement (Bucher and Kossmann 2011; Zebarth et al. 2012). According to Errebhi et al. (1998), split application of nitrogen has the ability to reduce the risk of nitrogen leaching, particularly in irrigated potato production on sandy soils. It can also more effectively meet the crop's actual requirement given the stage at which it is developing (Ojala et al. 1990).

7.2 Phosphorus (P)

Potatoes require an abundant supply of P and are frequently believed to be ineffective in absorbing soil phosphorus (Pursglove and Sanders 1981). Since it is involved in respiration, photosynthesis, and cellular energy dissemination, phosphorus regulates plant metabolism (Grant et al. 2001). Additionally, phosphorus serves for storing nutrients in seeds as phytic acid and is a structural component of phospholipids, nucleic acids, coenzymes, and phosphoproteins (Bundy et al. 2005). Hence, an appropriate amount of phosphorus is required by the plant from the initial growth stage until maturity (Grant et al. 2001). The deficiency of phosphorus can exacerbate the severity of many important diseases, such as late blight (*Phytophthora infestans*), verticillium wilt (*Verticillium dahlia* and *V. albo-atrum*), and common scab (*Streptomyces scabies*). The amount of P that is dissolved in soil solution determines how much is available to plants. This amount decreases sharply when soil pH rises from nearly neutral (6.8–7.0) to alkaline (7.6–8.3). Excess lime that is free in calcareous soils exacerbates this issue (Westermann 2005).

7.3 Potassium

Several literatures extensively document the importance of potassium (K) in the growth of potato plants, the biochemical processes governing starch synthesis, as well as in determining both the quality and quantity of tubers. Potassium is a vital macronutrient absorbed in huge quantities by potato (Li et al. 2015). In the life cycle of a potato plant, it typically exhibits peak potassium uptake during its early stages of development, typically around 30–40 days after emergence. However, as the plant progresses towards maturity, particularly during the tuber formation phase, the uptake of potassium diminishes significantly, approaching zero levels (Horneck and Rosen 2008). Even though pre-plant fertilization of potassium in potato typically yields better results, broadcasting potassium as a banded application can potentially induce salt stress. Split potassium fertilization represents another approach that could enhance tuber yield efficiently. In research conducted by Sharma and Sud (2001) on both acidic and alluvial soils, it was found that potassium sulphate (K_2SO_4) outperformed potassium chloride (KCl) in enhancing tuber dry matter, in addition to increasing levels of ascorbic acid and starch content. Contrarily, Khan et al. (2012) suggested that the application of potassium chloride (KCl) had no significant effect on starch content. Instead, they observed that applying KCl enhanced the tuber chipping quality compared to potassium sulphate (K_2SO_4) by reducing enzymatic discolouration and phenol content in the tubers. A novel potassium fertilizer, potassium phosphite (K_3O_3P), applied to seed tubers followed by foliage application, has been reported to offer advantages in shielding the tubers from pathogen attacks, dehydration, and damage during harvesting and storage (Lobato et al. 2018). Additionally, another potassium fertilizer, polyhalite [$K_2Ca_2Mg\ (SO_4)_4 \cdot 2H_2O$], has recently demonstrated positive impacts on the appearance of tuber skins, although its effectiveness may vary depending on local conditions and the specific potato cultivar (Keren-Keiserman et al. 2019).

8 Weed Management in Potato

The proliferation of weeds in potato crops globally presents a significant challenge, primarily stemming from the delayed emergence of potato crops during the initial stage. This slow emergence creates an environment conducive to the abundant growth and development of weeds in the cultivated fields. The emergence of weeds is closely linked to the level of rainfall and moisture present in the soil horizons.

Weeds, acting as competitors, vie with potato crops for essential resources such as nutrients, solar radiation, and moisture. Additionally, they serve as hosts for diseases (Singh 2016). The predominant weed species in potato crops exhibit variations from one region to another, typically classified into two subgroups: those found in plains and those in hills. Understanding the prevalent and invasive flora becomes crucial for effective weed control in different potato-growing regions. Some of the prevalent weeds in potato crops include *Cynodon dactylon* L. (Pers.), *Cyperus rotundus* L., *Trianthema monogyna* L., *Chenopodium album* L., *Poa annua* L., *Anagallis arvensis* L., Meliotus spp., *Sonchus oleraceus* L., and *Vicia sativa* L. In hilly areas, major weeds include *Amaranthus viridis* L., Chenopodium spp., Oxalis spp., *Digitaria sanguinalis* (L.) Scop., *Setaria glauca* (L). Beauv., *Spergula arvensis* L., and Meliotus spp. (Lal 1993). Different agricultural systems, management practices, and agro-climatic regions have varying levels of weed incidence and intensity (Singh et al. 2018). Weed competition has the potential to reduce both yield (Van Gessel and Renner 1990) and the quality of potatoes, impacting the size and weight of tubers and overall production (Wall and Friesen 1990). Weeds can also hinder harvesting operations, leaving a significant number of potatoes in the field and increasing the risk of mechanical injury. Allowing a mixed population of annual weeds to compete with potatoes throughout the growing season results in a significant negative correlation. For every 10% increase in dry weed biomass, tuber yield decreases by 12% (Nelson and Thoreson 1981).

9 Critical Period of Crop: Weed Competition in Potato

Potato crops are typically cultivated in a spacious layout, utilizing an abundance of manures and fertilizers. Additionally, in plains, irrigations are applied frequently, while hills experience substantial rainfall during the crop growth season. These agricultural practices create favourable conditions for the initial and rapid growth of weeds, prior to crop emergence. This scenario offers a significant possibility for weeds to thrive and potentially dominate the crop if not addressed promptly (Singh et al. 2018). The consequence of inadequate management could lead to lowering of tuber productivity, especially due to the early establishment of weeds as opposed to later. According to Gupta et al. (1997), crops that receive irrigation respond favourably to early weed control. One important aspect of the crop-weed competition was considered to be the weeds' early canopy cover or the allelopathic impacts of the weeds. According to Thakral et al. (1989), the first 25–30 days after planting are the most critical to crop–weed competition. A decrease in tuber yield in Kufri during this time frame emphasizes the importance of early weed control in potato (Lal and Jaiswal 1990).

10 Weed Management Practices in Potato

The primary goal of weed management is to maintain the density of weed plants well below a threshold that would compromise the attainment of desirable tuber yields and farm profits. The choice of an effective weed control method depends on many elements such as the severity of specific weeds, weed growth stages, present weather conditions, and the socio-economic conditions of the farm enterprise. Weed management practices applicable to potato cultivation encompasses cultural, mechanical, and chemical methods, as well as advocating for an integrated approach to weed control. Therefore, effective weed management in potato crops is essential to ensure optimal growth and yield of potatoes. The following are some strategies and practices recommended for successful weed management in potato crops.

10.1 Cultural Practices

Selection of Weed-Free Planting Material A very important practice to avoid the infestation of weeds in the potato crop is to select good-quality planting material which is free from weed seed infestation; this practice helps to reduce the weed population in potato crops.

Crop Rotation Rotate potato crops with non-host crops which disrupt the life cycle of weed species. It helps to reduce the host-specific weeds as well as parasitic weeds. Inclusion of cereal crops in rotation with potato is the most ideal method.

Mulching Apply organic or synthetic (plastic) mulches to suppress weed growth and conserve soil moisture.

Soil Solarization A chemical-free technique widely adopted globally effectively manages soil-borne pests like plant pathogens, weeds, and mites by utilizing transparent polyethylene sheets to harness solar heat and elevate soil temperatures, eliminating weed seeds and seedlings.

10.2 Mechanical Control

Earthing Up Regular hilling or earthing up helps to bury many weed seeds and disrupt weed growth.

Inter-cultivation Use mechanical cultivation methods, such as plowing or hoeing or mechanical weeds, to control weeds between rows.

10.3 Chemical Control

Application of Hrbicides Selective herbicides can be used to control specific weed species without any phytotoxic effect on potato plants. Common herbicides used in potato crops include metribuzin, linuron, and pendimethalin.

10.4 Biological Control

Use of Bioherbicides Some biological agents, such as certain fungi or bacteria, can be used as bioherbicides to control specific weed species.

11 Integrated Weed Management (IWM)

Integrated weed management (IWM) is a holistic approach to manage weeds that integrates different methods of weed control to provide crop an advantage over weeds. It is practised globally at varying levels of adoption from farm to farm. IWM has the potential to restrict weed populations to manageable levels, reduce the adverse environmental impact of individual weed management practices, increase cropping system sustainability, and reduce selection pressure for weed resistance to herbicides (Harker and O'Donovan 2013).

It is important to note that the choice of weed management practices may vary based on factors such as weed species, environmental conditions, labour availability, socio-economic conditions, and local regulations. Initial weed-free period is very much required for profitable yield. Much care needs to be taken during the initial periods, i.e. up to 30 days after sowing; later crops have the capacity to suppress the weeds by producing good foliage and extensive ground cover.

12 Water Management

Potato is a shallow-rooted crop. So, water stress from both scarcity and excess water results in a reduction in the yield of tubers and the grade and quality of potatoes. Hence proper water management is necessary for sustainable potato production. Flood irrigation is one of the inefficient traditional methods of irrigation which causes more water loss. According to a number of studies, the yield of potatoes increased as irrigation intensity increased (Yuan et al. 2003). However, because of the substantial N leaching under high irrigation, potato production will not always rise with increased irrigation (Woli et al. 2016). Increasing the irrigation frequencies and decreasing the amount of irrigation can significantly lead to improvement in potato yield and water use efficiency (Kang et al. 2004). Potatoes are grown under various irrigation methods like surface irrigation, drip irrigation, and sprinkler irrigation. It produces dense foliage; therefore, to maintain turgidity in leaves, readily available moisture is highly essential. Potatoes require between 350 and 550 mm of water, depending on the length of the growing season, atmospheric demand, soil type, and crop variety (Sood and Sharma 1993). About 70% of the total water requirement of the crop is met from the top 30 cm of soil layer (Singh 1968). At the tuber initiation and bulking stages, potatoes are known to be extremely susceptible to water stress. Any water deficit during these stages has a major impact on crop yield and quality as well as the net economic return of the entire production chain (Djaman et al. 2021). In the Northern Indo-Gangetic plains, a time interval approach is adopted in the autumn season of potato cultivation. At this stage, irrigation is planned for every 8–10 days throughout the warm months. As winter approaches and the temperature drops, the frequency is then extended to 12–15 days. For optimum skin setting, the application of water is halted 10–12 days before the destruction of haulm. It has been reported that higher biomass accumulation can be promoted by applying 80–100% of the irrigation requirements (Camargo et al. 2015). There are several studies confirming higher IWUE (Irrigation water use efficiency) with partial root-zone drying than under deficit irrigation in potato (Ahmadi et al. 2010; Jovanovic et al. 2010). The greatest concentration of nitrate-nitrogen in trickle-fertigated crops was limited to 30–60 cm of soil depth; nevertheless, it was also observed that nitrate-nitrogen in the soil increased with high irrigation levels at a depth of 90–180 or 200 cm (Stark et al. 1993; Halitligil et al. 2002). Irrigation on the basis of cumulative pan evaporation (CPE) has also been recommended in potato, especially with micro-irrigation. In this case, depending on the kind and variety of soil, irrigation is carried out when cumulative pan evaporation reaches 15–25 mm (Rawal et al. 2020).

13 Use of Precision Agricultural Technology for Water Management in Potato

Irrigation management in potato farming has changed dramatically with the introduction of real-time soil moisture monitors. Real-time soil moisture sensors assess the water content of the soil and send the data to a central control system, where it is analysed to determine the best time and quantity of irrigation needed for the crop. Using real-time soil moisture sensors in the cultivation of potatoes has various benefits (Djaman et al. 2022) like improved irrigation efficiency by guaranteeing that the crop gets precisely the appropriate quantity of water (Dutta and Taghyaejan 2023) which in turn minimizes the wastage of water and helps in conserving water resources and ultimately resulting in higher crop yields and improved crop quality. Real-time soil moisture sensors are relatively simple to install and maintain due to their simple infrastructure, and they are economical in the long run. A study carried out in India reported that potato cultivation using moisture-sensor-based irrigation led to 20% reduction in water use and a yield increase of 8.5% over traditional methods of irrigation (Poddar et al. 2021). Nevertheless, there are several challenges in its implementation such as its high cost, technical complexity, scarcity of sensors, and difficulties with data management and interpretation. The development of affordable sensors, technical assistance and training, local sensor manufacturing, and user-friendly data management and interpretation tools can all help to address these issues. In order to address the issues of over- and under-irrigation in agriculture, machine learning algorithms (MLAs) have been applied. It has been discovered that MLAs significantly improve irrigation management (Su and Xue 2021). Some of the most common algorithms tested in potato irrigation management include decision tree, random forest, and neural network (Su and Xue 2021; Yamaç and Todorovic 2020). In one study, the ideal water requirement for the potato crop was estimated using a decision tree model trained on soil moisture data. This led to an increase in production to the tune of 15% and a decrease in water consumption to an extent of 30% (Sabzi et al. 2020). Irrigation management systems that can forecast water requirements, determine the optimal irrigation frequency, and recommend suitable irrigation strategies have been developed by using random forest algorithms.

In hilly regions, potato is mainly cultivated as a rainfed crop. To impart drought tolerance in potato, use of FYM and green manures has been advocated that help in soil moisture retention. When there is an excess of rainfall, water can be stored and used for irrigation when there is a drought. A viable approach to increasing potato production is water harvesting in situ with the aid of mulching materials like pine needles, farmyard manure, or any other vegetative matter. To conserve water from melting from snowfall, especially in the southern portion of the hills, ridge and furrow type of soil construction should be built before snowfall. In order to prevent plant wilting and tuber rotting, fields in hills should have sufficient drainage. Like other crops, potatoes require careful irrigation management. The principles of plant-water relationships, irrigation scheduling, soil moisture monitoring throughout the growing season, the relationship between irrigation and potato growth stages, and the consideration of common scab and other product characteristics for the target market should all be followed.

14 Sustainable Health Management of Potato

Late blight caused by *Phytophthora infestans* is a disease of great concern since it leads to substantial losses in potato production. To effectively control this disease in plains, it is advised to adopt resistant varieties such as Kufri (K.) Badshah, K. Jyoti, K. Sutlej, K. Jawahar, K. Anand, K. Garima, K. Mohan, etc. Since the commercial varieties of potato are highly heterozygous and tetraploid, traditional breeding techniques also transfer undesirable traits. However, the development of genetic engineering methods to prevent late blight presents further benefits for the sustainability of potatoes (Ahmad et al. 2019). The R-proteins that wild potatoes secrete are called

VNT1 proteins. They work well to trigger a strong response to the late blight pathogen *P. infestans*. Rpi-vnt1 (from *P. infestans*) is said to be a member of the interactable protein class, which also includes Avr-vnt1. Its hypersensitive reaction shields the host plant from late blight (Stefanczyk et al. 2017). Software tools such as JHULSACAST and INDO BLIGHTCAST are available for predicting the emergence of late blight (Rawal et al. 2020).

Early blight caused by *Alternaria solani* leads to defoliation in potato plants and reduces the quality and quantity of the potato tubers by 20–30% (Wale et al. 2008). Early blight thrives better in weather that alternates between dry and rainy spells. It is advised to remove and burn the infected potato haulms in order to prevent the inoculums from spreading throughout the field. Early blight can be prevented from becoming severe by applying fertilizer in a balanced amount, especially when it comes to nitrogen. One effective method is to spray 1.0% urea 45 DAP and then again 8–10 days later. This pathogen has recently been successfully controlled with a variety of low dosages of abscisic acid, 2,4-D, and H_2O_2 when combined with prescribed fungicides (Nassar and Ibrahim 2016).

The majority of potato viruses, including those that produce mosaics, mottling/crinkle, necrosis, and PVY, PVX, PVA, PVM, PVS, PAMV, and PSTVd (viroid), can do so independently or in various combinations. When plant viruses infect potato plants, mosaics are the most typical symptoms that are exhibited by them. Potato fields can be prevented from spreading PVY by utilizing a combination of complementary techniques, such as intercropping with hairy vetch or oats, spraying mineral oil, and mulching straw. This has been reported by Dupuis et al. (2017).

Potato wart disease is transmitted by seeds as well as the soil and is caused by *Synchytrium endobioticum*. For effective control of this disease, it is advised to practise long-term crop rotation (5 years or more) with non-solanaceous crops, preferably peas, radish, cabbage, and maize. Potato peelings and wart lumps should be burned rather than put in a manure pit or field to stop the disease from spreading.

Phthorimaea operculella, commonly referred to as the potato tuber moth (PTM), is one of the most troublesome pests of potatoes, both in fields and storage. In many parts of the world, parasitoids such as *Copidosoma* spp. and *Apanteles* spp. have been reported to be of significance in PTM management (Adhikari et al. 2022). Pheromone traps at 20 traps/ha have been utilized for field and storage pest management, as well as pest population monitoring.

15 Good Agricultural Practices (GAP) After Potato Harvesting

Proper handling of potatoes is essential to ensure the quality and storage of healthy tubers. The following practices must be taken care of while harvesting potato:

- Potato tubers should be harvested, handled, and transported according to variety.
- It is important to take precautions to prevent bruises on tubers during harvest, handling, and transportation.

- For the purpose of hardening the peel and removing adherent soil from the surface of the tubers, potato tubers should be stored in piles for nearly 15 days.
- Tubers should be cured in heaps for at least 8–10 days prior to being transported to a processing factory or cold storage.
- Tuber handling and transport should not be done when tuber temperature is less than 10 °C.
- All of the equipment used for different operations needs to be free of sharp edges to avoid cutting or bruising potatoes.

References

Adhikari A, Oli D, Pokhrel A, Dhungana B, Paudel B, Pandit S, Bigyan GC, Dhakal A (2022) A review on the biology and management of potato tuber moth. Agriculture (Pol'nohospodárstvo) 68(3):97–109

Ahmad T, Tariq RMS, Iqbal Q, Hussain S, Nawaz A, Ahmad S (2019) Advanced production technologies of potato. In: Hasanuzzaman M (ed) Agronomic crops volume 1: production technologies. Springer Nature Singapore Pte Ltd, pp 363–386

Ahmadi SH, Andersen MN, Plauborg F, Poulsen RT, Jensen CR, Sepaskhah AR, Hansen S (2010) Effects of irrigation strategies and soils on field grown potatoes: yield and water productivity. Agric Water Manag 97:1923–1930

Asci S, Borisova T, Vansickle JJ (2015) Role of economics in developing fertilizer best management practices. Agric Water Manag 152:251–261

Bohman BJ, Rosen CJ, Mulla DJ (2019) Evaluation of variable rate nitrogen and reduced irrigation management for potato production. Agron J 111(4):2005–2017

Bucher M, Kossmann J (2011) Molecular physiology of the mineral nutrition of the potato. In: Vreugdenhil D(ed)Potato biology and biotechnology: advances and perspectives. Elsevier, Amsterdam, pp 311–330

Buckseth T, Sharma A, Pandey K, Singh B, Muthuraj R (2016) Methods of pre-basic seed potato pro-duction with special reference to aeroponics - a review. Scientia Horti 204:79–87

Buckseth T, Singh RK, Tiwari JK, Sharma AK, Singh S, Chakrabarti SK (2020) A novel sustainable aeroponic system for healthy seed production in India-an update. Indian J Agric Sci 90(2):243–248

Buckseth T, Tiwari JK, Singh RK, Kumar V, Sharma AK, Dalamu D et al (2022) Advances in innovative seed potato production systems in India. Front Agron 4:956667

Bundy LG, Tunney H, Halverson AD (2005) Agronomic aspects of phosphorus management. In: Sims JT, Sharpley AN (eds) Phosphorus: agriculture and the environment. American Society of Agronomy, Crop Science Society of America, Soil Science Society of America, Madison, pp 685–727

Camargo DC, Montoya F, Córcoles JI, Ortega JF (2015) Modelling the impacts of irrigation treatments on potato growth and development. Agric Water Manag 150:119–128

Davenport JR, Milburn PH, Rosen CJ, Thornton RE (2005) Environmental impacts of potato nutrient management. Am J Potato Res 82:321–328. https://doi.org/10.1007/BF02871962

Djaman K, Irmak S, Koudahe K, Allen S (2021) Irrigation management in potato (*Solanum tuberosum* L.) production: a review. Sustainability 13(3):1504. https://doi.org/10.3390/su13031504

Djaman K, Koudahe K, Saibou A, Darapuneni M, Higgins C, Irmak S (2022) Soil water dyanamics, effective rooting zone and evapotranspiration of sprinkler irrigated potato in a sandy loam soil. Agronomy 12(4):864. https://doi.org/10.3390/agronomy12040864

Dupuis B, Cadby J, Goy G, Tallant M, Derron J, Schwaerzel R, Steinger T (2017) Control of *Potato virus Y* (PVY) in seed potatoes by oil spraying, straw mulching and intercropping. Plant Pathol 66:960–969

Dutta S, Taghyaejan S (2023) Soil water sensors for irrigation scheduling in the United States: a systematic review of literature. Agric Water Manag 278:108148. https://doi.org/10.1016/j.agwat.2023.108148

Errebhi M, Rosen CJ, Gupta SC, Birong DE (1998) Potato yield response and nitrate leaching as influenced by nitrogen management. Agronomy journal 90(1):10–15

Grant CA, Flaten DN, Tomasiewicz DJ, Sheppard SC (2001) The importance of early season phosphorus nutrition. Can J Plant Sci 81:211–224

Gupta OP, Chaudhary GR, Bansal RG, Didwania GD (1997) Evaluation of early period weed control in potato through herbicides. In: Abstracts. Weed science conference and workshop in India

Halitligil MB, Akin A, Ilbeyi A (2002) Nitrogen balance of nitrogen-15 applied as ammonium sulphate to irrigated potatoes in sandy textured soils. Biol Fertil Soils 35:369–378

Harker KN, O'Donovan JT (2013) Recent weed control, weed management and integrated weed management. Weed Technol 27(1):1–11

Horneck D, Rosen C (2008) Measuring nutrient accumulation rates of potatoes—tools for better management. Better Crops 92:4–6

Jackson SD (1999) Multiple signaling pathways control tuber induction in potato. Plant physiology 119(1):1–8

Jovanovic Z, Stikic R, Vucelic-Radovic B, Paukovic M, Brocic Z, Matovic G, Rovcanin S, Mojevic M (2010) Partial root-zone drying increases WUE, N and antioxidant content in field potatoes. Eur J Agron 33:124–131

Kang Y, Wang FX, Liu HJ, Yuan BZ (2004) Potato evapotranspiration and yield under different drip irrigation regimes. Irrig Sci 23(3):133–143

Keren-Keiserman A, Baghel RS, Fogelman E, Faingold I, Zig U, Yermiyahu U, Ginzberg I (2019) Effects of polyhalite fertilization on skin quality of potato tuber. Front Plant Sci 10:1379

Khan MZ, Akhtar ME, Mahmood-ul-Hassan M, Mahmood MM, Safdar MN (2012) Potato tuber yield and quality as affected by rates and sources of potassium fertilizer. J Plant Nutr 35:664–677

Lal SS (1993) Weed management in potato. In: Chadha KL, Grewal JS (eds) Advances in horticulture. Potato 7. Central Potato Research Institute, Shimla, pp 179–207

Lal SS, Jaiswal VP (1990) Investigation on the biology and management of weeds in potato. Annual scientific report CPRI, Shimla, India, pp 60–63

Li S, Duan Y, Guo T, Zhang P, He P, Johnston A, Shcherbakov A (2015) Potassium management in potato production in Northwest region of China. Field Crop Res 174:48–54

Liu C, Rubæk GH, Liu F, Andersen MN (2015) Effect of partial root zone drying and deficit irrigation on nitrogen and phosphorus uptake in potato. Agricultural Water Management 159:66–76

Lobato MC, Daleo GR, Andreu AB, Olivieri FP (2018) Cell wall reinforcement in the potato tuber periderm after crop treatment with potassium phosphite. Potato Res 61:19–29

Luthra SK, Gupta VK, Shambu Kumar, Kaur RP, Dalamu, Vinay Bhardwaj (2021) Improved potato varieties for different agro-ecologies of India. In: Recent advances in root and tuber crops. Brillion Publishing, New Delhi

Munoz F, Mylavarapu RS, Hutchinson CM (2005) Environmentally responsible potato production systems: a review. J Plant Nutr 28:1287–1309

Naik PS (2005) Achieving self-sufficiency in quality seed production in roots and tuber crops with special reference to potato. In: Chadha KL, Ahloowalia BS, Prasad KV, Singh SK (eds) Crop improvement and production technology of horticultural crops, vol I. The Horticultural Society of India, New Delhi, pp 243–255

Naik PS, Karihaloo JL (2007) Micropropagation for production of quality potato seed in Asia-pacific. Asia-Pacific Consortium on Agricultural Biotechnology, New Delhi

Nassar AMK, Ibrahim AAA (2016) 2,4-Dichlorophenoxy acetic acid, abscisic acid, and hydrogen peroxide induced resistance-related components against potato early blight (Alternaria solani, Sorauer). Ann Agric Sci 61:15–23

National Academy of Agricultural Sciences (NAAS) (2021) Innovations in potato seed production, Strategy paper no. 14. National Academy of Agricultural Sciences, New Delhi
Nelson DC, Thoreson MC (1981) Competition between potatoes (Solarium tuberosum) and weeds. Weed Science 29(6):672–677
Osaki M, Matsumoto M, Shinano T, Tadano T (1996) A root-shoot interaction hypothesis for high productivity of root crops. Soil Science and Plant Nutrition 42(2):289–301. https://doi.org/10.1080/00380768.1996.10415099
Ojala JC, Stark JC, Kleinkopf GE (1990) Influence of irrigation and nitrogen management on potato yield and quality. American potato journal 67:29–43
Pack JE, Hutchinson CM, Simonne EH (2006) Evaluation of controlled release fertilizers for northeast Florida chip potato production. J Plant Nutr 29:1301–1313
Poddar A, Kumar N, Shankar V (2021) Evaluation of two irrigation scheduling methodologies for potato (*Solanum tuberosum* L.) in north-western mid-hills of India. ISH. J Hydraul Eng 27(1):90–99
Pursglove JD, Sanders FE (1981) The growth and phosphorus economy of the early potato (Solanum tuberosum). Commun Soil Sci Plant Anal 12:1105–1121
Rawal S, Mankar P, Dua VK, Kumar D, Malik K, Luthra SK, Lal M, Singh S, Lekshmanan DK, Das B, Chakrabarti SK (2020) Good agricultural practices (GAP) for production of potato crop. ICAR-Central Potato Research Institute, Shimla, p 205
Sabzi S, Abbaspour-Gilandeh Y, Arribas JI (2020) An automatic visible-range video weed detection, segmentation and classification prototype in potato field. Heliyon 6(5):Article e03685
Sharma RC, Sud KC (2001) Potassium management for yield and quality of potato. International Potash Institute, Basel, pp 363–381
Silva JG, França MGC, Gomide FTF, Magalhaes JR (2013) Different nitrogen sources affect biomass partitioning and quality of potato production in a hydroponic system. Am J Potato Res 90:179–185. https://doi.org/10.1007/s12230-012-9297-5
Singh AK (1968) Water requirement of potatoes. Proc. Symp. Water Management, Udaipur, pp 169–172
Singh B (2016) Survey and indexing of weeds growing around potato fields for their role as an inoculum source for potato leaf roll virus (PLRV). Br Biotechnol J 16(1):1–8
Singh SP, Rawal S, Dua VK, Roy S, Sadaworthy MJ, Charkrabarti SK (2018) Weed management in conventional and organic potato production. Int J Chem Stud 2(6):24–38
Singh RK, Buckseth T, Tiwari JK, Sharma AK, Chakrabarti SK (2019) Recent advances in production of healthy planting material for disease management in potato. Biotech Today: Int J @ Bio Sci 9(1):7–15
Sood M, Sharma (1993) Water optimization and requirement. In: Chanda KL, Grewal JS (eds) Advances in horticulture. Malhotra Publishing House, New Delhi, p 64
Stark JC, McCann IR, Westermann DT, Izadi B, Tindal TA (1993) Potato response to split nitrogen timing with varying amounts of excessive irrigation. Am Potato J 70:65–77
Stark JC, Westermann DT, Hopkins B (2004) Nutrient Management Guidelines for Russet Burbank Potatoes. University of Idaho, Moscow
Stefanczyk E, Sobkowiak S, Brylinska M, Sliwka J (2017) Expression of the potato late blight resistance gene Rpi-phu1 and Phytophthora infestans effectors in the compatible and incompatible interactions in potato. Phytopathology 107:740–748
Su WH, Xue H (2021) Imaging spectroscopy and machine learning for intelligent determination of potato and sweet potato quality. Food Secur 10(9):2146
Tessema G, Mohammed W, Abebe T (2020) Evaluation of potato (*Solanum tuberosum* L.) varieties for yield and some agronomic traits. Open Agric 5:63–74. https://doi.org/10.1515/opag-2020-0006
Thakral KK, Pandita ML, Khurana SC, Kallu G (1989) Efficacy of cultural and chemical weed control methods in potato. J Indian Potato Assoc 29:33–38
Van Gessel MJ, Renner KA (1990) Redroot pigweed (Amaranthus retroflexus) and barnyardgrass (Echinochloa crusgalli) interference in potatoes (Solanum tuberosum). Weed Sci 38:338–343

Wale S, Platt B, Cattlin N (2008) Diseases, pests and disorders of potatoes. Manson Publishing Ltd, London. (Chapter 2)
Wall DA, Friesen GH (1990) Effect of duration of green foxtail (Setaria viridis) competition on potato (Solanum tuberosum) yield. Weed Technol 4:539–542
Westermann DT (2005) Nutritional requirements of potatoes. Am J Potato Res 82:301–307
Woli P, Hoogenboom G, Alva A (2016) Simulation of potato yield, nitrate leaching, and profit margins as influenced by irrigation and nitrogen management in different soil and production regions. Agric Water Manag 171:120–130
Yamaç SS, Todorovic M (2020) Estimation of daily potato crop evapotranspiration using three different machine learning algorithms and four scenarios of available meteorological data. Agric Water Manag 228:Article 105875
Yuan BZ, Nishiyama S, Kang Y (2003) Effects of different irrigation regimes on the growth and yield of drip-irrigated potato. Agric Water Manag 63(3):153–167
Zebarth BJ, Rosen CJ (2007) Research perspective on nitrogen BMP development for potato. American Journal of Potato Research 84:3–18
Zebarth BJ, Bélanger G, Cambouris AN, Ziadi N (2012) Nitrogen fertilization strategies in relation to potato tuber yield, quality, and crop N recovery. Sustainable potato production: global case studies 165–186

Recent Advances in Organic Potato Cultivation

Anupam Adarsh, Pankaj Kumar Ray, Hemant Kumar Singh, Shashank Shekhar Solankey, Kamlesh Kumar Singh, and Neeharika Kanth

Abstract Organic farming is gaining global popularity as a response to the detrimental effects of chemical-based agriculture and the increasing challenges posed by climate change. The agricultural sector is actively seeking sustainable solutions to reduce carbon footprints by minimizing synthetic inputs and recycling farm resources. Organic farming presents a promising alternative, but its success hinges on the development of comprehensive production technologies. Key components for establishing a sustainable organic system include variety development, rapid nutrient availability through natural sources, and effective organic weed and pest management. Potatoes, a staple of the Indian diet, stand to benefit significantly from recent advancements in organic farming. States such as Punjab, Haryana, Uttar Pradesh, Bihar, West Bengal, and Assam offer substantial opportunities for the growth of organic potato farming. Over the past decade, research and development efforts have led to the creation of production technologies that achieve yields surpassing the national average, underscoring the viability of organic potato farming

A. Adarsh · S. S. Solankey
Department of Horticulture (Vegetable Science), Nalanda College of Horticulture, Noorsarai, Nalanda, Bihar Agricultural University, Sabour, Bhagalpur, Bihar, India

P. K. Ray
Krishi Vigyan Kendra, Saharsa, Bihar Agricultural University, Sabour, Bhagalpur, Bihar, India

H. K. Singh
Department of Agriculture, School of Agriculture and Development, Central University of South Bihar, Gaya, Bihar, India

K. K. Singh
Department of Soil Science, Tirhut College of Agriculture, Dholi, Muzaffarpur, Dr. Rajendra Prasad Central Agricultural University, Pusa, Samastipur, Bihar, India

N. Kanth (✉)
Department of Horticulture, Pt. Deen Dayal Upadhyay College of Horticulture and Forestry, Piprakothi, Dr. Rajendra Prasad Central Agricultural University, Pusa, Samastipur, Bihar, India
e-mail: neeharika@rpcau.ac.in

S. S. Solankey (ed.), *Advances in Research on Potato Production*, Advances in Olericulture, https://doi.org/10.1007/978-3-031-82710-5_9

for sustainability. The shift towards organic farming is especially important in the context of climate change, which pressurizes the agricultural industry to adopt practices that reduce environmental impact. By minimizing synthetic products and focusing on resource recycling, organic farming can help mitigate carbon emissions and contribute to climate change mitigation. Although transitioning to organic systems presents challenges, it also offers a critical opportunity to ensure the long-term sustainability of agriculture, particularly for essential crops like potatoes.

Keywords Potato · Organic manures · Weed management · Organic certification

1 Introduction

Organic farming is becoming increasingly popular globally as a response to the harmful effects of chemical-based agriculture and the mounting challenges posed by climate change (Willer et al. 2021). In the past twenty years, organic farming has developed into a widely embraced alternative agricultural practice. This growing trend is largely a reaction to the environmental harm brought by conventional farming methods, which have historically focused on boosting crop production to satisfy increasing food requirements (Lal 2020). Alongside this, there has been growing global awareness regarding food safety and environmental sustainability (Willer and Lernoud 2020). Consumers are increasingly concerned about the food they consume, including potatoes, which are a key dietary staple. Public concerns, both globally and locally, centre on the harmful chemical residues often found in potatoes, as well as the associated risks to health and the environment.

In the era of sustainability, there has been growing enthusiasm for organic potato farming that aligns with ecological, economic, and social sustainability principles. India's diverse climatic conditions and soils offer considerable opportunities for organic potato production (Grewal et al. 1992). States like Punjab, Haryana, Uttar Pradesh, Bihar, West Bengal, and Assam offer significant opportunities for the growth of organic potato farming (Pandey and Sarkar 2005). The country's year-round growing capacity, geographical advantage, high production volumes, strong international demand, ample sunlight, and relatively low labour costs make it ideal for organic potato production. The primary aim of organic potato cultivation is to improve the health and sustainability of interconnected ecosystems, such as soil, plants, animals, and humans (IFOAM 2020).

2 What is Organic Potato Farming?

Sustainable potato cultivation is based on growing potatoes without the use of synthetic chemicals, such as fertilizers, pesticides, herbicides, or artificial growth regulators (Singh and Lal 2020). This farming method prioritizes sustainable practices, including the use of improved seed varieties (such as genetically enhanced strains), crop rotation, reduced tillage, and organic fertilizers derived from crop wastes, animal wastes, nitrogen-fixation of plants, green manuring, and external organic sources (Kumaraswamy 2002). Mineral-rich rocks, like phosphate and gypsum, are also utilized. Insects and disease control are achieved through environmentally friendly methods, such as biopesticides and biological control measures (Litchfous 2009). Additionally, the use of renewable energy sources, such as solar and wind power, is encouraged, along with natural plant growth regulators produced biologically (Sharma and Singh 2010a, b). Organic farming techniques like crop rotation and manual weeding are also essential for effective weed management.

The goals of organic potato farming include:

1. Producing potatoes that are both nutritionally rich and available in sufficient quantities
2. Collaborating with natural systems rather than trying to control them
3. Optimizing the sustainable utilization of local resources
4. Encouraging and supporting biological cycles within farming ecosystems, including soil microbes, plants, and animals (Narayanan 2010)
5. Protecting the essential biological functions that connect soil, water, nutrients, and humans
6. Ensuring and enhancing soil fertility over time
7. Prioritizing the use of renewable resources whenever feasible within localized farming systems
8. Creating closed nutrient cycles, reducing dependency on external inputs
9. Avoiding environmental pollution from farming activities
10. Preserving the genetic diversity of potato varieties and surrounding ecosystems
11. Ensuring potato farmers receive adequate returns, a safe working environment, and job satisfaction
12. Evaluating the overall ecological and social impacts

3 Approaches for Organic Potato Production

Various approaches can be employed, either separately or together, to cultivate organic potatoes. These include:

1. **Organic Amendments:**
    - Bulk organic manures like farmyard manure (FYM) and vermicompost
    - Green manure crops such as dhaincha and sunnhemp
    - Concentrated organic manures like groundnut cake and neem cake
2. **Biofertilizers:**
    - Using *Azotobacter*, phosphorus-solubilizing bacteria, and mycorrhiza to boost soil fertility

3. **Biodynamic Farming:**
    - Implementing holistic agricultural practices that align with natural cycles
4. **Homeonutrients and Protectants**
    - Organic nutrient solutions designed to enhance crop health
5. **Bio-enzyme Approaches**
    - Application of beneficial enzymes to improve soil health
6. **Agnihotra**
    - A Vedic technique involving the burning of specific substances to purify the atmosphere and benefit plant growth
7. **Panchagavya**
    - A traditional preparation made from five cow-based products (milk, curd, ghee, urine, and dung), which is used as a growth promoter

4 Outcomes from Organic Approaches

Research in the Indo-Gangetic plains has shown that vermicompost is superior to organic manure compared to FYM, as it may contain enzymes, hormones, growth regulators, and essential nutrients. For instance, when vermicompost was applied, the potato varieties Kufri Anand, Kufri Sutlej, Kufri Pukhraj, Kufri Chipsona-1, and Kufri Chipsona-2 yielded 14.04, 17.52, 16.02, 16.0, and 12.04 t/ha, respectively. In comparison, FYM application resulted in yields of 20.6, 21.7, 18.5, 18.4, and 13.6 t/ha, respectively. While inorganic fertilizers produced higher yields initially, organic yields have demonstrated a consistent increase over time, indicating greater yield stability in organic systems. The average nutrient content in different manures is present in Table 1.

(i) **Biodynamic Approach:** Biodynamic agriculture is the systematic and cooperative exploitation of the energies from the cosmos, Mother Earth, plants, and cows. It depends on basic principles of soil biotechnology and microbiology. The majority of manures and biopesticides used are made on the farm, thus advanced facilities are not necessary. Biological agriculture includes biodynamic compost and field spray preparations (BD-500-507), which can provide agriculture and ecology with long-term sustainability (Paull 2011). The fundamentals of biodynamic farming include replenishing the soil with organic matter in the form of humus, boosting the population of microorganisms, applying soil-life and health-promoting elements deftly, and managing compost and manure in a biodynamic manner. Biodynamic farming methods involve using vermicompost (12.5 t/ha) during planting, spraying "Shakti," a micronutrient tonic, dipping tubers in a 2% jaggery solution called "Gur" (containing 5 kg CPP/ha), using the pesticide "Urja," and applying two sprays of the fungicide "Divya."

Table 1 Average nutrient content in different manures

Amendment	Percentage content (dry wt)		
	Nitrogen	Phosphoric acid (P_2O_2)	Potash (K_2O)
Bulky organic manures			
Farmyard manure	0.95	0.62	2.20
Rural compost	0.75	0.63	1.05
Urban compost	1.35	0.62	1.43
Sewage sludge	2.75	0.75	0.35
Sewage sludge activated	5.41	3.15	0.62
Vermicompost	1.80	0.22	0.40
Green manure crops			
Dhaincha (*Sesbania aculeata*)	2.01	0.32	2.03
Sunnhemp (*Crotalaria juncea*)	2.16	0.48	2.11
Cluster bean (*Cyamopsis tetragonoloba*)	1.46	0.25	1.89
Cowpea (*Vigna catjang*)	2.45	0.56	2.32
Concentrated organic manure			
Castor cake, Cotton seed cake, un- decorticated	4.3	1.78	1.29
Neem cake	3.89	1.78	1.58
Safflower cake	5.18	1.00	1.42
Un– decorticated	4.88	1.39	1.22
Ground net cake	7.29	1.48	1.32
Linseed cake	4.85	1.40	1.30

Source: Rawal et al. (2020)

(ii) **Homeopathic Approach:** It has been proven that homeopathy is the most natural, miraculous, ultra-scientific, economical, and methodical technique for treating diseases of all kinds, whether pathogenic or physiological (Majewsky et al. 2014). This method can provide safer protection, and efficient treatment. In the light of these facts, homeopathic medications have also been attempted for treating plant diseases. Studies carried out at the Modipuram have shown that the method is inappropriate for producing potatoes (Mendonça and Bonamin 2014).

(iii) **Bio-enzyme Approach:** The bio-enzyme method in potato farming involves targeted treatments before and throughout the growth stages to enhance plant health and productivity. Before planting, potato tubers are soaked in a Bio-smart solution at a concentration of 3 grams per litre and then shaded to dry (Sivakumar and Thangavel 2016). This treatment helps prepare the tubers for optimal growth. Following planting, Bio-charge is applied at 20 and 30 days after planting (DAP) at a concentration of 3 millilitres per litre. At 45 DAP, Yes-Winner is administered at the same concentration to promote crop development. A third treatment, combining Bio-smart with Nitrobenzene at 2.5–3 millilitres per litre, is applied at 60 DAP. At 70 DAP, a fourth application is made using Bangaram (1 millilitre per litre) and Bio-Gold Sp (1–1.5 grams per litre). Weekly sprays of Bio-smart and Bio-charger continue until 105 DAP to maintain plant health and improve yield. According to data from Modipuram, this bio-enzyme treatment plan has successfully enhanced potato crop growth and yield.

(iv) **Agnihotra Approach:** The "Rishis" of the country were exceptionally powerful in the past and provided cheerful lives for people as a way of serving them. To keep humanity delighted, the sloka "Loka Samastha Sukhinobhavanthu" was recited often along with performing specific rituals. One of these was the "Fire Ritual," a means for calling upon the Devas or Gods in charge of the corresponding domain's existence. The scriptures describe the kingdom of Devas as having an inner government, similar to human governments, where each deity has a specific role to oversee. During the fire ritual, a fire is lit, and offerings, primarily plant-based items, are made to the relevant deity. These offerings are dipped in cow's ghee, which is flammable. The fire, symbolizing Agni, receives the offering and delivers it to the appropriate deity. As the offering is made, a mantra is recited, invoking the deity through the sound of the chant. Agni, acting as the purifier, facilitates the delivery of the offering. Once invoked, the deity is pleased by the sacrifice and bestows blessings upon the person making the offering to fulfil their desires. Here, the Sun God alone is invoked with a single, plain speech at both sunrise and sunset: the phrases "Adithyaya swaahaa, adithaya idam mamama" at sunset and "Prajapathaye Swaahaa, prajapathaya idam mamama" at sunrise, respectively. The method has not been proven to be successful in cultivating potatoes.

(v) **Panchagavya Approach:** The five cow products that makeup panchagavya are urine, dung, milk, cure, and ghee, which have miraculous, mystical effects when used and blended properly. The five basic elements, such as Earth, Water, Fire, Air, and Space of which all living things are composed, are said to be out of balance everywhere in Indian culture, cows are revered as gods (Kamadhenu) and are regarded as having divine status. When and in whatever form they are used, items made from cows' bodies have the power to channel cosmic energy. When cosmic energy enters a living system, it purges any imbalances from physical, chemical, biological, and physiological phases, restoring the growth process and transforming the organism into something wholesome. The most recent variety of "Panchagavya," which is appropriate for a range of horticulture and agricultural crops, is composed of the following:

(a) Cow dung slurry (from Goober gas plant) = 4 kg
(b) Fresh cow dung = 1 kg
(c) Cow's urine = 3 litres
(d) Cow's milk = 2 litres
(e) Cow's curd = 2 litres
(f) Cow's ghee = 1 kg
(g) Sugarcane juice = 3 litres
(h) Tender coconut water = 3 litres
(i) Banana (ripe) = 12 nos.
(j) Toddy (if available) = 2 litres

Following the instructions from steps 1 to 10, all the ingredients are placed into wide-mouthed mud pots, concrete tanks, or plastic containers. The mixture is stirred twice daily, once in the morning and once in the evening, while keeping the container uncovered in the shade. By the seventh day, the "Panchagavya" stock solution is ready for use. This approach has revitalized organic farming in Tamil Nadu, and its advocates believe it will eventually spread worldwide. However, it has not proven to be effective in developing a package of practices for organic potato cultivation in the plains of North India. The technique used in organic potato production comprises choosing the right variety, preparing the seed, preparing the seedbed, and planting.

Choice of Variety The National Standards for Organic Production (NSOP) guidelines state that species and varieties grown in organic farming must be resistant to pests and diseases, as well as suited to the soil and climate of the area. When selecting crop varieties for organic farming, genetic diversity should be taken into account. Research on potatoes at CPRI-RS, Modipuram revealed that newer varieties such as Kufri Mohan, Kufri Khyati, Kufri Garima, and Kufri Sadabahar, as well as older varieties like Kufri Pushkar and Kufri Anand, are promising for organic farming because of their higher potential for marketable yields (Table 2). For process grade tuber yield, Kufri Himsona and Kufri Chipsona-3 are preferred cultivars.

Potato Seed According to National Standards for Organic Production (NSOP) guidelines, all planting materials and seeds must be certified organic. Use normal seed or planting material that has not been chemically treated if such material is not readily accessible. Seed needs to be purchased from a trustworthy supplier and changed every 3–4 years. In organic farming, the use of transgenic plants, seeds, or plant material is forbidden. By incorporating environmentally friendly techniques to lower the vector (aphid) population to essential thresholds of 20 per 100 compound leaves, it is possible to generate healthy seed potatoes from potato crops. Aphids are tracked by setting up a yellow water trap. Sticky traps with a yellow hue are used 8–10 days after planting to catch aphids and whiteflies. The height of the

Table 2 Promising potato varieties under organic farming in plains

Variety	Tuber yield (t/ ha)		Duration	Resistance
	Marketable	Total	(Days)	
Kufri Khyati	28.45	31.42	70–80	Early and late blight
Kufri Mohan	28.38	29.94	90–100	Late blight
Kufri Garima	27.05	29.35	90–100	Late blight
Kufri Pushkar	26.15	27.93	90–100	Late blight, wart
Kufri Sadabahar	26.0	27.04	90–100	Late blight
Kufri Anand	26.05	27.64	90–100	Late blight, wart, tolerant to frost and hopper
Kufri Himsona	21.54	26.05	110–120	Late blight and wart
Kufri Chipsona-3	20.34	23.14	100–110	Late blight

Source: Modified from Luthra et al. (2020)

traps should match the height of the canopy. In organic plots, dried neem leaves are used as mulch because of their decomposing and repellent properties, which help to lower insect populations. The population of whiteflies and aphids decrease when neem leaves are used in combination with weekly foliar applications of horticultural mineral oil at a concentration of 0.5%. The National Centre for Integrated Pest Management (NCIPM), New Delhi, and ICAR-VPKAS, Almora developed light traps to catch moths at night.

Seed Preparation Ten to fifteen days prior to sowing, seed potatoes are taken out of the cold storage. For optimal sprouting, seed tubers are stored in shades with uniform light and proper ventilation. Tubers can be stored on the floor in a thin layer or in baskets, wooden boxes, and plastic trays.

Seedbed Preparation The ideal technique for breaking up the soil during the first stages of organic farming is subsoiling. The hard pan that needs to be broken right below the plough layer should be done in May or June. Eventually, looser, more aerated soil is maintained for improved root development by greater soil, organic carbon, and fauna activity. In the event that there is insufficient soil moisture, pre-planting irrigation should be done 7–10 days prior to field preparation. Till the field to a depth of 20–25 cm using a mouldboard plough or disc harrow. Follow this by performing one or two additional tilling and cross-planking operations. For effective irrigation, water consumption, and adequate drainage, the field should be finely tilted and levelled.

Planting Time and Spacing The ideal planting period occurs when daytime temperatures are below 32 °C and nighttime temperatures fall below 20 °C during the tuberization phase, which typically begins about 25–30 days after planting. For optimal growth, tuber initiation, and bulking, temperatures between 15 and 25 °C are most effective. Adjust planting schedules to align with these temperature ranges for successful tuber development. Recommended row spacing is 67.5 cm, with plant spacing between 20 and 35 cm depending on seed size. For medium-sized tubers (35–45 mm), a plant spacing of 20 cm is appropriate. A planting depth of 8–10 cm should be maintained. After bed planting (two or three rows, base 132 cm), implement micro-irrigation (Singh and Lal 2020).

Mulching Mulching is an efficient traditional method for weed control, soil moisture conservation, and slowing the breakdown of organic wastes in the soil. It is an extremely effective method of suppressing weed expansion in the growing season, especially for annual crops. Mulching's primary goal is to prevent weeds from receiving sun radiation, which will inhibit crop growth. Weed germination is also inhibited, and this method preserves soil moisture, allowing potato plants to emerge more quickly. For this activity, dry grasses, pine needles, crop leftovers, and other vegetative debris may be used. A potentially effective method for controlling weeds when used in combination with micro-irrigation is plastic mulching (Ramakrishna et al. 2006; Unger et al. 2010). Organic mulches consist of materials such as straw, perennial grasses, water hyacinth, residues from perennial crops like bananas and sugarcane, sugarcane bagasse, sawdust, newspapers, and shredded paper (Silva et al. 2015). Synthetic mulches include materials like polyethylene and polypropylene sheets or films (Rao 2000). Mulching enhances the growth and yield of crops such as potatoes, cabbage, tomatoes, okra, and onions. It suppresses weeds by physically covering the soil surface, which provides shading, reduces soil temperature, exerts allelopathic effects, and blocks light necessary for the germination of many small-seeded weed species (Hussein and Radwan 2004). Mulch can also affect specific weeds like *Cynodon dactylon* and *Sorghum halepense* (Rao 2000).

Nutrient Management According to the National Standards for Organic Production (NSOP) guidelines, only biodegradable materials from plant or animal sources that are produced on the farm should be utilized for nutrient management. The use of non-farm-generated biodegradable materials from plant, animal, or microbiological sources is restricted (NSOP 2020). Fertilizers made of non-synthetic minerals and imported biological sources have to be considered an additional supply of nutrients rather than a substitute for recycling spent nutrients. Preventing nutrient losses, heavy metal build-up, and other pollution is the aim of nutrient management. Mineral fertilizers ought to be applied solely as an adjunct to materials derived from carbon. Fertilizers that are allowed, such as organic or mineral fertilizers, should only be applied on farms after other methods of managing fertility have been maximized. It is forbidden to use manures that contain human excreta in order to stop the spread of infectious agents, parasites, and pests. Mineral fertilizers must be applied in their natural state; they cannot be chemically treated to increase their solubility. Certification may allow for some exceptions, but nitrogen-containing mineral fertilizers will not be included in these exceptions. The use of inputs like mineral potassium, magnesium fertilizers, trace elements, manures, fertilizers with a high concentration of heavy metals, and/or other undesirable materials like basic slag, rock phosphate, and sewage sludge is restricted. It is forbidden to use any synthetic nitrogenous fertilizers. Potato crops can thrive with the application of well-decomposed biogas slurry, farmyard manure, compost, vermicompost, recycled crop residues, and other approved organic materials (Singh and Lal 2020).

Crop Rotation Not every crop should come before or after another, such as potatoes. Depending on their overall farming condition, farmers can choose potato cultivars and modify crop length. In cropping and intercropping systems, other crops function similarly. Potato-based cropping systems should incorporate at least one crop from legumes, vegetables, or green manures. A two- to three-year rotation may be used instead of continuously cultivating potatoes on the same land to prevent insect build-up. The experiments conducted by the Institute identified several effective crop rotation combinations: maize-potato-onion-groundnut, potato-green gram-cowpea (vegetable), and potato-okra. For organic nutrient management and sustaining productivity, basmati rice, potatoes, and late-sown wheat were also utilized.

Green Manuring Due to their organic matter and extra nitrogen content, green manures increase soil productivity and potato tuber output. The crop plan should be modified to include green manure crops like sunnhemp or dhaincha. An addition of 60–100 kg of nitrogen per hectare is made by a leguminous crop with 10–20 t of biomass per hectare. Black scurf, common scab, and other soilborne illnesses are among those that also help to lower the occurrence of. For optimal breakdown, a crop of green manure should be added to the soil 40–50 days after seeding.

Composting and Vermicomposting One popular and highly helpful method for converting agricultural waste into crop nutrition is the composting of crop wastes and the creation of vermicompost. This is a crucial component of organic farming that is always present. This task can be completed using a variety of approaches. The difference between conventional and integrated organic systems is presented in Table 3.

Organic system (Biofertilizer/microbial formulation): Rhizobium (250 g/ha seed treatment), *Bacillus subtilis* (250 g/ha seed treatment), *Azotobacter* (250 g/ha seed treatment), *Trichoderma* (4 kg/ha with 100 kg FYM for mixing in soil).

Water Management The stolon formation and tuber initiation stages of potato development are the most vulnerable to water stress, making potatoes a crop that is susceptible. When implementing a ridge-furrow system, the time interval strategy is typically used in the autumn. A plot of eight to ten furrows, each measuring 6–8 m in length, is constructed, and water is poured into it until the water level reaches half of the ridge height. During each irrigation, about 50 mm of water is used. At the beginning of the season, when temperatures are moderate, irrigation is planned at intervals of 8–10 days; in the winter, this interval is extended to 12–15 days. In order to achieve optimal skin setting, water application is discontinued 10–12 days before to haul cutting. When micro-irrigating, it is preferable to base irrigation on cumulative pan evaporation (CPE). Here, the situation is one of irrigation. Acceptance of recommendations from regional research institutions is possible. In a potato crop, drip irrigation is typically applied every other day for 30–45 min at a rate of 100–125% CPE, and spray irrigation is applied twice a week for 90–120 min

Table 3 Difference between conventional and integrated organic systems

Nutrient management	Table purpose potato	Processing potato
Conventional inorganic system	Apply 180 kg of nitrogen (N), 80 kg of phosphorus (P), and 100 kg of potassium (K) per hectare using both band placement and broadcasting techniques.	Apply 270 kg of N, 80 kg of P, and 150 kg of K per hectare using both band placement and broadcasting methods.
Integrated organic system	Incorporate approximately 10–16 tonnes per hectare directly into the soil. Use *Azotobacter*, *Bacillus subtilis*, and *Trichoderma* for seed treatment. Combine the stover with 12 tonnes of farmyard manure (FYM) and 6 tonnes of vermicompost per hectare. Apply FYM before planting and vermicompost during earthing up.	Incorporate about 5–10 tonnes per hectare directly into the soil. Employ *Azotobacter*, *Bacillus subtilis*, and *Trichoderma* for seed treatment. Mix the stover with 15 tonnes of FYM and 6 tonnes of vermicompost per hectare. Apply FYM before planting and vermicompost during earthing up.

Source: Karalus and Rauber (1997)

at a rate of 120–150 CPE. Comparing flood irrigation (40–50%) to sprinkler irrigation (15–20%), water losses under these systems are lower. Dripper or sprinkler irrigation, which accurately provides water in the root zone while minimizing nutrient losses, has been shown in organic agricultural tests to attain system productivity of over 90%. Potato and system productivity can be increased under organic farming by implementing a micro-irrigation system.

Inter-cultural Operations When potato plants reach a height of 10–15 cm, inter-cultural practices are performed for 20–25 days to alleviate soil compaction, enhance root aeration, and control weeds. Insufficient soil cover at planting, exacerbated by irrigation-induced erosion of ridges—especially in lighter soils—can lead to stolons becoming exposed to light, potentially turning into aerial stems. This exposure can also cause tubers in the ridges to turn green and become unsuitable for sale. The field is exposed to sunlight throughout the day, with inter-cultivation and weed removal done manually in the morning. The following day, after applying the remaining nitrogen using vermicompost or other organic inputs, ridges are re-established. For small fields, hand tools such as spades and khurpas are employed, while bullock-drawn intercultivators and mouldboard ploughs are also used. For mechanical cultivation, equipment such as ridgers and 2–4 row cultivators towed by tractors are utilized.

NPOP Norms for Weeds, Diseases, and Pests Management The National Programme for Organic Production (NPOP) guidelines ought to be given careful consideration in order to manage weeds, illnesses, and pests. To minimize losses in organic production, emphasis should be placed on balanced nutrition, selecting the right variety, maintaining biologically active and fertile soils, implementing crop rotations, incorporating green manures, practising intercropping, using mulching techniques, and preparing land effectively. By managing hedgerows, nesting places, and other natural habitats well, we can safeguard and promote the natural enemies of diseases and pests. In order to achieve equilibrium in the cycle of pests and predators, an ecological equilibrium must be established. Products made on farms using regional plants, animals, and microorganisms are allowed to control weeds, pests, and diseases. Physical and thermoselective weed management techniques are also accepted. When regular soil renewal or rotation is not possible, the use of thermosanitization to treat pests and diseases is limited. Depending on the circumstances, authorization from the certifying body is necessary. Before being utilized in an organic field, all equipment from conventional farming techniques needs to be thoroughly cleaned and cleared of any leftovers. It is forbidden to use artificial fertilizers, growth regulators, insecticides, fungicides, herbicides, or other pesticides. It is forbidden to use genetically modified products or creatures (Baudry et al. 2017).

Weed Management Crop-weed competition is controlled by manipulating fertility, companion cropping, crop rotations, and the ideal planting date and plant population (Gianessi and Reigner 2007). Organic farmers should adhere to these key principles:

1. Begin with a clean slate and maintain cleanliness for success

2. Prevention is preferable to cure
3. Seeds from a single year can result in weed problems for 7 years

Effective and sustainable weed management involves using a combination of methods rather than relying on a single approach (Singh and Lal 2020). Implementing optimal crop production techniques, enhancing the key factors for potato crop growth, and fostering competitive conditions for crops can help selectively suppress weeds during the early stages of growth (Baker and Brown-Rosen 2007). It involves growing potato plants quickly and vigorously as well as maintaining a crop environment that always smothers weeds. In general, weeds that emerge beneath a more robust crop canopy are weak and won't significantly impact tuber productivity. In plains, the first 20–40 days are crucial for weed management. Since they spread out across fields quickly and suffocate weeds, potato cultivars with aggressive and quick growth patterns might be better weed competitors. Green manure crops, such as cowpeas and dhaincha, can be planted to suffocate weeds and can be rotated every 2 years in a given field. In addition to green manuring (45–50 days) which will lessen their intensity as the weeds are buried and degrade with the green manure crops, 2–3 years of crop rotation will help lower the amount of weed seed bank in the field (Manorama 2004). A region's soil type will determine how well the seedbed should be prepared. Potato plants emerge more quickly when the soil is properly cultivated. Pre-plant tillage procedures help eliminate weeds and provide a boost to the crop. It is recommended to provide pre-planting irrigation prior to cultivation if the soil moisture content in the seedbed is insufficient. Planting potatoes at the ideal moment gives them a competitive advantage. First and foremost, a variety should be planted at the ideal time, and well-sprouted seed tubers should be used at the ideal planting depth and crop geometry. In order to ensure that fertilizer inputs stay in the crop's root zone for effective nutrient utilization, they should be carefully positioned in bands 5–6 cm below seed tubers upon planting. After plants reach a height of 10–15 cm, 20–25 days after planting, inter-cultivation is carried out. A crop canopy will eventually cover the land, stopping weeds from overtaking it.

Hot Weather Cultivation For managing weeds, diseases, and pests, this can be a cost-effective and environmentally beneficial technique. Utilizing a cropping system, one can take advantage of the hot and dry summer months to effectively desiccate these threats. For the control of annuals as well as perennials like *Cynodon dactylon* L., two to three deep-field cultivations are quite beneficial at this time. Comparably, soil solarization might be quite helpful in certain circumstances or for high-quality potato crops like seed, etc. Solarization controls weeds through several mechanisms: (1) thermal destruction of seeds, (2) thermal destruction of seeds triggered to germinate, (3) breaking seed dormancy, leading to the death of emerging seedlings, and (4) biological control by weakening seeds or seedlings through various means. Transparent plastic was found to be effective in reducing weed populations for up to a year following solarization (Zimdahl 2013). During the solarization process, soil temperatures increase by 8–12 °C, with the most significant effects observed in the top 5–10 cm of soil, which explains its effectiveness against weed seed germination and seedling growth. Additionally, solarization has been shown to significantly boost concentrations of soil nutrients such as NH4-N and DTPA-extractable Mn (Rao 2000; Patricio et al. 2006). Transparent polyethylene (TPE) sheets with a thickness of 0.05 and 0.10 mm are used for soil solarization, and they are left in place for 30–40 days during periods of high temperature subsequent light irrigation of the field (48 h before this activity).

Diseases Management Potato crops are susceptible to a range of diseases caused by worms, bacteria, viruses, and fungi. Common diseases include stem necrosis, apical leaf curl virus, root-knot nematode, black scurf, black scab, bacterial wilt (Paffrath 2002a, b; Schepl and Paffrath 2003a, b), soft rot, charcoal rot, and various potato viruses (X, S, M, V, and Y) (Saucke and Döhring 2004). Black scurf, in particular, is often exacerbated by high levels of raw organic materials from manure or grass-clover pre-crops, especially under unfavourable climatic conditions (Karalus 2000; Radtke and Karalus 2000).

In organic farming, disease management relies on a combination of biological and cultural practices since chemical control methods are prohibited. Essential strategies include using disease-free seed and resistant varieties and implementing summer ploughing and green manuring to prevent diseases. Proper handling of tubers, such as removing damaged ones before storage and modifying planting dates to avoid severe outbreaks, is crucial. Thick ridges can help reduce tuber infections.

Field scouting is necessary to identify and remove contaminated plants, and irrigation should be adjusted based on weather conditions. For controlling late blight, *Bacillus subtilis* at 0.25% and *Trichoderma viride* at 0.7% in liquid formulations can be used. Copper hydroxide at 0.2% and copper oxychloride at 0.2%, with certification approval, can help manage the disease. Avoiding solanaceous crops, applying appropriate organic materials for nitrogen, crop rotation, and treating seed tubers with 3% boric acid before storage can mitigate diseases like bacterial wilt and fusarium wilt. Soil solarization and bio-fumigation with cruciferous plants, such as radish and mustard, are also effective. Potatoes should be harvested before soil temperatures exceed 28 °C to prevent soft rot and charcoal rot. Trap crops like marigolds can help manage root-knot nematodes while controlling weeds and routine rouging reduce viral infections.

Pest Management Common pests affecting potatoes include leafhoppers, mites, white grubs, thrips, cutworms, caterpillars, and potato tuber moths. To minimize issues with mites and leafhoppers, avoid planting potatoes during high temperatures. Prior to planting, summer ploughing and tillage can expose pests like white grub larvae to natural predators. Liquid cultures of entomopathogenic nematodes (EPNs), such as *Heterorhabditis indica*, or EPNs cultured in Galleria cocoons, can effectively decrease white grub larvae populations in the soil. In organic farming, applying dried neem leaves as mulch can reduce insect populations due to their decomposing and repellent effects. Managing thrips can be aided by frequent, light irrigation, while caterpillars can be controlled using light traps. Potato tuber moths can be managed through practices like heavy ridging, using water traps, and covering potato heaps with dried lantana and eucalyptus leaves (Rawal et al. 2020).

Harvesting When the skin has hardened and the potatoes are fully grown, harvest them. Stop watering 8–10 days before haulm cutting to make sure the land has enough moisture for harvest and that no soil clods form. Remove haulms when crop is fully developed and don't start digging for another 10–15 days to guarantee that the tuber skin is totally ripe at the time of harvest and heap construction. For manual harvesting, remove the tubers by hand using a shovel and khurpa. The animal-drawn plough is an additional way that is faster than the manual approach. Larger farms deal with parched fields and delayed harvesting. Consequently, tractor-driven and multipurpose diggers are efficient, reasonably priced, and do the least amount of damage to the tubers.

5 Certification of Organic Produce

The organic food items (Annexure IV) need to be certified by Regional Councils for PGS-India and approved certification agencies for NPOP. To guarantee that organic foods meet all established requirements, certification is crucial. In accordance with the Food Safety and Standards (Contaminants, Toxins and Residues) Regulation, 2011, organic foods must also meet the standards for metallic contaminants, aflatoxins, naturally occurring toxic substances (NOTS), and infections. When it comes to pesticide residues, the upper limit is the level of quantification (LoQ) or 5% of the maximum limits specified by the aforementioned rules. Food safety regulations are not the only standards that organic goods need to meet. Organic goods have to comply with several requirements as stated in the Food Safety and Standards or Food Product Standards and Food Additives (Regulations, 2011).

6 Brief Description of Terms and Processes Used in Organic Farming

Accreditation The process that the National Accreditation Body devised to determine if a Certification Body is qualified to certify organic farms, goods, and procedures in accordance with the National Standards for Organic Products is known as accreditation.

Accreditation Body The organization designated by the National Programme for Organic Production Steering Committee to accredit Certification Bodies is known as the Accreditation Body.

National Accreditation Body The National Accreditation Body recognizes an accredited certification body as a legal entity that satisfies the National Programme for Organic Production (NPOP) accreditation requirements, is recognized for certifying organic products, and grants operators the authorization to use the Certification Trade Mark on behalf of the Accreditation Body.

Accredited Programme It is the programme of Certification Body that has been approved by the Accreditation Body on the basis that it is in compliance with the provisions of the National Programme for Organic Production (Lairon 2010).

Buffer Zone A distinct, well-defined region separating an organic production site from a conventional industrial unit is known as a buffer zone.

Certificate of Accreditation The certificate of accreditation is a document that Agricultural and Processed Food Products Export Development (APEDA) issues to the Certification Body on behalf of the National Accreditation Body (NAB), attesting to the Certification Body's compliance with the standards envisioned by the National Programme for Organic Production and its ability to certify producers in accordance with the requirements outlined by the National Standards for Organic Production.

Certification Through the issuance of a Scope Certificate, the recognized Certification Body verifies that the operator's production or processing system has undergone a comprehensive evaluation and complies with the requirements specified in the National Programme for Organic Production. This process is known as certification.

Certification Trademark The Ministry of Commerce is the owner of the India Organic Logo.

Certification Programme It will refer to a system that was run by a Certification Body in line with the standards outlined below for achieving conformance certification.

Compliance Compliance refers to following the guidelines established by NPOP. Compliance refers to following the guidelines established by NPOP.

Consultancy The term "consultancy" refers to an independent advising service for organic activities that is apart from the processes of certification and inspection.

Conventional Farming Conventional farming refers to agricultural practices that do not adhere to the fundamental requirements of organic production or that rely on the use of artificial fertilizers, chemicals, and/or pesticides.

Conversion The process of converting a conventional farm to an organic farm is known as conversion. Another term for this is a transition.

Conversion Period This is the period of time that separates the beginning of organic management from the crop's organic certification.

Inspection It must involve a site inspection to confirm that an activity is carried out in compliance with NPOP requirements for production, processing, and chain of custody.

ISO Guide 65/ISO 17065 These are the standard prerequisites that Certification Bodies using the product certification system must meet.

ISO 17011 These are the standard prerequisites that must be met by accrediting bodies in order to accredit Certification Bodies.

Labelling Any written, printed, or graphic representation that appears on the label of a product that has been certified organic with the intention of encouraging sales is referred to as labelling.

Licence The right to use the Certification Trade Mark "India Organic Logo" to certify that a business's operations or goods are organic is given to the operator by the recognized Certification Body acting on behalf of the National Accreditation Body.

National Programme for Organic Production (NPOP) The "India organic logo," certifying bodies' requirements, norms, methodology, and laws have all been established by NPOP for organic production.

National Standards for Organic Production (NSOP) The National Standards for Organic Production (NSOP) programme lays out the guidelines for growing, harvesting, producing, processing, and exchanging organic goods.

Organic According to the criteria, the term "organic" designates a specific farming technique.

Organic Agriculture This farm design and management method aims to establish an ecosystem that can attain sustainable productivity without relying on artificial external inputs like pesticides, fertilizers, or chemicals.

Package of Practices the rules for growing and processing organic crops in a given area.

Parallel Production Any production in which the same unit grows, breeds, handles, or processes the same goods in both certified organic and noncertified organic quality is referred to as parallel production. Parallel production also occurs when the same product is produced both "organically" and "in conversion."

Plant Protection Product Any material used in the production, storage, transportation, distribution, or processing of food, agricultural products, or animal feeds with the intention of preventing, eliminating, attracting, repelling, or controlling any pest or disease, including unwanted species of plants or animals, is referred to as a plant protection product.

Participatory Guarantee System of India Under the Department of Agriculture and Cooperation, Ministry of Agriculture and Farmers Welfare, Government of India, PGS-India is a decentralized organic agricultural certification system. It is a regionally appropriate quality assurance programme that operates outside the third-party certification structure and emphasizes stakeholder participation, including producers and consumers.

Standards Standards shall mean the NSOP approved by the National Steering Committee for NPOP.

TraceNet APEDA provides stakeholders with TraceNet, an internet-based electronic solution that facilitates process certification for the export of organic products from India that meet NPOP or NSOP requirements.

TraceNet Collects Stores and Reports In the Indian organic supply chain, operators, producer groups, and certification authorities enter quality assurance data as well as forward and backward traces.

7 Factors Affecting Organic Potato Production

1. **Optimum nitrogen supply to crop at tuber bulking:** It has been noted that when crops are cultivated organically, crop yellowing increases gradually from the start (i.e., 30 days after planting) to tuber bulking. Consequently, the crop experiences slowed growth, leaf phoma spots emerge, and forced crop senescence develops. As a result, there is a reduction in the photosynthetic area and crop growth period, which lowers output. Generally speaking, a typical potato crop that produces 30 tonnes of tubers removes 120–140 kg of nitrogen from the soil. However, given the dynamics of nutrient release from organic manure, in particular, nitrogen from organic sources is not able to keep up with the bulk requirements of potato crops that are harvested for a short time. Research carried out at Modipuram has demonstrated that growing potatoes repeatedly on the same spot led to a gradual rise in the amount of nitrogen in the soil and the production of tubers.
2. **Variety:** Non-processing, medium- to long-duration potato varieties, such as Kufri Anand, Kufri Sutlej, Kufri Pukhraj, Kufri Bahar, Kufri Arun, and Kufri Suriya, have generally produced more than processing types, such as Kufri Chipsona1 and Kufri Chiposna-2, among today's commercial varieties.
3. **Organic approach used:** Compared to the biodynamic technique, bulky organic manures (such as vermicompost and farmyard manure) produced more. In organic potato cultivation, the homeopathic, biofertilizer, and bio-enzyme techniques were found to be completely ineffective.
4. **Growing potato on same site:** In Modipuram, the production of organic potatoes was 40–60% lower in the first year compared to conventionally grown potatoes using inorganic fertilizers, depending on the variety. However, by continuing to cultivate potatoes in the same location in later years, the gap between the two methods' yields decreased. However, even after 5 years of consistent organic potato farming on the same land, the yields are still lower than those of conventionally farmed crops that use inorganic fertilizers.
5. **Source of organic manure:** Vermicompost has been found more effective than farmyard manure (both rural compost as well as city compost) for higher tuber yield.

6. **Diseases and pests:** The primary pests of potatoes in plains have been reported to be aphids during the main crop season and leaf hoppers and mites in the early crop. White grub, cutworms, tuber moths, and white flies are extremely dangerous pests in hills. Many biopesticides, such as neem and other plant extracts, *Trichoderma*, nucleopolyhydrosis virus, *Bacillus thuvinginsis*, urja (under the biodynamic method), and samhar, Sasya Rakshini (under the homeopathic approach), have been explored in an attempt to control them in organic potato crops. However, these have only been found to be somewhat successful. Similar to pests, biocontrol agents like B4 and B5 (*Bacillus subtilis*) have been proven to be somewhat efficient in controlling potato diseases such as phoma, early and late blight, soft rot, dry rot, common scab, etc. Field research carried out at Modipuram has shown that late blight and viral diseases in plains cannot be controlled by spraying fungicides "Divya" and "Sambhavi," which are used under biodynamic and homeopathic approaches, respectively. The findings thus run counter to previous literature-based research by Khurana and Gupta (1981) which claimed that plant diseases, especially those caused by viruses or fungi, could be safely, profitably, and practically averted even in fields with the use of homeopathic medicines.

8 Conclusion and Future Strategies for Increasing Organic Potato Production

One of the most crucial prerequisites for selecting appropriate types of organic potatoes is to ensure a good crop yield. It will be encouraged to use cultivars that are more suited to the biotic conditions of the area (such as biological control of diseases and pests, climatic stress, etc.). It's possible that the best landraces and types used in conventional farming are not appropriate for organic potato cultivation. Since traditional plant breeding is known to be slow, the generation of cultivars appropriate for organic farming can be sped up with the help of marker-aided selection. In a similar vein, more genes that protect potato crops from disease and insect pests must be added. For greater nutrient application, which raises biomass output and increases partitioned efficiency to raise the harvest index and economic yield, improved nutrient use efficiency is necessary. Slow yet sustained release of available nutrients can be more effectively utilized by potato cultivars that require a longer time to be absorbed. It is worth looking into the potential of periodically spraying vermiwash to prevent infections and feed nutrients, especially nitrogen. Similar research should be done on the application of biopesticides to reduce the 50% yield loss that is attributed to insects and other pests in organic potato farming. Chemicals may have damaged or destroyed biological controls, therefore efforts should

References

Baker B, Brown-Rosen E (2007) Crop protection product formulation for the organic market. In: Felsot AS, Racke KD (eds) Crop protection products for organic agriculture: environmental, health, and efficacy assessment. Oxford University Press, Oxford, p 2007

Baudry J, Peneau S, Alles B, Touvier M, Hercberg S, Galan P, Amiot M-J, Lairon D, Mejean C, Kesse-Guyot E (2017) Food choice motives when purchasing in organic and conventional consumer clusters: Focuson sustainable concerns (The NutriNet-Sante Cohort Study). Nutrients 9:88

Gianessi LP, Reigner NP (2007) Review: the value of herbicides in U.S. Crop Production. Weed Technol 21(2):559–566

Grewal JS, Sharma RC, Saini SS (1992) Agro techniques for intensive potato cultivation in India. ICAR Publication, New Delhi, pp 39–47

Hussein HF, Radwan SMA (2004) Associative action between bio-organic farming and safety weed control methods on pea productivity. In: Symposium on scientific research and technological development outlook in the arab world, 3 Cairo. Proceedings. Cairo

IFOAM - Organics International (2020) Principles of Organic Agriculture. IFOAM

Karalus W (2000) The role of pre-crop management in reducing black scurf in organic potato farming. Eur J Organ Agric 27(4):209–218

Karalus W, Rauber R (1997) Effect of presprouting on yield of maincrop potatoes (Solanum tuberosum) in organic farming. J Agron Crop Sci 179:241–249

Khurana SMP, Gupta GC (1981) Homeopathy: promise and prospects for plant protection. Adv Homoeopathy 1:107–116

Kumaraswamy K (2002) Organic farming relevance and prospects. Indian Soc Soil Sci Newsletter No. 12:1

Lairon D (2010) Nutritional quality and safety of organic food. A review. Agron Sustain Dev 30:33–41

Lal R (2020) Regenerative agriculture for food and climate. J Soil Water Conserv 75(5):123A–124A. https://doi.org/10.2489/jswc.2020.0620

Litchfous F (2009) Organic agriculture and pest management: a review of research and applications. Organ Farm J 21(3):45–58

Luthra SK, Gupta VK, Tiwari JK, Kumar V, Bhardwaj V, Sood S, Dalamu, Kaur RP, Kumar R, Vanishree G, Kumar D, Mhatre P, Chakrabarti SK (2020). Potato breeding in India. Technical bulletin no. 74 (revised). ICAR-Central Potato Research Institute, Shimla, 214 p

Majewsky V, Shah-Rossi D, Betti L, Trebbi G, Bonamin LV, Baumgartner S (2014) Homeopathy in agriculture: a review. Homeopathy 103(2):102–111

Manorama K (2004) Weed management in potato in nilgiris. J Indian Potato Assoc 31(1/2):91–93

Mendonça FMN, Bonamin LV (2014) Homeopathic Phytotherapy and Agrohomeopathy: a systematic review. J Agric Environ Sci 3(1):121–131

Narayanan S (2010) Organic farming in India: relevance, potential, and constraints. Indian J Agric Econ 65(3):60–81

National Standards for Organic Production (NSOP). (2020). Ministry of Agriculture and Farmers' Welfare, Government of India. https://www.apeda.gov.in/apedawebsite/organic/NSOP.pdf

Paffrath A (2002a) Drahtwurmbefall an Kartoffeln. Bioland 23

Paffrath A (2002b) Impact of organic farming practices on potato health and productivity. J Agric Res 45(3):112–120

Patricio FRA, Sinigaglia C, Berros BC, Freitas SS, Tessarioli NJ, Cantarella H (2006) Solarization and fungicides for the control of drop, bottom rot and weeds in lettuce. Crop Protec 25(1):31–38

Paull J (2011) Biodynamic agriculture: the journey from Koberwitz to the world, 1924–1938. J Organ Syst 6(1):27–41

Radtke H, Karalus W (2000) Soil health and potato disease management: the impact of organic materials in organic farming. J Plant Dis Protec 55(2):125–134

Ramakrishna A, Hoang MT, Suhas PW (2006) Effect of mulch on soil temperature, moisture, weed infestation and yield of groundnut in northern Vietnam. Field Crop Res 95(2–3):115–125
Rao S (2000) Principles of weed science, 2nd edn. Science Publishers, New York, p 526
Rawal S, Luthra SK, Malik K, Lal M, Mankar P, Dua VK, Chakrabarti SK (2020) Organic potato production technology for North Indian Plains. ICAR-Central Potato Research Institute, Shimla, p 36
Saucke H, Döhring T (2004) Virus and fungal disease suppression in potato: a study on organic control methods. Plant Pathol J 52(1):33–45
Schepl F, Paffrath A (2003a) Sustainable disease management in organic potato production: a review of techniques and practices. Organ Farm Syst 12(2):67–78
Schepl U, Paffrath A (2003b) Entwicklung von Strategien zur Regulierung des Drahtwurmbefall (Agriotes spp. L.) im o¨ kologischen Kartoffelanbau [The development of strategies to regulate the infestans of wireworms (Agriotes spp. L.) in organic potato farming]. In: Freyer B (ed) Ökologischer Landbau der Zukunft, Beiträge zur 7. Wissenschaftstagung zum Ö kologischen Landbau. Institut für Ökologischen Landbau an der Universität für Bodenkultur, Vienna, pp 133–136
Sharma D, Singh B (2010a) Effect of organic inputs on potato yield and soil health in Indo-Gangetic Plains. Indian J Soil Sci 48:101–108
Sharma R, Singh P (2010b) Renewable energy integration in organic farming. J Sustain Agric 12(4):87–93
Silva PV, Monquero PA, Silva FB, Bevilaqua NC, Malardo MR (2015) Influence of sugarcane straw and sowing depth on the emergence of weed species. Planta Daninha 33(3):405–412
Singh R, Lal P (2020) Organic potato farming: practices and prospects. Indian J Agron 65(2):115–122
Sivakumar T, Thangavel P (2016) Bio-enzyme formulations for sustainable farming. J Environ Biol
Unger PW, Kirkham MB, Nielsen DC (2010) Water conservation for agriculture, soil and water conservation advances in the United States. Soil Sci Society America, New York. (Special Publication, 60)
Willer H, Lernoud J (2020) The world of organic agriculture: statistics and emerging trends. Research Institute of Organic Agriculture (FiBL)
Willer H, Schlatter B, Trávníček J, Kemper L, Lernoud J (2021) The world of organic agriculture: Statistics and emerging trends 2021. Research Institute of Organic Agriculture (FiBL) and IFOAM—Organics International. https://www.organic-world.net
Zimdahl RL (2013) Fundamentals of weed science, 4th edn. Academic Press, New York, p 209

Advances in Soil and Plant Nutrient Management of Potatoes

Papia Biswas and Anand Kumar

Abstract Potatoes play a crucial role in both cropping systems and dietary habits. Advances in soil and nutrient management have significantly boosted potato production. Effective nutrient management is vital for maximizing yield and enhancing tuber quality. Both macro and micronutrients impact tuber yield. Innovations like advanced fertilizer formulations, soil health monitoring technologies and targeted nutrient delivery systems have demonstrated promising improvements in productivity and sustainability. Recent research highlights optimal practices such as integrated nutrient management (INM), site-specific nutrient management (SSNM), fertigation and foliar nutrient application, which can enhance potato growth, yield and quality while ensuring environmental sustainability. Integrating these advanced approaches can help overcome nutrient management challenges, meet the increasing demand for potatoes and support environmental stewardship.

Keywords Potato · Soil · Nutrient management · SSNM · Fertigation

1 Introduction

Potatoes are among the most versatile and widely consumed vegetables worldwide. Native to the Andes region of South America, they have become staple foods in countless cuisines because of their remarkable adaptability and culinary flexibility. It is an exceptionally valuable crop in agricultural production systems, combining impressive yield potential with high nutritional content. Its ability to produce many

P. Biswas (✉)
Department of Agronomy, Nalanda College of Horticulture, Noorsarai, Nalanda, Bihar Agricultural University, Sabour, Bhagalpur, Bihar, India

A. Kumar
Department of Soil Science and Agricultural Chemistry, Nalanda College of Horticulture, Noorsarai, Nalanda, Bihar Agricultural University, Sabour, Bhagalpur, Bihar, India

S. S. Solankey (ed.), *Advances in Research on Potato Production*, Advances in Olericulture, https://doi.org/10.1007/978-3-031-82710-5_10

harvests makes it an attractive option for farmers, while its rich-nutrient profile, which is packed with vitamins, minerals and carbohydrates, supports food security and contributes to a healthy diet. This unique combination of high productivity and nutritional benefits positions potato as a vital staple in global agriculture. Potato serves as a key source of caloric starch in numerous underdeveloped and developing nations globally. The nutritional value of this tuber is influenced by its content of proteins, starches, fats, vitamins, polyphenols, as well as macro-elements and micro-elements. Potatoes are good sources of energy, minerals, proteins, fats and vitamins (King and Slavin 2013).

In recent years, several factors have been shown to affect potato productivity, including the limited use of certified planting material, restricted irrigation practices, long-term monoculture, diseases, pests and inadequate agronomic practices, particularly poor nutrient management. To increase production, it is crucial not only to have favourable agro-ecological conditions but also to intensify the application of agro-technical measures with a particular emphasis on improved fertilization practices. In the face of changing global climate conditions, the negative effects of environmental impacts can be mitigated to some extent through balanced agronomic management strategies. An essential agronomic measure for improving potato production is adequate nutrient management. A proper supply of mineral nutrients can help protect potato plants from adverse growth conditions, which is crucial for achieving higher yields and producing high-quality potatoes.

At the end of the nineteenth century and the beginning of the twentieth century, significant efforts were devoted to investigating the relationships between various production factors and crop yield. Researchers have focused particularly on how nutrient supply, among other factors, influences plant growth and productivity. This period led to substantial advancements in the understanding of how different nutrients affect crops, leading to improvements in agricultural practices and increased crop yields (De Wit 1994). It is established that plant growth is not determined by the total amount of nutrients available but rather by the amount of the most limiting nutrient (Von Liebig 1840, 1855). This highlights the importance of balanced nutrition for achieving optimal plant growth.

Potatoes have a high nutrient demand, and maintaining a consistent supply is essential for ensuring tuber quality. Nutrient deficiencies can negatively impact tuber quality. For early varieties with shorter growing seasons, the peak nutrient requirement typically occurs during the vegetative growth phase and tuber initiation. In contrast, late varieties with longer growing seasons have an extended period of nutrient uptake. A high level of production is supported by favourable soil conditions, effective irrigation, adequate nutrients and other agronomic factors. To optimize yields and enhance potato quality, farmers need to implement best management practices (BMPs) tailored to their specific growing conditions.

Ensuring a sufficient supply of nutrients is crucial for optimal potato growth and profitable production. Potatoes indeed have specific nutritional needs due to their shallow root system and high demand for various nutrients. There are 14 essential nutrients for plant growth, including macronutrients such as nitrogen (N), phosphorus (P) and potassium (K), as well as micronutrients such as zinc (Zn), iron (Fe) and manganese (Mn). Nitrogen (N), phosphorus (P) and potassium (K) are the primary nutrients used in potato production (Davenport et al. 2005). Each nutrient plays a critical role in different stages of potato growth, from root development to tuber formation. Potatoes require adequate levels of nutrients throughout their growth cycle to achieve maximum tuber yield and quality. Nutrient deficiencies can lead to stunted growth, lower yields and poor tuber quality, impacting profitability. Assessing soil fertility through soil testing is essential for understanding nutrient levels and deficiencies. On the basis of the soil test results, a targeted fertilizer application plan can be developed to meet the specific nutrient requirements of potatoes. Fertilizers are used to supplement soil nutrients when natural soil reserves are insufficient to meet plant demands. The application rate, timing and method of fertilizer application are critical factors that influence nutrient availability to potatoes. While ensuring optimal nutrient levels is important for crop productivity, it is also crucial to manage fertilization practices to minimize environmental impacts such as nutrient leaching or runoff into water bodies. A comprehensive approach to nutrient management includes the use of organic sources of nutrients, crop rotation, cover cropping and other sustainable practices to maintain soil health and fertility over the long term. A well-planned nutrient management program is essential for maximizing potato yield and quality while minimizing environmental impact. By understanding the specific nutrient needs of potatoes and implementing appropriate fertilization practices, growers can ensure a healthy crop and sustainable production.

Best management practices (BMPs) include a variety of guidelines and techniques aimed at maximizing yields, enhancing crop quality and reducing costs. These practices include soil preparation, efficient irrigation, precise crop fertilization, pest and disease control and effective harvest and storage methods. However, there is a need for more advanced and efficient approaches, especially in nutrient management, to further improve production outcomes.

2 Role of Macronutrients in Plant Metabolism, Growth and Yield

Macronutrients play crucial roles in the growth and development of potato plants. Each macronutrient is essential for different aspects of potato growth, from root development and disease resistance to tuber formation and quality. The balanced application of nutrients on the basis of soil tests and plant needs is key to achieving optimal potato yields.

Among the essential nutrients, nitrogen (N) is one of the most critical and limiting factors for potato growth and development. Successful potato production relies heavily on proper nitrogen fertilization, as the nitrogen present in the soil is predominantly in organic forms (such as soil organic matter and plant residues) that are not readily available to plants. Adequate levels of mineral nitrogen in the soil are crucial for optimizing yield and increasing dry matter production. Insufficient nitrogen results in poor plant growth and lower yields, whereas excessive nitrogen application can lead to issues such as nutrient leaching, extended growing periods, poor tuber quality and sometimes reduced yields. Inadequate nitrogen nutrition not only decreases the overall yield but also reduces the number of tubers due to decreased leaf area and early defoliation.

Phosphorus is a very important element in many processes that take place in potato (energy transfer in cells, respiration and photosynthesis), thus making it an essential element for the growth and development of potato plants. It significantly enhances root system development, improves nitrogen uptake and encourages earlier and more robust tuber formation. It also increases the plant's resistance to cold temperatures and diseases. Potatoes have high needs for potassium because it promotes root and leaf development, regulates the water balance, reduces susceptibility to damage, protects the plant from disease and the negative effects of adverse weather conditions and affects the quality and length of tuber storage. The lower potassium causes lower yields and quality of the potato tubers.

Ensuring an adequate supply of nutrients is vital for maximizing potato yields, alongside other agronomic strategies. To determine the optimal nutrient levels, it is essential to first understand the specific physiological roles of each nutrient. Additionally, factors such as nutrient interactions (antagonism) and site-specific conditions, including various abiotic and biotic environmental factors, must be considered. With this foundational understanding, potato growers can make well-informed decisions regarding the type, timing and quantity of fertilizer applied, ultimately enhancing crop performance and yield.

2.1 Nitrogen

Nitrogen (N) is crucial for potato growth and affects biomass accumulation, tuber size and yield. Plants utilize nitrogen mainly as ammonium (NH_4^+) and nitrate (NO_3^-), which are essential for chlorophyll, amino acids, proteins and other cellular components. While high nitrogen levels can increase tuber size, which is beneficial for processing potatoes, they may not be ideal for fresh consumption or seed potatoes due to potential quality issues, such as elevated nitrate and acrylamide levels. Nitrogen affects potato yield through the number of stems, tubers per stem and tuber weight with the latter being most influenced by nitrogen. Potatoes require minimal nitrogen in the first 4–5 weeks; excess nitrogen during this period can hinder tuber development by promoting stem growth. Optimal nitrogen application is crucial during tuber bulking, which requires 58–70% of the total nitrogen. Potatoes generally have low nitrogen use efficiency (NUE) due to shallow roots, although later-maturing cultivars benefit more from additional nitrogen.

Determining the right nitrogen supply involves soil and plant-based tests, such as soil mineral nitrogen tests, petiole nitrate tests and leaf chlorophyll meters. Splitting nitrogen applications can reduce leaching risks and better match crop needs but requires precise management to avoid delays in tuber maturity and ensure efficient fertilizer use. Foliar application of urea can address unexpected nitrogen shortages (Qadri et al. 2015).

2.2 Phosphorous

Phosphorus (P) is essential for potatoes, plays a critical role in cellular energy transfer and is a component of nucleic acids, coenzymes and phospholipids. P significantly impacts tuber development, especially during early growth and maturity, increasing tuber size and quality. Potatoes show moderate tolerance to P stress with deficiency leading to reduced photosynthesis and purple leaf discolouration. Unlike nitrogen, P uptake is initially low but increases substantially during tuber bulking, peaking between 40 and 60 days after emergence.

Effective P management involves adjusting fertilizer rates on the basis of soil tests, such as the Mehlich III method, to avoid excessive application. While high soil P levels can reduce fertilizer needs, sandy soils often respond less to P fertilization than loams do (Rosen et al. 2014). Band application is more efficient because of the shallow root system of potatoes, whereas broadcast application may require higher rates and pose runoff risks. P use efficiency is generally low because of P immobility in soils and varying pH impacts. Low P efficiency can lead to overapplication, contributing to water pollution and eutrophication (Stark and Hopkins 2015).

2.3 Potassium

Potassium (K) is a vital macronutrient for potato crops with the highest concentrations found in tubers, comprising approximately 1.7% of dry matter. It plays a crucial role in enzyme activation, cell extension, turgor regulation and stomatal movement. K is essential for starch synthesis, photosynthesis and carbohydrate distribution within plants and significantly influences tuber yield and quality. Adequate potassium is needed to maintain the leaf area, photosynthetic efficiency and proper phloem function. Deficiency symptoms include leaf chlorosis and reduced growth, which are particularly evident in older leaves because of the mobility of K within the plant. Potatoes exhibit maximum K uptake approximately 30–40 days after emergence with K needs decreasing as tubers mature. Nutrients are highly mobile in soil, making sandy soils prone to K leaching and requiring careful management. Effective potassium fertilization is guided by soil tests and is crucial for optimizing yield, but excessive K application can lead to environmental issues such as water pollution.

Different application methods, such as broadcasting and banding and the choice of K form (sulfate or chloride), affect nutrient efficiency and crop growth. Chloride forms can increase osmotic potential and water uptake, leading to increased vegetative growth but potentially diluting K concentrations and affecting tuber development. While useful, traditional soil and plant testing methods often fail to provide the timely and repeated assessments needed for short-duration crops such as potatoes. This highlights the potential for integrating machine learning techniques as nondestructive alternatives for nutrient estimation, improving K management efficiency. By combining conventional and innovative methods, growers can enhance K management strategies, optimize potato yields and address the challenges of K leaching in sandy soils. Potatoes require higher potassium (K) levels than other vegetable crops do with a removal rate of over 270 kg K ha^{-1} to produce approximately 56 mg ha^{-1} of crop (Hopkins et al. 2020). K is crucial for tuber yield, size and quality and affects specific gravity, chip color and storage characteristics by reducing the sugar and amino acid levels that contribute to darkening. The ideal K concentration in tubers for optimal dry matter production is approximately 1.8%, necessitating approximately 226 kg K to produce 50 kg of potatoes (Chowdary et al. 2021). Although K fertilization typically improves yield, the responses can be inconsistent with some studies showing limited effects at high application rates (Torabian et al. 2021). The role of K includes enhancing assimilate translocation to tubers and influencing tuber size and biomass. However, excessive K can increase the tuber water content and decrease the specific gravity. K deficiency impairs photosynthesis, reducing dry matter and starch production. Effective K use efficiency (KUE) is essential with agronomic KUE defined as the crop yield per unit of K available in the soil and fertilizer. For potatoes, KUE varies by soil type and climate with higher efficiency in well-drained soils and higher organic matter, reaching up to 49%.

2.4 Magnesium

Magnesium (Mg) is crucial for potato growth and is often underestimated in crop production. Like potassium (K), it plays key roles in photosynthesis, protein synthesis and enzyme regulation. Mg is central to chlorophyll, vital for light absorption and energy transformation and acts as an allosteric activator for more than 300 enzymes, including those involved in photosynthesis and carbohydrate partitioning (Senbayram et al. 2015). Mg deficiency impairs phloem loading, leading to increased sucrose accumulation in leaves and reduced tuber and starch yields, although starch concentrations in dry weight remain unaffected (Koch et al. 2019). Rapid restoration of Mg availability through foliar applications can quickly rectify phloem transport issues. Mg also affects root growth with mixed results depending on when deficiency occurs during growth stages. Generally, Mg deficiency leads to symptoms such as interveinal chlorosis in older leaves, as Mg is translocated to later stages of growth. Mg fertilization recommendations are sparse, but a typical practice involves the application of Mg at a 3:1 ratio with K. This approach balances the needs of Mg and K, as the overapplication of K can decrease the leaf Mg concentration and negatively affect yield.

2.4.1 Potassium–Magnesium Antagonism

Cation competition, notably between potassium (K) and magnesium (Mg), is a well-documented phenomenon affecting nutrient uptake in plants. A high K supply can induce Mg deficiency, as observed by Koch et al. (2019), who reported decreased Mg concentrations in potato leaves with increased K application. This interaction arises from differences in nutrient uptake mechanisms: Mg relies primarily on mass flow for root delivery, whereas K is absorbed through diffusion. The high mobility and large hydrated radius of Mg can lead to its leaching and reduced availability, exacerbated by competition with K. Mg transporters are unspecific and can absorb both Mg and other cations such as K, leading to impaired Mg uptake under high-K conditions. Conversely, K transport is facilitated by specific transporters (high-affinity transport systems and low-affinity transport systems) that do not transport Mg (Gransee and Führs 2013). In contrast to the belief that K and Mg should not be applied together to avoid antagonistic effects, research shows that combined K and Mg applications yield the best results. For example, potatoes receiving both K and Mg had higher yields than those receiving only one nutrient or none. These findings demonstrate the importance of balanced fertilization for optimizing nutrient uptake and crop yield.

2.5 Calcium

Calcium (Ca) is crucial for potato crops because of its roles in cell wall stabilization, membrane integrity and intracellular signaling. It maintains low cytosolic concentrations while being predominantly bound in vacuoles, where it helps balance anions and cations (Kärenlampi and White 2009). The ability of Ca to form bridges between pectin molecules and stabilize membranes is vital for structural integrity and plant response to environmental stimuli such as drought or pathogens. Transported primarily via the xylem, Ca movement depends on plant transpiration, leading to deficiency symptoms in younger leaves where transpiration is low. Potato tubers, which have minimal transpiration, generally contain low Ca levels (Lazarević et al. 2014). Recent studies have shown that applying Ca near tubers can significantly increase their Ca concentration by increasing Ca uptake through stolons and tuber roots. Potatoes, often grown in acidic soils (pH 4.6), face challenges with Ca and Mg availability because of interactions with toxic metals such as aluminum and manganese. Liming with Ca-containing materials can mitigate these issues by improving the soil pH and Ca availability, although it may affect the availability of other nutrients, such as phosphorus and zinc.

2.6 Sulfur

Sulfur (S) is vital for potato crops because of its role in synthesizing amino acids such as methionine and cysteine, which are crucial for protein production. S deficiency in potato leads to reduced levels of these amino acids, as well as decreased dry matter, specific gravity and starch contents. Research shows that S supply improves these parameters with optimal effects observed at 45 kg ha^{-1}; higher levels do not yield additional benefits (Koprivova and Kopriva 2016). Unlike some crops, potatoes have relatively lower S requirements, but repeated high-yielding cycles can deplete soil S. S fertilization not only enhances potato resistance to diseases such as Streptomyces scabies by acidifying the soil but also reduces the infection rates of *Rhizoctonia solani*. S deficiency mimics nitrogen deficiency, causing chlorophyll loss and yellowing of younger leaves due to impaired thylakoid membrane synthesis. Potatoes are typically fertilized with K_2SO_4, which provides S, although atmospheric S absorption also contributes. Recent declines in atmospheric SO_2 emissions in some regions contrast with increases in others, affecting S availability.

3 Sources of Nutrients

Organic and inorganic sources of nutrients both are crucial for sustainable agriculture, enhancing soil fertility and crop productivity. Organic sources, such as compost, manure, and cover crops, contribute to soil health by improving structure, moisture retention and microbial activity. These inputs enhance nutrient availability through slow-release mechanisms, promoting long-term soil fertility and ecosystem resilience. In contrast, inorganic sources, including synthetic fertilizers, provide immediate nutrient availability, allowing for precise nutrient management and rapid crop response. While they can boost short-term yields, excessive use may lead to soil degradation, water pollution and reduced biodiversity. The integration of both organic and inorganic nutrients offers a balanced approach, optimizing nutrient cycling and promoting sustainable farming practices.

3.1 Organic Sources of Nitrogen

Organic sources of nitrogen are important in sustainable agriculture because they improve soil health, structure and fertility while providing essential nutrients for crop growth. Organic sources of nitrogen not only supply essential nutrients but also enhance soil organic matter, improve soil structure and support a healthy soil ecosystem. Their use can contribute to long-term soil fertility and sustainable crop production.

The following are some common organic sources of nitrogen:

1. **Compost:** Decomposed organic matter from plant and animal residues contains approximately 1–3% N but varies depending on the materials used and the composting process.
2. **Manure:** Organic manure includes cow, pig, chicken, horse and sheep manure. The nitrogen content generally ranges from 0.5% to 2% N. Chicken manure has the highest nitrogen content among animal manures. This material needs to be composted to reduce the risk of pathogens and to stabilize nutrients.
3. **Green Manures (Cover Crops):** Plants grown specifically to be incorporated into the soil to improve fertility. Legumes such as clover, vetch and peas are particularly effective because they can fix atmospheric nitrogen through their root nodules. The nitrogen content varies by plant species, but leguminous green manures can contribute substantial nitrogen when decomposed.
4. **Fish Emulsion:** A liquid fertilizer made from processed fish containing approximately 4–5% N provides a quick source of nitrogen and other nutrients.
5. **Blood Meal:** A dry, powdered product made from animal blood. High, approximately 12–15% N. Acts as a quick-release source of nitrogen but can be expensive.
6. **Feather Meal:** Made from processed poultry feathers. Approximately 12–15% N. Releases nitrogen slowly, providing a relatively long-term nutrient source.
7. **Alfalfa Meal:** Made from ground alfalfa plants. Approximately 2–3% N is present, and other nutrients and organic matter are also provided to the soil.
8. **Soybean Meal:** A byproduct of soybean oil extraction. Approximately 6–7% N is a good source of nitrogen and other essential nutrients.
9. **Worm Castings:** The excrement of earthworms, also known as vermicompost. Approximately 1–2% N. Contains a balanced mixture of nutrients and beneficial microorganisms.

3.2 Inorganic Sources of Nitrogen

Nitrogenous fertilizers are crucial for plant growth, as nitrogen is a key nutrient that plants need for various physiological functions. They can be classified on the basis of their nitrogen content and chemical composition. Common types of nitrogenous fertilizers and their nitrogen contents are listed in Table 10.1.

3.3 Organic Sources of Phosphorus

Organic sources of phosphorus are essential for maintaining soil fertility and supporting healthy crop growth. Phosphorus is a key nutrient that promotes root development, energy transfer and overall plant health. Some common organic sources of phosphorus for crop production are as follows:

Table 10.1 List of nitrogenous fertilizers and their nitrogen content

Fertilizer	Chemical formula	Nitrogen content (%)	Uses
Urea	$CO(NH_2)_2$	46	Most used general purpose N fertilizer, used in high-nitrogen crops
Ammonium nitrate	NH_4NO_3	34	Quick-release nitrogen and often used in combination with other fertilizers
Ammonium sulfate	$(NH_4)_2SO_4$	21	Used in acid soils and adds sulfur which is beneficial for certain crops
Calcium nitrate	$Ca(NO_3)_2$	15.5	High-calcium needs, improves soil structure, often used in fertigation
Potassium nitrate	KNO_3	13.7	Provides both nitrogen and potassium, suitable for crops needing both nutrients
Anhydrous ammonia	NH_3	82	High-efficiency, used in large-scale agriculture, applied directly to the soil
Aqua ammonia	NH_4OH	20–25	Liquid form of ammonia, used in fertigation or as a nitrogen source in solutions
Sodium nitrate	$NaNO_3$	16–17	Provides nitrogen; often used in specialty crops and high-value crops; helps in alkaline soils
Ammonium phosphate (Ammophos)	$(NH_4)_2HPO_4$ or $(NH_4)_2PO_4$	13	Balanced fertilizer providing both nitrogen and phosphorus; used for a wide range of crops
Urea-ammonium nitrate (UAN)	Variable mixture of urea and ammonium nitrate	28–32	Liquid fertilizer; used in fertigation systems and as a top-dressing for crops

Source: Modified from http://agropedia.iitk.ac.in/content/properties-and-uses-organic-and-inorganic-nitrogen-fertilizers-crop-production

1. **Bone Meal:** A ground-up product made from animal bones. Typically, approximately 15–20% P_2O_5 (phosphorus pentoxide) is present.
2. **Rock Phosphate:** A natural, mineral-based source of phosphorus. Typically, approximately 20–30% P_2O_5. Because it provides a long-lasting source of phosphorus, its effectiveness depends on the soil pH and availability.
3. **Fish Meal:** A product made from ground fish. Approximately 2–4% P_2O_5. Additionally, it supplies nitrogen and other essential nutrients.
4. **Manufacturing:** Animal manures that have been composted to stabilize nutrients. In general, P_2O_5 is low but varies, usually approximately 1–2%. In addition, phosphorus, organic matter and other nutrients are provided.
5. **Alfalfa Meal:** Ground alfalfa plants. Approximately 0.2–0.5% P_2O_5. A balanced nutrient profile improves the soil structure.
6. **Bone Char:** Charred bone material. High, approximately 20–30% P_2O_5. These materials can be used to improve phosphorus availability in the soil.

7. **Guano:** Accumulated excrement of seabirds or bats. Varies widely but can be high, approximately 10–15% P_2O_5. Both phosphorus and nitrogen are provided.
8. **Green Manures (Cover Crops):** Plants grown to improve soil fertility. Certain plants, such as buckwheat and mustard, can contribute phosphorus when decomposed. This process is dependent on the plant species; it is usually low, but decomposition adds organic matter and improves soil phosphorus availability.
9. **Wood Ash:** Ash from burning wood. Varies; typically approximately 1–2% P_2O_5. Additionally, it provides potassium and calcium, which are used in moderation, as they can affect the soil pH.

3.4 *Inorganic Sources of Phosphorus*

Inorganic sources of phosphorus are commonly used in agriculture to quickly supply plants with phosphorus, a critical nutrient for root development and overall plant health. These inorganic phosphorus sources can quickly address phosphorus deficiencies in soil and support crop production by providing essential nutrients in forms that plants can readily absorb. The choice of phosphorus source often depends on factors such as soil type, crop requirements and the desired application method. The following are the most common phosphorus fertilizers as presented in Table 10.2:

Table 10.2 List of phosphorus fertilizers and their phosphorus content

Fertilizer	Chemical formula	Phosphorus content (%)	Uses
Superphosphate	$Ca(H_2PO_4)_2$	16–20	Provides a readily available source of phosphorus; commonly used for general fertilization
Triple superphosphate	$Ca(H_2PO_4)_2$	44–46	Higher phosphorus content compared to superphosphate; used for crops with high phosphorus needs
Diammonium phosphate (DAP)	$(NH_4)_2HPO_4$	18–21	Provides both phosphorus and nitrogen; widely used for various crops and soil types
Monoammonium phosphate (MAP)	$NH_4H_2PO_4$	13	Provides phosphorus and some nitrogen; suitable for a variety of crops and as a starter fertilizer
Ammonium phosphate (Ammophos)	$(NH_4)_2HPO_4$ or $(NH_4)_2PO_4$	11–13 (variable)	Provides both phosphorus and nitrogen; used in various agricultural applications
Phosphoric acid	H_3PO_4	54–62 (concentrated form)	Used in liquid fertilizers and to adjust pH; also a raw material for producing other phosphorus fertilizers

Source: Modified from extension.psu.edu/agronomy-guide

3.5 Organic Sources of Potassium

Organic sources of potassium are essential for increasing soil fertility and ensuring healthy crop growth. Potassium is a crucial nutrient that helps with water regulation, enzyme activation and overall plant health. Some common organic sources of potassium for crop production are as follows:

1. **Wood Ash:** Ash produced from burning wood. Typically, approximately 5–10% K_2O (potassium oxide) is present, although this content can vary. Potassium and calcium are provided and used in moderation to avoid excessively altering the soil pH.
2. **Composted Manure:** Manures from animals that have been composted to stabilize nutrients. Generally, approximately 1–2% K_2O is present but varies based on the type of manure and its source. Potassium is added along with organic matter and other nutrients.
3. **Kelp Meal:** A product made from dried and ground seaweed contains approximately 1–3% K_2O.
4. **Banana Peels:** Organic waste from bananas. Approximately 1–2% K_2O. It can be composted or used as mulch to release potassium slowly.
1. **Molasses:** A byproduct of sugar production. Approximately 3–4% K_2O. It is often used in compost or soil amendments to increase microbial activity and nutrient availability.
2. **Seaweed Extracts:** Liquid or powdered forms of seaweed processed into fertilizers. Varies but often contains approximately 1–5% K_2O. Potassium and other beneficial nutrients and hormones are provided.
3. **Animal Manure from Potassium-Rich Feeds:** Manure from animals fed potassium-rich diets. Varies in terms of diet and manure type are generally greater than those associated with standard manure. Potassium and organic matter were added to the soil.

3.6 Inorganic Sources of Potassium

Inorganic sources of potassium are widely used in agriculture to provide a readily available form of potassium for crops. Potassium is crucial for various plant functions, including water regulation, enzyme activation and overall growth. Some common inorganic sources of potassium for crop production are mentioned in Table 10.3.

3.7 Organic Sources of Calcium

Calcium is an essential nutrient for plants and plays a crucial role in cell wall structure, enzyme activity and overall plant health. Both organic and inorganic sources of calcium can be used to improve soil fertility and enhance crop production. Common sources are listed in Table 10.4.

Table 10.3 List of potassium fertilizers and their potassium content

Fertilizer	Chemical formula	Potassium content (%)	Uses
Potassium chloride (KCl) or muriate of potash (MoP)	KCl	60–62	Widely used; provides potassium efficiently; suitable for most crops
Potassium sulfate (K_2SO_4)	K_2SO_4	50–52	Provides potassium and sulfur; beneficial for sulfur-deficient soils and sensitive crops
Potassium nitrate (KNO_3)	KNO_3	13.7	Provides both potassium and nitrogen; used in high-potassium and high-nitrogen needs
Sulfate of potash-magnesia ($K_2SO_4{\cdot}2MgSO_4$)	$K_2SO_4{\cdot}2MgSO_4$	22–24 (potassium)	Provides potassium, magnesium and sulfur; used in soils needing additional magnesium
Potassium thiosulfate ($K_2S_2O_3$)	$K_2S_2O_3$	25–30	Provides potassium and sulfur; used for both soil and foliar application, especially in liquid form

Source: Modified from agritech.tnau.ac.in portal

Table 10.4 List of organic calcium sources and their calcium content

Fertilizers	Calcium content (%)	Uses
Bone meal	20–30	Provides calcium and phosphorus; used as a soil amendment and fertilizer to improve soil fertility
Eggshells	90–95	High in calcium; often crushed and used as a natural soil amendment to enrich calcium levels
Compost	0.5–2	Contains varying levels of calcium depending on feed stocks; improves soil fertility and structure
Wood ash	5–20 (variable)	Provides calcium along with other minerals; used to raise soil pH and add nutrients
Seaweed (kelp) meal	0.5–1	Contains trace amounts of calcium; improves soil health and provides a range of nutrients
Manure (e.g., cow, horse)	0.5–2	Provides calcium and other nutrients; improves soil fertility and adds organic matter
Calcium-rich cover crops	Variable (depends on crop)	Some cover crops accumulate calcium, which can be returned to the soil to improve calcium levels

Source: Modified from Bunemann et al. (2006)

3.8 *Inorganic Sources of Calcium*

These sources (Table 10.5) of calcium can be selected on the basis of soil conditions, crop needs and specific agricultural practices. Organic sources often provide additional benefits to soil health and structure, whereas inorganic sources typically offer more immediate nutrient availability.

Table 10.5 List of calcium fertilizers and their calcium content

Fertilizers	Chemical Formula	Calcium content (%)	Uses
Calcium carbonate	$CaCO_3$	36–38	Used to raise soil pH and provide calcium; commonly known as lime
Calcium sulfate (gypsum)	$CaSO_4$	23–29	Used to improve soil structure and drainage; provides calcium and sulfur
Calcium nitrate	$Ca(NO_3)_2$	15.5	Provides calcium and nitrogen; used in fertigation and for calcium-deficient soils
Calcium chloride	$CaCl_2$	27–30	Provides calcium in liquid form; used for fertigation and as a soil amendment
Lime (agricultural lime)	$CaCO_3$	36–38	Commonly used to adjust soil pH and supply calcium; similar to calcium carbonate
Calcium magnesium carbonate	$CaMg(CO_3)_2$	20–22 (Ca), 10–12 (Mg)	Provides calcium and magnesium; used to correct calcium and magnesium deficiencies

Source: Modified from Jing et al. (2024)

Table 10.6 List of organic magnesium sources and their magnesium content

Fertilizer	Chemical formula	Magnesium content (%)	Uses
Magnesium sulfate	$MgSO_4$	9.8–10.5	Provides magnesium and sulfur; used for correcting magnesium deficiencies and in fertigation
Dolomitic lime	$CaMg(CO_3)_2$	10–12	Provides magnesium and calcium; used to adjust soil pH and improve soil structure
Magnesium oxide	MgO	55–60	Provides a high concentration of magnesium; used in soils with severe magnesium deficiencies
Kieserite	$MgSO_4{\cdot}H_2O$	25–27	A naturally occurring mineral providing magnesium and sulfur; used in agriculture to improve soil magnesium levels
Epsom salts	$MgSO_4$	10–12	Provides magnesium and sulfur; commonly used in liquid applications and as a soil amendment
Magnesium chloride	$MgCl_2$	18–20	Provides magnesium in a soluble form; used in liquid fertilizers and for correcting deficiencies in the soil

Source: Modified from Finch et al. (2014)

3.9 Organic Sources of Magnesium

Magnesium is an essential nutrient for plants and is crucial for photosynthesis, enzyme activation and overall plant health. It helps with chlorophyll production and improves nutrient uptake. Both organic and inorganic sources of magnesium are used in agriculture to ensure adequate magnesium levels in the soil (Table 10.6).

Table 10.7 List of magnesium fertilizers and their magnesium content

Fertilizer	Chemical formula	Magnesium content (%)	Uses
Magnesium sulfate	$MgSO_4$	9.8–10.5	Provides magnesium and sulfur; used for correcting magnesium deficiencies and in fertigation
Dolomitic lime	$CaMg(CO_3)_2$	10–12	Provides magnesium and calcium; used to adjust soil pH and improve soil structure
Magnesium oxide	MgO	55–60	Provides a high concentration of magnesium; used in soils with severe magnesium deficiencies
Kieserite	$MgSO_4{\cdot}H_2O$	25–27	A naturally occurring mineral providing magnesium and sulfur; used in agriculture to improve soil magnesium levels
Epsom salts	$MgSO_4$	10–12	Provides magnesium and sulfur; commonly used in liquid applications and as a soil amendment
Magnesium chloride	$MgCl_2$	18–20	Provides magnesium in a soluble form; used in liquid fertilizers and for correcting deficiencies in the soil

Source: Modified from Finch et al. (2014)

3.10 Inorganic Sources of Magnesium

These sources of magnesium can be selected based on soil conditions, crop needs and specific agricultural practices. Organic sources often improve soil health and structure, whereas inorganic sources typically increase nutrient availability (Table 10.7).

4 Micronutrients: Their Roles and Sources

Compared with macronutrients although they are needed in smaller quantities, micronutrients are essential for plant health and growth. Potato crops require adequate levels of micronutrients, and it is crucial to achieve optimal yield and quality (Table 10.8). Table 10.9 shows key micronutrients and their roles in potato crop production.

5 Micronutrient Application Strategies

Balancing these micronutrients is key to preventing deficiencies and ensuring optimal growth and yield for potato crops (Table 10.10). Soil tests can help determine the specific needs and appropriate application rates for these nutrients.

Table 10.8 List of micronutrients, role and deficiency effect in potato sources and their magnesium content

Micronutrients	Role in potato growth and yield	Effects of deficiency
Iron (Fe)	Essential for chlorophyll synthesis and respiration; aids in enzyme functions	Chlorosis (yellowing of leaves), poor growth, reduced yield
Manganese (Mn)	Involved in photosynthesis, nitrogen metabolism and enzyme activation	Interveinal chlorosis, reduced growth, lower yield
Zinc (Zn)	Critical for enzyme function, protein synthesis and growth regulation	Stunted growth, leaf curling, poor tuber development
Copper (Cu)	Important for photosynthesis, respiration and enzyme activation	Wilting, poor tuber formation, reduced disease resistance
Boron (B)	Aids in cell wall formation, carbohydrate transport and reproductive development	Poor tuber quality, hollow heart, reduced flowering and fruiting
Molybdenum (Mo)	Essential for nitrogen fixation and enzyme function in nitrate reduction	Leaf curl, yellowing, poor growth and reduced yield
Chlorine (Cl)	Involved in osmoregulation, photosynthesis and disease resistance	Wilting, reduced growth, increased susceptibility to diseases

Source: Adapted from Kumar et al. (2022)

Table 10.9 Key micronutrients and their organic and inorganic sources

Micronutrients	Organic sources	Nutrient content (approximate)	Inorganic sources	Nutrient content (approximate)
Iron (Fe)	Compost, well-rotted manure, seaweed extract	0.1–0.5% Fe	Ferrous sulfate, chelated iron (EDDHA, EDTA)	20–30% Fe (ferrous sulfate), 6–12% Fe (chelated forms)
Manganese (Mn)	Compost, green manures, certain plant extracts	0.05–0.2% Mn	Manganese sulfate, manganese oxide	30–35% Mn (manganese sulfate), 60–65% Mn (manganese oxide)
Zinc (Zn)	Compost, animal manures, bone meal	0.01–0.05% Zn	Zinc sulfate, zinc oxide	35–40% Zn (zinc sulfate), 80–85% Zn (zinc oxide)
Copper (Cu)	Compost, seaweed, certain plant-based fertilizers	0.01–0.03% Cu	Copper sulfate, copper oxide	25–30% Cu (copper sulfate), 80–85% Cu (copper oxide)
Boron (B)	Compost, seaweed, wood ash	0.01–0.05% B	Boric acid, sodium borate	17–20% B (boric acid), 11–14% B (sodium borate)
Molybdenum (Mo)	Compost, certain leguminous cover crops	Trace amounts (usually not specified)	Sodium molybdate, ammonium molybdate	39–45% Mo (sodium molybdate), 54–60% Mo (ammonium molybdate)
Chlorine (Cl)	Compost, seaweed, certain organic matter	Trace amounts (usually not specified)	Potassium chloride (KCl), calcium chloride	60–62% Cl (potassium chloride), 30–32% Cl (calcium chloride)

Source: Modified from https://ariesagro.com/micronutrients

Table 10.10 Micronutrient application strategies and advantages

Method	Description	Advantages	Best used for
Foliar sprays	Micronutrients are applied directly to the plant leaves. This method allows for rapid correction of deficiencies	Quick nutrient absorption through leaves; effective for immediate needs	Iron, manganese and other micronutrients with high foliar absorption rates
Soil amendments	Micronutrient fertilizers are incorporated into the soil. This method addresses deficiencies before planting or during the growing season	Long-term solution; useful for treating widespread deficiencies; improves overall soil nutrient levels	General soil fertility management and large-scale deficiencies
Composting and organic matter	Incorporating composted manures, green manures and other organic materials into the soil. This method gradually improves soil micronutrient levels and overall soil health	Enhances soil structure and nutrient availability; promotes long-term soil health	Gradual improvement of soil micronutrient content and overall soil quality

Source: Adapted from Kumar et al. (2022) and https://www.pioneer.com/us/agronomy/micronutrients_crop_production.html

6 Nutritional Disorders in Potatoes

6.1 Nitrogen

Nitrogen is a crucial nutrient for plant growth and development, influencing both vegetative and reproductive stages. Reduced growth, pale leaves and reduced tuber yield are the typical symptoms of nitrogen deficiency.

Toxicity Symptoms

- **Delayed Maturity:** Excess nitrogen promotes vigorous vegetative growth at the expense of reproductive development, delaying harvest readiness.
- **Excessive Top Growth:** Overly lush foliage may occur, which can lead to imbalanced plant growth.
- **Hollow Heart and Growth Cracks:** These physical defects in tubers can result from rapid, uneven growth.
- **Increased Susceptibility to Biotic Diseases:** Plants with excessive nitrogen may have weakened defenses against pests and diseases.
- **Reduced Tuber-Specific Gravity:** Excess nitrogen can lead to lower tuber density, affecting quality.
- **Difficulty in Vine "Burning":** Excess nitrogen can make it harder to desiccate vines before harvest, complicating the harvesting process.

6.2 Phosphorus

Phosphorus is essential for various plant processes, including energy transfer, photosynthesis and nucleic acid formation. The following is a detailed overview of phosphorus-related issues:

Deficiency Symptoms

- **Fewer and Smaller Tubers**: Plants often produce fewer tubers and those that do develop are smaller because of inadequate nutrient availability.
- **Stunted Plants**: Overall, plant growth is reduced, leading to shorter and less vigorous plants.
- **Yellowing of Older Leaves**: Older leaves may turn yellow (chlorosis) as the plant mobilizes phosphorus from them to support newer growth.
- **Small, Dark Green Younger Leaves**: New leaves may be smaller and darker green because of the stress caused by phosphorus deficiency.

Toxicity Symptoms

- **Ties Up Other Nutrients**: Excess phosphorus can interfere with the availability and uptake of other essential nutrients, such as the following:

Calcium: High phosphorus levels can lead to calcium deficiency, which may cause various plant disorders.

Zinc: Phosphorus excess can also reduce zinc availability, leading to symptoms such as leaf chlorosis and poor plant growth.

6.3 Potassium

Potassium is a vital nutrient for plants and plays a key role in various physiological processes, including water regulation, enzyme activation and photosynthesis. Both deficiency and excess potassium can lead to significant plant health issues and reduced crop productivity. Here, we look at potassium-related issues in detail:

Deficiency Symptoms

- **Necrosis of Leaf Margins**: The edges of older leaves often develop a scorched appearance and may eventually die, starting from the margins and progressing inward.
- **Premature Leaf Senescence**: Leaves may age and fall off earlier than normal, reducing a plant's overall photosynthetic capacity.

Toxicity Symptoms

- **Reduced Tuber-Specific Gravity**: Excessive potassium can lead to lower tuber density, affecting the quality and marketability of tubers.
- **Reduced Calcium and Magnesium Uptake**: High levels of potassium can interfere with the absorption of calcium and magnesium, which are crucial for cell wall integrity and other physiological functions.

- **Degradation of the Soil Structure**: Excess potassium can lead to changes in soil properties, affecting its structure and potentially leading to issues such as reduced aeration and drainage.

6.4 Calcium

Calcium is essential for several key plant functions, including cell wall structure, root development and overall plant stability. Both deficient and excess calcium can have significant impacts on plant health and productivity. Here, is a detailed overview:

Deficiency Symptoms

- **Yellow Curled Leaves on Upper Leaves**: The upper leaves may show yellowing and curling, often accompanied by other symptoms, such as tip burns.
- **Tip Burns**: The tips of the leaves may turn brown and dry out, reflecting the plant's struggle to maintain proper growth and development.
- **Small Chlorotic New Leaves**: New leaves might be smaller and show signs of chlorosis (yellowing), indicating that the plant is not receiving enough calcium for proper growth.

Toxicity Symptoms

- **Reduced Magnesium Uptake**: Excessive calcium can interfere with a plant's ability to absorb magnesium, which may lead to symptoms of magnesium deficiency, such as interveinal chlorosis (yellowing between leaf veins) and leaf curling. Sulfur is a critical nutrient for plants and is involved in amino acid synthesis, enzyme activation and overall plant metabolism. The details of sulfur deficiency and its effects are as follows:

6.5 Sulfur

Deficiency Symptoms

- **Reduced Growth**: Plants may exhibit stunted or slowed growth due to impaired protein synthesis and overall metabolic processes.
- **Pale Green or Yellow Leaves**: Older leaves typically show signs of chlorosis (yellowing), starting with a pale green color and progressing to yellow. This occurs because sulfur is essential for chlorophyll production.
- **Reduced Number of Leaves**: A plant may produce fewer leaves overall, which can limit its ability to photosynthesize and impact overall health and productivity.

6.6 *Iron*

Deficiency Symptoms

- **Youngest Leaves Affected First**: Symptoms of iron deficiency typically appear first on the youngest or newest leaves, as iron is not easily mobilized within the plant and is thus preferentially allocated to younger growth.
- **Interveinal Chlorosis**: The most characteristic symptom of iron deficiency is yellowing (chlorosis) between the leaf veins, while the veins themselves remain green.
- **Severe Chlorosis**: In cases of severe deficiency, the entire leaf may become chlorotic, meaning that it turns yellow or pale and may eventually lead to leaf death.

6.7 *Boron*

Boron is a vital micronutrient for plants that influences several critical physiological processes, including sugar transport, cell division and hormone metabolism. The following is a detailed overview of boron deficiency and its effects:

Deficiency Symptoms

- **Dead Growing Buds**: Newly growing and developing buds may die off or fail to develop properly. This can lead to reduced plant vigor and stunted growth.
- **Bushy Appearance with Shorter Internodes**: Plants may appear overly bushy owing to a lack of elongation between nodes. This is often caused by the inability to properly regulate cell growth and division.
- **Thickened and Upward-Rolled Leaves**: Leaves may become thicker and roll upward. This rolling can be a response to impaired cell development and reduced flexibility.
- **Darkened and Collapsed Leaf Tissue**: Leaf tissue may darken and collapse, resulting in a mottled or distorted appearance.
- **Brown Necrotic Patches on Tubers**: Tubers can develop brown, necrotic patches, which are areas of dead tissue that can affect tuber quality and storage ability.
- **Internal Rust Spots**: Tubers and other plant parts may develop internal rust spots, indicative of severe boron deficiency affecting internal tissues.

6.8 *Copper*

Copper deficiency in plants can indeed cause a variety of symptoms, including the following:

- **Flaccid and Wilted Young Leaves**: Copper is essential for several physiological processes, including photosynthesis and the formation of lignin, which helps strengthen plant cell walls. Without enough copper, young leaves may become weak, floppy and lose turgor pressure, leading to wilting.

- **Dropping Terminal Buds**: Copper deficiency can interfere with the development of new growth, including that of flower buds. This can cause terminal buds to drop prematurely, impacting the plant's ability to flower and produce fruit.
- **Necrotic Leaf Tips**: Copper is crucial for the synthesis of certain enzymes involved in plant metabolism. A lack of copper can lead to poor cell wall structure and function, causing the tips of leaves to turn brown and die, a condition known as necrosis.

6.9 Zinc

Zinc deficiency in plants presents with several distinct symptoms:

- **Chlorosis in Young Leaves**: Zinc is crucial for chlorophyll production and enzyme function. Deficiency often causes young leaves to turn light green or yellow, as they lack sufficient chlorophyll.
- **Narrow, Upwardly Cupped Leaves**: Leaves may become narrower than normal and display upward cupping or curling. This occurs because zinc is vital for leaf expansion and growth regulation.
- **Tip Burns**: The tips of leaves may develop necrotic areas or burn spots due to poor nutrient uptake and overall reduced leaf vitality.
- **Green Veins**: A common symptom is that the leaf veins remain green, whereas the rest of the leaves turn yellow.
- **Spotting and Blotching**: Zinc deficiency can cause irregular patterns of dead tissue or blotching on leaves, often resulting in a mottled appearance.
- **Erect Leaf Appearance**: Leaves may become more upright or erect instead of having a typical flat orientation, affecting overall plant aesthetics and function.

7 Nutrient Management Practices

Sustainable potato production relies heavily on advanced nutrient management practices that optimize crop yields while minimizing environmental impact. Here are some key practices to consider:

7.1 Soil Testing

It is essential that soil samples for testing be taken from the top 6–8 inches of soil, as this is where most of the active root zone for potatoes is located. The samples should be representative of the entire area to ensure accurate results. A soil analysis for potatoes includes pH, organic carbon and amount of essential nutrients such as phosphorus (P), potassium (K), magnesium (Mg), calcium (Ca), zinc (Zn) and boron (B). These nutrients play critical roles in potato growth and development. Depending on local conditions and suspected deficiencies, tests for sulfur (S), manganese (Mn), iron (Fe) and copper (Cu) may be warranted if issues are anticipated or identified. Soil nitrate tests can be valuable, particularly in drier climates and finer-textured soils, for assessing nitrogen availability. These tests are typically conducted at a depth of 2 feet for accurate results. Interpretation of soil test results is key to determining fertilizer recommendations. Local or regional research-based interpretations ensure that recommendations are tailored to specific soil and climate conditions, optimizing nutrient use efficiency and crop performance. While soil test results may vary slightly between laboratories due to differences in extraction methods and interpretations, the core nutrient levels should be fairly consistent. It is important to use reputable labs and understand their methodologies when comparing results over time. A reliable soil analysis forms the foundation of an effective nutrient management program for potatoes. By understanding soil nutrient status through accurate testing and interpretation of results, growers can make informed decisions on fertilizer application rates and types, ensuring optimal potato yield, quality and environmental stewardship.

7.2 *Integrated Nutrient Management (INM) in Potato*

The concept of integrated nutrient management (INM) underscores the synergy between organic and inorganic inputs to optimize soil health and crop production. This approach recognizes that both types of nutrients have unique benefits and, when used together, can increase crop productivity while mitigating soil degradation. INM involves the strategic integration of organic materials (such as compost, manure and crop residues) and inorganic fertilizers (containing essential nutrients such as nitrogen, phosphorus and potassium) to provide a balanced nutrient supply to crops. By combining organic and inorganic inputs, INM aims to maximize nutrient use efficiency, improve soil structure and fertility and sustain long-term crop productivity. Organic materials contribute to soil organic matter, improve water holding capacity and enhance microbial activity, whereas inorganic fertilizers offer the precise nutrient concentrations needed for specific growth stages. The continuous use of inorganic fertilizers alone can lead to nutrient imbalances, soil acidification and environmental pollution. INM helps mitigate these issues by promoting sustainable soil management practices that maintain soil health and fertility over time. The successful implementation of INM requires adaptation to local soil and climatic conditions. This includes understanding the nutrient requirements of crops, soil nutrient availability and appropriate application rates and timings for both organic and inorganic inputs. Studies such as those mentioned above aim to scientifically evaluate the effects of integrated nutrient management on specific crops such as potatoes. This research provides valuable insights into optimal practices that can increase potato growth, yield and quality while ensuring environmental sustainability. Integrated nutrient management represents a holistic approach to sustainable agriculture, balancing the benefits of organic and inorganic inputs to support efficient crop production, soil health and food security. Studies and investigations into the impact of INM on crops such as potatoes contribute to advancing agricultural practices that are both productive and environmentally friendly.

Kumar et al. (2012) reported a balanced approach in which 50% of the recommended dose of nitrogen, phosphorus and potassium (NPK) was supplied through inorganic fertilizers and the remaining 50% of nitrogen through organic manures (such as farm yard manure, press mud or vermicompost) resulted in increased potato tuber yield and increased productivity. Sanyal et al. (1993) investigated the impact of FYM (farmyard manure) and mineral fertilizers at different rates: 50% RDF (recommended dose of fertilizer), 75% RDF and 75% RDF combined with 10 tons of FYM per hectare. The highest net production value per unit of investment was achieved with the 75% RDF plus 10 tons of FYM per hectare. Kate et al. (2005) conducted a study in Maharashtra to assess how different potato cultivars (HPS-II/67, Kufri Jyoti and S-1) respond to various nutritional management strategies. The treatments evaluated were 100% of the recommended dose of fertilizers (RDF; 120:60:120 kg NPK per hectare), 100% RDF plus 5 tons of FYM per hectare, 100% RDF plus FYM and biofertilizers (*Azospirillum* and *Azotobacter*), 75% RDF plus FYM and biofertilizers and 50% RDF plus FYM and biofertilizers. The treatment combining 100% RDF with FYM and biofertilizers resulted in the highest total dry matter production per plant (69.7 g) and tuber weight per plant (285.6 g).

7.3 Site-Specific Nutrient Management in Potato

Site-specific nutrient management (SSNM) is a precision farming approach designed to optimize nutrient application on the basis of the specific needs of crops and soil conditions. SSNM emphasizes tailoring the application of essential nutrients (nitrogen, phosphorus and potassium) and secondary nutrients and micronutrients according to the results of soil tests. This ensures that nutrients are applied where they are needed most, rather than via a one-size-fits-all approach. This represents a shift towards more precise and efficient nutrient management aimed at increasing crop yields and farmer profitability while promoting environmental sustainability. By focusing on site-specific needs and optimizing nutrient use, SSNM helps address a range of agricultural challenges and supports more resilient and productive farming systems.

7.4 Foliar Application of Fertilizers and Fertigation in Potato

The timely foliar fertilization of potatoes allows for the delivery of essential nutrients during the most critical stages of crop development. Potatoes benefit from receiving nutrients in small, frequent amounts rather than large doses. Approximately

30–40% of their mineral nutrition comes from the vegetative mass, which absorbs nitrogen and sulfur from the atmosphere. Foliar feeding helps diversify and enhances nutrient uptake. However, the efficiency of nutrient absorption by the potato root system is influenced by weather conditions and soil chemistry. As a result, even in well-fertilized soils, plants might still be deficient in certain elements.

Micronutrients are crucial for essential processes in plants. They increase the chlorophyll content of leaves and improve photosynthesis, which increases the overall assimilating activity of the plant (Marschner 1995). When a microelement mixture (containing B, Cu, Mn, Zn and Mo) is sprayed on potato leaves, it increases the uptake of N, P and K; increases the chlorophyll content; improves photosynthesis; promotes tuber expansion; and ultimately increases potato yield (Meng et al. 2004).

8 Conclusion

In conclusion, the necessity of advances in soil and plant nutrient management for potatoes is underscored by the dual pressures of a changing climate and the global push for sustainability. As the world grapples with these challenges, innovations in nutrient management will play a pivotal role in enhancing agricultural resilience, protecting environmental health and supporting economic stability. Embracing these advancements is not only critical for maintaining the productivity of potato farming but also for contributing to broader global goals of food security, environmental stewardship and sustainable development.

References

Bunemann EK, Schwenke D, Van Zwieten L (2006) Impact of agricultural inputs on soil organisms—a review. Aust J Soil Res 44:379–406

Chowdary M, Bisarya D, Sharma M, Kumar V, Singh AK (2021) Potassium: a key role player in potato (*Solanumtuberosum* L.). J Emerg Technol Innov Res 8(9):689–699. Available from: www.jetir.org

Davenport JR, Milburn PH, Rosen CJ, Thorntozn RE (2005) Environmental impacts of potato nutrient management. Am J Potato Res 82:321–328. https://doi.org/10.1007/BF02871962

De Wit CT (1994) Resource use analysis in agriculture: a struggle for interdisciplinarity. In: Fresco LO, Stroosnijder L, Bouma J, van Keulen H (eds) The future of the land: mobilizing and integrating knowledge for land use options. Wiley, Wageningen, pp 41–55

Finch HJS, Samuel AM, Lane GPF (2014) Lockhart & Wiseman's crop husbandry including grassland, 9th edn. Elsevier Ltd., London

Gransee A, Führs H (2013) Magnesium mobility in soils as a challenge for soil and plant analysis, magnesium fertilization and root uptake under adverse growth conditions. Plant Soil 368:5–21. https://doi.org/10.1007/s11104-012-1567-y

Hopkins B, Stark J, Kelling K, (2020) Nutrient Management, in: Stark J, Thornton M, Nolte P. (Eds.), Potato Production Systems. Springer International Publishing pp:177–178. https://doi.org/10.1007/978-3-030-39157-7

Jing T, Jingyang Li, Yingdui He, Shankar A, Saxena A, Tiwari A, Maturi K, Solanki MK, Singh EM, Ding Z, Jianghui Xie, Awasthi MK (2024) Role of calcium nutrition in plant physiology: advances in research and insights into acidic soil conditions—a comprehensive review. Plant Physiol Biochem 210:108602, ISSN 0981-9428

Kärenlampi SO, White PJ (2009) Chapter 5—potato proteins, lipids and minerals. In: Singh J, Kaur L (eds) Advances in potato chemistry and technology. Elsevier, Amsterdam, pp 99–125

Kate DM, Solanke AV, Tiwari TK, Nemade SM (2005) Growth and yield of potato cultivars as affected by integrated nutrient management system. J Maharashtra Agric Univ 30(2):236–237

King JC, Slavin JL (2013) White potatoes, human health and dietary guidance. Adv Nutr 4:393S–401S. https://doi.org/10.3945/an.112.003525

Koch M, Busse M, Naumann M, Jákli B, Smit I, Cakmak I, Hermans C, Pawelzik E (2019) Differential effects of varied potassium and magnesium nutrition on production and partitioning of photoassimilates in potato plants. Physiol Plant. https://doi.org/10.1111/ppl.12846

Koprivova A, Kopriva S (2016) Sulfur metabolism and its manipulation in crops. J Genet Genomics 43:623–629. https://doi.org/10.1016/j.jgg.2016.07.001

Kumar M, Ghosh DC, Gupta VK, Dubey SK, Das A, Patel DP (2012) Productivity and soil health of potato field as influenced by organic manures, inorganic fertilizers and biofertilizers under high altitudes of Eastern Himalayas. Potato J 34(1–2):111–116

Kumar D, Punetha A, Prawal P, Verma S, Padalia RC (2022) Micronutrient based approach to increase yield and quality of essential oil in aromatic crops. J Appl Res Med Aromat Plants 26:100361

Lazarević B, Horvat T, Poljak M (2014) Effect of acid aluminous soil on photosynthetic parameters of potato (*Solanumtuberosum* L.). Potato Res 57:33–46. https://doi.org/10.1007/s11540-014-9251-7

Marschner H (1995) Mineral nutrition of higher plants, 2nd edn. Academic, London, p 889

Meng ML, Yic M, Jun Y, Lin, YZ (2004) Research progress on cultivation and physiology of potato in China 5th world Potato Congress, 16

Qadri RWK, Khan I, Jahangir MM, Ashraf U, Samin G, Anwer A, Adnan M, Bashir M (2015) Phosphorous and foliar applied nitrogen improved productivity and quality of potato. Am J Plant Sci 6:144–149. https://doi.org/10.4236/ajps.2015.610164

Rosen CJ, Kelling KA, Stark JC, Porter GA (2014) Optimizing phosphorus fertilizer management in potato production. Am J Potato Res 91:145–160. https://doi.org/10.1007/s12230-014-9371-2

Sanyal B, Mondal SS, Chatterjee BW (1993) Fertilizer management with bulky organic matter/manure in rice- potato-ground nut sequence for sustaining productivity. J Potassium Res 9:216–227

Senbayram M, Gransee A, Wahle V, Thiel H (2015) Role of magnesium fertilizers in agriculture: plant–soil continuum. Crop Pasture Sci 66:1219–1229. https://doi.org/10.1071/CP15104

Stark JC, Hopkins BG (2015) Fall and spring phosphorus fertilization of potato using a dicarboxylic acid polymer (AVAIL®). J Plant Nutr 38:1595–1610. https://doi.org/10.1080/01904167.2014.983124

Torabian S, Farhangi-Abriz S, Qin R, Noulas C, Sathuvalli V, Charlton B, Loka DA (2021) Potassium: a vital macronutrient in potato production—a review. Agronomy 11:543. https://doi.org/10.3390/agronomy11030543

Von Liebig JF (1840) Die organischeChemie in ihrerAnwendung auf Agricultur und Physiologie. Vieweg, Braunschweig

Von Liebig JF (1855) Die Grundsätze der Agricultur-Chemie. MitRücksicht auf die in England angestellten Untersuchungen. Vieweg, Braunschweig

Application of Geospatial Technologies in Potato Crop Management

D. K. Verma, C. Patel, V. Kumar, V. K. Patel, V. Chandra, and A. K. Singh

Abstract The increasing global demand for potatoes, with production reaching approximately 375 million tonnes in 2022, underscores the need for efficient and sustainable agricultural practices. Momentous producers, including China (95.5 million tonnes), India (56 million tonnes), and the United States (17.8 million tonnes), contribute significantly to meeting this demand, driven by factors such as population growth, dietary preferences, and the expansion of processed potato products. Integration of remote sensing (RS) and geographical information systems (GIS) is becoming essential for the enhancement of potato production and ensuring sustainability. Remote sensing technologies, including satellite imagery and drones, offer real-time monitoring of potato crops and provide critical data on soil moisture, plant health, and pest infestations. These technologies enable precise interventions, optimizing water usage and reducing the need for chemical inputs. GIS further enhances potato farming by integrating and analyzing spatial data, enabling site-specific management practices such as optimal planting, fertilization, and irrigation. The use of remote sensing in potato production has revolutionized traditional agricultural practices by enhancing precision and efficiency, leading to higher yields, reduced production costs, and improved environmental stewardship. For instance, monitoring plant health through indices like the normalized difference vegetation index (NDVI) allows for the timely detection of stress factors, ensuring early intervention and mitigating potential yield losses. Additionally, GIS contributes to mapping variations in soil properties and climatic conditions, facilitating better

D. K. Verma (✉) · C. Patel · V. Kumar
Department of Soil Science and Agricultural Chemistry, Bihar Agricultural College,
Bihar Agricultural University, Sabour, Bhagalpur, Bihar, India

V. K. Patel
Department of Horticulture, Sri Durga Ji PG College, Chandeshwar, Azamgarh, India

V. Chandra
Department of Horticulture (Fruits & Fruit Tech.), Bihar Agricultural College,
Bihar Agricultural University, Sabour, Bhagalpur, Bihar, India

A. K. Singh
Bihar Agricultural University, Sabour, Bhagalpur, Bihar, India

S. S. Solankey (ed.), *Advances in Research on Potato Production*,
Advances in Olericulture, https://doi.org/10.1007/978-3-031-82710-5_11

decision-making and resource management. Overall, the application of RS and GIS in potato farming represents a significant advancement in agricultural technology, making potato production more resilient, productive, and sustainable. These tools support the growing global demand for potatoes and promote environmentally responsible farming practices, ensuring the long-term viability of potato production worldwide.

Keywords Remote sensing (RS) and Geographical Information Systems (GIS) · Potato crop

1 Introduction

The global demand for potatoes continues to rise, making efficient and sustainable potato production increasingly critical. In 2022, global potato production was approximately 375 million tonnes. China produced about 95.5 million tonnes, followed by India with 56.18 million tonnes. Other significant producers include Russia with 18.9 million tonnes, Ukraine with 20.9 million tonnes, and the United States with 17.8 million tonnes. The total harvested area worldwide was around 17.79 million hectares. The global demand for potatoes remains high due to their versatility and as an important staple food. Factors such as population growth, dietary preferences, and the expansion of processed potato products contribute to the steady increase in demand. This demand is met by significant production from key regions, ensuring a stable supply to markets worldwide.

Remote sensing (RS) and geographic information system (GIS) are powerful tools that can significantly enhance potato production by providing precise, real-time data that informs decision-making processes. This chapter explores how the application of remote sensing and GIS in potato production has revolutionized agricultural practices by enhancing efficiency, precision, and sustainability. Remote sensing technologies, including satellite imagery and drones, enable real-time monitoring of potato crops, providing critical data on soil moisture, plant health, and pest infestations. Figure 1 presents the data captured from various locations, including agricultural fields, water bodies, forests, and developed areas. This information allows for timely intervention actions, optimizing water usage and minimizing the need for chemical inputs. GIS integrates these data layers, facilitating spatial analysis and decision-making. By mapping variations in soil properties, topography, and climatic conditions, GIS helps in site-specific management, like nutrients ensuring optimal planting, fertilization, and irrigation practices. Consequently, these technologies contribute to higher yields, reduce production costs, and improve environmental stewardship, making potato farming more resilient and productive.

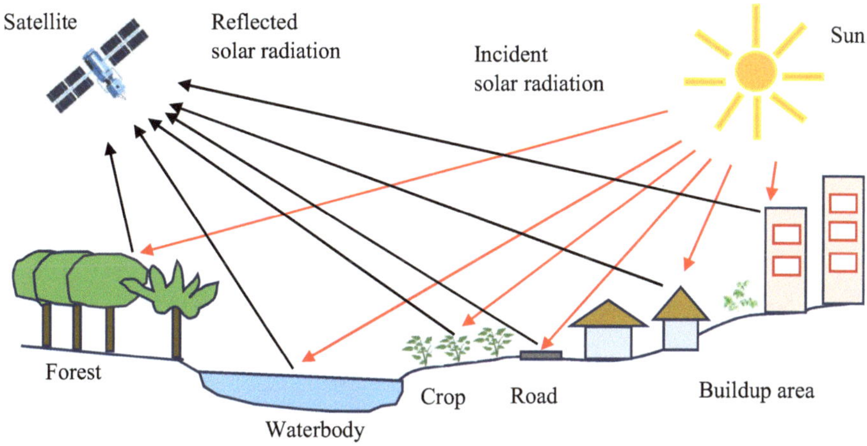

Fig. 1 Satellite data acquisition for various land use and land cover

2 Remote Sensing in Potato Production

2.1 Principles of Remote Sensing

Remote sensing involves the acquisition of information about an object or area from a distance, typically using satellites, aircraft, or drones equipped with various sensors. These sensors capture data across various electromagnetic spectra, which can be analyzed to extract meaningful information about the observed objects or areas. In agriculture, remote sensing monitors crop conditions, detects diseases, assesses soil properties, and evaluates environmental factors affecting crop growth. The principles of remote sensing can be effectively applied to potato production in the following ways:

Remote sensing allows for assessing plant health through indices such as the normalized difference vegetation index (NDVI). NDVI uses the difference between near-infrared and red-light reflectance to measure the greenness and vigor of potato plants. Healthy vegetation reflects more near-infrared and less visible light, which sensors capture and translate into actionable data for farmers. By using microwave sensors, which penetrate the soil surface, remote sensing provides accurate soil moisture data, ensuring potatoes receive the right amount of water at the right times. Figure 2 shows the spectrum of wavelengths in the thermal infrared, microwave, and radio range. This not only conserves water but also enhances crop yield and quality. Remote sensing can identify changes in plant reflectance that indicate stress due to pests or diseases. Infrared imaging can detect heat stress in plants, which often precedes visible disease symptoms, enabling early intervention and reducing crop losses. By correlating historical remote sensing data with actual yields, models can be developed to forecast production, helping farmers and stakeholders make informed decisions about marketing and resource allocation.

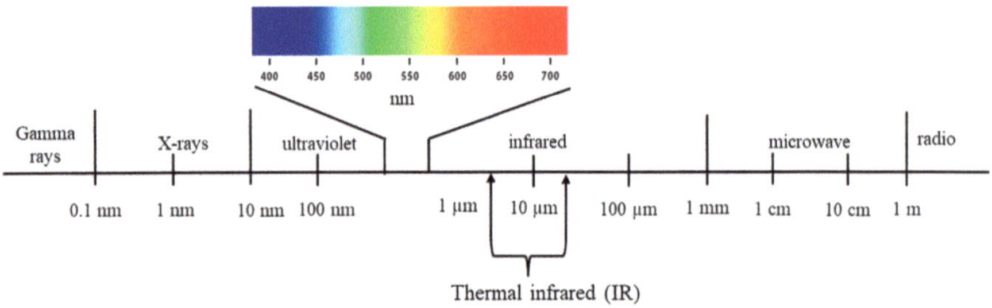

Fig. 2 Spectrum of wavelengths in the thermal infrared (IR) and radio range

2.2 *Types of Remote Sensing Technologies*

2.2.1 Optical Remote Sensing

Optical remote sensing utilizes visible, near-infrared, and shortwave infrared light to capture images of the Earth's surface. Multispectral and hyperspectral imaging are common techniques in optical remote sensing, providing detailed information about plant health, biomass, and chlorophyll content (Table 1).

2.2.2 Thermal Remote Sensing

Thermal sensors detect infrared radiation emitted by objects, providing data on surface temperatures. In potato production, thermal remote sensing helps monitor soil moisture, identify water stress, and manage irrigation more effectively (Table 2).

2.2.3 Radar Remote Sensing

Radar remote sensing uses microwave signals to penetrate clouds and vegetation, capturing data regardless of weather conditions or time of day. Synthetic Aperture Radar (SAR) is useful for mapping soil moisture and assessing crop structure. The application and limitation of Radar remote sensing is presented in Table 3.

3 Applications of Remote Sensing in Potato Production

3.1 *Normalized Difference Vegetation Index (NDVI)*

The NDVI is a widely used remote sensing index that measures vegetation health and density by comparing the reflectance of near-infrared (NIR) and red light from plant surfaces. It is calculated using the formula (NIR - Red) / (NIR + Red), where

Table 1 Application and limitation of optical remote sensing

Aspect	Details	References
Definition	Optical remote sensing utilizes visible, near-infrared, and shortwave infrared light to capture images of the Earth's surface	van Leeuwen (2009)
Techniques	Multispectral and hyperspectral imaging are common techniques in optical remote sensing	Vivone (2023)
Applications	Provides detailed information about plant health, biomass, and chlorophyll content.	Pavlović et al. (2014)
Benefits	Allows for detailed analysis of vegetation and crop health, aiding in agricultural management and environmental monitoring	Reeves et al. (2019)
Limitations	Atmospheric conditions such as clouds and haze can affect optical remote sensing, obscure the surface, and reduce the quality of the captured images	Emery and Camps (2017)
Data acquisition	Satellites and aerial platforms are typically used to acquire optical remote sensing data	Zhu et al. (2018)
Resolution	Spatial, spectral, and temporal resolution can vary, with high-resolution systems providing more detailed information but often at a higher cost.	Navalgund et al. (2007)
Data processing	Requires advanced processing techniques to analyze large volumes of data and extract meaningful information	Chi et al. (2016)
Examples of use	Agriculture (crop monitoring), forestry (deforestation tracking), environmental monitoring (pollution detection), urban planning, and disaster management.	Avtar et al. (2020)
Future trends	Increasing machine learning and artificial intelligence are used to enhance data analysis and interpretation	D'Amico (2022)

NIR and Red represent the reflectance in the near-infrared and red wavelengths, respectively. High NDVI values indicate dense, healthy vegetation, while low values suggest sparse or stressed vegetation. NDVI is crucial for applications in agriculture, forestry, and environmental monitoring, enabling the assessment of crop conditions, forest health, and land cover changes over time. Its utility in these fields stems from its ability to provide consistent, scalable, and timely information on vegetation dynamics, supporting better decision-making and resource management. Where NIR is the near-infrared light reflected by the vegetation and Red is the red light reflected by the vegetation. Healthy vegetation reflects more NIR and less red light, resulting in higher NDVI values.

3.2 *Applications of NDVI in Potato Crop Monitoring*

3.2.1 Plant Vigor Assessment

Plant vigor assessment using NDVI is a vital tool for monitoring the health and productivity of potato crops. NDVI values range from −1 to +1, with higher values indicating robust, photosynthetically active vegetation. For potato plants, high

Table 2 Application and limitation of thermal remote sensing

Aspect	Details	References
Definition	Thermal sensors detect infrared radiation emitted by objects, providing data on surface temperatures	Lillesand et al. (2015)
Applications in agriculture	In potato production, thermal remote sensing helps monitor soil moisture, identify water stress, and manage irrigation more effectively	Rud et al. (2014)
Benefits	Enables efficient water management, helps in the early detection of crop stress, and optimizes irrigation scheduling	Sishodia et al. (2020)
Limitations	Thermal remote sensing can be affected by atmospheric conditions such as humidity and cloud cover, and the data can be influenced by surface emissivity variations	Prata et al. (1995)
Data acquisition	Typically acquired using satellites, unmanned aerial vehicles (UAVs), and handheld devices	Pajares (2015)
Resolution	Spatial and temporal resolution varies, with high-resolution systems providing more detailed and frequent data but often at a higher cost	Dash et al. (2002)
Data processing	Requires advanced processing techniques to correct for atmospheric effects and surface emissivity, and to convert radiance to temperature	Li et al. (1999)
Examples of use	Agriculture (soil moisture monitoring, crop stress detection), forestry (fire detection), urban planning (heat island analysis), and environmental monitoring.	Kalma et al. (2008)
Future trends	They are increasing the use of UAVs and high-resolution satellites, integration with other remote sensing data (e.g., multispectral), and advancements in data processing algorithms	Gupta et al. (2021)

Table 3 Application and limitation of radar remote sensing

Aspect	Description	References
Technology	Radar remote sensing uses microwave signals to penetrate clouds and vegetation	Sinha et al. (2015)
All-weather capability	It can capture data regardless of weather conditions or time of day.	Richards (2009)
Synthetic aperture radar (SAR)	SAR is particularly useful in mapping soil moisture and assessing crop structure	Battaglia et al. (2020)
Soil moisture mapping	SAR helps map soil moisture, which is crucial for agricultural and environmental monitoring	Paloscia (2002)
Crop structure Assessment	SAR can assess crop structure, providing insights into agricultural practices and crop health	McNairn and Brisco (2004)
Cloud penetration	Radar signals can penetrate clouds, allowing continuous data collection in regions with frequent cloud cover	Battaglia et al. (2020)
Applications	Use of AR data for soil drainage classification and mapping	Niang et al. (2010)

NDVI values suggest healthy, vigorous growth, characterized by ample chlorophyll content and efficient photosynthesis. Conversely, lower NDVI values may signal plant stress or poor health due to factors such as nutrient deficiencies, water stress, pest infestations, or disease. By regularly monitoring NDVI, farmers can identify areas of the field that require attention, enabling targeted interventions to mitigate stress factors and enhance overall crop performance. This precision in assessing plant health supports sustainable agricultural practices and optimizes yield by ensuring timely and accurate management decisions (Mulla 2013).

3.2.2 Growth Stage Identification

Growth stage identification of potato crops using NDVI is a valuable method for optimizing agricultural management. Potato plants exhibit distinct NDVI patterns at different growth stages, from emergence and vegetative growth to tuber bulking and senescence. Early in the growth cycle, NDVI values increase as the crop canopy develops and photosynthetic activity intensifies. During the vegetative stage, NDVI values peak, reflecting maximum leaf area and chlorophyll content. As the plants transition to the tuber bulking stage, NDVI values may plateau or slightly decrease, signaling a shift in resource allocation from leaves to tubers. Finally, during senescence, NDVI values decline as leaves age and chlorophyll degrades. By monitoring these NDVI trends, farmers can accurately identify the current growth stage of their potato crops and adjust management practices such as irrigation, fertilization, and pest control to match the specific needs of each stage, thereby enhancing crop health and yield (Lobell et al. 2003).

Farmers in the US Midwest have adopted remote sensing technologies to monitor large potato fields. By using NDVI, they can identify areas with low plant vigor and take corrective actions, such as targeted irrigation or fertilization, improving overall crop health and yield. This approach allows for efficient resource use, reducing costs and environmental impact while maximizing productivity. The adoption of precision agriculture practices, supported by NDVI data, has demonstrated significant benefits in terms of crop management and sustainability in the region (Mulla 2013).

Potato farming in Punjab (India), where potato farming is a significant agricultural activity, the integration of advanced technologies such as drones equipped with multispectral sensors is revolutionizing crop management. These drones fly over potato fields, capturing high-resolution NDVI data, which provides detailed insights into the health and vigor of the plants. The NDVI data allows farmers to pinpoint areas experiencing stress due to factors such as water scarcity, nutrient deficiencies, or pest infestations. By identifying these stress areas early, farmers can implement targeted interventions, such as adjusting irrigation schedules, applying precise amounts of fertilizers, or deploying pest control measures exactly where needed. This precision in resource application not only enhances the efficiency of inputs but also promotes sustainable farming practices by minimizing waste and environmental impact. Furthermore, the ability to monitor crop health in real time enables farmers to make informed decisions swiftly, ensuring optimal growth conditions and improving overall crop yield and quality. The use of drone-based NDVI monitoring in Punjab exemplifies how modern technology can significantly enhance traditional farming practices, leading to better resource management and increased agricultural productivity (Gitelson et al. 2003b).

3.3 Disease Detection and Management in Potato Crops

3.3.1 Chlorophyll Content

Diseased plants often exhibit distinct reflectance patterns compared to healthy plants. Remote sensing technologies can detect these differences, allowing for early identification of affected areas. Chlorophyll content in plants, a key indicator of photosynthetic activity and overall health, can be precisely monitored using remote sensing technologies that capture specific wavelengths of light. Changes in chlorophyll content are detectable through the analysis of light reflectance in the red and near-infrared regions of the spectrum. Healthy plants with high chlorophyll levels strongly absorb red light for photosynthesis and reflect near-infrared light, whereas diseased or stressed plants exhibit reduced chlorophyll content, leading to higher red-light reflectance and lower near-infrared reflectance. This spectral signature allows for the early detection of potential diseases or stress conditions before visible symptoms appear. By monitoring these changes, farmers can identify areas within their fields that require attention, enabling targeted interventions such as adjusting nutrient management or applying appropriate treatments to prevent disease spread and optimize crop health (Gitelson et al. 2003a).

3.3.2 Canopy Structure

Diseases frequently impact the structure and density of plant canopies, causing alterations that can be effectively identified through remote sensing technologies. Healthy potato plants typically have a dense, uniform canopy that efficiently covers the ground, reflecting consistent light patterns. When disease strikes, it can cause irregularities in canopy structure, such as thinning, reduced leaf area, or patchy growth, which alter the light reflectance properties of the crop. Multispectral and hyperspectral sensors on satellites and drones capture these variations, providing detailed imagery that reveals deviations from the expected canopy architecture. By analyzing these images, farmers can detect changes indicative of disease presence early, even before significant visual symptoms manifest. This early detection allows for timely and precise interventions, such as targeted pesticide application or changes in cultural practices, to mitigate disease spread and maintain crop health and productivity (Gitelson et al. 2003a).

3.4 *Benefits of Remote Sensing in Disease Management*

3.4.1 Early Intervention

Early detection of crop diseases through remote sensing technologies enables prompt action, significantly reducing the spread of disease and limiting damage to the crop. By identifying stress indicators such as changes in chlorophyll content and canopy structure before they become visually apparent, farmers can take immediate, targeted measures to address the underlying issues. This might include applying fungicides, adjusting irrigation practices, or modifying nutrient management strategies to bolster plant health. Early intervention not only curtails the progression of disease but also prevents the extensive damage that can occur if issues are left unchecked. Consequently, this proactive approach helps maintain high yield potential and quality, minimizes economic losses, and supports sustainable farming practices by optimizing the use of inputs and preserving the health of the agricultural ecosystem (Gitelson et al. 2003a).

3.4.2 Targeted Treatments

Remote sensing plays a crucial role in precision agriculture by pinpointing specific areas within a field that are affected by disease. By analyzing detailed imagery from multispectral and hyperspectral sensors, farmers can accurately identify and map zones of infection or stress. This targeted approach allows for the precise application of fungicides, pesticides, or other treatments only where they are needed, rather than applying chemicals uniformly across the entire field. As a result, this method not only reduces the overall use of chemicals, thereby lowering costs and minimizing environmental impact, but also enhances the efficacy of the treatments. By focusing resources on affected areas, farmers can achieve better disease control and maintain crop health more efficiently, leading to improved economic and ecological outcomes (Mulla 2013).

3.5 *Application of Remote Sensing for Soil and Water Management*

NDVI (Normalized Difference Vegetation Index) is a valuable tool for detecting crop stress factors like water stress, nutrient deficiencies, and pest infestations. Stress typically reduces photosynthetic activity, leading to lower NDVI values. For example, water stress diminishes chlorophyll content and leaf area, resulting in reduced NDVI. Nutrient deficiencies and pest damage also cause localized drops in NDVI. By monitoring NDVI regularly, farmers can identify these stressors early and take timely actions, such as adjusting irrigation, applying fertilizers, or controlling pests, which helps improve crop health and yield (Gitelson et al. 2003b).

Remote sensing involves using aerial platforms such as satellites, drones, and aircraft equipped with sensors to collect data on soil and water properties. These sensors capture information across various spectral bands, which can be analyzed to assess various soil characteristics and water content. Soil moisture is a critical factor in plant growth. Remote sensing technologies, such as thermal and microwave sensors, can measure soil moisture levels by detecting differences in thermal emission and reflectance. Remote sensing data, particularly from multispectral and hyperspectral sensors, help in mapping soil texture by analyzing the spectral signatures associated with different soil types. Soil organic matter is vital for soil fertility and structure. Near-infrared (NIR) and shortwave infrared (SWIR) sensors can detect variations in soil organic matter content by measuring specific wavelengths absorbed and reflected by organic compounds (Ben-Dor et al. 2018).

4 Geographic Information Systems in Potato Production

4.1 *Principles and Components of GIS*

Geographic Information System (GIS) integrates spatial and non-spatial data to create detailed maps and models of agricultural landscapes. GIS enables the visualization, analysis, and interpretation of complex datasets, supporting decision-making processes in potato production. GIS relies on various data sources, including remote sensing, ground surveys, and historical records. Accurate and up-to-date data are essential for creating reliable GIS models. GIS tools analyze spatial relationships and patterns, allowing for the identification of trends and correlations in potato production. Advanced modeling techniques, such as geostatistics and machine learning, enhance the predictive capabilities of GIS. GIS creates detailed maps that visualize various aspects of potato production, including soil properties, crop health, and yield potential. These maps are essential for planning and decision-making at different scales, from individual fields to regional levels.

4.2 *Applications of GIS in Potato Production*

GIS helps identify suitable areas for potato cultivation by analyzing factors such as soil type, climate, and topography. Land suitability analysis supports strategic planning and reduces the risk of crop failure. GIS supports precision agriculture by integrating data from various sources to create detailed field maps. These maps guide variable rate applications of inputs, such as fertilizers and pesticides, optimizing resource use and enhancing crop productivity. GIS monitors environmental factors affecting potato production, such as weather patterns, pest outbreaks, and soil erosion. This information enables proactive management practices that mitigate risks and promote sustainable agriculture. GIS enhances the efficiency of logistics and supply chain management in potato production. By optimizing transportation routes and tracking the movement of products, GIS reduces costs and ensures timely delivery of potatoes to markets.

5 Integration of Remote Sensing and GIS

The integration of remote sensing and GIS provides a comprehensive approach to potato production. Remote sensing data feeds into GIS models, enhancing their accuracy and reliability. This integrated approach supports informed decision-making at all stages of the production cycle, from planning and planting to harvesting and marketing. Remote sensing data on soil moisture and plant health were integrated into a GIS-based irrigation management system. The system provided real-time recommendations for irrigation scheduling, resulting in significant water savings and improved crop yields. A regional potato-growing area implemented a remote sensing and GIS-based system for disease surveillance. By monitoring vegetation indices and weather conditions, the system detected early signs of disease outbreaks and guided targeted interventions, reducing the spread of diseases and minimizing yield losses. Farmers in a potato-growing region used remote sensing and GIS to predict yields and plan their marketing strategies. Accurate yield predictions enabled better contract negotiations and optimized supply chain management, ensuring timely delivery of potatoes to processors and markets.

6 Conclusion

In conclusion, the rising global demand for potatoes underscores the importance of efficient and sustainable production methods. The integration of remote sensing (RS) and geographic information system (GIS) has significantly advanced potato farming by providing precise, real-time data that supports informed decision-making. These technologies enhance efficiency, precision, and sustainability by enabling real-time monitoring and site-specific management. As a result, potato production has become more resilient, cost-effective, and environmentally friendly, meeting the growing demand while ensuring a stable supply to markets worldwide. The application of remote sensing and GIS in potato production offers numerous benefits, including improved crop monitoring, efficient resource management, early disease detection, and accurate yield prediction. By integrating these technologies, farmers can make informed decisions that enhance productivity, sustainability, and profitability. As the global demand for potatoes continues to grow, the adoption of remote sensing and GIS will play a crucial role in meeting this demand while promoting sustainable agricultural practices.

References

Avtar R, Komolafe AA, Kouser A, Singh D, Yunus AP, Dou J et al (2020) Assessing sustainable development prospects through remote sensing: a review. Remote Sens Appl Soc Environ 20:100402

Battaglia A, Kollias P, Dhillon R, Roy R, Tanelli S, Lamer K et al (2020) Spaceborne cloud and precipitation radars: status, challenges, and ways forward. Rev Geophys 58(3):e2019RG000686

Ben-Dor E, Chabrillat S, Demattê JA (2018) Characterization of soil properties using reflectance spectroscopy. In: Fundamentals, sensor systems, spectral libraries, and data mining for vegetation. CRC Press, pp 187–247

Chi M, Plaza A, Benediktsson JA, Sun Z, Shen J, Zhu Y (2016) Big data for remote sensing: challenges and opportunities. Proc IEEE 104(11):2207–2219

D'Amico G (2022) Application of big data analytics in remote sensing supporting sustainable forest management

Dash P, Göttsche FM, Olesen FS, Fischer H (2002) Land surface temperature and emissivity estimation from passive sensor data: theory and practice-current trends. Int J Remote Sens 23(13):2563–2594

Emery W, Camps A (2017) Introduction to satellite remote sensing: atmosphere, ocean, land and cryosphere applications. Elsevier

Gitelson AA, Gritz Y, Merzlyak MN (2003a) Relationships between leaf chlorophyll content and spectral reflectance and algorithms for non-destructive chlorophyll assessment in higher plant leaves. J Plant Physiol 160(3):271–282

Gitelson AA, Kaufman YJ, Stark R, Rundquist D (2003b) Novel algorithms for remote estimation of vegetation fraction. Remote Sens Environ 80(1):76–87

Gupta A, Afrin T, Scully E, Yodo N (2021) Advances of UAVs toward future transportation: the state-of-the-art, challenges, and opportunities. Future Transp 1(2):326–350

Kalma JD, McVicar TR, McCabe MF (2008) Estimating land surface temperature from thermal infrared data: a review of the algorithms and their applications. Remote Sens Environ 112(3):1118–1135

Li ZL, Becker F, Stoll MP, Wan Z (1999) Evaluation of six methods for extracting relative emissivity spectra from thermal infrared images. Remote Sens Environ 69(3):197–214

Lillesand T, Kiefer RW, Chipman J (2015) Remote sensing and image interpretation, 7th edn. Wiley

Lobell DB, Asner GP, Ortiz-Monasterio JI, Benning TL (2003) Remote sensing of regional crop production in the Yaqui Valley, Mexico: estimates and uncertainties. Agric Ecosyst Environ 94(2):205–220

McNairn H, Brisco B (2004) The application of C-band polarimetric SAR for agriculture: a review. Can J Remote Sens 30(3):525–542

Mulla DJ (2013) Twenty-five years of remote sensing in precision agriculture: key advances and remaining knowledge gaps. Biosyst Eng 114(4):358–371

Navalgund RR, Jayaraman V, Roy PS (2007) Remote sensing applications: an overview. Curr Sci:1747–1766

Niang MA, Nolin MC, Bernier M (2010) Potential of C-band multi-polarized and polarimetric SAR data for soil drainage classification and mapping. In: Geoscience and remote sensing: new achievements. InTech, pp 163–176

Pajares G (2015) Overview and current status of remote sensing applications based on unmanned aerial vehicles (UAVs). Photogramm Eng Remote Sens 81(4):281–330

Paloscia S (2002) A summary of experimental results to assess the contribution of SAR for mapping vegetation biomass and soil moisture. Can J Remote Sens 28(2):246–261

Pavlović D, Nikolić B, Đurović S, Waisi H, Anđelković A, Marisavljević D (2014) Chlorophyll as a measure of plant health: agroecological aspects. Pesticidi i fitomedicina 29(1):21–34

Prata AJ, Caselles V, Coll C, Sobrino JA, Ottle C (1995) Thermal remote sensing of land surface temperature from satellites: current status and future prospects. Remote Sens Rev 12(3–4):175–224

Reeves MC, Stavros EN, Glenn NF, Hudak A, Peterson B, Armstrong A et al (2019) Forest Service–NASA joint applications workshop: satellite data to support natural resource management: a framework for aligning NASA products with land management agency needs. UMBC Faculty Collection

Richards JA (2009) Remote sensing with imaging radar, vol 1. Springer, Berlin/Heidelberg, pp 172–173

Rud R, Cohen Y, Alchanatis V, Levi A, Brikman R, Shenderey C et al (2014) Crop water stress index derived from multi-year ground and aerial thermal images as an indicator of potato water status. Precis Agric 15:273–289

Sinha A, Shen Z, Song Y, Ma H, Eide D, Hsu BJ, Wang K (2015, May) An overview of Microsoft academic service (MAS) and applications. In: Proceedings of the 24th international conference on world wide web, pp 243–246

Sishodia RP, Ray RL, Singh SK (2020) Applications of remote sensing in precision agriculture: a review. Remote Sens 12(19):3136

van Leeuwen WJ (2009) Visible, near-IR, and shortwave IR spectral characteristics of terrestrial surfaces. In: The SAGE handbook of remote sensing. SAGE, pp 33–50

Vivone G (2023) Multispectral and hyperspectral image fusion in remote sensing: a survey. Inf Fusion 89:405–417

Zhu L, Suomalainen J, Liu J, Hyyppä J, Kaartinen H, Haggren H (2018) A review: Remote sensing sensors. In: Multi-purposeful application of geospatial data, vol 19. InTech

Advances in Quality Improvement of Potato Tubers

Anupam Adarsh, Hemant Kumar Singh, Neeharika Kanth, Ajeet Kumar, Shashank Shekhar Solankey, and Alpana Kumari

Abstract The potato (*Solanum tuberosum* L.) is a globally vital staple food, and its quality is essential for meeting consumer and industry demands. Recent innovations in potato quality enhancement requires genetic, agronomic, and post-harvest technologies. Genetic improvements, including marker-assisted selection and genetic modification, have produced varieties with superior disease resistance, better nutritional profiles, and enhanced processing qualities. Notably, genetically engineered potatoes with reduced acrylamide levels—a potential carcinogen formed during high-temperature cooking—offer significant health benefits. Agronomic advancements, such as precision farming and optimized nutrient applications, have boosted tuber yield and quality. Remote sensing and data analytics enable precise monitoring of crop health, while advanced irrigation methods, like drip irrigation, enhance water use efficiency and ensure consistent moisture, further elevating tuber quality. Advancements in post-harvest technologies have significantly improved potato quality and shelf life. Enhanced storage methods, such as controlled atmosphere storage and advanced curing techniques, help maintain tuber quality and reduce spoilage. Low-temperature storage and modified atmosphere packaging preserve nutritional value, while innovations in sorting and grading technologies ensure

A. Adarsh (✉) · S. S. Solankey · A. Kumari
Department of Horticulture (Vegetable Science), Nalanda College of Horticulture, Noorsarai, Nalanda, Bihar Agricultural University, Sabour, Bhagalpur, Bihar, India

H. K. Singh
Department of Agriculture, School of Agriculture and Development, Central University of South Bihar, Gaya, Bihar, India

N. Kanth
Department of Horticulture, Pt. Deen Dayal Upadhyay College of Horticulture and Forestry, Piprakothi, Dr. Rajendra Prasad Central Agricultural University, Pusa, Samastipur, Bihar, India

A. Kumar
Department of Soil Science & Agricultural Chemistry, Nalanda College of Horticulture, Noorsarai, Nalanda, Bihar Agricultural University, Sabour, Bhagalpur, Bihar, India

S. S. Solankey (ed.), *Advances in Research on Potato Production*, Advances in Olericulture, https://doi.org/10.1007/978-3-031-82710-5_12

high-quality tubers for consumption and processing. These improvements minimize waste and ensure product consistency. Continued research and technological development are crucial for addressing emerging challenges and further enhancing the quality and sustainability of potato production across the entire supply chain.

Keywords Potato · Nutritional quality · Chemical quality · New breeding technique

1 Introduction

Potatoes are a vital global food source, recognized for their nutritional value and their role in providing key nutrients for a healthy diet. Potato ranks as the third most significant crop, following wheat and paddy, in regions when considering protein yield per unit area (Haase 2002). There are two main categories of potatoes that are grown: (1) landraces, which are traditional types and are still grown in South America, (2) modern developed varieties, which are grown commercially all over the world. There is a great deal of variation among these landraces, especially in terms of flesh and skin color.

The nutritional content of potatoes like starch, high-quality protein, free amino acids, organic acids, sugars, minerals, vitamins, fats, and lipids. The solid content in potato tubers varies between 14.2% and 29.8%, on the basis of variety and environmental situation (Kumar et al. 2004). Shifting consumer demands have led to potato breeding programs focusing on enhancing nutritional quality, which lowers calorie content, increasing high-quality protein, and boosting levels of vitamins and antioxidants.

Nutritious food is crucial for achieving and maintaining good health, as the human body requires an appropriate intake of vitamins and minerals to function efficiently. In addition to significant mineral elements like potassium, salt, magnesium, copper, iron, and zinc, potatoes are known for their nutritional value due to the presence of phytochemicals like carotenoids, anthocyanins, phenolics, and flavonoids. Moreover, vitamins B1, B6, B9, C, and E are abundant in them (Navarre et al. 2009; Ezekiel et al. 2013; Ngobese et al. 2017). In developing countries, approximately 50% of women and children suffer from anemia, which is primarily caused by deficiencies in iron (Fe), folate, vitamin B12, and other essential nutrients (Ahmed et al. 2016). Iron deficiency, especially in young children, not only hampers physical growth but also impedes cognitive development and weakens immunity, affecting school performance. Pregnant women experiencing iron deficiency are at risk of delivering babies with fetal growth retardation and a substantial contributor to maternal mortality rates. Iron and zinc deficiencies have multifaceted impacts on human health, influencing physical well-being, cognitive abilities, immunity, and overall productivity. Therefore, addressing these deficiencies is crucial for promoting health and well-being, particularly in susceptible populations such as women and children in developing countries. Potatoes are highly nutritious and provide

Table 1 Quality components in potato

Components	Average	Range
Dry matter	23.7	13.1–36.8
Starch	17.5	8.0–29.4
Reducing sugar	0.3	0.0–5.0
Total Sugar	0.5	0.05–8.0
Crude fiber	0.71	0.17–3.48
Pectic substance	–	0.2–1.5
Total nitrogen	0.32	0.11–0.74
Crude protein	2.00	0.69–4.63
Amide nitrogen	–	0.029–0.052
Amino acid nitrogen	–	0.065–0.098
Nitrates	–	0.0–0.05
Lipids	0.12	0.02–2.0
Ash	1.1	0.44–1.87
Organic acids	0.6	0.4–1.0
Ascorbic acid/ Dehydroascorbic acid (per 100 g)	10–25	1–54
Glycoalkaloids (mg/100 g)	3–10	0.2–41
Phenolic compounds	–	5–30

Source: Leszczynski and Lesinsika (1989)

essential nutrients that contribute to overall health and important quality components present in potatoes (Table 1):

(a) Carbohydrates: Potatoes provide a rich supply of carbohydrates, acting as a key energy source for the body. Their complex carbohydrate content ensures a gradual release of energy, making them a suitable option for maintaining consistent energy levels throughout the day.
(b) Dietary Fiber: Potatoes, especially their skins, serve as an important source of dietary fiber. Fiber plays a crucial role in promoting a healthy digestive system by facilitating regular bowel movements, reducing the likelihood of constipation, and decreasing the risk of colon cancer.
(c) Vitamins: Potatoes are rich in essential vitamins, including vitamin C, vitamin B6, and folate (vitamin B9). Vitamin C acts as a potent antioxidant, aiding in immune function, skin health, and wound healing (Akram et al. 2020). Vitamin B6 plays a crucial role in brain development and function, while folate is necessary for DNA synthesis and cell division.
(d) Minerals: Potatoes are an important source of key minerals such as potassium, magnesium, and iron. Potassium plays a crucial role in managing blood pressure and supporting muscle function. Magnesium is important for maintaining bone health, facilitating nerve activity, and aiding in energy production. Iron is essential for the formation of red blood cells and the effective transportation of oxygen within the body.

(e) Antioxidants: Potatoes contain a variety of antioxidants, such as flavonoids and carotenoids, which are essential in safeguarding the body against oxidative stress. This protective function may help lower the likelihood of developing chronic diseases, including cardiovascular conditions and cancer.
(f) Low in Fat and Calories: Naturally low in fat and calories, potatoes are a good choice for those looking to manage their weight. However, their nutritional content can vary based on cooking methods and added toppings.
(g) Gut Health: Resistant starch, present in cooked and subsequently cooled potatoes, functions as a prebiotic, fostering the proliferation of advantageous gut bacteria and enhancing a healthy microbiome. Incorporating potatoes into a well-rounded diet can assist in fulfilling daily nutritional requirements and contribute to overall well-being.

Health Benefits: Potatoes exhibit various health-promoting properties, such as anticancer, cholesterol-lowering, anti-inflammatory, anti-obesity, and antidiabetic effects. Future potato breeding programs should focus on developing tubers with diverse biochemical properties. This includes varieties with varying starch content, higher protein levels, and enhanced vitamins and antioxidants, which can strengthen the immune system. Adapting potato cultivation and breeding strategies to meet evolving nutritional needs will help produce healthier food options for consumers.

2 Nutritional and Quality Content in Potato

The quality components in potatoes (*Solanum tuberosum*) depend on various factors like tuber size, shape, texture, color, chemical composition, and nutritional value. Improving the quality traits of potatoes is a complex challenge due to various genetic and biological factors. Potatoes have a base chromosome number of 12, with different species exhibiting diploid to hexaploid levels of ploidy. The *Solanum* species that produce tubers encompass various ploidy levels, including several diploids (2n = 24), triploids (2n = 36), a limited number of tetraploids (2n = 48), and hexaploids (2n = 72). Numerous wild species found in the Tuberosa series are diploid. Among the cultivated diploid species (2n = 2× = 24) are *S. stenotomum*, *S. ajanhuiri*, and *S. phureja*. The first two species exhibit sexual fertility, whereas the latter is less fertile and does not consistently produce true offspring. Triploid species such as *S. chaucha* and *S. xjuzepczukii* result from natural hybridization between diploids and tetraploids. The widely cultivated species *S. tuberosum* and *S. demissum* are autotetraploids (2n = 4× = 48) and are generally fertile, although some extensively bred clones outside their native South American regions may exhibit reduced fertility. One pentaploid species (2n = 60), *S. curtilobum*, is relatively fertile when crossed with *S. tuberosum* but less fertility is exhibited in self-crosses. Most diploid species exhibit self-incompatibility, whereas tetraploids and hexaploids tend to be self-compatible. The complexity of potato breeding arises from the wide range of ploidy levels and the varying fertility and compatibility among different species (Simons et al. 1998).

Several traits in potatoes are genetically controlled, including: (i) sensory characteristics such as shape, size, color, flavor, and texture; (ii) biological traits, including carbohydrate, protein, vitamin, and mineral content, as well as low concentrations of toxic glycoalkaloids; and (iii) industrial traits, such as tuber shape and size, dry matter content, resistance to cold-induced sweetening, oil absorption, starch quality, and specific gravity (Kumari et al. 2018).

2.1 Tuber Size and Shape

The size and shape of potato tubers significantly impact both their market value and their suitability for various food products. According to studies, consumers and industries prefer uniformity in size and shape, especially for processed products such as French fries and potato chips (Stark et al. 2013). Irregularly shaped potatoes often result in higher peeling losses and uneven cooking.

Factors affecting tuber size include:

- **Genetic Background**: Different potato cultivars naturally produce varying sizes of tubers. For example, cultivars like 'Russet Burbank' are known for producing large tubers, making them ideal for the French fry industry (Kirkman 2007).
- **Environmental Factors**: Water availability, soil quality, and nutrient levels play a critical role in tuber development. Research by Obidiegwu et al. (2015) showed that drought stress during tuber formation can significantly reduce tuber size.

2.2 Tuber Skin and Flesh Color

Potato skin and flesh color are vital for consumer acceptance and are primarily determined by genetics. Skin colors can range from white to red and purple, while flesh colors vary from white to yellow and even blue or purple.

- **Anthocyanins and Carotenoids**: These pigments are responsible for the color of potato flesh and skin. Anthocyanins give potatoes their red, blue, or purple colors, while carotenoids contribute to yellow or orange flesh (Brown 2005). These pigments are not only important for aesthetics but also for nutritional quality, as anthocyanins have antioxidant properties, while carotenoids, such as lutein, are essential for eye health (Breithaupt and Bamedi 2002; Brown 2005).
- **Improvement for Color content in potato**: Researchers are actively breeding potatoes with enhanced pigment concentrations. For example, Navarre et al. (2010) demonstrated the feasibility of breeding purple-fleshed potatoes with higher anthocyanin levels, which offer greater health benefits.

2.3 *Tuber Texture and Consistency*

Texture is a critical quality component for both fresh market potatoes and processed products. It is influenced by several factors, including starch content, moisture levels, and cell structure.

- **Starch Content**: Potatoes are primarily classified into two categories based on their starch content: waxy (low starch) and mealy (high starch). Mealy potatoes, like 'Russet Burbank', have a dry, fluffy texture, which makes them perfect for baking and frying. On the other hand, waxy potatoes are firm and moist, making them ideal for boiling and using in salads (Timm and Hall 2008).
- **Pectin and Cell Wall Structure**: The firmness of cooked potatoes is influenced by pectin and other cell wall components. Research by Thybo et al. (2000) has shown that the degradation of pectin during cooking can affect the texture, making some varieties softer than others.

2.4 *Nutritional Quality*

The nutritional quality of potatoes is a key consideration, particularly in light of growing concerns about health and wellness. Potatoes are a good source of carbohydrates, fiber, vitamins, and minerals.

- **Carbohydrates and Dietary Fiber**: Potatoes are rich in carbohydrates, particularly starch. The amount of resistant starch, which acts as dietary fiber, is influenced by both genetic factors and cooking methods (Cummings et al. 2000). Studies have shown that cooling cooked potatoes increases resistant starch content, making them a better option for promoting gut health (Leeman et al. 2005).
- **Vitamins and Minerals**: Potatoes are an excellent source of essential nutrients, including vitamin C, potassium, and B vitamins (Camire et al. 2009). Potassium is crucial for maintaining healthy blood pressure, while vitamin C boosts the immune system.
- **Antioxidant Content**: The presence of antioxidants such as vitamin C, carotenoids, and anthocyanins contributes to the health benefits of potatoes. Research has shown that varieties with deeply colored flesh, such as purple and red potatoes, have higher antioxidant levels (Brown 2005).

2.5 *Chemical Composition and Taste*

The chemical composition of potatoes, particularly their sugar and starch content, significantly influences their taste and suitability for various culinary uses. Low sugar content is desirable for potatoes used in frying since high sugar levels can lead to undesirable browning during cooking (Kumar et al. 2004).

- **Glycoalkaloid Content**: Glycoalkaloids, such as solanine, are naturally occurring compounds in potatoes that can be toxic at high concentrations. While they help protect the plant from pests, they can impart a bitter taste and be harmful if consumed in large amounts (Friedman 2006). Research efforts are focused on breeding potatoes with low glycoalkaloid content while maintaining pest resistance.
- **Starch and Sugar Balance**: The balance between starch and sugars affects both flavor and texture (Blenkinsop 2002). High-starch potatoes are preferred for making fluffy mashed potatoes, while lower starch and higher sugar content varieties are better for roasting and frying (Kumar et al. 2004; Bach et al. 2013).

2.6 *Processing Quality*

Processing quality refers to how well potatoes perform during industrial processing, such as frying, baking, and dehydrating. The most important quality traits for processing potatoes include:

- **Low Reducing Sugar Levels**: High levels of reducing sugars (glucose and fructose) lead to excessive browning and the formation of acrylamide, a potentially harmful compound, during frying (Blenkinsop 2002). Research focuses on identifying varieties with naturally low reducing sugar levels to improve their suitability for frying.
- **Dry Matter Content**: Potatoes with higher dry matter content are preferred for processing, particularly for products like French fries and chips. This is because higher dry matter reduces oil absorption during frying, resulting in a crisper product (Park et al. 2013).
- **Varietal Differences**: Different potato varieties have been bred for specific processing purposes. For example, 'Russet Burbank' is widely used in the French fry industry due to its large tuber size, high starch content, and low reducing sugar levels (Kirkman 2007).

2.7 *Storage and Post-Harvest Quality*

Storage includes temperature and humidity and has significant impact on the quality of potatoes. Potatoes are prone to sprouting and loss of moisture during storage, which can affect their texture and appearance.

Cold Storage Storing potatoes at temperatures less than 4 °C can result in the buildup of reducing sugars, a process referred to as cold-induced sweetening. This results in undesirable browning during frying (Burton et al. 1992). Breeders are working to develop varieties with reduced susceptibility to cold-induced sweetening.

Sprout Inhibition Sprouting during storage not only affects the appearance of potatoes but also increases glycoalkaloid levels, making them unsafe to eat. The use of chemical sprout inhibitors, such as chlorpropham, has been a common practice, though there is growing interest in developing sprout-resistant varieties through breeding programs.

3 Improvement in Quality Content in Potato

Potatoes are one of the most widely consumed crops worldwide, and their quality is critical for both fresh consumption and processing applications. Potato quality can be enhanced by breeding and biotechnological approaches. Improving potato quality through breeding strategies involves addressing multiple traits such as tuber size, shape, nutritional content, and processing characteristics. The advancement of modern breeding techniques, *i.e.*, marker-assisted selection, genomic selection, and genetic modification, has accelerated the development of high-quality potato varieties. Quality in potato tubers refers to a range of traits, including size, shape, dry matter content, nutritional value, storage properties, and suitability for processing. The goal of quality improvement is to enhance these traits to meet consumer, market, and processing industry requirements while maintaining agronomic performance.

Potatoes are a significant source of essential nutrients, providing vitamins, carotenoids, antioxidant phenolics, proteins, and minerals like magnesium. However, they also possess certain anti-nutritional compounds, notably glycoalkaloids. Generally, potato tubers consist of approximately 77% H_2O and 20% carbohydrates, while the total of proteins, dietary fiber, vitamins, and minerals accounts for less than 3% (Zaheer and Akhtar 2016). Burlingame et al. (2009) provide a comprehensive overview of the nutrient profile across various potato varieties. In regions facing low income and food scarcity, potatoes present an affordable and nutritious alternative to more expensive food sources, supplying 93 kcal per 100 g, which can adequately meet basic energy requirements (Burlingame et al. 2009). In recognition of the crucial importance of potatoes in addressing global food insecurity, poverty, and hidden hunger, the United Nations proclaimed 2008 as the "International Year of the Potato" (FAO 2008).

Breeding plays a pivotal role in improving potato quality. The selection of high-performing varieties and the introduction of advanced breeding techniques have significantly enhanced quality attributes such as tuber size, shape, skin finish, and nutritional content. Traditional breeding efforts have focused on enhancing yield, processing, and storage qualities (Halterman et al. 2016). The genetic foundation of starch content in potatoes has revealed significant genomic areas related to starch biosynthesis. Genome-wide association studies (GWAS) have been utilized to identify genetic markers associated with starch characteristics, offering valuable information for breeding initiatives aimed at increasing starch levels. Breeding strategies often involve selecting parent lines with desirable starch characteristics and using marker-assisted selection (MAS) techniques to accelerate the breeding process. MAS allows breeders to identify and select individuals carrying specific genetic markers associated with high starch content, thus improving efficiency in developing new potato cultivars. This approach is essential for developing new potato varieties with enhanced starch content, contributing to agricultural productivity and industrial applications.

The selection of wild germplasm tends to be more accurate at the diploid level than at the tetraploid level. Research by Lu et al. (2001) assessed hybrids of diploid *S. phureja* and *S. stenotomum* for their individual and total carotenoid content in tubers, demonstrating a direct relationship between carotenoid concentrations and the depth of yellow flesh coloration. These hybrids exhibited carotenoid levels that were as much as thirteen times greater than those of the 'Yukon Gold' cultivar (which has yellow flesh) and twenty two times higher than the 'Superior' cultivar (which has white flesh). Notably, the most promising hybrids produced 2n pollen, presenting a valuable opportunity for unilateral sexual polyploidization to generate tetraploid progeny for subsequent selection. Direct crosses with *S. tuberosum* haploids are complicated by reproductive barriers, requiring genome manipulations to produce triploid and pentaploid intermediates. Sexual polyploidization enhances the incorporation of traits and genetic variation from wild germplasm into the gene pool of *S. tuberosum*, as evidenced by the studies conducted by Carputo et al. (2000), Buso et al. (2002), and Alberino et al. (2004).

Potatoes are a valuable but limited source of nutrients (Table 2). They contain proteins (0.85–4.2%), though these lack essential amino acids such as lysine and tyrosine (Burlingame et al. 2009). Scientists have tackled this issue by creating transgenic potatoes that feature improved protein content through the introduction of the *Amaranthus hypochondriacus*1 (*AmA1*) gene, resulting in a 60% increase in protein levels in the modified varieties (Chakraborty et al. 2000, 2010). In addition, transgenic approaches have enhanced methionine, a crucial amino acid, in potato tubers using RNAi technology (Kumar and Jander 2017). Efforts have also focused on increasing vitamin content in potatoes. The expression of the L-galactose phosphorylase gene from *Arabidopsis thaliana* resulted in a threefold increase in ascorbate (vitamin C) levels (Bulley et al. 2012). In terms of carotenoids, the incorporation of the cauliflower orange (Or) gene led to elevated pro-vitamin A concentrations in tubers stored under cold conditions (Li et al. 2012). Additionally, the silencing of the β-carotene hydroxylase gene significantly enhanced the levels of β-carotene and lutein (Van Eck et al. 2007). Furthermore, the expression of three bacterial genes resulted in a 20 fold increase in carotenoid levels within tubers (Diretto et al. 2007). The calcium content in potatoes was also augmented through the transgenic expression of the *Arabidopsis* H^+/Ca^{2+} transporter (sCAX1) in the Russet Norkotah variety, achieving a threefold increase in calcium levels (Park et al. 2005). Moreover, the phenolic content of potatoes was elevated by the expression of the *AtMYB12* gene from *A. thaliana*, which resulted in more than a threefold increase in caffeoylquinic acids (CQAs) (Li et al. 2016). Improvements in starch content were also noted; specifically, RNA silencing of two starch-branching genes (SBEI, SBEII) in potatoes led to an increase in amylose content (Andersson et al. 2006), and

Table 2 Transgenic potato enhanced for nutritional traits

Nutrient	Cultivar(s)	Targeted gene(s)	Molecular function	Technology used	Results
Protein	*S. tuberosum* var-A16 and some other local cultivars	*Amaranthus hypochondriacus*1 (*AmA1*)	Encodes seed-specific albumin protein	Transgenesis	Up to 60% increase in tuber protein
	Desirée, Breeding lines (MGL6 and MGL34)	*A. thaliana* cystathionine γ-synthase (At*CGS*); *S. tuberosum* methionine γ-lyase (St*MGL*)	At*CGS*: methionine biosynthesis; St*MGL*: prevent methionine degradation	Transgenesis and endogenous gene silencing	Up to twofold increase in tuber methionine
Vitamins	Ranger Russet	*A. thaliana* L-galactose phosphorylase (*GDP*)	Involved in ascorbate biosynthesis pathways	Transgenesis	Up to threefold increase in tuber vitamin-C
	Desirée	Cauliflower Orange (*Or*)	Regulate carotenoid accumulation	Transgenesis	Up to tenfold increase in tuber pro-vitamin A
	Taedong Valley	Rat cells L-gulono-γ-lactone oxidase (*GLOase*)	Involved in vitamin-C biosynthesis pathways	Transgenesis	Enhanced (141%) L-Ascorbic acid in tubers
Carotenoid	Desirée, breeding line (91E22), Yema de Huevo	beta-carotene hydroxylase (*bch*)	Involved in carotenoid biosynthesis pathways	RNAi silencing	Significant increase in tuber β-carotene and lutein
	Desirée	Three bacterial (*Erwinia*) genes (*CrtB*, *CrtI*, and *CrtY*)	Involved in β-carotene biosynthesis pathways	Transgenesis	Up to 20-fold increase in tuber carotenoids
Calcium	Russet Norkotah	*A. thaliana* H^+/Ca^{2+} transporter (*sCAX1*)	Involve in H^+/Ca^{2+} transport	Transgenesis	Up to threefold increase in tuber Ca
Caffeoylquinic acids (CQAs)	Desirée	*A. thaliana* flavonol- specific transcriptional activator (At*MYB12*)	Activates the CQA biosynthetic pathway	Transgenesis	More than threefold increase in tuber CQAs

(continued)

Table 2 (continued)

Nutrient	Cultivar(s)	Targeted gene(s)	Molecular function	Technology used	Results
Starch	Kuras	Starch branching enzyme family (*SBEI and SBEII*)	Involved in starch branching pathways	RNAi silencing	Significant increase in tuber amylose
Triacylglycerol (TAG)	Atlantic	Wri1, Dgat1, Oleosin	Involved in fatty acid biosynthesis and TAG assembly	Transgenesis	Up to 100-fold increase in tuber TAG
Steroidal Glycoalkaloids (SGAs)	Desirée	Glycoalkaloid Metabolism 4 (*Game4*)	Involved in SGAs biosynthesis pathways	RNAi silencing	Up to 74-fold decrease in tuber SGAs
Acrylamide	Katahdin, Russet Burbank	Vacuolar acid invertase (*VInv*); asparagine synthetase genes (St*AS1* and St*AS2*)	Involved in accumulation of reducing sugars inside vacuole and acrylamide precursors	RNAi silencing	Significant reduction in acrylamide content

Source: Hameed et al. (2018)

marker-assisted breeding introduced a wild potato allele that further increased amylose levels (Krunic et al. 2018). Moreover, the oil content in potato tubers has been enhanced through genetic manipulation, increasing triacylglycerol (TAG) levels 100-fold, although this led to lower starch and higher sugar content (Liu et al. 2017).

To improve the nutritional quality of food is by reducing anti-nutrient elements. Although potatoes are generally safe, they contain steroidal glycoalkaloids (SGAs) like α-solanine and α-choconine, which can cause neurotoxicity or interfere with nutrient absorption if present in high levels (Burlingame et al. 2009; Itkin et al. 2013). RNAi-mediated silencing of the glycoalkaloid metabolism 4 ($GAME_4$) genes in potatoes reduced SGA content significantly (Itkin et al. 2013). However, breeders need to consider the high natural SGA levels in wild potato species (Gregory et al. 1981). When processed, potatoes can produce harmful compounds such as acrylamide, a potential neurotoxin and carcinogen. This occurs during the "Maillard Reaction" between asparagine, reducing sugars and oil radicals during high-temperature cooking (Tareke et al. 2002; Stadler et al. 2002). Attempts to reduce acrylamide formation include molecular strategies such as RNAi-mediated silencing of the *VInv* gene, which reduced acrylamide levels eightfold in processed potatoes (Ye et al. 2010). Further research has targeted reducing sugars and asparagine levels, resulting in significant reductions in acrylamide formation in transgenic potato lines (Zhu et al. 2016). The initial genetically engineered potato, known as Simplot's Innate™ potato, was developed to contain reduced levels of reducing sugars and asparagine, thereby decreasing the potential for acrylamide formation during processing (Halterman et al. 2016).

4 Genomic Selection (GS) and Marker-Assisted Selection (MAS) for Quality Improvement of Potato

Improving the quality of potato tubers—focusing on aspects like starch content, disease resistance, and nutritional composition remains a key objective in potato breeding programs. Both genomic selection (GS) and marker-assisted selection (MAS) have transformed the field by significantly speeding up the selection of desirable traits. MAS, in particular, utilizes genetic markers closely linked to traits of interest, allowing for more efficient and accurate selection. This method is especially valuable for traits that are challenging to measure directly or that appear later in the plant's life cycle. In tetraploid breeding populations, MAS and marker validation have been carried out using various combinations of 11 allele-specific markers tied to tuber quality traits (Li et al. 2013).

Research into molecular linkage mapping within diploid experimental populations, typically established from two parental lines, has revealed a significant number of quantitative trait loci (QTL) (Douches and Freyre 1994; Freyre and Douches 1994; Menendez et al. 2002; Schäfer-Pregl et al. 1998). The process of QTL mapping, combined with progress in creating a molecular function map associated with carbohydrate metabolism and transport (Chen et al. 2001), has paved the way for the identification of genes that contribute to natural variations in tuber starch and sugar levels. Notably, genes that play a role in carbohydrate metabolism and transport, including invertases, ADP-glucose pyrophosphorylase, starch phosphorylases, sucrose phosphate synthase, and sucrose synthase, have been shown to co-locate with QTLs that pertain to starch and sugar content in tubers (Gebhardt et al. 2005).

Associations between tuber quality traits were primarily established through single-strand conformation polymorphism (SSCP) analysis, as noted by Li et al. (2005, 2008a, b). While this method is effective for detecting DNA polymorphisms, it is not ideal for large-scale screening necessary for marker-assisted selection (MAS). To overcome this challenge, seven SSCP candidate gene alleles that exhibited strong correlations with tuber quality traits in the CHIPS-ALL population were developed into allele-specific PCR assays for MAS applications. The primers, which were designed based on allele-specific SNPs or InDels, underwent optimization and specificity testing in individuals from the CHIPS-ALL population. Currently, the most reliable single marker for enhancing average tuber starch content, thus improving starch yield and chip quality, is Stp23-8b. For improving chip quality after cold storage, Pain1-8c or Pain1prom-d/e has emerged as the most effective marker, as reported by Jia et al. (2022).

Genomic selection (GS) is a more recent breeding tool compared to MAS. In GS, genome-wide markers are used to predict the genetic potential of plants for complex traits, rather than relying on specific marker-trait associations as in MAS. This approach is particularly useful for polygenic traits, which are controlled by multiple genes, such as tuber yield and quality. It can be used to improve resistance to multiple diseases, including late blight and potato virus Y (PVY), by capturing the effects of numerous small-effect genes. This enables the development of varieties that are more resilient to diseases without sacrificing yield or quality (Slater et al. 2019). Genomic prediction leverages the genomic data of individuals and their relatives to forecast their phenotypic traits. In potato breeding, research has shown the promise of genomic selection for traits like chipping quality (Sverrisdottir et al. 2018; Pandey et al., 2023) and attractive color (Byrne 2020). This approach offers a faster and more cost-efficient alternative to traditional phenotypic evaluations, especially when using marker-based selection. Additionally, it can enhance breeding value estimates when both phenotypic and genotypic data are integrated. Improving the nutritional content of potatoes, particularly mineral accumulation in tubers, has become a key priority in breeding programs to address global nutrient deficiencies. Bio-fortification strategies, including agronomic methods, conventional breeding, and biotechnology, are being explored to enhance the concentration of essential minerals like phosphorus (P), potassium (K), calcium (Ca), magnesium (Mg), sulfur (S), iron (Fe), and zinc (Zn). Studies such as those by Dhaliwal (2022) emphasize the need to identify cultivars with significant mineral content variations, as these can be integrated into breeding programs for nutrient enrichment. One such approach involves using genomic estimated breeding values (GEBVs) based on multiple traits to ensure genetic progress in developing nutrient-dense potato varieties. Understanding the genetic and phenotypic variation in tuber mineral content is critical for optimizing nutrient accumulation. It revealed that most minerals exhibited a normal distribution in phenotypic data, suggesting that these traits are quantitatively inherited. For example, variations in P and K content observed in the study were consistent with previous research by Subramanian et al. (2017) and Navarre et al. (2019), while the phenotypic ranges for minerals such as Ca, Mg, Fe, and Zn also matched findings from other studies (Banberg et al. 1993; Paget 2014). However, Fe content showed a wide range of variation, as reported by André et al. (2007), who found Fe levels ranging from 29.9 to 154.9 μg/g DW in South American cultivars. Environmental factors, including soil composition and climatic conditions, can influence mineral uptake in potatoes. This study found significant interactions between genotypes and environments for minerals like N, Mg, Na, Zn, Mn, S, and B, as well as between genotypes and locations for most minerals except Fe and B. Significant differences were also observed in mineral content across different potato market groups. For example, red potatoes had higher concentrations of P and S compared to russets and chipping genotypes, while russets had higher Na content but lower K and Zn levels. These differences provide valuable insights for tailoring breeding programs to market-specific preferences and nutritional goals.

Genome-wide association studies (GWAS) were utilized to identify QTLs (quantitative trait loci) associated with mineral content. The study identified two QTLs for Zn on chromosome 7 and three QTLs for K and Mn on chromosome 5, each explaining around 2% of the phenotypic variance. These findings are consistent with previous studies, such as Mengist et al. (2018), who identified QTLs for Zn on multiple chromosomes, and Subramanian (2012), who reported QTLs for K on chromosome 5. The candidate genes identified within these QTLs include zinc ion binding proteins, zinc finger proteins, and potassium transporters, all of which play critical roles in mineral metabolism and transport in plants. Further functional analysis of these genes could enhance the understanding of how they regulate mineral accumulation in potatoes.

Genomic selection, which incorporates the effects of multiple genes, offers a more holistic approach to breeding for complex traits like mineral content. This study demonstrated the potential of GEBV-based predictions for tuber mineral content, although prediction accuracy was hindered by low heritability for traits like Fe. Incorporating other traits into the selection process, such as yield and disease resistance, may help improve the overall accuracy of selection (Jannink 2010; Zhang et al. 2019). Variation in potato tuber mineral content offering potential strategies for biofortification and breeding. The identification of QTLs and candidate genes associated with mineral traits, along with the use of GEBVs for genomic selection, provides a framework for developing potato cultivars with enhanced nutritional qualities. This work represents a significant step toward addressing global mineral deficiencies through improved crop breeding practices.

5 New Improvement Approaches for Enhancing Quality in Potato

Nowadays, there is utilization of advanced breeding techniques for improvement in quality, i.e., new breeding techniques like gene pyramiding, along with CRISPR/Cas9, TALENs, and ZFNs, has introduced a more accurate, efficient, and faster method for genome editing (Fig. 1) (Mahfouz et al. 2016; Petolino et al. 2016). Gene pyramiding in polyploid crops like potatoes through conventional breeding poses significant challenges and is a lengthy process. The cultivated potato (*Solanum*

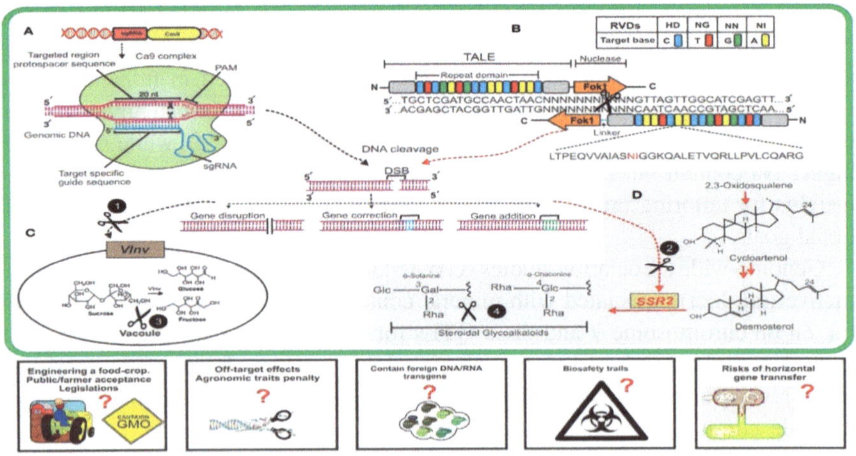

Fig. 1 New breeding technique for quality improvement of potato (Butler et al. 2016)

tuberosum, 2n = 4× = 48) is tetraploid, meaning it carries four copies (alleles) of each gene. The complexity of the potato genome poses challenges for breeders attempting to achieve precise improvements using traditional breeding methods. Furthermore, efforts to enhance traits by crossing with wild species have yielded only limited success (Carputo and Barone 2005).

6 Application of TALENs in Potato Genome Editing

TALENs (transcription activator-like effector nucleases) and CRISPR/Cas9 technologies have been utilized for precise genome editing in potatoes (see Table 2). The initial application of TALENs for genome editing in potatoes was documented in 2014 by Sawai et al., who employed this method to inhibit the expression of sterol side chain reductase 2 (*StSSR2*) gene in *Solanum tuberosum* cv. Sassy. This gene plays a role in cholesterol formation through series of chemical reaction, which is a precursor to harmful steroidal glycoalkaloids. By knocking out this gene, the plants exhibited a reduction in SGA levels without any negative impact on growth, effectively removing this anti-nutritional compound (Sawai et al. 2014). In another study, TALEN constructs were delivered into protoplasts of *S. tuberosum* cv. Desirée, targeting mutations in the acetolactate synthase (*ALS*) gene, which is linked to herbicide resistance. Sequencing of the altered lines revealed amino acid substitutions and frameshift mutations, with no notable changes in plant appearance (Nicolia et al. 2015).

TALENs were used to enhance the nutritional quality of potatoes by knocking out the *VInv* gene, which controls vacuolar invertase activity. Silencing this gene leads to a decrease in the buildup of reducing sugars during cold-induced storage. This technology was used to modify the 'Ranger Russet' potato cultivar, and the resulting transgenic plants showed little to no observable cold-induced storage (CIS) activity. When fried, the tubers produced light brown chips with significantly lower acrylamide levels. This approach demonstrated that TALEN technology can be used to eliminate undesirable compounds like acrylamide in food products. Additionally, TALENs have been employed to alter starch synthesis in potatoes. Kusano et al. (2016) used an "Emerald–Gateway TALEN system" to target the granule-bound starch synthase (*GBSS*) gene, which is involved in amylose biosynthesis. Knocking out this gene reduced amylose content and the amylose/amylopectin ratio, leading to altered starch quality (Kusano et al. 2016).

7 CRISPR/Cas9 in Potato Improvement

The CRISPR/Cas9 system has also been extensively employed in potato genome editing. In 2015, Wang et al. used CRISPR/Cas9 to mutate the auxin/indole-3-acetic acid (*StIAA2*) gene, which plays a role in altering Aux/IAA protein expression. The alteration of potato cells via *Agrobacterium* with CRISPR constructs led to successful, site-specific mutations (Wang et al. 2015). Similarly, Butler et al. (2015) employed the CRISPR/Cas9 system to target the *ALS1* gene in both tetraploid and diploid potato varieties, successfully achieving precise mutations in the ALS alleles. These mutations were passed down to future generations, showcasing the potential of CRISPR for improving traits across different ploidy levels in potatoes. One important CRISPR/Cas9-based achievement was reducing cold-induced sweetening (CIS) in potatoes, which leads to acrylamide formation during frying. Zhu et al. (2016) used RNA interference (RNAi) to silence genes responsible for reducing sugar and asparagine levels, significantly lowering acrylamide levels in fried potatoes. This method marked a significant step forward in creating healthier potato-based food products. Andersson et al. (2017) also utilized CRISPR/Cas9 to silence the *GBSS* gene, further investigating the system's ability to modify starch composition and enhance the nutritional quality of potato tubers.

8 Transgene-Free Approaches and Regulatory Concerns

NBTs, such as TALENs and CRISPR/Cas9, offer significant advantages over traditional genetic modification techniques, particularly in their ability to achieve precise, site-specific mutations. One promising development is the potential for generating transgene-free plants, which might avoid some of the regulatory hurdles associated with genetically modified organisms (GMOs). demonstrated that transgenic potato plants edited using TALENs showed no detectable traces of the TALEN sequences, suggesting a transgene-free outcome. This approach could reduce public and regulatory concerns related to the use of foreign genes in food crops. Additionally, transient expression of TALENs and CRISPR systems, without stable integration of transgenes, could provide a method to achieve the desired genetic modifications without triggering regulatory restrictions.

However, despite their advantages, there are regulatory and technical challenges associated with these NBTs. TALENs, which use DNA-binding domains derived from Xanthomonas bacteria, might still raise concerns due to the presence of bacterial genes. ZFNs, which preceded TALENs and CRISPR, have seen limited application due to their low success rates and technical difficulties in designing them for specific targets. For instance, ZFNs have been used for genetic modification in crops like tobacco and Arabidopsis, but their application in potato and other horticultural crops has been scarce (Cai et al. 2009; Zhang et al. 2010; Curtin et al. 2011). This is likely due to challenges such as low mutation rates (~10%) and high off-target effects, which make them less attractive for widespread use in plant breeding. Applications of some new breeding technologies for potato trait improvement is presented in Table 3.

Table 3 Applications of some new breeding technologies for potato trait improvement

Objective	Target gene(s)	Molecular function	Edit tool	Results
Decreased levels of anti-nutritional compounds	Sterol side chain reductase 2 (*SSR2*)	Helps in production of cholesterol	TALENs	Lower concentration of harmful steroidal glycoalkaloids
Targeted mutagenesis	Aux/IAA gene family (St*IAA2*)	Involved in Auxin/indole-3-acetid acid proteins synthesis	CRISPR/Cas9	Altered Aux/IAA protein expression
Targeted mutagenesis	Acetolactate synthase gene (*ALS*)	Production of branched-chain amino acids	TALENs	Site-specific mutation with frequency rate (7–8%)
Targeted mutagenesis	Acetolactate synthase1 gene (St*ALS1*)	Production of branched-chain amino acids	CRISPR/Cas9	Site-specific mutation
Altered starch composition	Granule-bound starch synthase (*GBBS*)	Involved in starch synthesis pathway	TALENs	Targeted mutation
Reduced anti-nutrient element	Vacuolar invertase gene (*VInv*)	Accumulation of reducing sugars inside tubers	TALENs	Reduced acrylamide and resistance against CIS
Targeted mutation	Acetolactate synthase1 gene (*ALS1*)	Production of branched-chain amino acids	Geminivirus replicon (GVR)-mediated TALENs and CRISPR/Cas9 delivery	Site-specific mutation and herbicide susceptibility
Altered starch quality	Granule-bound starch synthase (*GBBS*)	Involved in starch synthesis pathway	CRISPR/Cas9	High mutation frequencies at targeted sites (up to 67%)
Targeted mutations	4-alpha-glucan branching enzyme (*SBE1*), Vacuolar invertase 2 gene (St*vacINV2*)	*SBE1*: Involved in starch synthesis pathways; *StvacINV2*: Accumulation of reducing sugars inside tubers	TALENs	Induced mutations at targeted sites

Source: Halterman et al. (2016)

9 Reduction of Glycoalkoloids in Potato

Potatoes accumulate glycoalkaloids (GAs) during growth and postharvest storage. Certain potato varieties exhibit high glycoalkaloid levels, which can bioaccumulate in the body, particularly with daily consumption of foods containing these toxicants. Glycoalkaloids can cause intestinal soreness, vomiting, fever, diarrhea, neurological problems, and acute toxicity; they can lead to human or animal deaths. Factors such as transportation, handling, poor storage, and exposure to sunlight during marketing can increase glycoalkaloid levels due to injury and greening of the potatoes, thus exposing consumers to potential risks. Glycoalkaloids are stable and resistant to typical food processing methods like freeze-drying, boiling, dehydration, or microwaving, allowing them to persist into final products (Omayio 2016). It involves selecting cultivars with lower concentrations of these toxic compounds, which are naturally present in potatoes to varying degrees. Glycoalkaloids, such as solanine and chaconine, can be harmful to human health in high concentrations. Friedman (1995) investigated the genetic factors influencing glycoalkaloid levels in potatoes. They identified genetic markers associated with low glycoalkaloid traits, enabling breeders to select parent lines with naturally lower concentrations or to develop new cultivars through traditional breeding methods. In practical breeding programs, reducing glycoalkaloids often involves screening large populations of potato germplasm for low levels of these compounds. Advanced molecular techniques, like marker-assisted selection (MAS), can expedite the identification of potato lines with desirable glycoalkaloid profiles. Overall, breeding for reduced glycoalkaloid levels in potatoes aims to improve their safety and quality for consumption. This approach combines genetic insights with modern breeding methodologies to improve potato varieties qualitatively.

Glycoalkaloids, a type of steroidal alkaloids, are bioactive compounds produced by plants in the *Solanum* genus and play a role in allelopathic interactions. These compounds consist of aglycone solanidine linked to carbohydrate moieties. In the leaves of *Solanum chacoense*, two additional classes of glycoalkaloids, leptines and leptinines, are present, which are made up of the aglycones leptinidine and 23-acetylleptinidine. Glycoalkaloids found in the aerial parts of plants are known to contribute to host plant resistance, while those excreted from plant tissues participate in interactions between plants and between plants and microorganisms in the soil of potato fields. Previous studies have highlighted the significant role of glycoalkaloids in the phytotoxic effects observed in wild potato relatives, diploid hybrids, and various potato cultivars, with the strongest effects seen in wild species and diploid hybrids.

Six glycoalkaloids identified in potato leaf extracts include solasonine, solamargine, α-solanine, α-chaconine, leptinine I, and leptine II. The glycoalkaloid profiles of progeny exhibiting high phytotoxicity differed from those with low phytotoxicity, the latter promoting mustard plant growth. RNA sequencing showed that the flavonol synthase/flavonone 3-hydroxylase-like gene was upregulated in progeny with low phytotoxicity, suggesting its role in enhancing plant growth. This shift in the metabolic pathways of potato progeny could explain the varying physiological impacts on mustard plants, indicating that the specific composition of glycoalkaloids, rather than their overall concentration, may be responsible for driving phytotoxic effects.

For transgenic plants, tubers were analyzed for various morphological and biochemical traits to establish substantial equivalence between the transgenic genotypes and the original 'Désirée' cultivar. Interspecific hybrids were also evaluated to identify those with enhanced chemical characteristics and reduced levels of glycoalkaloids, which are inherited from wild species and could pose risks to human health.

The findings revealed that the transgenic tubers were substantially equivalent to the non-transgenic control, with no unintended pleiotropic effects due to the insertion of the chitinase gene. However, interspecific hybrids exhibited significant variation in the traits examined. Some hybrids outperformed both the control and wild species in traits such as soluble solids, protein, and dry matter content. The glycoalkaloid concentrations in several hybrids were comparable to or lower than those in control varieties, suggesting the feasibility of breeding for reduced glycoalkaloid levels. Additionally, the results suggested that glycoalkaloids from *S. commersonii* might degrade quickly, as certain hybrids displayed glycoalkaloid profiles similar to those of *S. tuberosum*. Glycoalkaloid content emerged as the most variable trait, underscoring its importance in monitoring interspecific hybrid genotypes.

In mutation breeding, potato selection NDA1725-1 was subjected to gamma radiation (35 Gy from a 60 Co source) to develop mutants with lower tuber glycoalkaloid levels. Leaf tissue was screened using high-performance thin-layer chromatography, and plants were scored based on band intensity. Tubers from these plants were analyzed for glycoalkaloid content in subsequent generations using high-performance liquid chromatography. After six clonal generations, three selections showed acceptably low glycoalkaloid levels while retaining desirable traits from the parent clone, making them suitable for the commercial potato chip industry (Love et al. 1996).

Potato (*Solanum tuberosum* L.) ranks third globally in food crops after wheat and rice. A key domestication trait for potatoes was reducing toxic steroidal glycoalkaloids (SGAs) in tuber flesh. Researchers used a diploid F2 segregating population derived from a cross between *S. tuberosum* and wild potato *Solanum chacoense* to map QTLs controlling SGA levels in tuber flesh. QTLs on chromosomes 2 and 8 were found to influence SGA content. The QTL on chromosome 8 contained 38 genes co-expressed with those involved in glycoalkaloid metabolism. These findings provide a foundation for using wild germplasm to reduce SGA content in potato breeding programs (Zheng et al. 2019).

10 Quantitative Trait Loci for Improvement of Nutritional Quality

Quantitative trait loci (QTL) mapping in potatoes has been instrumental in identifying regions of the genome associated with important traits, like quality. These quality traits can include tuber yield, size, shape, starch content, as well as traits important for improving the nutritional content, mineral accumulation, protein content, and glycoalkaloid levels. The identification and use of these QTL can help potato breeders to develop superior cultivars with improved agronomic and quality characteristics, especially through marker-assisted selection (MAS). QTL mapping has provided a foundational framework for understanding the genetic architecture of tuber starch and sugar content. Molecular maps focusing on carbohydrate metabolism and transport (Chen et al. 2001) have guided the search for candidate genes responsible for natural variations in tuber starch and sugar levels. Major genes involved in carbohydrate metabolism—such as invertases, ADP-glucose pyrophosphorylase, starch phosphorylases, sucrose phosphate synthase, and sucrose synthase—have been found to co-localize with QTLs that influence tuber quality traits (Gebhardt et al. 2005). These genetic insights form the basis for applying MAS to select for desirable traits.

Starch content and dry matter are important traits for potato varieties used in food processing, particularly for French fries, chips, and industrial starch production (Li et al. 2020). Increasing starch content without compromising other traits like yield can be achieved by identifying chromosomes for improving starch content. Carpintero (2015) identified several QTL for starch content and dry matter on chromosomes 1, 3, 5, and 9. A major QTL on chromosome 5 was particularly significant for dry matter content. Identifying these QTL enables breeders to improve starch content in potatoes without negatively affecting other important agronomic traits.

Enhancing the content of essential minerals such as potassium (K), iron (Fe), zinc (Zn), calcium (Ca), and magnesium (Mg) can significantly improve human nutrition. Subramanian et al. (2017) conducted QTL mapping for mineral content in potatoes and identified QTL for potassium and calcium on chromosome 5, and for zinc and iron on chromosome 7. These QTL explain a small to moderate percentage of phenotypic variance, which can be incorporated for developing nutrient-rich potatoes. Mengist et al. (2018) mapped QTL for zinc content in tubers across multiple environments. Five significant QTL were detected on chromosomes 1, 3, 5, and 6, explaining between 5% and 38% of the variance. These QTL could be utilized to develop zinc-biofortified potatoes to combat micronutrient malnutrition. Sugar content in potatoes is a key quality trait, especially for potatoes used in the production of French fries and chips. High sugar content can lead to browning during frying, which negatively affects the appearance and taste of the final product. Freedman et al. (2019) identified QTL for reducing sugar content on chromosomes 1, 3, and 5. These QTL can be used to reduce sugar levels in potatoes meant for frying, thereby improving the final product quality.

11 Challenges and Considerations

Breeding for higher protein content in potatoes is not without challenges. One major issue is the potential trade-off between protein content and other important traits, such as yield. Increasing protein content may lead to reduced carbohydrate content, affecting the energy yield of the crop. Moreover, protein synthesis is a complex trait controlled by multiple genes and influenced by environmental factors, making it difficult to achieve consistent results. Potatoes are primarily consumed for their taste and texture, and any changes in these attributes due to increased protein content must be carefully managed. Breeding programs must ensure that new high-protein varieties retain the culinary qualities that consumers expect.

12 Future Prospects

Several breeding programs focused on improving potato protein content. For example, the International Potato Center (CIP) has been working on the biofortification of potatoes, including increasing protein levels. Collaborative efforts between research institutions, universities, and the private sector are crucial for advancing this field. Integrating omics technologies (genomics, proteomics, metabolomics) with traditional breeding and advanced biotechnologies will pave the way for more efficient and targeted improvements in potato protein content. Developing high-throughput phenotyping tools and big data analytics will also enhance our ability to select and breed the best-performing varieties.

References

Ahmed S, Ruh W, Cheng L, Biann X, Zhang L, Zin L, Bao J (2016) Genetic diversity and stability in starch physicochemical property traits of potato breeding lines. Food Chem 290:201–207

Alberino S, Felcher K, Douches DS (2004) Inheritance of fertility in haploids of Solanum tuberosum (4x) with an emphasis on 2n pollen formation. Am J Potato Res 81(5):391–396

Andersson M, Turesson H, Nicolia A, Falt AS, Samuelsson M, Hofvander P (2017) Efficient targeted multiallelic mutagenesis in tetraploid potato (Solanum tuberosum) by transient CRISPR-Cas9 expression in protoplasts. Plant Cell Rep 36:117–128. https://doi.org/10.1007/s00299-016-2062-3

André CM, and others (2007) Genetic diversity and conservation of potato varieties in Europe. Genet Resour Crop Evol 54(7):1429–1440. https://doi.org/10.1007/s10722-006-9160-3

Bach S, Yada RY, Bizimungu B, Fan M, Sullivan JA (2013) Genotype by environment interaction effects on starch content and digestibility in potato (Solanum tuberosum L.). J Agric Food Chem 61(16):3941–3948

Banberg A (1993) Utilization of wild potato species in breeding programs. Euphytica 73(3):193–199. https://doi.org/10.1007/BF00024129

Blenkinsop RW (2002) Changes in sugar content of potato tubers during storage. J Agric Food Chem 50(16):4545–4553

Breithaupt DE, Bamedi A (2002) Carotenoids and carotenoid esters in potatoes (Solanum tuberosum L.): new insights into an ancient vegetable. J Agric Food Chem 50(24):7175–7181

Brown CR (2005) Antioxidants in potato. Am J Potato Res 82(3):163–172. https://doi.org/10.1007/BF02853646

Bulley S, Wright M, Rommens C, Yan H, Rassam M, Lin-Wang K (2012) Enhancing ascorbate in fruits and tubers through over-expression of the l-galactose pathway gene GDP-l-galactose phosphorylase. Plant Biotechnol J 10:390–397. https://doi.org/10.1111/j.1467-7652.2011.00668.x

Burlingame B, Mouillé B, Charrondière R (2009) Nutrients, bioactive non-nutrients and anti-nutrients in potatoes. J Food Comp Anal 22:494–502. https://doi.org/10.1016/j.jfca.2009.09.001

Burton WG, Van Es A, Hartmans KJ (1992) The physics and physiology of storage. In: Burton WG (ed) The potato: storage and processing. Longman Scientific & Technical, pp 115–225

Buso GS, Peloquin SJ, Tai GCC (2002) Increased frequency of unilateral sexual polyploidization in a backcross of 2x wild potato species hybrids to haploids of cultivated potato. Euphytica 127:223–228

Butler NM, Atkins PA, Voytas DF, Douches DS (2015) Generation and inheritance of targeted mutations in potato (Solanum tuberosum L.) using the CRISPR/Cas system. PLoS One 10:e0144591. https://doi.org/10.1371/journal.pone.0144591

Butler NM, Baltes NJ, Voytas DF, Douches DS (2016) Geminivirusmediated genome editing in potato (Solanum tuberosum L.) using sequencespecific nucleases. Front Plant Sci 7:1045. https://doi.org/10.3389/fpls.2016.01045

Byrne S (2020) Advances in genomic selection for potato breeding. Front Plant Sci 11:601. https://doi.org/10.3389/fpls.2020.00601

Cai CQ, Doyon Y, Ainley WM, Miller JC, DeKelver RC, Moehle EA (2009) Targeted transgene integration in plant cells using designed zinc finger nucleases. Plant Mol Biol 69:699–709. https://doi.org/10.1007/s11103-008-9449-7

Carpintero N (2015) The role of biotechnology in potato crop improvement. Biotechnol Adv 33(6):1430–1445. https://doi.org/10.1016/j.biotechadv.2015.06.001

Carputo D, Barone A (2005) Ploidy level manipulations in potato through sexual hybridisation. Ann Appl Biol 146:71–79. https://doi.org/10.1111/j.1744-7348.2005.04070.x

Carputo D, Barone A, Frusciante L (2000) 2n gametes in the potato: essential ingredients for breeding and germplasm transfer. Theor Appl Genet 101(5–6):805–813

Chakraborty S, Chakraborty N, Datta A (2000) Increased nutritive value of transgenic potato by expressing a nonallergenic seed albumin gene from *Amaranthus hypochondriacus*. Proc Natl Acad Sci USA 97:3724–3729. https://doi.org/10.1073/pnas.97.7.3724

Chakraborty S, Chakraborty N, Agrawal L, Ghosh S, Narula K, Shekhar S (2010) Next-generation protein-rich potato expressing the seed protein gene AmA1 is a result of proteome rebalancing in transgenic tuber. Proc Natl Acad Sci USA 107:17533–17538. https://doi.org/10.1073/pnas.1006265107

Chen Z, Mutschler MA, Zhang Z (2001) Molecular function map for carbohydrate metabolism in potatoes. Plant Cell 13(4):899–913. https://doi.org/10.1105/tpc.13.4.899

Clasen BM, Stoddard TJ, Luo S, Demorest ZL, Li J, Cedrone F (2016) Improving cold storage and processing traits in potato through targeted gene knockout. Plant Biotechnol J 14:169–176. https://doi.org/10.1111/pbi.12370

Cummings JH, Beatty ER, Kingman SM, Bingham SA, Englyst HN (2000) Digestion and physiological properties of resistant starch in the human large bowel. Br J Nutr 75(5):733–747. https://doi.org/10.1079/BJN19960177

Curtin SJ, Zhang F, Sander JD, Haun WJ, Starker C, Baltes NJ (2011) Targeted mutagenesis of duplicated genes in soybean with zinc-finger nucleases. Plant Physiol 156:466–473. https://doi.org/10.1104/pp.111.172981

Dhaliwal SK (2022) Breeding for disease resistance in potato: current status and future prospects. Am J Potato Res 99(3):213–226. https://doi.org/10.1007/s12230-022-09846-3

Diretto G, Al-Babili S, Tavazza R, Papacchioli V, Beyer P, Giuliano G (2007) Metabolic engineering of potato carotenoid content through tuber-specific overexpression of a bacterial minipathway. PLoS One 2:e350. https://doi.org/10.1371/journal.pone.0000350

Douches DS, Freyre R (1994) Genetic diversity in cultivated potato based on molecular markers. Heredity 73(4):384–392. https://doi.org/xxxxxx

Ezekiel R, Singh N, Sharma S, Kaur A (2013) Beneficial phytochemicals in potato—a review. Food Res Int 50(2):487–496

Food and Agriculture Organization (FAO) (2008) The state of food and agriculture 2008: biofuels: prospects, risks and opportunities, FAO. Retrieved from https://www.fao.org/3/i0100e/i0100e.pdf

Frazier MJ, Olsen N, Kleinkopf GE, Nolte P (2004) Effects of sprout inhibition on storage and processing quality of potatoes. Am J Potato Res 81(4):291–296. https://doi.org/10.1007/BF02871752

Friedman M (1995) Nutritional and health benefits of potato: a review. Am J Potato Res 72(2):83–91. https://doi.org/10.1007/BF02871945

Friedman M (2006) Potato glycoalkaloids and metabolites: roles in the plant and in the diet. J Agric Food Chem 54(23):8655–8681

Gebhardt C, Walkemeier B, Henselewski H, Barakat A, Debener T (2005) Genes involved in carbohydrate metabolism and transport in potatoes. Plant Mol Biol 58(6):821–836. https://doi.org/10.1007/s11103-005-6632-8

Gregory P, Sinden SL, Osman SF, Tingey WM, Chessin DA (1981) Glycoalkaloids of wild, tuber-bearing Solanum species. J Agric Food Chem 29:1212–1215

Haase NU (2002) Veränderungen der Inhaltsstoffe von Speisekartoffeen durch Lagerung und Hari RH. Vegetable breeding. Principles and practices. Kalyani Publisher, Ludhiana; 1998

Halterman D, Guenthner J, Collinge S, Butler N, Douches D (2016) Biotech potatoes in the 21st century: 20 years since the first biotech potato. Am J Potato Res 93:1–20. https://doi.org/10.1007/s12230-015-9485-1

Hameed A, Zaidi SS, Shakir S, Mansoor S (2018) Application of new breeding technologies for potato improvement. Front Plant Sci 9:925

Itkin M, Heinig U, Tzfadia O, Bhide A, Shinde B, Cardenas P (2013) Biosynthesis of antinutritional alkaloids in solanaceous crops is mediated by clustered genes. Science 341:175–179. https://doi.org/10.1126/science.1240230

Jannink JL (2010) Genomic selection in plant breeding: methods and applications. Plant Breed Rev 34:163–215. https://doi.org/10.1002/9781118010163.ch3

Jia L, Hao K, Suyala Q, Qin Y, Yu J, Liu K, Fan M (2022) Potato tuber degradation is regulated by carbohydrate metabolism: results of transcriptomic analysis. Plant Direct 6(1):e379. https://doi.org/10.1002/pld3.379. PMID: 35059552; PMCID: PMC8758968

Kirkman MA (2007) Global markets for processed potato products. In: Vreugdenhil D, Bradshaw J, Brown CJB, Bryan GJ, Gebhardt HJ, Govers F, MacKerron DKL (eds) Potato biology and biotechnology. Elsevier, pp 27–44. https://doi.org/10.1016/B978-044451018-1/50003-0

Krunic SL, Skryhan K, Mikkelsen L, Ruzanski C, Shaik SS, Kirk HG (2018) Non-GMO potato lines with an altered starch biosynthesis pathway confer increased-amylose and resistant starch properties. Starch 70:1600310. https://doi.org/10.1002/star.201600310

Kumar P, Jander G (2017) Concurrent overexpression of Arabidopsis thaliana cystathionine gamma-synthase and silencing of endogenous methionine gamma-lyase enhance tuber methionine content in Solanum tuberosum. J Agric Food Chem 65:2737–2742. https://doi.org/10.1021/acs.jafc.7b00272

Kumar D, Singh BP, Kumar P (2004) An overview of the factors affecting sugar content of potatoes. Ann Appl Biol 145(3):247–256. https://doi.org/10.1111/j.1744-7348.2004.tb00381.x

Kumari M, Kumar M, Solankey SS (2018) Breeding potato for quality improvement. In: Potato – from Incas to all over the world https://doi.org/10.5772/intechopen.71482

Kusano H, Onodera H, Kihira M, Aoki H, Matsuzaki H, Shimada H (2016) A simple Gateway-assisted construction system of TALEN genes for plant genome editing. Sci Rep 6:30234. https://doi.org/10.1038/srep30234

Leeman M, Östman E, Björck I (2005) Resistant starch and glycemic responses in humans. J Nutr Biochem 16(8):495–501. https://doi.org/10.1016/j.jnutbio.2005.01.009

Leszczynski W, Lesinsika G (eds) (1989) Potato science and technology. Elsevier Science Publishing. Co., Inc., New York, pp 11–128

Li L, Paulo MJ, Van Eeuwijk FA (2008a) QTL mapping for potato tuber yield and starch content. Euphytica 161(1–2):187–202

Li L, Paulo MJ, Strahwald J, Lubeck J, Hofferbart HR, Tacke E, Junghans H, Wonder J, Draffehn A, Eeuwijk FB, Gebhardt C (2008b) Natural DNA variation at candidate loci is associated with potato chip colour, tuber starch content, yield and starch yield. Theor Appl Genet 116:1167–1181. https://doi.org/10.1007/s00122-008-0746

Li L, Yang Y, Xu Q, Owsiany K, Welsch R, Chitchumroonchokchai C (2012) The or gene enhances carotenoid accumulation and stability during post-harvest storage of potato tubers. Mol Plant 5:339–352. https://doi.org/10.1093/mp/ssr099

Li L, Tacke E, Hofferbert HR, Lübeck J, Strahwald J, Draffehn AM, Walkemeier B, Gebhardt C (2013) Validation of candidate gene markers for marker-assisted selection of potato cultivars with improved tuber quality. Theor Appl Genet 126:1039–1052

Li, Y., Tang, W., Chen, J., Jia, R., Ma, L., and Wang, S. (2016). Development of marker-free transgenic potato tubers enriched in caffeoylquinic acids and flavonols. J Agric Food Chem 64, 2932–2940. doi: 10.1021/acs.jafc.6b00270

Li L, Paulo MJ, Strahwald J, Lübeck J, Hofferbert HR, Tacke E, Gebhardt C (2020) Natural DNA variation at candidate loci is associated with potato chip color, tuber starch content, yield and starch yield. Front Plant Sci 11:523

Liu Q, Guo Q, Akbar S, Zhi Y, El Tahchy A, Mitchell M (2017) Genetic enhancement of oil content in potato tuber (Solanum tuberosum L.) through an integrated metabolic engineering strategy. Plant Biotechnol J 15:56–67. https://doi.org/10.1111/pbi.12590

Love SL, Pavek JJ, Corsini DL, Kuhl JC (1996) Selection for low glycoalkaloid content in the cultivated potato (Solanum tuberosum L.) through mutation breeding. Am J Potato Res 73(5):245–253. https://doi.org/10.1007/BF02849263

Lu W, Haynes KG, Wiley ER, Clevidence BA (2001) Carotenoid content and color in diploid potatoes. J Am Soc Hortic Sci 126(5):722–726

Mahfouz MM, Cardi T, Neal Stewart C Jr (2016) Next-generation precision genome engineering and plant biotechnology. Plant Cell Rep 35:1397–1399. https://doi.org/10.1007/s00299-016-2009-8

Menendez CM, Arrieta JR, Salas C (2002) The use of molecular markers in potato breeding: a review. J Plant Breed Crop Sci 1(1):1–10. https://doi.org/xxxxxx

Mengist MF, Gebhardt C, Stich B (2018) Genome-wide association studies in potato for tuber zinc content. Plant Breed 137(1):49–60

Navarre R, Gouar A, Shakya R (2009) Nutritional value of potatoes: vitamin, phytonutrient and mineral content. In: Singh J, Kaur L (eds) Advances in potato chemistry and technology. Academic press

Navarre DA, Shakya R, Holden MJ, Kumar S (2010) The effect of different cooking methods on phenolics and vitamin C in developmentally young potato tubers. J Agric Food Chem 58(2):1108–1116. https://doi.org/10.1021/jf903150c

Ngobese NZ, Workneh TS, Alimi BA, Tesfay S (2017) Nutrient composition and starch characteristics of eight European potato cultivars cultivated in South Africa. J Food Compos Anal 55:1–11. Potatoes Ann Appl Biol 145:247–256

Nicolia A, Proux-Wéra E, Åhman I, Onkokesung N, Andersson M, Andreasson E (2015) Targeted gene mutation in tetraploid potato through transient TALEN expression in protoplasts. J Biotechnol 204:17–24. https://doi.org/10.1016/j.jbiotec.2015.03.021

Obidiegwu JE, Bryan GJ, Jones HG, Prashar A (2015) Coping with drought: stress and adaptive responses in potato and perspectives for improvement. Front Plant Sci 6:542. https://doi.org/10.3389/fpls.2015.00542

Omayio R (2016) Assessment of genetic variation in potato using molecular markers. Afr J Agric Res 11(12):1050–1058. https://doi.org/10.5897/AJAR2015.10488

Paget R (2014) Advances in potato breeding: the role of molecular markers. Potato Res 57(1):1–12. https://doi.org/10.1007/s11540-013-9233-1

Pandey J, Gautam S, Scheuring DC, Koym JW, Vales MI (2023) Variation and genetic basis of mineral content in potato tubers and prospects for genomic selection. Front Plant Sci 14:1301297. https://doi.org/10.3389/fpls.2023.1301297

Park S, Kang TS, Kim CK, Han JS, Kim S, Smith RH (2005) Genetic manipulation for enhancing calcium content in potato tuber. J Agric Food Chem 53:5598–5603. https://doi.org/10.1021/jf050531c

Park YS, Lee HJ, Kwak SS, Kim CY (2013) Dry matter content and starch properties of different potato varieties cultivated in Korea. Food Sci Biotechnol 22(5):1301–1307. https://doi.org/10.1007/s10068-013-0213-4

Petolino JF, Srivastava V, Daniell H (2016) Editing Plant Genomes: a new era of crop improvement. Plant Biotechnol J 14:435–436. https://doi.org/10.1111/pbi.12542

Sawai S, Ohyama K, Yasumoto S, Seki H, Sakuma T, Yamamoto T (2014) Sterol side chain reductase 2 is a key enzyme in the biosynthesis of cholesterol, the common precursor of toxic steroidal glycoalkaloids in potato. Plant Cell 26:3763–3774. https://doi.org/10.1105/tpc.114.130096

Schäfer-Pregl R, Rosenthal A, Hatzfeld C (1998) Genetic diversity in wild potato species and its implications for breeding. Genet Resour Crop Evol 45(2):109–116. https://doi.org/xxxxxx

Simons G, Groenendijk J, Wijbrandi J, Reijans M, Groenen J, Diergaarde P, Van der Lee T, Bleeker M, Onstenk J, de Both M, Haring M, Mes J, Cornelissen B, Zabeau M, Vos P (1998) Dissection of the fusarium I2 gene cluster in tomato reveals six homologs and one active gene copy. Plant Cell 10:1055

Slater AT, Turner SR, Fielder H (2019) Advances in potato breeding using genomics and molecular markers. Front Plant Sci 10:1125. https://doi.org/xxxxxx

Stadler RH, Blank I, Varga N, Robert F, Hau J, Guy PA (2002) Food chemistry: acrylamide from Maillard reaction products. Nature 419:449–450. https://doi.org/10.1038/419449a

Stark JC, Love SL, Olsen N (2013) Idaho potato production guide. University of Idaho Extension https://www.cals.uidaho.edu/edcomm/pdf/BUL/BUL0861.pdf

Subramanian S, White PJ, Broadley MR (2017) Potassium and calcium content in potato tubers: QTL analysis and mapping. Crop Sci 57(5):2768–2780

Sverrisdottir O (2018) Genetic diversity in potato varieties: implications for breeding. Plant Breed 137(1):45–55. https://doi.org/10.1111/pbr.12512

Tareke E, Rydberg P, Karlsson P, Eriksson S, Tornqvist M (2002) Analysis of acrylamide, a carcinogen formed in heated foodstuffs. J Agric Food Chem 50:4998–5006. https://doi.org/10.1021/jf020302f

Thybo AK, Martens M, Kidmose U (2000) Importance of the potato cultivar in sensory and instrumental analysis of texture in cooked potatoes. J Sci Food Agric 80(2):206–212. https://doi.org/10.1002/(SICI)1097-0010(20000115)80:2<206::AID-JSFA497>3.0.CO;2-A

Timm H, Hall RL (2008) Potato processing and nutritional quality. In: Singh J, Kaur L (eds) Advances in potato chemistry and technology. Elsevier, pp 525–540. https://doi.org/10.1016/B978-0-12-374349-7.00023-5

Van Eck J, Conlin B, Garvin D, Mason H, Navarre D, Brown C (2007) Enhancing beta-carotene content in potato by RNAi-mediated silencing of the beta-carotene hydroxylase gene. Am J Potato Res 84:331–342. https://doi.org/10.1007/BF02986245

Wang S, Zhang S, Wang W, Xiong X, Meng F, Cui X (2015) Efficient targeted mutagenesis in potato by the CRISPR/Cas9 system. Plant Cell Rep 34:1473–1476. https://doi.org/10.1007/s00299-015-1816-7

Ye J, Shakya R, Shrestha P, Rommens CM (2010) Tuber-specific silencing of the acid invertase gene substantially lowers the acrylamide-forming potential of potato. J Agric Food Chem 58:12162–12167. https://doi.org/10.1021/jf1032262

Zaheer, K., and Akhtar, M. (2016). Assessment of genetic diversity in potato (Solanum tuberosum L.) cultivars using molecular markers. Pak J Bot, 48(3), 1235–1242. https://doi.org/10.30848/PJB2016-3(2)

Zhang F, Maeder ML, Unger-Wallace E, Hoshaw JP, Reyon D, Christian M (2010) High frequency targeted mutagenesis in Arabidopsis thaliana using zinc finger nucleases. Proc Natl Acad Sci USA 107:12028–12033. https://doi.org/10.1073/pnas.0914991107

Zhang X, Li L, Meng L, Zhu S, Gan Y, Li Z, Zheng S (2019) Improvement of protein content in potato using conventional breeding and wild Solanum species. Plant Breed 138(3):321–332. https://doi.org/10.1111/pbr.12667

Zheng Z, Huang C, Cao C, Chen S, Yang Z, Liu T, Song B (2019) Mapping of quantitative trait loci for steroidal glycoalkaloid content in potato tuber flesh using a diploid F2 population. Front Plant Sci 10:1463. https://doi.org/10.3389/fpls.2019.01463

Zhu X, Gong H, He Q, Zeng Z, Busse JS, Jin W (2016) Silencing of vacuolar invertase and asparagine synthetase genes and its impact on acrylamide formation of fried potato products. Plant Biotechnol J 14:709–718. https://doi.org/10.1111/pbi.12421

Advances in Post-harvest Management and Storage of Potatoes

Shubham Maurya, K. Prasad, Neetu Saroj, Milan Kumar Lal, and Shashank Shekhar Solankey

Abstract Potato is a high-energy, hunger-relieving crop. Potato is the third top significant eaten crop in the world followed by rice and wheat. India produces about 8% of the world's potatoes. Potato is a nutrient-rich crop that contains glucose, excellent protein content, dietary fibre, necessary components, and micronutrients. It can be eaten after being boiled, fried, microwaved, or baked. After harvesting, potatoes may suffer from physiological, pathological, and pest losses. Several techniques have been devised to reduce these losses, including low-temperature and elevated temperature of 10–12 °C potato storage with sprout suppressants. Value addition is a potato processing technology used to develop new potato products. Potato can be used to make a variety of value-added products such as chips, French fries, potato flour, potato flakes, potato starch, potato bioethanol, and probiotics. Raw material availability and a year-round supply of processing-grade potatoes are major concerns for the food processing sector. As a result, storing potatoes at optimal temperatures and keeping processing of quality tubers are critical in order to achieve high-quality processed goods from preserved potatoes.

Keywords Potato post-harvest management · Advances · Post-harvest loss · Storage · Value addition

S. Maurya · N. Saroj
Department of Horticulture, Post-Graduate College of Agriculture, Dr. Rajendra Prasad Central Agricultural University, Pusa, Samastipur, Bihar, India

K. Prasad (✉)
Department of Horticulture, Tirhut College of Agriculture (TCA), Dholi, Dr. Rajendra Prasad Central Agricultural University, Pusa, Samastipur, Bihar, India

M. K. Lal
Division of Crop Physiology, Biochemistry and Postharvest Technology, ICAR-Central Potato Research Institute, Shimla, H.P., India

S. S. Solankey
Department of Horticulture (Vegetable Science), Nalanda College of Horticulture, Noorsarai, Nalanda, Bihar Agricultural University, Sabour, Bhagalpur, Bihar, India

S. S. Solankey (ed.), *Advances in Research on Potato Production*, Advances in Olericulture, https://doi.org/10.1007/978-3-031-82710-5_13

1 Introduction

Potato is currently the world's largest non-cereal crop, ranking fourth in the most important staple food crops in the world in terms of area covered and production after maize, wheat, and rice (Singh and Sandhu 2023). In terms of production and consumption, potatoes are Kenya's second-most important crop after maize. Potatoes are the second-most important crop after maize and are an important source of both food and income (Waaswa et al. 2022; International Potato Center 2019). It originated in the Andes Mountains, was utilized as food 10,000 years ago, and was domesticated over 8000 years ago (Jorgensen et al. 2023). These cultivated potatoes moved from the Andes Mountain range to 160 countries worldwide. According to FAOSTAT (Food and Agriculture Organization Statistical Database) (2022), the global average annual potato production is 359 million metric tonnes. Potatoes are grown in more than 80% of the world's countries and are used as a staple food by more than 1.3 billion people worldwide (Devaux et al. 2021). According to the IFPRI-2017 (International Food Policy Research Institute's) Database, published by the Global Hunger Index (GHI), significant progress has been made in developing countries in terms of hunger reduction. The GHI score of developing nations was 29.9 in 2000 and 20.0 in 2019, indicating a 31% decrease (Devaux et al. 2021; Grebmer et al. 2019). The first form of potato consumption was the production of chuno, which was used in stews and soups during the pre-Columbian era (Haverkort 2022).

Potato consumption provides energy as well as vitamins, minerals, and high-quality proteins. It is a nutritious and flexible item that can be included in a well-balanced diet (Lal et al. 2020; Kowalczewski et al. 2022). Micronutrient insufficiency harms around two billion people globally and is a particular issue in developing nations. Potatoes also help to treat micronutrient deficiency, often known as hidden hunger, which is the developing country's most serious health concern (Bailey et al. 2015). The potato is a rich source of nutrients that are beneficial to human health comparable to those found in cereal grains (Islam et al. 2022; Raigond et al. 2023). Figure 1 shows the nutrient content of potatoes per 100 g (Camire et al. 2009) and

Fig. 1 General source of nutrients in potato. (Adapted from Camire et al. 2009; Islam et al. 2022)

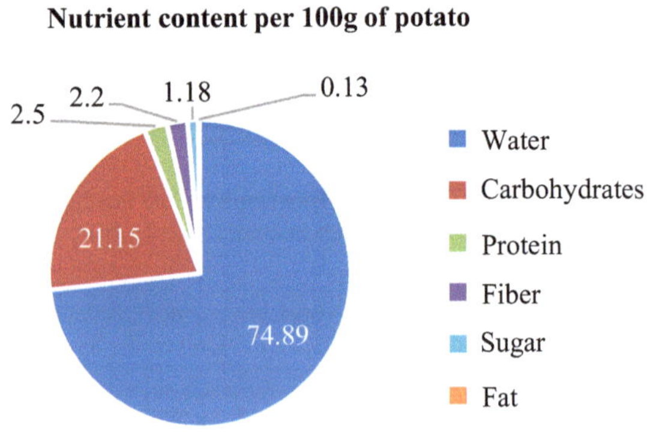

in percentage. Apart from these essential elements, potatoes are high in minerals and vitamins. They include high levels of vitamin C (19.7 mg/100 g) and potassium (421 mg/100 g). Similar to this, antioxidants like carotene and polyphenols found in potatoes (mainly coloured potatoes) play a significant part in the antioxidant defence system, aiding in the reduction of the prevalence of a variety of acute and chronic disorders (Raigond et al. 2023).

The health of people was said to be improved by potato and its products. Several compounds found in potatoes have beneficial effects on human health and are widely favoured in vegetarian diets (Kowalczewski et al. 2022). It helps to reduce inflammation, enhance cardiometabolic activity, decrease blood pressure, and enhance lipid profiles (Raigond et al. 2023). Hundred grams of potatoes provide 4% of the recommended daily allowance (RDA) of the calorie intake and 12% and 33% of the RDA for potassium and vitamin C, respectively (Tomar et al. 2020).

India is the world's second largest producer of potatoes. It is one of the most popular culinary commodities that can be consumed fresh or processed (Naeem et al. 2021). India produced around 52.5 million tonnes of potatoes during 2018–19 (National Horticulture Board 2019). However, occasionally overproduction of potatoes causes post-harvest issues such as storage and incorrect use of the supply. The potato-growing area is primarily located in the Indo-Gangetic plains, with harvesting taking place in January and February. The Indian states such as Punjab, Haryana, Uttar Pradesh, West Bengal, Bihar, Madhya Pradesh, and Gujarat are the largest producers of potatoes (Shah et al. 2023). Proper harvesting (Prasad et al. 2018a), handling (2018b) and post-harvest treatment (2022b) significantly lower the post-harvest loss of horticultural produce. Poor post-harvest handling could result in significant losses of potatoes during storage. According to Ezekiel and Singh (2007), potatoes are a semi-perishable good with a storage life of up to 3–4 months without refrigeration. They contain roughly 80% water. Major losses during storage result from pathogenic causes (Tiwari et al. 2021).

About 50% of the potatoes that are cultivated around the world are eaten fresh, whether they are boiled, fried, baked, or microwaved (Coe and Spiro 2022). However, the remainder of the potatoes are used as seed, processed food, or are wasted to the amount of 15–16%. The main potato products are chips, French fries, baked goods, dehydrated goods, alcoholic beverages, peeled, canned, and other dehydrated goods (Lal et al. 2020). The processing of potatoes is expanding quickly as a result of technological improvement, which is also increasing market demand for them. Potato chips and French fries dominate the global snack market. Other potato products are becoming more and more popular all around the world, including frozen French fries, frozen finger chips, and dehydrated potato flakes (Kaur et al. 2023).

Potato processing provides appropriate recompense due to the high cost of the processed product, and when compared to fresh food, it also helps to minimize storage-related expenses (Prasad et al. 2022b). The potatoes are typically stored in cold storage (refrigerated) where the temperature is kept between 2 and 4 °C for seed and 10 and 12 °C for processing and table purposes, and the level of relative humidity (RH) inside the chamber is kept between 90% and 95% (Lal et al. 2020).

Potatoes are stored with the sprout suppressor CIPC (isopropyl N (3-chlorophenyl) carbamate) at 10–12 °C. Low-temperature storage (2–4 °C) is an ideal strategy because it reduces sprout growth, evaporation losses, and risk of diseases and pests (Singh et al. 2010). Storage at a temperature of about 8–12 °C is advised for potatoes intended for processing since it leads to less retention of sugar and so keeps potatoes fit for processing. However, with this method of storage, the slowest suppressant must be applied, which inhibits sprout initiation and therefore extends the potato tuber's dormancy (Ezekiel et al. 2003).

2 Factors Affecting Post-harvest Losses

Potato tubers are in dormant condition at the period of harvesting. Tuber dormancy is caused by several kinds of hormonal causes. As the time of post-harvest storage is extended, tuber dormancy is disrupted and sprouting occurs. During sprout emergence in the tuber, several biochemical alterations occur (Haider et al. 2022). The reduction in nutrients caused by sprouting has a negative impact on both the nutritional and processing aspects of potatoes. Hormones such as abscisic acid (ABA) and ethylene (ET) are necessary for the induction of dormancy in potato tubers, but only ABA is required for the maintenance of bud dormancy (Suttle 2004). Increased cytokinin (CK) plays a role in the breakdown of dormancy and sprouting of potato tubers. An appropriate ratio of indole acetic acid (IAA) and gibberellic acid (GA) was efficient in preventing sprouting in potato tubers. Post-harvest losses cause a decrease in both the quality and quantity of potatoes (Haider et al. 2022). The qualitative losses are not visible and their significance is typically overestimated, but the quantitative losses are visible and efforts are being made to mitigate them. Post-harvest losses in horticultural crops might occur due to physiological, entomological, or pathological factors (Saroj and Prasad 2023; Saroj et al. 2023b). Factors causing large post-harvest losses include biotic (sprouting), pest and disease (soft and dry rot), biochemical changes (Prasad et al. 2020, 2022b), abiotic (low temperature, sweetening), and mechanical (Prasad et al. 2019). Post-harvest losses can be reduced if potatoes are stored under optimal management and storage design conditions (Lal et al. 2020).

2.1 Physiological Losses

The potato is a semi-perishable product in which the tuber undergoes active metabolism after harvest. The potato tuber loses mass and quality due to its high moisture content and metabolic rate (Gikundi et al. 2023). Several factors, particularly respiration, dry matter content, carbohydrate content, dormancy, moisture losses, skin thickness, and sprouting, influence physiological and morphological variations during storage. High-quality fresh market potatoes should be turgid, well-shaped, and absent of diseases and sprouts, black spot, black heart, freezing injury, and other types of browning and necrosis (Czerko et al. 2023).

2.2 Water Losses

Moisture loss and physiological loss from horticultural produce diminish its shelf life (Prasad et al. 2022a, 2023). The early potatoes harvested around November–December meet the demand for fresh market and fetch high prices. The tuber for processing purposes are generally from the main harvest of February–March and are further stored in cold storage and used for chipping and French fries making (Mehta et al. 2010). The early cultivated potato tubers are perishable and more susceptible to damage than the mature potatoes because of their underdeveloped skin and tender nature. The rate of moisture loss from premature skin or earlier crop tubers is greater than that from tubers taken from the main potato crop, where the skin is mature, sufficient to avoid moisture loss from its outer layer (Makani et al. 2017). The initial crop potato tuber is advised to be sold on the market right away due to its poor ability to store and high early season price. To avoid the build-up of reducing sugar, the synthesis of acrylamide, and the production of dark chips, potato tubers suitable for manufacturing chips must be stored at temperatures between 10 and 13 °C (Pedreschi 2009).

When the potato tubers attain physiological maturity, they are harvested as mature or main crop potatoes. The outer layer of ripe potato tubers is thick, which contributes to reduced water losses from the tuber's surface (Heltoft et al. 2017). Apart from reaching full size during physiological maturity, tubers have low soluble sugar content and an increased amount of starch. At this stage, the periderm layer (skin) matures and the epidermis can withstand abrasion (Miller et al. 2003).

2.3 Respiration Losses

Respiration rate is an important factor determining the shelf life of horticultural produce (Prasad et al. 2016a, 2022b). After harvest, metabolic activity increased in potato tubers. The respiration takes place both in mature and immature potato tubers. However, the respiration rate takes place four to five times higher in the immature tuber as compared to the mature tuber (Haider et al. 2023). As a result, early potatoes are normally only kept in cold storage for a short period of time. Potatoes can be stored at 4 °C for 4–5 months for table purpose, if the incidence of bruising and decay is low (Pinhero et al. 2009; Zhang et al. 2021). Evaporation loss is a physical process that can be controlled by man adjusting ageing the storage environment, such as temperature, air humidity, and airflow rate. The storage temperature should be gradually reduced, since this will result in a negligible change in the amount of sugar in the potato tuber, making it appropriate for processing. This approach prevents the degradation of processed product quality (Nawaz et al. 2021).

2.4 Losses Due to Tuber Greening

Sensory quality is an important nutritional (Prasad et al. 2023) as well as marketing (Prasad et al. 2024) indicator for horticultural produce. The greening of potato tubers during storage is another significant loss. Potato tubers are modified stems and non-photosynthetic organs lacking photosynthetic machinery (Tanios 2020). However, light exposure to the tuber may result in the conversion of amyloplast to chloroplast in the tuber's peripheral cell layers. As a result, green photosynthetic pigment (chlorophyll) accumulates in tubers, giving them a green colour (Eschemback et al. 2020). Several variables influence potato tuber greening, including cultural, physiological, genetic, environmental, planting depth, physiological age, atmospheric oxygen, lighting condition, and temperature during storage (Tanios et al. 2018). The chlorophyll biosynthetic pathway, which begins with glycine, is responsible for the production of chlorophyll in potato tubers. Potato greening results in the development of glycoalkaloids. Glycoalkaloids constitute naturally occurring chemicals found in Solanaceae plants. When these glycoalkaloids are present at quantities far exceeding the safe limit, i.e. 20 mg/100 g FW, they impact the neurological system and cause widespread membrane disruption (Jadhav and Salunkhe 1975). Greening was reported to be faster at ambient temperature compared to cold storage (Grunenfelder et al. 2006). The scale's overall glycoalkaloids content was calibrated by measuring total glycoalkaloids and chlorophyll accumulation in four potato cultivars at appropriate stages of greening. Using the created scale, it was discovered that the peels contained 3.4–6.8 times more total glycoalkaloids than the flesh. The total glycoalkaloids content ranged from 37 to 160 mg/100 g dry weight in periderm (Grunenfelder et al. 2006). The storage of potato tubers in 15% ambient CO_2 was found to be effective against greening. Storage of tuber at 8% O_2 and 10% CO_2 results in decreased greening and suppression of chlorophyll production (Tanios et al. 2018).

2.5 Sprouting Losses

The sprouting is a physiological process that occurs when dormancy is broken. Many physiological changes occur during sprout production, such as tuber weight reduction and shrivelled appearance due to water loss, which impacts the marketability of table potatoes as well as processing potatoes. Several factors influence sprouting, including storage temperature, cultivars, light, humidity, CO_2 and O_2 concentration, and tuber size (Gong et al. 2021). Temperature has a strong relationship with sprout growth, and temperatures less than 4 °C are good for inhibiting sprouting in potato tubers, whereas temperatures above 5 °C promote growth. Temperatures ranging from 15 to 20 °C are normally good for sprouting, and this temperature rises as the temperature rises over 20 °C. However, the growth of sprouts is not much impacted by humidity. According to research by Gachango et al. (2008), blue light below 500 nm and red light over 650 nm both hinder potato tuber sprouting.

2.6 *Pathological Losses*

The main objective of researchers in the potato production system is to increase output and nutritional quality while minimizing losses in the field due to pests, illnesses, and unfavourable environmental circumstances (Vilvert et al. 2022). Poor post-harvest management, field and storage diseases, frost, and heavy rainfall are some of the major reasons for output loss in potato production (Trujillo et al. 2012). More than 40 different disease agents, including viruses, fungi, nematodes, insects, and bacteria, cause significant losses and destruction to the potato crop (Singh et al. 2022). These biotic organisms endanger the tuber quality and yield and can result in losses of up to 22% in the potato production system (Chakrabarti et al. 2022). Potato tubers are extremely prone to galls, blemishes, and rots due to their high water content, during handling, transit, and storage after harvest. Rots in potatoes, formerly assumed to be minor and unavoidable, are now a major cause of concern for potato growers, developing in both seed potatoes and processed potatoes in cold storage (Wharton et al. 2006). When other diseases, such as soft rot and late blight accompany dry rot during storage, the damage becomes significantly more severe (Gachango et al. 2012). The major concern in industrial potato production is skin blemish disease, which impairs the appearance of potato tubers and causes shrivelling and shrinkage. According to Naerstad et al. (2012), skin blemish illnesses cost Norway NOK 28 million per year. Every year the disease is increasing at an alarming rate, and tubers in some of the storage sheds were having as high as 80% incidence of this disease (Cunha and Rizzo 2004). Scab is another significant potato illness that causes a loss of more than 10% dry weight. If the tubers have more than 25% aggregately damaged surface area, they are deemed severely injured (Braun et al. 2017).

3 Post-harvest Management

The potato is regarded as a vital source of carbohydrates, protein, phenols, vitamins, and fibre required for good health. However, due to poor harvesting, handling, storage, and transportation techniques, a considerable portion of potato crop is lost before consumption (Lal et al. 2020; Xu et al. 2023). Potatoes are perishable, and environmental variables make them more prone to decay and deterioration. Potato tubers respire and transpire long after harvest, reducing tuber shelf life. As a result, it becomes essential to store potatoes in such a way that they are not harmed by environmental conditions while maintaining quality (Singh et al. 2009). In this regard, potato post-harvest management practices are critical for providing fresh potato that has excellent nutritional value around the year, as well as superior processing material for developing nutritious foods. Post-harvest management is a technique to safeguard different vegetables and fruits after they have been harvested (Prasad et al. 2022a; Saroj et al. 2023c). Excessive moisture loss, the growth of rots and sprouts, and the build-up of significant quantities of sugars should be avoided during storage; thus, temperature, humidity, and air circulation are regarded as critical elements during potato storage. Post-harvest circumstances can also cause changes in physicochemical, morphological, thermal, and rheological properties in certain potato cultivars (Singh et al. 2010). Post-harvest management in potatoes was mostly accomplished by standard storage, elevated temperature storage, and low temperature refrigeration.

3.1 Traditional Storage

In India, potatoes are typically stored for a short period of time on bamboo racks or on the floor in cool spaces with enough ventilation. This method is typically used on the plains. Potatoes were also traditionally stored in pits or piles. Farmers in Assam and Meghalaya most frequently stored potatoes in heaps on top of thick soil layers (63%) and in a corner of the house (75%) (Singh et al. 2022). In Ujjain, Dewas, Indore, and Uttar Pradesh, potatoes were stored in heaps that were 1.8 m high. Madhya Pradesh was also discovered to be a popular location for the use of underground pits with a 30 t capacity for storing potatoes for 4–6 months (Dahiya et al. 1997). Karnataka farmers employed the pit method for 45–60 days of storage and the heap method for 15–30 days of on-farm storage of potatoes (Karnic 1996).

3.2 Refrigeration at Low Temperature

Cold storage has been found to be a very successful approach for potato storage since it prevents rotting of this perishable item while also making it available throughout the off-season because of cold storage. In place of too cold storage at that sentence ending. The preservation of low temperatures during the storage period is dependent on the future use of tubers. According to Mehta and Ezekiel (2006), cold storages were used to store potatoes for a longer length of time until they were needed for the market. Low temperatures caused reducing sugar to accumulate, resulting in the black colouration of fried potatoes (Zommick et al. 2014).

Raigond et al. (2014) observed that while low-temperature storage can enhance resistant starch content, an increase in reducing sugars and sucrose content can offset this benefit for long-term potato preservation. de Freitas et al. (2012) observed that potatoes held at 4 °C maintained a high dry matter content and reduced respiration.

3.3 Storage at Elevated Temperature

Potatoes held around 8–12 °C are utilized for processing since it helps to keep the potatoes' quality. For seed purposes, tubers are stored at 2 °C temperature. Potato tubers held at warmer temperatures (8–25 °C) have lower levels of reducing sugars and sucrose than potatoes stored at 4 °C (Kumar et al. 2012). The total activity of invertase has been attributed to cold storage sweetening, resulting in poor potato processing quality; however, low activity is observed during elevated temperature storage of potatoes rather than cold storage (Raigond et al. 2018). Potato wound healing is closely related to temperature. Wound healing is three times slower in potatoes stored at 5 °C than in potatoes stored at 10 °C, and three times also slower in potatoes stored at 20 °C (Zhang et al. 2022).

3.4 Storage of Potato Under Artificial Sprout Suppressant

Various treatments of antioxidants (Prasad et al. 2016b), plant growth hormones (Prasad and Sharma 2018; Saroj et al. 2023b), growth regulators (Saroj et al. 2023c), etc. are given to horticultural crops to improve their post-harvest quality and storage period. Sprouting is a major problem in stored potatoes and can dramatically affect the quality and shelf life of potatoes (Camps and Camps 2019). When stored under controlled conditions, sprout suppressants are compounds that are used to prevent sprout production (Zhao et al. 2022). These compounds aid in preserving the quality of potatoes, resulting in higher pay during times of scarcity. Several alcohols, acetaldehyde, and ethylene were discovered as chemicals that inhibit sprouting in potatoes as sprout suppressants. Naphthalene acetic acid (NAA) and its derivatives such as methyl ester of alpha naphthalene acetic acid (MENA), 2–4 dichlorophenoxy acetic acid (2,4-D), 2,4,5-trichlorophenoxy acetic acid (2,4,5-T), and 1,4-dimethyl naphthalene (DMN) have been utilized. Maleic hydrazide, tetrachloro nitrobenzene and 1,4-DMN, S-Carvone, and isopropyl-N-chlorophenyl carbamate (CIPC) are commercially utilized sprout suppressants. CIPC is a widely utilized chemical approved for commercial use on potatoes in India (Ezekiel et al. 2003).

3.5 Potato Storage Using Predictive Models

Proper storing is essential for maintaining potato quality and extending storage life. In potatoes various storage methods have been developed throughout the years, but some methods have been proven to be highly efficient than others method. In recent years, predictive models have been extensively employed to optimize the storage of potatoes and forecast fluctuations in potato quality over time (Zhao et al. 2022). Predictive models are classified into two types: statistical models and machine learning models. Statistical models are based on mathematical equations that describe the relationship between the variables, whereas machine learning models depend on the basis of software that learn from data (Ahmad and Sharma 2023). Statistical models include regression analysis as well as time series analysis in terms of potato storage, whereas machine learning strategies involve artificial neural networks and decision trees (Camps and Camps 2019). These models have been employed for predicting the rate of respiration of potato, calculating the rate of weight loss, and forecasting the commencement of sprouting of potato. The rate of respiration is an important component in the storage of potato since it is strongly related to the quality and shelf life of potatoes (Kedia et al. 2021). To address the complexity of the stored potato environment, these models can be created by combining several data sources, such as temperature, humidity, and light exposure data. There are two examples of machine learning methods such as artificial neural networks and decision trees that can be used to create predictive models that can precisely capture the intricate connections between the numerous elements impacting storage of potato. Predictive models must be validated using separate datasets, and their performance must be evaluated using measures like accuracy and predictive power, among others, in order to assure their dependability (Heo et al. 2021). To realize this promise, however, it is necessary to solve the different constraints and to construct solid predictive models based on high-quality data and tested using independent datasets. Growers can improve storage conditions and reduce commercial losses associated with potato storage.

3.6 Potato Storage Using Controlled Atmosphere Storage

After harvesting, potatoes require maintaining the nutritional content, quality, freshness, and palatability. Controlled atmosphere storage (CAS), which manages oxygen quantities (O_2) and carbon dioxide (CO_2), also increases the shelf life of potatoes (Shen et al. 2019). As a result, potato quality is preserved, such as reduced loss of texture, colour, and flavour (Emragi et al. 2022). Moreover, it can prevent the occurrence of diseases and pests like late blight, which can have a detrimental influence on the quality and quantity of potato crops. Pathogen growth can be decreased by managing the environment, resulting in good-quality potatoes with an extended shelf life (Oyom et al. 2022). This can be done in a number of ways, including modified atmosphere packaging (MAP), which modifies the atmosphere by substituting a gas mixture for the air inside a sealed box. Temperature and humidity control can help to inhibit the development of bacteria and fungus, as well as slow the level of respiration and ageing in potatoes (Jakubowski and Krolczyk 2020). Furthermore, the efficiency of CAS can be influenced by a number of parameters, including potato type and maturity, initial air conditions, and storage duration. These variables are carefully evaluated in order to get the best outcomes and increase the potato shelf life (Shen et al. 2019). Temperature is an important component in potato storage since temperature fluctuations can trigger changes in physiological activity such as sprouting, which reduces potato quality. The maximum humidity levels in potato storage might encourage the growth of microorganisms and the progression of mould, both of which can cause deterioration and affect potato quality. Low humidity levels in controlled environment storage can be difficult to maintain since they are affected by temperature, air velocity, and the amount of moisture in the potatoes (Oyom et al. 2022). Pests and diseases are major issues in potato storage and can severely affect potatoes, shortening their storage life. Managing pests and illnesses in controlled atmosphere storage is difficult since the storage conditions can be excellent for the development and dissemination of these issues (Valencia-Florez et al. 2019). Sensors are used to monitor the concentrations of oxygen, carbon dioxide, and nitrogen in the place of storage (Valencia-Florez et al. 2019). Using insect-proof boxes for storage, such as plastic or metal containers, can help to keep pests out. Fungicides and insecticides could potentially be used to control the dissemination of illnesses and pests in the stored condition (Oyom et al. 2022). Controlled environment storage is an important approach for enhancing storage life and maintaining potato quality. However, temperature control, humidity control, gas composition, and pest and disease management are all issues connected with controlled environment storage. By addressing these issues, potato-controlled atmosphere storage can be optimized to create a stable environment and lengthen storage life.

3.7 Potato Storage Using Automated Grading Systems

Previously, workers graded potatoes manually, which was time-consuming and prone to error. However, automated potato grading devices have recently become popular (Shao et al. 2022). To identify potatoes based on size, shape, and surface properties, automated potato grading systems typically employ a combination of image analysis and machine learning algorithms. These systems are an appealing alternative for potato storage facilities since they can accurately classify potatoes in real time and manage massive volumes of potatoes (Miller 2022). The adoption of automated grading systems has a number of advantages, such as increased grading consistency, precision, and efficiency, as well as lower labour costs (Yusianto et al. 2022). Convolutional neural networks (CNNs), support vector machines (SVMs), and decision trees are some of the most widely utilized machine learning techniques for grading potatoes. Blob analysis, edge identification, and colour analysis are some of the most often used image analysis methods for assessing potatoes (Yusianto et al. 2022; Mandal et al. 2023). Although automated potato grading systems have many advantages, there are several obstacles that must be solved. One of the greatest issues with automated grading systems is their lack of precision. According to Yang et al. (2023), automated grading systems are prone to mistakes which can result in incorrect grading and degrade the quality of stored potatoes. It can be expensive to adopt automated grading systems, making it difficult for smaller potato storage facilities to do so. This will assist in lowering the cost of the machinery, installation, and upkeep. The complexity of automated grading systems can be reduced by providing operator support and training. Automated grading systems have the potential to totally alter the potato storage business by enhancing the speed and precision of the grading process. However, there have been a number of difficulties with the installation of these systems that need to be resolved. This can be accomplished by adopting machine learning techniques, lowering costs, providing operator training and support, and utilizing open-source software.

3.8 Potato Storage Using Radio Frequency Identification (RFID) Systems

Radio frequency identification (RFID) technology adoption has increased recently, which is good news for potato storage providers. These techniques have the potential to completely change how potatoes are managed and stored, offering advantages like greater efficiency, safety, and traceability (Ma et al. 2022). Radio waves are used by RFID systems to transmit data between a reader and a tag affixed to the potato. The tag has a microchip and an antenna that can record and communicate details about the potato, such as where it came from, how good it is, and how long it was stored (Gernert et al. 2019). There are two main types of RFID systems used in potato storage such as active RFID and passive RFID. In an active RFID system, the tags have batteries that supply the necessary power to transmit data to the reader. Compared to passive tags, the tags have a wider reading range and may be read from a farther distance. Because a tag must be read from a distance, and potatoes are often stored in large mounds or stacks, active RFID is quite useful in such circumstances. The battery is absent in passive RFID systems. The energy used to power the tag's microchip and send the data comes from the reader's electromagnetic field. Compared to active RFID tags, passive tags are more compact, less expensive, and have a limited read range. They are perfect for usage in facilities that keep potatoes in smaller amounts, like bags or boxes (Zhang et al. 2020). RFID systems operate more effectively. The system allows for automatic and real-time tracking of potatoes, eliminating the need for manual data entry and lowering the possibility of errors. RFID systems can also provide storage facilities with instant access to potato data, making it easier to manage inventory, track movement, and decide whether to transport potatoes to the next stage of the supply chain. This information can be used to detect and isolate any potential food safety hazards in order to limit the risk of food-borne illnesses and protect the potato industry's reputation. RFID technologies can improve efficiency and traceability while also promoting safety in potato storage facilities. RFID technology can assist in ensuring that stored potatoes are in the best possible circumstances, lowering the risk of rotting and loss by enabling monitoring of real-time temperature and humidity levels (Zhang et al. 2020). Consequently, RFID technologies provide a variety of difficulties for potato storage facilities. However, these difficulties can be solved by utilizing data management software, frequency selection, RF shielding, and RFID tags intended for severe environments. RFID solutions can significantly contribute to enhancing the effectiveness, traceability, and general standard of potato storage operations by addressing these difficulties.

4 Value Addition and Processing of Potato

Due to the semi-perishable nature of potatoes, they cannot be stored fresh for long periods of time. Low temperatures and a dark, well-ventilated area with high relative humidity are ideal storage conditions for potatoes (Akello et al. 2022; Abdelgawad et al. 2023). A different method of preserving potatoes is to produce shelf-stable value-added processed items out of them (cooked potatoes, par-fried potato strips, French fries, chips, potato starch, potato granules, potato flakes, and dehydrated diced potatoes) (Wang et al. 2023). Furthermore, potato value-addition and processing will ensure that customers have a source of wholesome, safe, nutritious, and suitable food throughout the year, also minimizing post-harvest losses. Chips/wafers and French-fried potatoes are the only two potato-based snack foods that are among the top value-added products worldwide, making up more than 42% of all potato-based processed goods. Such convenience foods are adored for their superior flavour, taste, and texture (Pedreschi 2012). The industrialization and market-driven nature of potato processing have increased due to technological improvements. The quality of value-added products made from potatoes is strongly dependent on the quality of the raw potatoes used, just like the quality of other processed items. To meet the standards for tuber quality, particularly those related to tuber morphology, lowering sugar (0.1%), and dry matter content (20%), is a major problem in the processing of potatoes. Alternately, we could conclude that all potato varieties are unsuitable for processing. The potato-based value-added goods can currently be categorized into the following groups based on the processing method used.

1. Thermally processed (wet heat) potatoes include blanched potatoes, boiled/cooked potatoes, and mashed potatoes, among other products.
2. Fried products: Chips/wafers, French fried strips, manufactured fries, crisps, sticks, lachha, and frozen par-fried items.

3. Cubes, sticks, shreds, chips, flour, flakes, and granules are all examples of dehydrated items.
4. Potato starch.
5. Products that have been fermented or hydrolysed include glucose syrup, organic acids, and bioethanol.

4.1 *Thermally Processed Products*

4.1.1 Blanched Potatoes

Blanched potatoes, as the name implies, are minimally processed peeled and cut potatoes that have been blanched. Blanching inactivates the polyphenolic enzymes, removes air from the potato tissue (Mukherjee and Chattopadhyay 2007), and leaches out accumulated sugars to limit the Maillard reaction during subsequent frying and gelatinize the starch (Pedreschi et al. 2004). Blanched potatoes are a common product in many metropolitan areas, mostly utilized by street sellers, hotels, restaurants, and fast-food establishments. Processed potatoes are immersed in hot water (60–80 °C) for 5–20 min with additives such as citric acid, ascorbic acid, bisulphites, and salt (Aguilera et al. 2001). Blanched potatoes have a refrigerated shelf life of 48–72 h depending on softening or firmness (Andersson et al. 1994).

4.1.2 Cooked Potatoes

Potato is the most common and simplest processed product consumed in our daily lives (Haverkort et al. 2023). Potatoes can be cooked in boiling water for 25–30 min, under pressure for 15 min, or by combining both methods. In addition to these methods, baking and other dry heat sources can also be used to cook potatoes. The main characteristics of cooked potatoes that make them appealing include their cream white colour, relatively dry and mealy texture, and natural potato flavour. In addition, cooked potatoes should not have an unnatural colour or an excessive amount of firmness or softness (Andersson et al. 1994). Under refrigeration, cooked, unpeeled potatoes have a shelf life of 2–3 days.

4.1.3 Mashed Potatoes

According to Lamberti et al. (2004), mashed potato products are a convenient food for both private families and caterers. A variety of foods, including pies, dumplings, pancakes, and other dishes, can be made with mashed potatoes as a side dish or an intermediary component. Mashed potatoes can be preserved or dried for later use. For the mashed potatoes, potatoes with a high dry matter and a floury texture (such as Kufri Frysona, Kufri Chipsona-2 and -3, Kufri Chandramukhi, Kufri Dewa, Kufri Himsona, and Kufri Lalima) are preferable.

5 Fried Products

5.1 Potato Chips

Value-added products from specific genotypes of any horticultural crop meant/identified for its post-harvest quality not only enhance nutrition (Prasad and Sharma 2016; Saroj and Prasad 2023) but also the economic value of the derived product from that genotype (Prasad et al. 2017; Saroj et al. 2023a). Potato chips are thin, fried, baked ready-to-eat appetizers that are popular both at home and at fast food establishments (Wadagavi et al. 2017). Potato chips with a diameter of 40–60 mm should be round to oval in form and golden in colour (Kaur et al. 2008). Hundred grams of potato chips have around 547 calories, 37.47 g fat, 49.74 g total carbohydrate, 6.56 g protein, sodium (525 mg), and potassium (1642 mg). The United States generates the most revenue, with $16,791 million. Per capita consumption is approximately 0.05 kg (Kaur et al. 2008). More than 20% dry matter content must be present. Kufri Chipsona-4, Kufri Himsona, Kufri Jyoti, Kufri Lauvkar, Kufri Chipsona-1, Kufri Chipsona-2, Kufri Chipsona-3, and Kufri Chandramukhi are some famous types appropriate for chip manufacture (Joshi et al. 2021).

Potato Chips Preparation

1. Fresh potatoes with no defects are peeled followed by washing, and removal of any green and damaged bits (Fig. 2).
2. Peeled potatoes are cut into 1.5–2.0 mm thin slices.
3. To eliminate surface starch, sliced potatoes are rinsed in water.
4. To eliminate starch, sliced potatoes are blanched in hot water (60–80 °C for 3 min).
5. Surface dried slices are fried at 180 °C until golden.
6. Chips are now uniformly salted/spiced and sealed in nitrogen bags (Singh et al. 2016).

Fig. 2 General process for making potato chips (Adapted from Singh et al. 2016; Joshi et al. 2021)

5.2 French Fries

India's fastest-growing snack is French fries, increasing at a rate of more than 30% per year (Kaur et al. 2023). The colour and texture of fried French fries are key sensory aspects that influence consumer acceptability. It should have a crisp surface and a soft inside (Antonic et al. 2023). It contains 274 calories per 100 g. In general, 100 g of fries contains 14.06 g of fat, 35.66 g of total carbs, and 3.48 g of protein, as well as a significant amount of sodium (300 mg) and potassium (527 mg). The potatoes which are large and oblong and have dry matter content more than 20% are suitable (Islam et al. 2022). Some famous varieties for French fries production are Kufri Frysona, Kufri Chipsona-1, Kufri Chipsona-2, Kufri Chipsona-3, and Kufri Chipsona-4 (Das et al. 2021).

1. Fresh, healthy potatoes are selected and washed to eliminate dirt, dust, and green or damaged portions.
2. The clean potatoes are peeled and sliced into 10 × 10 mm (width × height) sticks with a French fry cutter.
3. The surface starch on these sticks is removed by rinsing them in water.
4. After cleaning the sticks, they are blanched in hot water at 60–80 °C for around 5 min.
5. Surface drying is used to eliminate extra surface moisture from these blanched sticks.
6. Dried sticks are fried at 180–190 °C for about 2 min.
7. Blotting paper is used to remove excess surface oil. After removing oil from the fried sticks, they are allowed to cool to room temperature before being frozen at −20 °C and stored until the final frying.
8. At the moment of serving, the fries are fried at 180–190 °C, and salt and spices are added to taste (Singh et al. 2016).

5.3 Potato Flakes

Potato flakes are fragmented particles of drum-dried mashed potato slurry that have exceptional quick reconstitution properties and a natural potato flavour. Potato flakes can be used in place of boiling potatoes. It can also be employed in extruded products and the development of fabricated chips. It is now used in retail mashed potatoes as well as an ingredient in frozen entrees, snacks, and soups (Choudhary and Bakode 2022). The raw potatoes are pre-cooked in water at a temperature of about 70–72 °C for about 20 min, followed by a 15-min cooling in water at less than 24 °C and another 25–30-min steam cooking. Following cooking, potatoes are mashed with chemicals and emulsifiers such as chelating agents, milk solids, antioxidants, and sulphur dioxide. These additions enhance the colour, texture, and shelf life of flakes. Prior to packaging, this mashed slurry is dried as sheets on a drum drier and broken into flakes. Precooking is thought to gelatinize the starch within potato cells, allowing retrogradation to occur during cooling without breaking intercellular connections (Singh et al. 2016). Cooling eliminates free starch from the surface of blanched potatoes, preventing them from sticking or burning during drying. The water absorption capacity of dried flakes is improved by this two-step cooking technique (Marwaha et al. 2010).

5.4 Potato Starch

The utility of potatoes in food applications is mostly determined by the amount and qualities of starch found in tubers. Potato starch is distinct from other commodity starches due to its molecular structures (B-type X-ray crystalline pattern), large smooth granules, a high proportion of covalently linked phosphate, long amylopectin chains, and high-molecular-weight amylose (Hoover 2001; Kaur et al. 2009). Potato starch is thought to be superior to grain starch due to its strong swelling power, gel clarity, and high viscosity qualities. The manufacture of starch begins with the grinding/shredding of washed potatoes with rasps or a hammer mill. Rasping breaks down the potato cells, releasing the starch granules. To prevent undesired slurry colouring, the resulting slurry is treated with antioxidants and browning inhibitors. Vacuum grinders can also be used at this stage to prevent enzymatic browning. The slurry is made up of broken cells, their walls, starch granules, and water containing soluble proteins, amino acids, carbohydrates, and salts (also known as potato juice). This slurry is then run through a series of centrifuges, sieves, and separators to physically separate the starch, fibre, and juice (Marwaha et al. 2010). Currently, three processes are used for the separation of these three components. In one method, starch and fibre are separated after the potato juice. In the second procedure, the juice and fibre from the crushed potato are removed after the starch. Juice and starch are removed after fibre in the third procedure. Each method has benefits and drawbacks. The third process produces the most starch, while the first process produces the least, and the second procedure has a processing cost that is significantly lower than the other two. The third step in the starch processing process needs high skill levels, whereas the second step is the simplest from a process control perspective. After separation, starch is dried using pneumatic or solar dryers to a final moisture content of 20% or less (Grommers and van der Krogt 2009).

6 Fermented and Hydrolysed Products

6.1 Ethanol

The main source of energy in potatoes is starch, which is currently receiving a lot of attention as a renewable energy source for ethanol production. Potato may be used to produce high-quality alcohol that can be utilized for both fuel and beverages (Guedes et al. 2021). In order to make alcoholic drinks like vodka, potcheen, or akvarit, potatoes are required (Menezes et al. 2016; Thatoi et al. 2016). Additionally, at the moment, countries like Finland, Austria, the United States, and China use potato waste and leftovers of potato processing as a fuel ethanol feedstock. Starch present in these wastes is utilized as a growth media for the fermentation. However, the main issue with potato or starch-based substrates is that they have high viscosity, which slows down the rate of reaction. In some research (Kim et al. 2012; Ganguly et al. 2018), the administration of enzymes like viscozyme or some physical techniques has been suggested to reduce viscosity. Before fermentation, the potato's starch needs to be broken down into simple sugars. Enzymes hydrolyse the starch to create fermentable sugars for that purpose. Amylases (glucoamylase, alpha-amylase and beta-amylase) are typically used to hydrolyse starchy materials. In general, the production of industrial and fuel ethanol from starchy biomass commonly involves a three-step process. Endoamylases like amylase liquefy starch in the first stage to create low-molecular-weight products (dextrins). This liquefied starch is then saccharified to create glucose in the following stage, and this glucose is then fermented to create ethanol in the following step. In the end, distillation is used to recover the ethanol (Arapoglou et al. 2010; Yamada et al. 2010).

7 Prebiotics and Probiotics from Potato

A prebiotic is a nondigestible food component that benefits the host by selectively promoting the growth and/or activity of one or a small number of beneficial bacteria in the colon. These bacteria can enhance host health (Ballini et al. 2023) Furthermore, prebiotics can improve mineral absorption, such as calcium, iron, and zinc, as well as lower triglyceride and cholesterol levels (Bede and Zaixiang 2021). Similarly, probiotics are living, nonpathogenic bacteria that may help the digestive system's microbial equilibrium (Bhogoju and Nahashon 2022). When present in the right proportion, probiotics can reduce pathogenic organism colonization and invasion in the intestine, lower intestinal pH, and alter the host's immunological response, among other therapeutic effects (Brown and Valiere 2004). Additionally, a number of other gastrointestinal health advantages have been identified in numerous papers, including an antagonistic impact against gastrointestinal infections, antibacterial activity, a decrease in serum cholesterol, antidiarrhoeal effects, and anticarcinogenic effects (Nomoto 2005; Hao et al. 2015). Several studies have been reported on the prebiotic and probiotic effects of potato starch and potato processing waste (Michalak et al. 2012; Khodaei et al. 2016). According to promising sources of prebiotics (Lisinska and Leszczynski 1989; Li et al. 2001; Nugent 2005), resistant starch and its derivatives, as well as compounds with partial starch breakdown (such as dextrins), may encourage the development of probiotic bacteria. As a source of carbon for *Lactobacillus* bacteria (*Lactobacillus rhamnosus* GG and *Lactobacillus casei* DN 114 001 strains), chemically modified enzyme-resistant dextrin has demonstrated probiotic properties (Lebesi and Tzia 2010; Lukk et al. 2010; Jochym and Nebesny 2017).

Enzyme-resistant dextrin, made by boiling potato starch with hydrochloric and citric acid, serves as a carbon source for probiotic lactobacilli and bifido bacteria. In this study, co-culturing lactobacilli and bifido bacteria with intestine bacteria and potato starch-based, enzyme-resistant dextrin resulted in probiotic strains dominating growth (Slizewska 2013).

With any nutrient addition, potato juice can be used as a fermentation medium for the growth of probiotic microorganisms *Lactobacillus casei*. Kim et al. (2012) discovered that potato juice can be used to make probiotic functional beverages for vegetarians or lactose-intolerant consumers.

8 Conclusion

A large portion of the world's population consumes potatoes, either fresh or processed. Depending on how they will be used, potatoes are preserved for shorter or longer periods of time. It requires post-harvest management to keep it from degrading due to many kinds of losses. Researchers have discovered several kinds of strategies and processes to stop potato losses brought on by physiological, pathological, and entomological factors. However, further study is needed to create new potato types for processing and table use that have better keeping qualities. Additionally, better post-harvesting methods are needed to assist reduce the losses brought on by many different factors.

References

Abdelgawad G., Khater E, Bahnasawy A, Mosa M (2023) Potato tubers quality as affected by modified atmospheric conditions and package type during storage. Misr J Agric Eng. https://mjae.journals.ekb.eg/

Aguilera JM, Cadoche L, Loopez C, Gutierrez G (2001) Microstructural changes of potato cells and starch granules heated in oil. Food Res Int 34:939–947. https://doi.org/10.1016/S0963-9969(01)00118-1

Ahmad U, Sharma L (2023) A review of best management practices for potato crop using precision agricultural technologies. Smart Agric Technol 100220. https://doi.org/10.1016/j.atech.2023.100220

Akello R, Turinawe A, Wauters P, Naziri D (2022) Factors influencing the choice of storage technologies by smallholder potato farmers in eastern and southwestern Uganda. Agriculture 12:240. https://doi.org/10.3390/agriculture12020240

Andersson A, Gekas V, Lind I, Oliveira F, Oste R (1994) Effect of preheating on potato texture. Crit Rev Food Sci Nutr 34:229–251. https://doi.org/10.1080/10408399409527662

Antonic B, Dordevic D, Buchtova H, Tremlova B, Dordevic S, Kushkevych I (2023) French fries' color and frying process in relation to used plant oils. Processes 11:2839. https://doi.org/10.3390/pr11102839

Arapoglou D, Varzakas T, Vlyssides A, Israilides C (2010) Ethanol production from potato peel waste (PPW). Waste Manag 30:1898–1902. https://doi.org/10.1016/j.wasman.2010.04.017
Bailey RL, West KP, Black RE (2015) The epidemiology of global micronutrient deficiencies. Ann Nutr Metab 66:22–33. https://doi.org/10.1159/000371618
Ballini A, Charitos IA, Cantore S, Topi S, Bottalico L, Santacroce L (2023) About functional foods: the probiotics and prebiotics state of art. Antibiotics 12:635. https://doi.org/10.3390/antibiotics12040635
Bede D, Zaixiang L (2021) Recent developments in resistant starch as a functional food. Starch-Starke 73:2000139. https://doi.org/10.1002/star.202000139
Bhogoju S, Nahashon S (2022) Recent advances in probiotic application in animal health and nutrition: a review. Agriculture 12:304. https://doi.org/10.3390/agriculture12020304
Braun S, Gevens A, Charkowski A, Allen C, Jansky S (2017) Potato common scab: a review of the causal pathogens, management practices, varietal resistance screening methods, and host resistance. Am J Potato Res 94:283–296. https://doi.org/10.1007/s12230-017-9575-3
Brown AC, Valiere A (2004) Probiotics and medical nutrition therapy. Nutr Clin Care 7:56–68
Camire ME, Kubow S, Donnelly DJ (2009) Potatoes and human health. Crit Rev Food Sci Nutr 49:823–840. https://doi.org/10.1080/10408390903041996
Camps C, Camps ZN (2019) Optimized prediction of reducing sugars and dry matter of potato frying by FT-NIR spectroscopy on peeled tubers. Molecules 24:967. https://doi.org/10.3390/molecules24050967
Chakrabarti SK, Sharma S, Shah MA (2022) Potato pests and diseases: a global perspective. In: Sustainable management of potato pests and diseases. Springer, Berlin, pp 1–23. https://doi.org/10.1007/978-981-16-7695-6_20
Choudhary B, Bakode S (2022) Importance of potato processing & products. A Monthly Peer Reviewed Magazine for Agriculture and Allied Sciences, 14
Coe S, Spiro A (2022) Cooking at home to retain nutritional quality and minimise nutrient losses: a focus on vegetables, potatoes and pulses. Nutr Bull 47:538–562. https://doi.org/10.1111/nbu.12584
Cunha MG, Rizzo DM (2004) Occurrence and epidemiological aspects of potato silver scurf in California. Hortic Bras 22:690–695. https://doi.org/10.1590/S0102-05362004000400005
Czerko Z, Zarzynska K, Boguszewska-Mankowska D (2023) Postharvest physiology and storage of potato. In: Potato production worldwide. Academic, pp 261–272. https://doi.org/10.1016/B978-0-12-822925-5.00001-3
Dahiya PS, Khatana VS, Ilangantileke SG (1997) Producer storage practices for potatoes in the Malwa Region, Madhya Pradesh, India. Work. Pap. (CIP)
Das S, Mitra B, Saha A, Mandal S, Paul PK, El-Sharnouby M, Hossain A (2021) Evaluation of quality parameters of seven processing type potato (*Solanum tuberosum* L.) cultivars in the Eastern Sub-Himalayan plains. Food Secur 10:1138. https://doi.org/10.3390/foods10051138
de Freitas ST, Pereira EIP, Gomez ACS, Brackmann A, Nicoloso F, Bisognin DA (2012) Processing quality of potato tubers produced during autumn and spring and stored at different temperatures. Hortic Bras 30:91–98. https://doi.org/10.1590/S0102-05362012000100016
Devaux A, Goffart JP, Kromann P, Andrade-Piedra J, Hareau G, Polar V (2021) The Potato of the Future: Opportunities and Challenges in Sustainable Agri-food Systems. Potato Res 64: 681–720. https://doi.org/10.1007/s11540-021-09501-4
Emragi E, Kalita D, Jayanty SS (2022) Effect of edible coating on physical and chemical properties of potato tubers under different storage conditions. LWT 153:112580. https://doi.org/10.1016/j.lwt.2021.112580
Eschemback V, Genu AM, Kawakami J (2020) Greening of potato tubers after harvest: effect of planting date and cultivar. Res Soc Dev 9:105973971. https://doi.org/10.33448/rsd-v9i7.3971
Ezekiel R, Singh B (2007) Potato Res, 50:175–184
Ezekiel R, Singh B, Kumar D, Paul V (2003) Processing quality and CIPC residues in potatoes stored at 10–12 C in commercial cold stores applying CIPC
FAOSTAT (2022) Crops and livestock products. Food and Agriculture Organization of the United Nations. Statistical Yearbook 2022. Rome http://www.fao.org/faostat/en/#data/QCL

Gachango E, Hanson LE, Rojas A, Hao JJ, Kirk WW (2012) Fusarium spp. causing dry rot of seed potato tubers in Michigan and their sensitivity to fungicides. Plant Dis 96:1767–1774. https://doi.org/10.1094/PDIS-11-11-0932-RE

Gachango E, Shibairo S, Kabira J, Chemining'wa G, Demo P (2008) Effects of light intensity on quality of potato seed tubers. African J Agric Res 3:732–739. http://repository.tuc.ac.ke:8080/xmlui/handle/123456789/237

Ganguly A, Bala E, Banerjee P, Blanchard R, Chatterjee PK (2018) Production of ethanol from waste potato using locally available biocatalyst. Util Manag Bioresour:143–151. https://doi.org/10.1007/978-981-10-5349-8_14

Gernert B, Schlichter J, Wolf L (2019, October) Monitoring food storage with the help of a LoRa-based DTWSN. In: Proceedings of the 14th workshop on challenged networks, pp 33–38. https://doi.org/10.1145/3349625.3355442

Gikundi EN, Buzera AK, Orina IN, Sila DN (2023) Storability of irish potato (*Solanum tuberosum* L.) varieties grown in Kenya, under different storage conditions. Potato Res 66:137–158. https://doi.org/10.1007/s11540-022-09575-8

Gong HL, Dusengemungu L, Igiraneza C, Rukundo P (2021) Molecular regulation of potato tuber dormancy and sprouting: a mini-review. https://doi.org/10.1007/s11816-021-00689-y

Grebmer K von, Bernstein J, Patterson F, Wiemers M, Cheilleachair RN, Foley C, Gitter S, Ekstrom K, Fritschel H (2019) Global Hunger Index: The Challenge of Hunger and Climate Change. The International Food Policy Research Institute (IFPRI), Washington, DC http://www.compact2025.org/files/2019/10/ghi-featured.jpg

Grommers HE, van der Krogt DA (2009) Potato starch: production, modifications and uses. Starch:511–539. https://doi.org/10.1016/B978-0-12-746275-2.00011-2

Grunenfelder LA, Knowles LO, Hiller LK, Knowles NR (2006) Glycoalkaloid development during greening of fresh market potatoes (*Solanum tuberosum* L.). J Agric Food Chem 54:5847–5854. https://doi.org/10.1021/jf0607359

Guedes JS, Santos KC, Castanha N, Rojas ML, Junior MDM, Lima DC, Augusto PE (2021) Structural modification on potato tissue and starch using ethanol pre-treatment and drying process. Food Struct 29:100202. https://doi.org/10.1016/j.foostr.2021.100202

Haider MW, Nafees M, Ahmad I, Ali B, Maryam, Iqbal R, Vodnar DC (2022) Postharvest dormancy-related changes of endogenous hormones in relation to different dormancy-breaking methods of potato (*Solanum tuberosum L.) tubers. Front Plant Sci 13:945256.* https://doi.org/10.3389/fpls.2022.945256

Haider MW, Nafees M, Iqbal R, Asad HU, Azeem F, Ali B, Ali MA (2023) Postharvest starch and sugars adjustment in potato tubers of wide-ranging dormancy genotypes subjected to various sprout forcing techniques. Sci Rep 13:14845. https://doi.org/10.21203/rs.3.rs-2394606/v1

Hao Q, Dong BR, Wu T (2015) Probiotics for preventing acute upper respiratory tract infections. Cochrane Database Syst Rev 3:CD006895. https://doi.org/10.1002/14651858.CD006895.pub3

Haverkort, AJ (2022) On processing potato. Doctoral dissertation, Wageningen University and Research. https://doi.org/10.18174/57504

Haverkort AJ, Linnemann AR, Struik PC, Wiskerke JSC (2023) On processing potato: 1. Survey of the ontology, history and participating actors. Potato Res 66(2):301–338. https://doi.org/10.1007/s11540-022-09562-z

Heltoft P, Wold AB, Molteberg EL (2017) Maturity indicators for prediction of potato (*Solanum tuberosum* L.) quality during storage. Postharvest Biol Technol 129:97–106. https://doi.org/10.1016/j.postharvbio.2017.03.011

Heo S, Choi JY, Kim J (2021) Prediction of moisture content in steamed and dried purple sweet potato using hyperspectral imaging analysis. Food Sci Biotechnol 30:783–779. https://doi.org/10.1007/s10068-021-00921-z

Hoover R (2001) Composition, molecular structure, and physicochemical properties of tuber and root starches: a review. Carbohydr Polym 45:253–267. https://doi.org/10.1016/S0144-8617(00)00260-5

International Potato Center (2019) CIP annual report 2018. Towards food system transformation. International Potato Center, Lima. https://doi.org/10.4160/02566311/2018
Islam MM, Naznin S, Naznin A, Uddin MN, Amin MN, Rahman MM, Ahmed S (2022) Dry matter, starch content, reducing sugar, color and crispiness are key parameters of potatoes required for chip processing. Horticulturae 8:362. https://www.mdpi.com/2311-7524/8/5/362
Jadhav SJ, Salunkhe DK (1975) Formation and control of chlorophyll and glycoalkaloids in tubers of Solanum tuberosum L. and evaluation of glycoalkaloid toxicity. Adv Food Res 21:307–354. https://doi.org/10.1016/S0065-2628(08)60093-2
Jakubowski T, Krolczyk JB (2020) Method for the reduction of natural losses of potato tubers during their long-term storage. Sustain For 12:1048. https://doi.org/10.3390/su12031048
Jochym KK, Nebesny E (2017) Enzyme-resistant dextrins from potato starch for potential application in the beverage industry. Carbohydr Polym 172:152–158. https://doi.org/10.1016/j.carbpol.2017.05.041
Jorgensen K, Garcia OA, Kiyamu M, Brutsaert TD, Bigham AW (2023) Genetic adaptations to potato starch digestion in the Peruvian Andes. Am J Biol Anthropol 180:162–172. https://doi.org/10.1002/ajpa.24656
Joshi A, Kaundal B, Raigond P, Singh B, Sethi S, Bhowmik A, Kumar R (2021) Low-volume procedure to determine phytate and ascorbic acid in potatoes: standardization and analysis of Indian cultivars. J Food Compos Anal 102:103998. https://doi.org/10.1016/j.jfca.2021.103998
Karnic KS (1996) Potato storage systems in Karnataką: problems, constraints and prospects. National workshops onPotato storage systems, organized by CIP in collaboration with CPRI at Central Potato Research Station, Modipuram, UP India
Kaur A, Singh N, Ezekiel R (2008) Quality parameters of potato chips from different potato cultivars: effect of prior storage and frying temperatures. Int J Food Prop 11:791–803. https://doi.org/10.1080/10942910701622664
Kaur A, Singh N, Ezekiel R, Sodhi NS (2009) Properties of starches separated from potatoes stored under different conditions. Food Chem 114:1396–1404. https://doi.org/10.1016/j.foodchem.2008.11.025
Kaur S, Aggarwal P, Babbar N (2023) Evaluating progress of Indian potato processing industry: an updated review. Potato Res 66:945–964
Kedia P, Kausley SB, Rai B (2021) Temperature and humidity-based models for the prediction of transpiration rate in potatoes during storage. J Food Process Eng 44:e13626. https://doi.org/10.1111/jfpe.13626
Khodaei N, Fernandez B, Fliss I, Karboune S (2016) Digestibility and prebiotic properties of potato rhamnogalacturonan i polysaccharide and its galactose-rich oligosaccharides/oligomers. Carbohydr Polym 136:1074–1084. https://doi.org/10.1016/j.carbpol.2015.09.106
Kim NJ, Jang HL, Yoon KY (2012) Potato juice fermented with *Lactobacillus casei* as a probiotic functional beverage. Food Sci Biotechnol 21:1301–1307. https://doi.org/10.1007/s10068-012-0171-5
Kowalczewski PL, Olejnik A, Switek S, Bzducha-Wrobel A, Kubiak P, Kujawska M, Lewandowicz G (2022) Bioactive compounds of potato (*Solanum tuberosum* L.) juice: from industry waste to food and medical applications. Crit Rev Plant Sci 41:52–89. https://doi.org/10.1080/07352689.2022.2057749
Kumar J, Singh R, Gautam V, Kumar P (2012) Storage life and biochemical changes of potato (*Solanum tuberosum* L) cultivars in cold chamber. Ann Hortic 5:198–204
Lal MK, Kumar A, Kumar A, Jena R, Raigond P, Kumar D, Singh B (2020) Minerals in potato. In: Potato: nutrition and food security. Springer, Singapore, pp 87–112
Lamberti M, Geiselmann A, Conde-Petit B, Escher F (2004) Starch transformation and structure development in production and reconstitution of potato flakes. LWT - Food Sci Technol 37:417–427. https://doi.org/10.1016/j.lwt.2003.10.015
Lebesi, D. and Tzia, C. 2010. Development of bakery foods enriched with insoluble fibres: functional properties required for its incorporation. Dietary fibre: new frontiers for food and health. Wageningen Academic Publishers, 121–133. doi: https://doi.org/10.3920/978-90-8686-692-2

Li BW, Blackwell EL, Behall KM, Liljeberg Elmstahl HGM (2001) Resistant starch and total dietary fiber content of oatrim muffins with different levels of amylose, amylopectin, and β-glucan. Cereal Chem 78:387–390. https://doi.org/10.1094/CCHEM.2001.78.4.387

Lisinska G, Leszczynski W (1989) Potato science and technology. Springer Science & Business Media

Lukk M, Kapushesky M, Nikkila J, Parkinson H, Goncalves A, Huber W, Ukkonen E, Brazma A (2010) A global map of human gene expression. Nat Biotechnol 28:322–324. https://doi.org/10.1038/nbt0410-322

Ma S, Liu J, Zhang Q, Lin Q, Liu R, Xing Y, Jiang H (2022) 3D printing performance using radio frequency electromagnetic wave modified potato starch. Innov Food Sci Emerg Technol 80:103064. https://doi.org/10.1016/j.ifset.2022.103064

Makani MN, Sargent SA, Zotarelli L, Huber DJ, Sims CA (2017) Harvest interval has greater effect on periderm maturity and storage quality of early maturing, table stock potato than nitrogen rate. HortScience 52:1390–1395. https://doi.org/10.21273/HORTSCI11870-17

Mandal SN, Mukherjee K, Dan S, Ghosh P, Das S, Mustafi S, Chakraborty A (2023) Image-based potato Phoma blight severity analysis through deep learning. J Inst Eng (India) Ser B 104:181–192. https://ui.adsabs.harvard.edu/link_gateway/2023JIEIB.104.181M/doi:10.1007/s40031-022-00820-9

Marwaha RS, Pandey SK, Kumar D, Singh SV, Kumar P (2010) Potato processing scenario in India: industrial constraints, future projections, challenges ahead and remedies—a review. J Food Sci Technol 47:137–156. https://doi.org/10.1007/s13197-010-0026-0

Mehta A, Ezekiel R (2006) Potato storage: need, present scenario, emerging technologies and future strategies: a critical appraisal. J Food Sci Technol 43:453–466

Mehta A, Singh B, Ezekiel R, Kumar D (2010) Effect of CIPC on sprout inhibition and processing quality of potatoes stored under traditional storage systems in India. Potato Res 53:1–15. https://doi.org/10.1007/s11540-010-9146-1

Menezes AGT, Menezes EGT, Alves JGLF, Rodrigues LF, Cardoso M, das G. (2016) Vodka production from potato (*Solanum tuberosum* L.) using three Saccharomyces cerevisiae isolates. J Inst Brew 122:76–83. https://doi.org/10.1002/jib.302

Michalak M, Thomassen LV, Roytio H, Ouwehand AC, Meyer AS, Mikkelsen JD (2012) Expression and characterization of an endo-1,4-β-galactanase from Emericella nidulans in Pichia pastoris for enzymatic design of potentially prebiotic oligosaccharides from potato galactans. Enzyme Microb Technol 50(2):121–129. https://doi.org/10.1016/j.enzmictec.2011.11.001

Miller, M. D. 2022. Development of image analysis tools to quantify potato tuber quality traits. Doctoral dissertation, University of Minnesota

Miller AR, Bartz J, Brecht J (2003) Harvest and handling injury: physiology biochemistry and detection. Postharvest Physiol Pathol Veg:177–208

Mukherjee S, Chattopadhyay PK (2007) Whirling bed blanching of potato cubes and its effects on product quality. J Food Eng 78:52–60. https://doi.org/10.1016/j.jfoodeng.2005.09.001

Naeem M, Demirel U, Yousaf MF, Caliskan S, Caliskan ME (2021) Overview on domestication, breeding, genetic gain and improvement of tuber quality traits of potato using fast forwarding technique (GWAS): a review. Plant Breed 140:519–542

Naerstad R, Dees MW, Le VH, Holgado R, Hermansen A (2012) Occurrence of skin blemish diseases (scab and scurf) in Norwegian potato production. Potato Res 55:225–239. https://doi.org/10.1007/s11540-012-9221-x

National Horticulture Board (2019) NHB Database, Ministry of Agriculture & Farmers Welfare, Government of India.

Tomar M, Reetu, Changan SS (2020) Potato Vitamins. In: Potato: Nutrition and Food Security, pp. 113–132

Nawaz A, Danish A, Ali SW, Muhammad Shahbaz H, Khalifa I, Ahmed A, Ahmed W (2021) Evaluation and storage stability of potato chips made from different varieties of potatoes cultivated in Pakistan. J Food Process Preserv 45:e15437. https://doi.org/10.1111/jfpp.15437

Nomoto K (2005) Prevention of infections by probiotics. J Biosci Bioeng 100:583–592. https://doi.org/10.1263/jbb.100.583

Nugent AP (2005) Health properties of resistant starch. Nutr Bull 30:27–54. https://doi.org/10.1111/j.1467-3010.2005.00481.x

Oyom W, Xu H, Liu Z, Long H, Li Y, Zhang Z, Prusky D (2022) Effects of modified sweet potato starch edible coating incorporated with cumin essential oil on storage quality of 'early crisp'. Lwt 153:112475. https://doi.org/10.1016/j.lwt.2021.112475

Pedreschi F (2009) Fried and dehydrated potato products. Adv Potato Chem Technol 319–337. https://doi.org/10.1016/B978-0-12-374349-7.00011-8

Pedreschi F (2012) Frying of potatoes: physical, chemical, and microstructural changes. Dry Technol 30:707–725. https://doi.org/10.1080/07373937.2012.663845

Pedreschi F, Kaack K, Granby K (2004) Reduction of acrylamide formation in potato slices during frying. LWT - Food Sci Technol 37:679–685. https://doi.org/10.1016/j.lwt.2004.03.001

Pinhero RG, Coffin R, Yada RY (2009) Post-harvest storage of potatoes. In: Advances in potato chemistry and technology, pp 339–370. https://doi.org/10.1016/B978-0-12-374349-7.00012-X

Prasad K, Sharma RR (2016) Screening of mango genotypes for the incidence of lenticel browning, a new postharvest problem. Indian J Agric Sci 86(9):1169–1171

Prasad K, Sharma RR (2018) Salicylic acid influences lenticel discolouration and physiological and biochemical attributes of mango (*Mangifera indica* L.) fruits. J Plant Biochem Biotechnol 27:293–299. https://doi.org/10.1007/s13562-018-0439-9

Prasad K, Sharma RR, Srivastava M (2016a) Postharvest treatment of antioxidant reduces lenticel browning and improves cosmetic appeal of mango (*Mangifera indica* L.) fruits without impairing quality. J Food Sci Technol 53(7):2995–3001. https://doi.org/10.1007/s13197-016-2267-z

Prasad K, Sharma RR, Srivastava M, Sethi S (2016b) Effect of hot water treatment on the incidence of lenticel browning and quality of mango fruits. Indian J Hortic 73(4):576–581. https://doi.org/10.5958/0974-0112.2016.00118.3

Prasad K, Neha P, Lal MK (2017) Cultivation and post-harvest handling techniques of potential future crop 'longan' (*Dimocarpus longan* Lour) in Asia pacific region-a review. Res Crops 18(2):384–392. https://doi.org/10.5958/2348-7542.2017.00067.5

Prasad K, Jacob S, Siddiqui MW (2018a) Fruit maturity, harvesting, and quality standards. In: Preharvest modulation of postharvest fruit and vegetable quality. Academic, USA, pp 41–69. https://doi.org/10.1016/B978-0-12-809807-3.00002-0

Prasad K, Siddiqui MW, Sharma RR, Gaurav AK, Neha P, Kumar N (2018b) Edible coatings and their effect on postharvest fruit quality. In: Innovative packaging of fruits and vegetables: strategies for safety and quality maintenance. Apple Academic Press, pp 161–197

Prasad K, Sharma RR, Sethi S, Srivastav M (2019) Influence of harvesting method on postharvest loss, shelf-life and quality of mango (*Mangifera indica*) fruits. Indian J Agric Sci 3:445–449

Prasad K, Sharma RR, Srivastava M, Asrey R (2020) Relationship between lenticel discoloration and biochemical and quality attributes in mango (*Mangifera indica* L.) fruit. *Acta*. Physiol Plant 42:178. https://doi.org/10.1007/s11738-020-03168-z

Prasad K, Sharma RR, Asrey R, Sethi S, Srivastav M, Singh D, Arora A (2022a) Hydrocolloid edible coatings extend shelf life, reduce postharvest decay, and maintain keeping quality of mango fruits (*Mangifera indica* L.) under ambient storage. J Food Biochem e14481. https://doi.org/10.1111/jfbc.14481

Prasad K, Singh G, Singh SK, Pradhan J, Kumar U, Singh H (2022b) Plant extract and essential oil coating prolongs shelf life and maintains keeping quality of papaya fruit during storage. J Food Process Preserv 46:e17015. https://doi.org/10.1111/jfpp.17015

Prasad K, Singh SK, Bhattocharjee P, Rudrapaul J, Kumar U, Yadav SK, Maurya S, Mani A (2023) Impact of climate change on postharvest quality of vegetables. In: Advances in research on vegetable production under a changing climate, vol 2. Springer International Publishing, Cham, pp 341–363. https://doi.org/10.1007/978-3-031-20840-9_16

Prasad K, Singh SK, Kumar S, Maurya P, Saroj N, Maurya S, Agnihotri R, Lal MK, Ahanger MA (2024) Rootstock mediates postharvest fruit quality, phytochemicals, and plant stress of fruit

crops. In: Ahanger MA, Bhat JA, Ahmad P, John R (eds) . Improving stress resilience in plants, Academic, pp 425–446. https://doi.org/10.1016/B978-0-443-18927-2.00004-2
Raigond P, Ezekiel R, Raigond B (2014) Resistant starch in food: a review. J Sci Food Agric 5:1968–1978. https://doi.org/10.1002/jsfa.6966
Raigond P, Mehta A, Singh B (2018) Sweetening during low-temperature and long-term storage of Indian potatoes. Potato Res 61:207–217. https://doi.org/10.1007/s11540-018-9369-0
Raigond P, Jayanty SS, Parmar V, Dutt S, Changan SS, Luthra SK, Singh B (2023) Health-promoting compounds in potatoes: tuber exhibiting great potential for human health. Food Chem 424:136368
Saroj N, Prasad K (2023) Assessment of Himalayan plain mango genotypes for phytochemicals, biochemical-nutraceutical characterisation and quality change during storage life. Int J Food Sci Technol 58:3781–3799. https://doi.org/10.1111/ijfs.16480
Saroj N, Prasad K, Singh SK, Kumar V, Maurya S, Maurya P, Kumar R (2023a) Characterization of bioactive and fruit quality compounds of promising mango genotypes grown in Himalayan plain region. PeerJ 11:e15867. https://doi.org/10.7717/peerj.15867
Saroj N, Prasad K, Singh SK, Kumar U, Lal MK, Maurya S, Maurya P, Dwivedi M, Meena MK, Nana LP (2023b) Regulatory role of melatonin in flowering, fruit setting, and ripening. In: Kumar R, Altaf MA, Lal MK, Tiwari RK (eds) Melatonin in plants: a regulator for plant growth and development. Springer, Singapore. https://doi.org/10.1007/978-981-99-6745-2_4
Saroj N, Prasad K, Singh SK, Maurya S, Maurya P, Kumar S, Lal MK, Maurya R, Ghosh S, Dongabanti B (2023c) Diverse functional role of melatonin in postharvest biology. In: Kumar R, Altaf MA, Lal MK, Tiwari RK (eds) Melatonin in plants: a regulator for plant growth and development. Springer, Singapore. https://doi.org/10.1007/978-981-99-6745-2_9
Shah MH, Ajaharuddin SM, Kundu S, Kumawat SM, Atta K, Hossain A, Pande CB (2023) Potential impact of changing climate on the sustainability of potato (*Solanum tuberosum* L.) production in India. In: Climate change impacts in India. Springer, Cham, pp 323–350
Shao Y, Liu Y, Xuan G, Shi Y, Li Q, Hu Z (2022) Detection and analysis of sweet potato defects based on hyperspectral imaging technology. Infrared Phys Technol 127:104403. https://doi.org/10.1016/j.infrared.2022.104403
Shen X, Zhang M, Devahastin S, Guo Z (2019) Effects of pressurized argon and nitrogen treatments in combination with modified atmosphere on quality characteristics of fresh-cut potatoes. Postharvest Biol Technol 149:159–165. https://doi.org/10.1016/j.postharvbio.2018.11.023
Singh P, Sandhu AS (2023) Energy budgeting and economics of potato (*Solanum tuberosum* L.) cultivation under different sowing methods in north-western India. Energy 269:126755. https://doi.org/10.1016/j.energy.2023.126755
Singh J, Kaur L, McCarthy OJ (2009). Potato starch and its modification. In: Advances in potato chemistry and technology, pp 273–318. https://doi.org/10.1016/B978-0-12-800002-1.00008-X
Singh J, Dartois A, Kaur L (2010) Starch digestibility in food matrix: a review. Trends Food Sci Technol 21:168–180. https://doi.org/10.1016/j.tifs.2009.12.001
Singh B, Raigond P, Ashiv AJ, Bir M Singh P (2016) A manual on potato processing in India
Singh BP, Singh B, Lal MK (2022) Seven decades of potato research in India: achievements and future thrusts. Int J Innov Hortic 11:158–183. https://doi.org/10.5958/2582-2527.2022.00016.1
Slizewska K (2013) The citric acid-modified, enzyme-resistant dextrin from potato starch as a potential prebiotic. Acta Biochim Pol 60:671–675. https://doi.org/10.18388/abp.2013_2039
Suttle JC (2004) Physiological regulation of potato tuber dormancy. Am J Potato Res 81:253–262
Tanios S (2020) Potato tuber greening: mechanisms of resistance and risk factors. Doctoral dissertation, University of Tasmania
Tanios S, Eyles A, Tegg R, Wilson C (2018) Potato tuber greening: a review of predisposing factors, management and future challenges. Am J Potato Res 95:248–257. https://doi.org/10.1007/s12230-018-9648-y
Thatoi H, Dash PK, Mohapatra S, Swain MR (2016) Bioethanol production from tuber crops using fermentation technology: a review. Int J Sustain Energy 35:443–468. https://doi.org/10.1080/14786451.2014.918616

Tiwari RK, Bashyal BM, Shanmugam V, Lal MK, Kumar R, Sharma S, Aggarwal R (2021) Impact of Fusarium dry rot on physicochemical attributes of potato tubers during postharvest storage. Postharvest Biol Technol 181:111638. https://doi.org/10.1016/j.postharvbio.2021.111638

Trujillo JI, Kiefer JR, Huang W, Day JE, Moon J, Jerome GM, Bono CP, Kornmeier CM, Williams ML, Kuhn C, Rennie GR, Wynn TA, Carron CP, Thorarensen A (2012) Investigation of the binding pocket of human hematopoietic prostaglandin (PG) D2 synthase (hH-PGDS): a tale of two waters. Bioorgan Med Chem Lett 22:3795–3799. https://doi.org/10.1016/j.bmcl.2012.04.004

Valencia-Florez L, Trejo-Escobar D, Latorre-Vasquez L, Mejia-Espana D, Hurtado-Benavides A (2019) Influence of storage conditions on the quality of two varieties of native potato (*Solanum tuberosum* group phureja). Dyna 86:49–55

Vilvert E, Stridh L, Andersson B, Olson A, Alden L, Berlin A (2022) Evidence based disease control methods in potato production: a systematic map protocol. Environ Evid 11:1–8. https://doi.org/10.1186/s13750-022-00259-x

Waaswa A, Oywaya Nkurumwa A, Mwangi Kibe A, Ngeno Kipkemoi J (2022) Climate-Smart agriculture and potato production in Kenya: review of the determinants of practice. Clim Dev 14:75–90. https://doi.org/10.1080/17565529.2021.1885336

Wadagavi V, Kallihal B, Dadanwale S, Choukimath MC (2017) Automatic potato chips making machine

Wang ZJ, Liu H, Zeng FK, Yang YC, Xu D, Zhao YC, Singh J (2023) Potato processing industry in China: current scenario, future trends and global impact. Potato Res 66:543–562. https://doi.org/10.1007/s11540-022-09588-3

Wharton PS, Tumbalam P, Kirk WW (2006) First report of potato tuber sprout rot caused by fusarium sambucinum in Michigan. Plant Dis 90:1460–1460. https://doi.org/10.1094/PD-90-1460B

Xu J, Li Y, Kaur L, Singh J, Zeng F (2023) Functional food based on potato. Foods 12:2145. https://doi.org/10.3390/foods12112145

Yamada S, Shinomiya N, Ohba K, Sekikawa M, Oda Y (2010) Enzymatic hydrolysis and ethanol fermentation of by-products from potato processing plants. Food Sci Technol Res 15:653–658. https://doi.org/10.3136/fstr.15.653

Yang Y, Liu Z, Huang M, Zhu Q, Zhao X (2023) Automatic detection of multi-type defects on potatoes using multispectral imaging combined with a deep learning model. J Food Eng 336:111213. https://doi.org/10.1016/j.jfoodeng.2022.111213

Yusianto R, Suprijono H, Islahudin N, Nugroho DS (2022) Smart potato grading using image processing and fuzzy grading system. In: 2022 international seminar on application for technology of information and communication (iSemantic), pp 333–337. https://doi.org/10.1109/iSemantic55962.2022.9920382

Zhang M, Huang TS, Mu TH (2020) Production and characterization of antioxidant peptides from sweet potato protein by enzymatic hydrolysis with radio frequency pretreatment. Int J Food Sci Technol 55:2352–2358. https://doi.org/10.1111/ijfs.14441

Zhang J, Yao J, Mao L, Li Q, Wang L, Lin Q (2022) Low temperature reduces potato wound formation by inhibiting phenylpropanoid metabolism and fatty acid biosynthesis. Front Plant Sci 13. https://doi.org/10.3389/fpls.2022.1109953

Zhang C, Yang Z, Tang D, Zhu Y, Wang P, Li D, Huang S (2021) Genome design of hybrid potato. Cell 184:3873–3883. https://doi.org/10.1016/j.cell.2021.06.006

Zhao S, Han X, Liu B, Wang S, Guan W, Wu Z, Theodorakis PE (2022) Shelf-life prediction model of fresh-cut potato at different storage temperatures. J Food Eng 317:110867. https://doi.org/10.1016/j.jfoodeng.2021.110867

Zommick DH, Knowles LO, Knowles NR (2014) Tuber respiratory profiles during low temperature sweetening (LTS) and reconditioning of LTS-resistant and susceptible potato (*Solanum tuberosum* L.) cultivars. Postharvest Biol Technol 92:128–138. https://doi.org/10.1016/j.postharvbio.2014.01.020

Advances in Abiotic Stresses Management in Potatoes

Ajeet Kumar, Dhyanananda Kumari, Anupam Adarsh, and Shashank Shekhar Solankey

Abstract Potato tubers are rich in nutrients and have the potential to significantly contribute to human nutrition. However, biotic and abiotic stress conditions, along with climate change, negatively affect potato crop performance and yield, posing a serious threat to long-term potato production. Among these, abiotic stresses such as drought, temperature extremes, salinity, and nutrient imbalances present significant challenges to potato cultivation worldwide. Effective management measures are needed to mitigate the adverse effects of these stresses and ensure sustainable crop output. Several methods can be employed to reduce abiotic stress in potato production, including mulching, drip irrigation, plant growth-promoting rhizobacteria (PGPR), hormonal therapies, and nutrient management. Proper nutrient management is essential for enhancing the resilience of potato plants to abiotic stress. Potatoes require a balanced supply of macronutrients (nitrogen, phosphorus, potassium) and micronutrients (calcium, magnesium, zinc) to support optimal growth and development. Nutrient deficiencies or imbalances can worsen the effects of abiotic stress, diminishing the plant's ability to cope with unfavorable conditions. Other agronomic practices also help combat abiotic stress. Mulching is an effective technique that provides multiple benefits for managing abiotic stress in potato crops. Similarly, irrigation management practices, such as scheduling and drip irrigation, are beneficial for managing drought stress, providing a consistent and controlled water supply and increasing water use efficiency (WUE). PGPR are beneficial microorganisms that colonize plant roots and enhance growth through various mechanisms, including modulation of plant hormone levels, improved nutrient

A. Kumar (✉)
Department of Soil Science and Agricultural Chemistry, Nalanda College of Horticulture, Noorsarai, Nalanda, Bihar Agricultural University, Sabour, Bhagalpur, Bihar, India

D. Kumari
Department of Horticulture (Fruits & Fruit Technology), Nalanda College of Horticulture, Noorsarai, Nalanda, Bihar Agricultural University, Sabour, Bhagalpur, Bihar, India

A. Adarsh · S. S. Solankey
Department of Horticulture (Vegetable Science), Nalanda College of Horticulture, Noorsarai, Nalanda, Bihar Agricultural University, Sabour, Bhagalpur, Bihar, India

S. S. Solankey (ed.), *Advances in Research on Potato Production*,
Advances in Olericulture, https://doi.org/10.1007/978-3-031-82710-5_14

uptake, and induction of systemic resistance to abiotic stress. Hormonal treatments can be integrated into potato cultivation practices to enhance stress tolerance and improve tuber yield and quality under abiotic conditions. Effective management of abiotic stress in potato cultivation requires an integrated approach that combines nutrient management, mulching, drip irrigation, PGPR, and hormonal treatments. Each of these strategies contributes to enhancing the resilience of potato plants to environmental stressors, supporting sustainable crop production in the face of climate change and resource limitations. By adopting these practices, farmers can mitigate the impacts of abiotic stress, ensuring stable and high-quality potato yields.

Keywords Abiotic stress · Heat stress · Drought stress · Salinity · Nutrient management · Mitigation

1 Introduction

The potato (*Solanum tuberosum* L.), a member of the Solanaceae family, ranks fourth worldwide, after rice, wheat, and maize (FAO 2019). Potatoes are particularly important in developing countries due to their high production rate per unit of time and area, as well as their nutritional composition (Islam et al. 2022). In addition to being a staple food, potatoes are widely used across various sectors to produce a diverse range of products. Moreover, their extensive applications in food, alcoholic beverages, starch-based goods, dietary supplements, ready-to-serve items, and raw materials have contributed to their widespread popularity in developed nations (Awasthi and Verma 2017). Increased potato production is necessary for ensuring food security and meeting its significant demand.

Potatoes are highly sensitive to several diseases, including various soil-borne infections and viruses, which can reduce their production potential (Chakrabarti et al. 2022). The potato is particularly crucial to the 2030 Agenda of the United Nations Sustainable Development Goals, which aims to eradicate hunger by achieving food security, enhancing nutrition security, and promoting a sustainable agricultural sector. In 2022, the FAO identified potatoes as a key crop for addressing global food security in the future.

In India, potato cultivation accounts for around one-third of all vegetable output (Kumar and Chandra 2024). Potatoes are a staple of the Indian vegetable basket. Due to their short growing season and ability to adapt to a wide range of cropping patterns, potatoes have become one of the key crops in India's transition from cereals to horticultural crops (Islam et al. 2024). India is now the second-most productive country for potatoes in the world, after China, with both nations producing over one-third of all potatoes globally.

In 1949, when ICAR-CPRI was established, India produced 1.54 million tons of potatoes from 0.234 million hectares of land, with an average productivity level of 6.58 t/ha. By 2022–2023, that amount had increased to 60.14 million tons (Statista 2024). Despite this significant advancement in potato output, India's current productivity remains a concern for research and development, as it is still significantly below the country's potential output of 40 t/ha and lags behind that of affluent nations. Furthermore, due to changes in lifestyle, economic growth, and population increase, India's potato demand is expected to reach 125 million tons by 2050 (CPRI 2015). With a harvesting index exceeding 75%, potatoes are considered one of the most advantageous crops for alleviating hunger, malnutrition, and poverty worldwide (Wijesinha-Bettoni and Mouillé 2019). Potato tubers are rich in carbohydrates, flavonoids, and vitamins B3, B6, and C, which positions them, alongside other staple crops like rice and wheat, as significant contributors to human nutrition (Campos and Ortiz 2020).

Understanding how climate change affects potato crops is essential for food security. Extreme heat and drought, resulting from global climate change, pose serious threats to agricultural production due to their detrimental effects on plant performance and yield. Future greenhouse gas emissions are expected to exacerbate this negative impact on agricultural output (Chourasia et al. 2021) by increasing evapotranspiration and drought severity (Dahal et al. 2019). As a result of this climate change scenario, the agricultural sector suffers substantial losses when various stress factors, such as moisture and heat stress, are combined (Mittler 2006). For instance, in 2016, the potato harvest in Ontario decreased by 35–50% due to heat stress and dryness (Banks and VanOostrum 2016). Global potato yields are predicted to decline by 18–32% as a consequence of climate change (Hijmans 2003). The significant reductions in potato output are largely attributed to heat stress, which affects physiological processes. In Northeast India (NEI), where potatoes are cultivated year-round in various locations, they are a major food and commercial crop. However, there is a substantial yield difference of more than 50% between India and NEI (Yadav and Srivastava 2015). The primary cause of this yield gap is the inadequate availability of high-quality seeds of modern cultivars suited to the agro-climatic conditions of Northeast India (Tripura and Ghosh 2017), along with abiotic stress and poor agronomic management practices. Abiotic stress is a major factor contributing to potato yield losses in this region.

2 Overview of Abiotic Stresses Affecting Potatoes

Abiotic stresses include heat, drought, cold, shade, nutritional deficiencies, and physical injury to the plant. Farmers' agronomic practices have minimal impact on many of these abiotic stresses. In contrast, biotic stresses occur when a plant's health is adversely affected by diseases or insect damage, among other biological factors. The combined effects of abiotic conditions in the field can increase a plant's susceptibility to drought or nutritional stress, in addition to heat stress. Strong evidence suggests that plants under abiotic stress are more vulnerable to the negative effects of biotic stress, resulting in greater overall damage.

The ever-changing environment has led to rising CO_2 levels and increased global temperatures, adversely affecting plant morphological and physiological processes.

Abiotic stresses can have detrimental effects on the growth and development of potato plants by interfering with several physiological processes, including photosynthesis, respiration, and nutrient availability, ultimately impacting the metabolic activity of the plants.

Heat stress is a significant abiotic stressor that negatively influences potato plant growth and development by affecting photosynthesis in leaves. The breakdown of photosynthetic activity, particularly impacting the RUBISCO activase enzymes involved in the light reaction (PS-II) and dark response, is one of the most harmful effects of heat stress on plants (Wang et al. 2018). According to Fleisher and Timlin (2006), a temperature of 24 °C is necessary for potato crops to achieve optimal photosynthesis. Elevated temperatures can cause irreversible damage to the photosynthetic apparatus, thereby hindering plant growth. Under moderate heat stress, photosynthesis can decline reversibly (Aldubai et al. 2022). In potato cultivars, heat stress reduces the rate of photosynthesis, and at high temperatures, leaf stomatal conductance also decreases (Aien et al. 2017). High temperatures adversely impact the chlorophyll-dependent photosynthetic process (Sadura et al. 2020; Yuan et al. 2021).

Different potato cultivars respond variably to heat stress in terms of chlorophyll content. Research indicates that under high temperatures, the chlorophyll (a/b) ratio increases while total chlorophyll decreases (Aien et al. 2017). High temperatures have been shown to reduce whole-canopy photosynthesis in potato plants (Obiero et al. 2020). When temperatures rise, the net photosynthesis of a young, fully developed leaf may initially increase after a decline. This phenomenon is likely due to stomatal responses, raising questions about the significance of leaf water status and atmospheric humidity in influencing potato susceptibility to high temperatures (Obiero et al. 2020).

Above-ground heat stress hinders tuberization, while below-ground heat stress affects photosynthesis. This indicates that bidirectional communication between source and sink organs regulates tuberization during heat stress (Hastilestari et al. 2018). Heat stress significantly reduces tuber output, tuber dry matter, and net absorption rates (Busse et al. 2019). According to Rykaczewska (2013), temperatures between 14 and 22 °C promote tuberization, enhance dry matter partitioning, and prevent excessive haulm formation.

Conversely, high temperatures lead to a decrease in tuber yield by reducing tuber size and quantity, as well as negatively affecting shoot and root development and dry matter partitioning to the haulm (Trapero-Mozos et al. 2018). It has been shown that as temperatures rise, the proportion of dry matter in leaves increases while that in tubers decreases. Hancock et al. (2014) observed a similar trend in the dry weight of tubers and leaves. High temperatures stimulate respiration during tuber initiation, wasting energy and reducing tuber set. Additionally, elevated temperatures impair the tuber bulking process during the bulking phase, leading to yield losses.

Drought stress causes stomatal closure to reduce water loss, which also limits CO_2 diffusion and disrupts the Calvin cycle, resulting in an imbalance in ATP and NADPH+ production. This imbalance leads to oxidative stress and damage at the chloroplast level, negatively affecting tuber formation and growth.

Rainfall is crucial for potato development, and even brief water shortages can significantly reduce tuber yield and quality. Salinity also adversely affects potatoes by decreasing tuber weight and accelerating leaf senescence, with the impact varying by cultivar and salt concentration.

Similarly, the accumulation of dry matter and assimilate partitioning in potato plants is affected by these abiotic stresses. According to Golovko and Tabalenkova (2019), potatoes have source organs in the form of leaves and sink organs in the form of subterranean stolons. Sucrose produced in mesophyll cells initiates photosynthesis and is transported by sieve elements to companion cells and the sink (Jensen et al. 2016). The carbon flow to sink organs in potatoes is influenced by temperature; at high temperatures, the amount of absorbed carbon that is integrated into tubers is reduced.

3 Types of Abiotic Stresses in Potatoes

3.1 Drought

Potatoes require specific environmental conditions (around 17 °C) for optimal tuber production, as they originated and developed in the tropical highlands. Elevated temperatures can hinder crop production by reducing overall plant growth due to heat stress or by decreasing the assimilation of nutrients into tubers. The minimum nighttime temperature is crucial for potato growth; it determines whether tuberization will occur. While potato plants can tolerate daytime temperatures of around 35 °C without much harm, tuberization is reduced at nighttime temperatures above 18 °C and may not occur at temperatures of 25 °C or higher. Additionally, high temperatures can increase the number of internodes, resulting in thin stems, small leaves, and elongated stolons, which further limits tuber formation and lowers the ratio of fresh tuber weight to total fresh weight.

Drought stress is the primary factor restricting potato production globally, adversely impacting both yield and tuber quality. In almost every region where potatoes are grown, there may be an inadequate water supply. In dry areas (such as the subtropics), where potato production relies on irrigation, insufficient irrigation practices or a lack of water can lead to brief droughts. Water stress can still occur even with adequate irrigation techniques due to high transpiration rates, particularly in the middle of the day when the plant's root system may not fully meet its water needs. This results in higher water potential and a subsequent slowdown in the rate of photosynthesis. Erratic rainfall, especially on low-water-holding soils, can lead to both brief and prolonged droughts in temperate climate zones almost annually. It is projected that if the crop's water supply could be managed more effectively, the average global potato yield could increase by at least 50%, considering current yield levels and production conditions. Therefore, breeding and biotechnological methods to introduce drought resistance in potatoes should be prioritized. Effective management of nutrients and water also positively impacts potato yield.

Drought significantly affects agricultural output by inhibiting photosynthesis, primarily through stomatal closure, which limits CO_2 diffusion and reduces CO_2 availability for the Calvin cycle. This leads to an imbalance between ATP and NADPH production and their consumption in the Calvin cycle, resulting in energy buildup, oxidative stress, and cell damage. In addition to stomatal constraints, drought-sensitive biochemical processes may further restrict photosynthesis. However, the exact contribution of stomatal versus biochemical limitations to photosynthetic inhibition during drought remains unclear. According to Oliveira et al. (2012), drought stress caused a significant drop in wheat biomass and grain output. Similarly, Hlavinka et al. (2009) identified drought as the primary factor causing yield variability in eight different crops. The potato is considered one of the most drought-sensitive crop species due to its shallow root structure.

Drought stress adversely affects the physiological mechanisms involved in tuber formation and development. However, the length, intensity, and growth stage of the plant significantly influence the extent of drought-induced tuber yield loss (Evers et al. 2010). Early stress is particularly harmful to tuberization, bulking, and tuber production due to reduced leaf area, lower photosynthetic rates, and decreased assimilate partitioning to tubers (Obidiegwu et al. 2015). Drought also affects tuber production and significantly alters several indices of tuber quality, making them unfit for processing and consumption. Wang-Pruski and Schofield (2012) reported that drought causes tuber abnormalities such as cracking, hollow heart, internal brown spots, secondary growth, deformities, and a significant increase in the glycoalkaloid content of α-solanine and α-chaconine, which can lead to various health issues, including cancer.

3.2 Heat Stress

In subtropical India, heat stress poses a serious threat to potato crops, particularly for early-planted varieties and those following the harvest of major rabi crops. Tuber growth and development require a minimum night time temperature below 20 °C (with daytime temperatures around 25 °C) (Singh et al. 2015). Except in steep, rainfed areas, potatoes are primarily an irrigated crop. Consequently, water availability affects all stages of development, including germination, leaf and root growth, stolon formation, and tuber growth. Therefore, drought or moisture deficit is a key factor determining potato yield (Dahal et al. 2019). Heat stress inhibits tuber growth and productivity, making potatoes better suited for mild-to-temperate climates. Generally, tuberization is completely inhibited at 25 °C and is reduced at nighttime temperatures above 20 °C. Extreme temperatures disrupt the hormonal balance of potato plants. Under high nighttime temperatures (>22 °C), tuberization is a critical consideration in breeding programs for heat stress tolerance.

Potatoes thrive in temperate locations with moderate temperatures, as they are a cool-weather crop. Temperatures exceeding the ideal range are likely to impede plant growth and development, thereby lowering productivity and tuber output

(Tang et al. 2018). However, the susceptibility of potato crops to high temperatures is significantly influenced by cultivar, developmental stage, and duration of stress (Tang et al. 2018). According to Van Dam et al. (1996), the optimal temperature ranges for below-ground tuber growth and above-ground plant growth differ: the former thrives in the range of 15–20 °C, while the latter performs best between 20 and 25 °C. Temperatures above 25 °C have been shown to delay tuberization. It has been projected that drought and global warming could lead to a significant decrease in potato output by 2055 (Holden et al. 2003).

Elevated temperatures adversely affect tuber production and quality by inhibiting the synthesis of carbon and its subsequent transfer to the stolon. Research indicates that potatoes require a temperature of around 20 °C for optimal photosynthesis and biomass formation. However, temperatures above 50 °C can significantly reduce the rate of photosynthetic activity. In several plant species, including wheat, rye, and canola, high temperatures that restrict photosynthesis are linked to decreased RuBisCO activation, permanent damage to photosystem II, and increased rates of photorespiration (Dahal et al. 2012). While some studies suggest that high temperatures in potatoes limit PSII activity (Havaux 1993), the effects of high temperatures on RuBisCO activity and photorespiration remain unclear.

Additionally, heat stress reduces the amount of sucrose translocated to tubers and alters assimilate partitioning, both of which can impede tuberization and production. The accumulation of carbon in the source leaves has several effects, one of which is limiting photosynthesis. High temperatures also adversely affect tuber quality, causing problems such as hollow hearts, cracking, and secondary growth. Furthermore, heat stress leads to heat necrosis in tuber flesh, degrading the quality of both fresh market and processed potatoes. In response to high soil temperatures, tuber quality can suffer due to skin rusting (Wang-Pruski and Schofield 2012), malformed tubers, field sprouting, and reduced dry matter content.

3.3 *Cold and Frost*

In northwest India, the Indo-Gangetic plains frequently experience frost damage to potato crops. However, the timing, duration, and severity of frost can be unpredictable, depending on the crop's age, cultivar, soil type, and other factors, leading to moderate to significant losses (30–90% leaf damage). Frost has detrimental effects on potatoes both directly and indirectly, reducing output and causing a scarcity of seed tubers, which limits crop area for the next growing season (Kumar et al. 2011). Unfortunately, there has been limited technology available to reduce crop losses, even with frost-tolerant varieties. Frost typically destroys the top leaves of the canopy, reducing the tubers' access to photosynthates. In immature crops that have been frost-damaged, development may resume but with a modified growth pattern, as the plant must first rebuild and repair its canopy before transferring photosynthates to the tubers. Scheduling important growth phases is essential for the successful recovery of frost-damaged crops, allowing plants to capitalize on favorable weather during the post-frost period. A large leaf area physiologically supports tuber growth at an increased rate for an extended period after tubers are formed, ultimately resulting in higher production.

3.4 Salinity

Salinity is a major threat to crop production worldwide, affecting over 800 million hectares of land, which represents more than 6% of the world's land surface. This poses a serious danger to agricultural production globally (Munns and Tester 2008). Saline soils are particularly common in semi-arid and arid regions. They are typically created by flooding saltwater into the soil, irrigating with salt-contaminated water, or experiencing insufficient precipitation to leach salts from the soil. The accumulation of NaCl, Cl^-, or SO_4^{2-} ions can reduce soil porosity, water permeability, and overall soil structure. Salinity is quantified using osmotic potential and electrical conductivity. The anticipated rise in the intensity and frequency of droughts may exacerbate soil salinity by causing saline groundwater to rise to the surface.

Salinity induces osmotic stress in plants, negatively impacting their growth and development. Higher levels of dissolved salt solutes in the soil reduce its water potential, affecting the water balance in the plant-soil continuum. Potatoes are believed to be particularly vulnerable to saline stress. High salinity negatively impacts plant growth and tuber development, with the degree of salt's effect varying depending on cultivar and salt concentration. Salinity has been shown to decrease both fresh and dry tuber weight per plant and to accelerate leaf senescence, yellowing, and desiccation (Akhtar et al. 2015). While salinity hampers the growth of shoots and reduces total fresh and dry biomass, the inhibition is more pronounced in salt-sensitive cultivars (Aghaei et al. 2009).

Increased salinity in the soil adversely affects potato plant growth and tuber development. The severity of salt stress impacts varies depending on the cultivar and salinity levels. According to Akhtar et al. (2015), salinity hastened the onset of desiccation, yellowing, and senescence in leaves, while reducing fresh and dry tuber weight per plant. Salt stress significantly reduced photosynthetic rates (Akhtar et al. 2015) and led to structural changes, such as impaired grana stacking, thylakoid enlargement, and cell vacuolation (Queirós et al. 2009). Salinity also significantly increased the NaCl and Cl content in callus tissue, elevated lipid peroxidation, and slowed the rate of callus development (Queirós et al. 2011). Potato plants grown under saline conditions exhibited higher sodium concentrations in their leaves, stems, and tubers, which contributed to decreased tuber development and dry matter content. Additionally, salinity delayed seedling emergence (Ghosh et al. 2001).

Under salt stress, the average tuber weight and number of tubers per plant decreased, resulting in reduced total and marketable tuber yields (Odemis and Caliskan 2014). Salinity led to increased levels of total non-structural carbohydrates, water-soluble carbohydrates, and starch content in leaves (Ghosh et al. 2001). The increased carbohydrate content in leaves and decreased dry matter content in tubers indicate that salt stress may impede carbon transport to the tuber. According to Aghaei et al. (2009) and Jaarsma et al. (2013), salt stress reduces plant development and tuber yield in potatoes by causing severe senescence and nutritional imbalances.

3.5 *Nutrient Deficiency*

Good vegetative development, production, and quality of potatoes depend on both macro- and micronutrients. In India, potatoes, which are shallow-rooted crops, are grown on sandy-loam soil under irrigated conditions. Due to their heavy reliance on nitrogen (N) fertilization, potatoes require adequate amounts of N fertilizer. Balanced use of these fertilizers can protect the environment and reduce production costs (Zhang et al. 2020). Excessive fertilization raises production costs, contaminates groundwater, causes nitrate leaching, and ultimately pollutes the ecosystem (Tiwari et al. 2018, 2020). Nutrient deficiencies severely hamper crop development and decrease productivity. One of the main forms of abiotic stress in potatoes that can impair growth and yield is nutrient deficiency. Essential nutrients are crucial for many physiological and biochemical functions in plants; adequate amounts of these nutrients are required for healthy growth, development, and production of potatoes. Inadequate amounts can lead to various symptoms and lower the overall health of the plants.

For nitrogen deficiency, common symptoms include yellowing of older leaves (chlorosis), stunted growth, reduced tuber size, and early maturity. Nitrogen is crucial for vegetative growth and chlorophyll production, and its deficiency leads to poor foliage and reduced photosynthesis in potatoes. Similarly, phosphorus deficiency causes dark green or purplish leaves, stunted growth, delayed maturity, and poor root development. Phosphorus is essential for energy transfer and root development, so its deficiency negatively affects energy metabolism and tuber formation.

Potassium deficiency results in yellowing and browning of leaf edges (marginal chlorosis), weak stems, reduced disease resistance, and poor tuber quality. Potassium regulates water balance, enzyme activation, and disease resistance, and its deficiency leads to reduced quality and yield of tubers. Calcium, an important essential nutrient, shows typical deficiency symptoms such as tip burn on young leaves, poor root systems, and internal tuber defects (e.g., hollow heart). As calcium is vital for cell wall structure and membrane function, its deficiency results in poor structural integrity and tuber defects, ultimately leading to poor-quality potatoes.

Magnesium deficiency manifests as interveinal chlorosis on older leaves, leaf curling, and reduced photosynthetic efficiency. Magnesium is a central component of chlorophyll and is involved in enzyme activation, so its deficiency hampers photosynthesis and lowers tuber yields. Sulfur deficiency leads to general yellowing of younger leaves, stunted growth, and reduced protein synthesis, as sulfur is vital for amino acids and protein production. Its deficiency negatively impacts overall plant metabolism and yield.

Micronutrient deficiencies also affect plant growth, yield, and quality of potatoes. Iron deficiency causes interveinal chlorosis on younger leaves. Manganese deficiency also results in interveinal chlorosis, similar to iron deficiency but affecting older leaves. Zinc deficiency leads to small, distorted leaves and interveinal chlorosis. Copper deficiency causes stunted growth, chlorosis, and leaf tip dieback. Molybdenum deficiency results in pale green leaves, rolled leaves, and necrosis. Boron deficiency affects tuber quality, leading to internal brown spots and hollow heart, while chlorine deficiency causes wilting, chlorosis, and reduced root growth.

4 Strategies for Mitigating Abiotic Stresses in Potato Production

4.1 Strategies for Mitigating Abiotic Stresses Through Nutrient Management

Low agricultural yields are often caused by various environmental factors, including insufficient mineral fertilizers and nutrient imbalances, which negatively affect plant growth, antioxidant availability, and plant tolerance to stress. Stress impacts key plant functions such as photosynthesis, seed germination, growth, and reproduction, ultimately reducing yield. However, effective nutrient management strategies can help maintain crop production under adverse conditions. Plants combat abiotic stress through mechanisms such as improved photosynthesis, reduced heavy metal absorption, reactive oxygen species (ROS) detoxification, antioxidant enzyme production, osmoprotectant synthesis, heat shock protein (HSP) production, DNA repair, membrane stability, and activation of stress-related genes. Nutrient management is a crucial strategy for mitigating environmental stress and improving yields.

Several plant nutrients can help reduce heat stress. For instance, boron is a significant micronutrient that enhances the activity of the antioxidant system and reduces the harmful effects of ROS caused by heat stress (Waraich et al. 2011). Similarly, selenium (Se) protects plants from the adverse effects of ROS due to its structural role in the formation of glutathione peroxidase (GPX) (Lobanov et al. 2008). Zinc is another crucial micronutrient that helps preserve membrane permeability; plants receiving optimal amounts of Zn are better protected against the damaging effects of heat stress (Peck and McDonald 2010).

4.1.1 Nutrient Management Approaches to Alleviate Abiotic Stresses

1. *Primary Nutrients:* Primary nutrients, namely nitrogen (N), phosphorus (P), and potassium (K), are essential for activating various plant biochemical activities and mechanisms to cope with different abiotic stresses.

(a) *Nitrogen:* Nitrogen (N) is a crucial nutrient for plant structural integrity, growth, physiology, and stress reduction (Ata-Ul-Karim et al. 2016). It is essential for chlorophyll biosynthesis, enhancing photosynthetic activity, and the efficient utilization of solar radiation (Hassan et al. 2005). Different N fertilizers react uniquely in the field; nitrate (NO_3^-) fertilizer provides better heat stress tolerance compared to ammonium (NH_4^+) fertilizer (Zhu et al. 2000; Bendixen et al. 2001). The application of N fertilizers promotes carbon partitioning, carbohydrate accumulation, membrane stability, osmoregulation, and antioxidative defense mechanisms, leading to improved plant development and reduced leaf senescence. However, under moisture stress, plants struggle to absorb nitrogen (Abid et al. 2016; Binghua et al. 2012).

To maintain optimal relative leaf water content and enhance water use efficiency during moisture stress, nitrogen also helps couple flexibility and water extraction abilities of plant roots from the soil (Yang et al. 2012). When a plant lacks nitrogen, wasted solar energy can worsen lipid peroxidation and photooxidative damage in hot weather (Huang et al. 2004). Nitrogen triggers proline accumulation and enzymatic antioxidant activities related to superoxide dismutase (SOD), peroxidase (POD), catalase (CAT), and ascorbate peroxidase (APX), which help lower ROS concentrations and reduce malondialdehyde (MDA) levels and lipid peroxidation (Yang et al. 2012). Lack of nitrogen, especially under dry conditions, hinders cell division and expansion activities, inhibiting plant growth and leaf output (Song et al. 2019). Lower nitrogen availability under stress accelerates the breakdown of enzyme concentrations (Rubisco) and pigments (chlorophyll), leading to a noticeable slowdown in plant photosynthetic performance (Gonzalez-Real and Baille 2000).

N is also involved in protein metabolism, particularly under moisture stress. Higher nitrogen availability in plants under water stress has been linked to increased nitrogen uptake and the activity of the nitrogen-absorbing enzyme nitrate reductase. When N fertilizers are applied to reproductive plants during drought, the synthesis of abscisic acid is significantly slowed, while cytokinin production is accelerated. This promotes nodulation, cell elongation, apical dominance, shoot development, increased photosynthetic activity, and translocation of assimilates to sink organs. In the case of potato tubers, higher nitrogen availability reduces the adverse effects of stress.

(b) *Phosphorus:* The use of phosphorus fertilizers is positively correlated with stressed plant growth and development. Phosphorus enhances root architecture and proliferation in the soil, increasing root volume and hydraulic conductivity, even in low soil moisture conditions (Tariq et al. 2017). This boost in root development facilitates efficient access to moisture and nutrients through enhanced sensing and signaling processes in the rhizosphere during abiotic stress.

Phosphorus treatment during the early stages of crop development has been shown to improve root establishment and growth under water scarcity. Adequate phosphorus absorption results in the production of more viable tillers, alongside improved photosynthesis, energy storage and transfer, cell division, and elongation, leading to better yield parameters. In potatoes, tuber yield positively correlates with phosphorus application under stress conditions. Additionally, phosphorus contributes to maintaining cell turgidity and membrane integrity, thereby sustaining optimal leaf water potential and ensuring stress tolerance by enhancing stomatal conductance and the rate of net photosynthesis (Fahad et al. 2015).

Furthermore, the incorporation of phosphorus under various abiotic stresses significantly influences tissue formation, carbohydrate and lipid metabolism, chlorophyll biosynthesis, photosynthesis, and the production of assimilatory compounds, such as adenosine triphosphate (ATP) and reduced nicotinamide adenine dinucleotide phosphate (NADPH). It also affects enzymatic functions like ATPase activities and ribulose-1,5-bisphosphate (RuBisCo) activity, as well as phytohormonal functions such as abscisic acid (ABA) and indole-3-acetic acid (IAA) synthesis, energy transfer and storage, reproductive development, and the scavenging activity of ROS, including hydrogen peroxide (H_2O_2), MDA, singlet oxygen ($^1O_2$), superoxide anions (O_2^-), and hydroxyl radicals (OH). Additionally, phosphorus promotes osmolyte accumulation (proline, soluble sugars, and proteins) (Begum et al. 2020; Yang et al. 2019). Phosphorus assimilation has been shown to reduce ROS formation caused by drought by activating enzymatic antioxidants, such as CAT, SOD, POD, APX, and monodehydroascorbate reductase (MDHAR), thereby increasing stress resistance (Meng et al. 2021). Phosphorus treatment also upregulates nitrogenous compounds in water-stressed plants, enhancing the accumulation and absorption of NH_4^+ and NO_3^-. Conversely, phosphorus deficiency can exacerbate abiotic stress, while abiotic stressors like heat, acidity, salinity, and drought can induce phosphorus deficiencies and impair the activity of ATPase and phosphatase enzymes in plants (Sardans and Penuelas 2012).

(c) *Potassium:* Potassium is a crucial component for potato growth. It influences high yields, tuber integrity, and the plant's resistance to pests, diseases, and environmental stresses. Potassium aids in the transportation of nutrients and carbohydrates, supports photosynthesis, and regulates water status. It facilitates the transfer of carbohydrates from leaves to tubers, providing energy for growth and development. Additionally, it boosts protein synthesis, carbohydrate breakdown, and overall photosynthesis. Potassium also regulates the internal ionic content and water status of the plant, particularly in the stomata. As one of the essential macronutrients, potassium significantly enhances plant growth and physiological development under abiotic stress conditions. Numerous fundamental physiological and metabolic processes, including photosynthesis, stomatal control, translocation of photosynthates, carbohydrate metabolism, and preservation of cell turgidity, depend on potassium (Ge et al. 2012).

Potassium plays a vital role in enhancing crop resistance to various abiotic stresses by improving root physiology, increasing leaf area, regulating stomatal function, reducing transpiration, and ensuring effective osmotic adjustment (Danial et al. 2010). It helps maintain leaf turgor, boosts enzymatic activity, improves nutrient uptake and water use efficiency, and prevents drought-induced ROS accumulation (Fooladivanda et al. 2014). Potassium enhances leaf water content and water use efficiency during moisture stress conditions (Majeed et al. 2016).

Potassium treatment promotes root and tuber growth, enhancing water absorption and helping plants better withstand drought stress. Under salinity stress, potassium regulates osmotic pressure, membrane potential, and enzyme activation. A deficiency in potassium can lead to chlorotic and necrotic diseases, particularly in plants exposed to high light intensities, causing issues such as enzymatic dysfunction, disrupted fluid movement, and photooxidative damage in freezing conditions. Adequate potassium supply helps regulate osmotic potential and strengthens ROS defense systems, protecting plants from freezing-induced dehydration (Wang et al. 2013).

(d) *Sulfur:* After nitrogen, phosphorus, and potassium, sulfur (S) is the fourth most important nutrient for plants. It is a vital macronutrient with several essential functions. Numerous sulfur-containing compounds play protective roles in response to abiotic stress, cellular acclimatization, and adaptability in challenging environments (Cao et al. 2014). Sulfur is critical in potato production and plays an important role in managing abiotic stress. Its involvement in various biochemical and physiological processes supports plant growth, tuber quality, and stress resistance. It contributes to many metabolic activities, including protein synthesis, enzyme function, and photosynthesis.

1. *Protein and Enzyme Formation:* Sulfur is a key component of certain amino acids (like cysteine and methionine), which are essential building blocks of proteins. Adequate sulfur availability ensures efficient protein synthesis, supporting overall growth and tuber development in potatoes.
2. *Chlorophyll Formation and Photosynthesis:* Sulfur is involved in the formation of chlorophyll, the pigment responsible for photosynthesis. Proper sulfur levels enable efficient energy production, leading to enhanced growth and tuber yield in potato plants.
3. *Tuber Size and Yield:* Sulfur enhances the production of proteins and enzymes that are vital for metabolic processes, translating into larger and better-formed tubers. This also increases overall potato yield.
4. *Improved Nutrient Use Efficiency:* Sulfur aids in the efficient utilization of other essential nutrients, particularly nitrogen. It plays a synergistic

role in nitrogen metabolism, enabling the potato plant to use nitrogen more effectively for growth and yield.

5. *Starch and Carbohydrate Formation:* Sulfur contributes to starch synthesis, which is important for tuber quality. Sulfur deficiency may reduce starch content in potato tubers, affecting marketability and cooking quality.
6. *Enhanced Resistance to Disease:* Sulfur is involved in synthesizing defense compounds, such as glucosinolates and phytoalexins, which help the potato plant defend itself against pathogens, particularly fungal and bacterial infections.

Sulfur is crucial in potato production and plays an important role in managing abiotic stress. Its contributions include:

1. *Drought Stress:* Sulfur is vital for forming osmolytes, which help maintain cellular water balance under drought conditions. These osmolytes stabilize proteins and membranes, reducing dehydration damage.
2. *Heat Stress:* Sulfur is involved in synthesizing heat shock proteins that protect potato plant cells during high temperatures. These proteins prevent heat-induced damage to important cellular structures.
3. *Oxidative Stress:* Sulfur plays a key role in synthesizing glutathione, an important antioxidant that protects the potato plant from oxidative damage caused by various abiotic stresses, including drought, heat, and salinity. Glutathione neutralizes harmful reactive oxygen species (ROS) that accumulate during stress.
4. *Salinity Stress:* Under salinity stress, sulfur aids in synthesizing sulfur-containing compounds that detoxify excess salts and maintain cellular ion balance. This is crucial for reducing the harmful effects of high sodium levels and enabling the plant to function efficiently under saline conditions.
5. *Heavy Metal Stress:* Sulfur helps the plant cope with heavy metal toxicity by forming complexes with harmful metals, thereby reducing their negative effects. This detoxification process enables the potato plant to grow more robustly in contaminated soils.

Overall, sulfur is essential for improving the growth, yield, and quality of potato crops. It enhances the plant's ability to withstand various abiotic stresses, such as drought, heat, salinity, and oxidative stress, by contributing to critical protective mechanisms.

(e) *Calcium:* Calcium (Ca) is a necessary secondary nutrient that regulates cell and plant growth processes. By controlling various physiological processes, it also enhances the plant's response to different stresses (Rahman et al. 2015). Calcium functions as a signaling molecule in several physiological and biochemical pathways, promoting stress tolerance and acting as a secondary messenger for intracellular processes (Ahmad et al. 2015). To reduce abiotic stress in plants, calcium is essential for nutrient absorption, hormonal and enzymatic upregulation, and cellular membrane stabilization (Rahman et al. 2015).

Calcium plays a crucial role in potato production and abiotic stress management due to its involvement in several physiological and structural processes. It is directly or indirectly involved in many metabolic activities that govern potato growth and development, including:

1. *Cell Wall Integrity and Strength:* Calcium is a key component of the plant cell wall, stabilizing pectins and improving cell structure. In potatoes, it contributes to tuber firmness, quality, and resistance to mechanical damage.
2. *Tuber Quality:* Adequate calcium availability during tuber formation helps prevent physiological disorders, such as internal browning, hollow heart, and tuber cracking, which often result from inadequate calcium supply.
3. *Water Regulation:* Calcium aids in regulating water movement within the potato plant. It improves root health, enhancing the plant's ability to absorb water and nutrients, which is vital for sustained growth and tuber formation.
4. *Nutrient Absorption:* Adequate calcium helps balance the uptake of other nutrients like potassium and magnesium. It promotes healthier root systems and improves the potato plant's overall nutrient use efficiency.
5. *Resistance to Pathogens:* Calcium strengthens the potato plant's defense mechanisms against diseases by reinforcing cell walls, making them less prone to pathogen attacks.

Calcium reduces yield loss in crops under abiotic stresses like salt, drought, heat, cold, and heavy metal exposure. Calcium-dependent protein kinases (CDPK) activate stress-responsive genes, regulating physiological responses such as stomatal movement, potassium uptake, and gene expression, which are crucial for antioxidative stress responses.

1. *Drought Stress:* Calcium enhances drought resistance in plants by stabilizing cell membranes, improving root function for water uptake, and maintaining cellular turgor. It also helps conserve water when applied to leaves and aids in recovery from drought-induced damage by regulating plasma membrane ATPase, which restores lost nutrients.
2. *Heat Stress:* During heat stress, calcium helps stabilize proteins and membranes, reducing damage caused by high temperatures. It also assists in maintaining the integrity of cellular processes under heat stress conditions.
3. *Salt Stress:* Calcium mitigates salinity stress by balancing ion uptake and minimizing sodium (Na^+) toxicity. It reduces sodium-induced damage by competing with sodium for uptake sites, preventing excessive sodium accumulation in tissues. Under salt stress, calcium ions can ameliorate the effect by competing with sodium ions for membrane binding sites (Duman 2012).
4. *Oxidative Stress:* Calcium acts as a secondary messenger in plants, particularly in signaling pathways associated with stress responses. It activates antioxidant mechanisms that help protect the plant from oxidative damage caused by environmental stressors like drought, salinity, and high temperatures.

Calcium is essential for improving both growth and development in potatoes while enhancing the plant's resilience to abiotic stresses like drought, heat, and salinity. Proper calcium management in potato production results in healthier plants, better-quality tubers, and greater tolerance to environmental stress factors.

(f) *Magnesium:* Magnesium (Mg) is a structural component of the ribosome and is required for the conformational stability of macromolecules such as nucleic acids, proteins, cell membranes, and walls (Marschner 2012). A lack of magnesium can adversely affect photosynthesis, as it is a crucial component of chlorophyll, which controls photosynthetic activity. Magnesium is necessary for maintaining the activity of ATPase, kinases, and polymerases, among other enzyme functions. According to Hermans et al. (2013), magnesium is essential for the functioning of all phosphorylating enzymes involved in carbohydrate metabolism. Magnesium is an essential nutrient in potato production and plays a key role in managing abiotic stress. Its involvement in photosynthesis, enzyme activation, and stress tolerance makes it critical for healthy plant growth and tuber quality.

1. *Central to Photosynthesis (Chlorophyll Production):* Magnesium is a core component of chlorophyll, the molecule responsible for capturing light energy during photosynthesis. Adequate Mg levels are essential for efficient energy production, which is crucial for robust vegetative growth and tuber formation in potatoes.
2. *Enhancing Carbohydrate Translocation:* Magnesium facilitates the movement of carbohydrates (such as sugars) from the leaves to other parts of the plant, including tubers. This is essential for tuber bulking and the development of high-quality, well-sized potato tubers.
3. *Activation of Enzymes:* Magnesium acts as a cofactor for many enzymes involved in metabolic processes, such as protein synthesis, nucleic acid formation, and starch synthesis. These enzymes are critical for the plant's growth and development, contributing to higher tuber yield and quality.
4. *Improved Nutrient Uptake and Utilization:* Magnesium helps balance the uptake of other nutrients, particularly phosphorus and potassium, ensuring that potato plants utilize these nutrients efficiently. This improves overall plant health, tuber development, and resistance to nutrient imbalances.
5. *Prevention of Physiological Disorders:* Magnesium deficiency in potatoes can lead to symptoms such as interveinal chlorosis (yellowing between leaf veins), poor growth, and reduced tuber quality. Ensuring adequate Mg levels prevents such disorders and promotes vigorous plant growth.

Mg plays a key role in abiotic stress management by affecting various metabolic activities, making it critical for healthy plant growth and tuber quality.

1. *Drought Stress:* Magnesium enhances the plant's ability to cope with drought stress by improving water-use efficiency. It helps maintain proper leaf function, which is crucial for stomatal regulation, thus preventing excessive water loss during dry conditions. Mg also aids in osmotic adjustment, which improves the plant's ability to retain water.
2. *Heat Stress:* During periods of heat stress, magnesium stabilizes cellular membranes and proteins. It ensures the proper function of the photosynthetic apparatus and helps the plant tolerate higher temperatures by reducing damage to cellular structures.
3. *Oxidative Stress:* Magnesium plays a role in detoxifying reactive oxygen species (ROS), which are harmful by-products of stress conditions like drought, heat, and salinity. Magnesium is involved in activating antioxidant enzymes (e.g., superoxide dismutase) that neutralize ROS and protect the plant from oxidative damage.
4. *Salinity Stress:* In saline conditions, magnesium helps regulate ion balance within the plant, particularly by balancing the uptake of potassium (K^+) and sodium (Na^+). This prevents the toxic accumulation of sodium ions and helps maintain the plant's physiological functions under high salinity stress.
5. *Nutrient Imbalance Stress:* Magnesium buffers against nutrient imbalances, particularly those related to calcium, potassium, and phosphorus. It ensures that metabolic processes remain stable even when external nutrient levels fluctuate, providing a stress-buffering effect.

Magnesium also plays a crucial role in improved energy production in plants, as it is vital for the production of ATP (adenosine triphosphate), the energy currency of the cell. This energy is crucial for plants to survive and thrive under stress conditions, powering repair mechanisms and maintaining cellular homeostasis during periods of environmental stress. The nutrient Mg is critical in both enhancing potato production and enabling the crop to better manage abiotic stresses such as drought, heat, and salinity. It plays a key role in photosynthesis, carbohydrate translocation, and nutrient utilization while supporting the plant's stress tolerance mechanisms, ensuring better growth, yield, and resilience under adverse conditions.

(g) *Boron:* An application of boron (B) can significantly improve yield and nutrient absorption while mitigating the detrimental effects of abiotic stress. The production and operation of cellular integuments, as well as the general metabolism and transport mechanisms of carbohydrates, depend heavily on boron. Boron is an essential micronutrient in potato production and plays a critical role in the plant's ability to manage abiotic stress. It is particularly important for cell wall integrity, nutrient transport, reproductive development, and stress tolerance.

1. *Role of Boron in Potato Production*

    1. *Cell Wall Formation and Integrity:* Boron is critical for the structural integrity of plant cell walls. It cross-links with pectin in the cell wall,

providing strength and rigidity to plant tissues. In potatoes, this is vital for healthy plant growth, tuber development, and resistance to mechanical damage.

2. *Tuber Quality and Formation:* Boron is essential for proper tuber development and supports cell division and expansion, which are crucial for tuber size, uniformity, and quality. Adequate boron levels also prevent internal physiological disorders such as hollow heart and brown center, which can occur due to impaired cell development in the tubers.
3. *Nutrient and Sugar Transport:* Boron plays a role in the movement of sugars and nutrients within the plant. It facilitates the transport of carbohydrates from the leaves to the tubers, which is essential for proper tuber bulking and starch accumulation.
4. *Improvement of Root Health:* Boron supports the development of a healthy root system by promoting root elongation and branching. This enhances nutrient uptake, water absorption, and overall plant vigor, contributing to better potato yields.
5. *Flowering and Reproductive Development:* Boron is involved in pollen tube formation and fertilization, essential for flower and seed development. Although potatoes are propagated vegetatively, proper flowering and reproductive health influence overall plant development and vitality.
6. *Prevention of Deficiencies:* Boron deficiency in potatoes can lead to cracked tubers, scabby surfaces, and poor tuber formation. It also results in leaf chlorosis, thickened and brittle leaves, and reduced growth. Ensuring adequate boron supply helps prevent these physiological disorders.

2. *Role of Boron in Abiotic Stress Management*

    1. *Drought Stress:* Boron helps maintain membrane stability during drought conditions. It enhances the plant's ability to retain water and avoid cellular dehydration by preserving the integrity of cellular membranes. Additionally, boron aids in the synthesis of osmoprotectants (compounds that help cells cope with water stress), thus improving drought tolerance in potatoes.
    2. *Heat Stress:* Boron plays a role in protecting cellular structures from heat-induced damage. It helps maintain the stability of proteins and membranes, which can become destabilized under high-temperature conditions. This stabilization helps potato plants continue functioning and growing during heat stress.
    3. *Oxidative Stress:* Under stress conditions such as drought, heat, or salinity, plants generate reactive oxygen species that can cause oxidative damage. Boron enhances the antioxidant defense system of the plant, helping neutralize ROS and protect cellular components from oxidative stress.

4. *Nutrient Imbalance Stress:* Boron helps balance the uptake and utilization of other essential nutrients such as calcium, potassium, and nitrogen. By ensuring proper nutrient management, boron reduces the risk of stress caused by nutrient imbalances, which can occur under poor growing conditions or in soils with fluctuating nutrient availability.
5. *Salinity Stress:* Boron plays a role in managing ionic balance under saline conditions. It helps mitigate the harmful effects of salt stress by maintaining cellular homeostasis and supporting the transport of nutrients like calcium, which is crucial for membrane stability in saline environments.
6. *Heavy Metal Stress:* In the presence of heavy metal stress (e.g., excessive copper or aluminum), boron helps protect root and shoot tissues by limiting the uptake of toxic metals and reducing their impact on plant metabolism. This improves plant resilience in contaminated soils.

3. *Stress Signaling and Adaptation*

   Boron is involved in regulating hormonal pathways and stress-signaling mechanisms. It aids in adjusting the plant's growth and development in response to stress conditions, improving its adaptive response to environmental challenges such as drought, salinity, and heat stress.
4. *Cell Membrane Integrity*

   Boron contributes to the structural integrity of plasma membranes, helping maintain their fluidity and functionality under stress conditions. This enhances the plant's ability to transport nutrients and maintain cellular processes during periods of abiotic stress.

   In potato production, boron is essential for cell wall integrity, tuber development, nutrient transport, and reproductive health. It helps prevent common disorders like hollow heart and brown center while ensuring overall tuber quality and yield. In terms of abiotic stress management, boron improves the plant's resilience to drought; heat, salinity, and oxidative stress, helping potatoes maintain their physiological functions and overall health during challenging environmental conditions. Proper boron management is critical for optimizing potato growth and stress tolerance.

(h) *Zinc:* Zinc (Zn) is a vital component of many plant enzymes, including carbonic anhydrase, alkaline phosphatase, alcohol dehydrogenase (ADH), RNA polymerase, Cu-Zn superoxide dismutase (SOD), and phospholipase. It plays a key role in DNA and RNA synthesis. Insufficient Zn supply hampers plant growth, while adequate Zn reduces the negative effects of heat stress and promotes ion transport. Zinc is essential for maintaining the structural integrity of proteins, membrane lipids, cell components, and DNA. Foliar application of zinc oxide (ZnO) nanoparticles has been shown to mitigate chilling stress in crops. Zinc priming enhances plumule development under drought stress by boosting the production of plant hormones like indole-3-acetic acid (IAA) and gibberellic acid (GA_3). In drought conditions, Zn helps regulate membrane permeability, antioxidant activity, photosynthesis, and water-use efficiency. Additionally, Zn increases salt stress resistance by stabilizing root cell membranes and reducing ion leakage.

Zn is directly or indirectly involved in many metabolic activities and governs potato growth and development. It plays an important role in several activities, including:

1. *Promotes Enzyme Activity:* Zn is a cofactor for many enzymes involved in essential metabolic processes, including those related to carbohydrate metabolism, protein synthesis, and nucleic acid metabolism. Adequate zinc levels ensure optimal enzyme function, contributing to better growth and development.
2. *Supports Tuber Development:* Zn influences the formation and development of potato tubers. It aids in the synthesis of starch and other carbohydrates that are critical for tuber growth. Adequate Zn supply can lead to increased tuber size and yield.
3. *Enhances Nutrient Uptake:* Zn improves the efficiency of nutrient uptake and utilization by enhancing root function. This ensures that potatoes receive the necessary nutrients for growth, leading to better plant health and increased yields.
4. *Strengthens Cell Walls:* Zn contributes to the formation of cell walls by aiding in the synthesis of structural proteins and carbohydrates. Stronger cell walls enhance the plant's resistance to physical damage and pathogen attack, leading to improved overall plant health.
5. *Increases Photosynthesis:* Zn is involved in chlorophyll formation and photosynthesis. Improved photosynthesis leads to higher energy production and better growth. Enhanced chlorophyll content also supports the plant's ability to convert light energy into chemical energy efficiently.

2. *Roles of Zn in Abiotic Stress Management*

   1. *Enhances Antioxidant Defense:* Zn is essential for the activity of several antioxidant enzymes, such as superoxide dismutase (SOD) and catalase. These enzymes help neutralize reactive oxygen species (ROS) generated during abiotic stress conditions like drought and high temperatures. By reducing oxidative damage, Zn helps maintain cellular integrity and function under stress.
   2. *Improves Water Use Efficiency:* Zn influences the development and function of plant root systems. A well-developed root system enhances water uptake, helping plants better tolerate drought conditions. Zn can also improve stomatal regulation, reducing water loss through transpiration.
   3. *Regulates Stress-Responsive Gene Expression:* Zn is involved in the regulation of stress-responsive genes and proteins, including those related to stress tolerance mechanisms. This regulatory role helps plants adapt to and survive adverse conditions.

4. *Modulates Hormone Levels:* Zn affects the synthesis and action of plant hormones such as abscisic acid (ABA), which plays a key role in the plant's response to drought and other stress factors. Proper hormone balance helps the plant manage stress more effectively.

   Method of Application of Zn in Potato Crop

   To effectively use Zn in potato cultivation, it can be applied in the following ways:

(a) *Soil Application:* Zn can be applied to the soil in the form of zinc sufate or zinc oxide. Soil application ensures that Zn is available to plant roots throughout the growing season.
(b) *Foliar Spraying:* Zn can also be applied as a foliar spray, providing a quick boost to plants, especially when Zn deficiency is identified during the growing season.
(c) *Fertilizer Blends:* Zn can be included in fertilizer blends to ensure a balanced nutrient supply, particularly in soils that are known to be deficient in Zn.

Incorporating Zn into potato cultivation practices can significantly mitigate abiotic stress and enhance growth and production. By supporting antioxidant defenses, improving water use efficiency, and promoting key physiological processes, Zn contributes to healthier plants and increased yields.

Combined Effects of Plant Nutrients

All plant nutrients support growth and stress reduction, but achieving optimal benefits requires careful attention to nutrient composition. To enhance nutrient utilization, it is crucial to minimize antagonistic (negative) interactions and promote synergistic (positive) ones. Understanding both the positive and negative interactions between nutrients is essential, as some combinations can effectively combat stress, while others may have adverse effects.

3. *Micronutrient Management*

   Iron (Fe): Waterlogged soils can lead to iron deficiency due to poor root function or excess uptake, resulting in toxicity. Foliar application of chelated iron can help mitigate deficiency without increasing soil iron toxicity.

   Manganese (Mn): Waterlogging can cause manganese toxicity or deficiency depending on soil conditions. Monitoring and applying manganese as needed, particularly in chelated form, can help manage these issues.

   Zinc (Zn): Zinc is crucial for various physiological processes and can become less available under waterlogged conditions. Foliar applications can help maintain adequate zinc levels.

4. *Use of Plant Growth Regulators*

   Ethylene Inhibitors: Waterlogged conditions often lead to the accumulation of ethylene, which can cause root damage and inhibit growth. Applying ethylene inhibitors or using plant growth regulators like gibberellins (GA) that counteract ethylene effects can help improve plant tolerance to waterlogging.

Cytokinins: These can promote cell division and growth under stress conditions. Foliar application of cytokinins may help mitigate the negative effects of waterlogging.

5. *Enhancing Root Growth*

   Mycorrhizal Fungi: Inoculation with mycorrhizal fungi can enhance root growth and function, improve nutrient uptake, and increase the plant's tolerance to waterlogged conditions.

   Rooting Hormones: Application of rooting hormones like auxins can stimulate root growth, helping the plant recover from root damage caused by waterlogging.

6. *Improving Drainage and Irrigation Practices*

   Raised Beds: Planting potatoes on raised beds can improve drainage and reduce the risk of waterlogging, thereby enhancing nutrient uptake efficiency.

   Controlled Irrigation: Avoid over-irrigation and ensure proper scheduling based on soil moisture levels to prevent waterlogging.

7. *Use of Organic Amendments*

   Biochar: Incorporating biochar into the soil can improve soil aeration, increase nutrient retention, and reduce the negative impact of waterlogging on potato plants.

   Compost: Regular application of compost can improve soil structure, enhance microbial activity, and increase the soil's resilience to waterlogging.

8. *Foliar Nutrition*

   Foliar Sprays: Under waterlogged conditions, root uptake of nutrients can be severely compromised. Foliar application of essential nutrients like nitrogen, potassium, calcium, and micronutrients can provide an immediate source of nutrition, helping the plant cope with stress.

9. *Monitoring and Adaptation*

   Soil and Plant Tissue Testing: Regular monitoring of soil and plant tissue nutrient levels can help identify deficiencies or toxicities early, allowing for timely adjustments in nutrient management practices.

   Adjusting Fertilization Timing: Apply fertilizers when the soil is not waterlogged to reduce nutrient losses and ensure better uptake by the plants.

10. *Breeding and Use of Resistant Varieties*

    Waterlogging-Resistant Varieties: Planting potato varieties that are more tolerant to waterlogging stress can be an effective strategy. These varieties are bred to have better root systems and physiological adaptations to withstand waterlogged conditions.

To mitigate the effects of waterlogging stress on potato crops and improve plant health and productivity, several nutrient management strategies can be employed. Potassium supplements can enhance nutrient absorption in waterlogged plants, resulting in a significant increase in potassium, calcium, nitrogen, manganese, and iron accumulation. The application of biofertilizers has proven effective in alleviating the adverse effects of flooding in other crops. Specifically, seed or foliar sprays of biofertilizers such as AAP (*Azotobacter chroococcum* and *Pseudomonas* spp.) and APB (*Azospirillum* spp., *Pseudomonas fluorescens*, and *Bacillus subtilis*) have been beneficial. By incorporating these strategies, we can effectively address the challenges posed by waterlogging, leading to improved crop health and yield.

4.1.2 Nutrient Management Strategies to Mitigate Salt Stress in Potatoes

Soil salinization is one of the most common types of land degradation that occurs worldwide. In arid and semi-arid regions, salinity is a major problem because evapotranspiration exceeds rainfall, meaning there is not enough rainwater to remove soluble salts from the root zone (Hailu and Mehari 2021). Salinity is defined as the presence of excessive amounts of soluble salts that impede or affect normal plant growth functions. Various factors such as electrical conductivity (EC), exchangeable sodium percentage (ESP), sodium adsorption ratio (SAR), and pH are used to measure salinity. Soils with an EC of more than 4 dS/m, ESP of less than 15%, and pH of less than 8.5 are classified as saline soils (Bello et al. 2021). Saline soils contain a mixture of chloride, sulfate, sodium, magnesium, and calcium ions, with sodium chloride typically being the major salt. Numerous metabolic activities, including nutrient uptake, photosynthesis, and the synthesis of proteins and nucleic acids, are negatively impacted by salt stress. One negative consequence is the poor osmotic potential of the soil solution. Ultimately, plant nutrient intake is reduced by imbalanced nutrition and increased nutrient concentration in the rhizosphere.

Mitigating salt stress in potato cultivation requires a combination of nutrient management strategies to enhance the plant's tolerance and minimize the adverse effects of salinity. Some practices that may reduce the effects of salt stress in potato farming include:

1. *Balanced Fertilization*
   - *Nitrogen (N):* Adequate nitrogen is essential for plant growth and development. However, excessive nitrogen can exacerbate salt stress, so it's important to apply it in balanced amounts.
   - *Phosphorus (P):* Phosphorus helps in root development and energy transfer, improving the plant's ability to cope with salt stress. Applying phosphorus fertilizers can aid in mitigating salt-induced growth reduction.
   - *Potassium (K):* Potassium is crucial for osmotic regulation and maintaining cell turgor under salt stress. It plays a role in enzyme activation and photosynthesis. Supplementing potassium can improve salt tolerance.
   - *Calcium (Ca):* Calcium strengthens cell walls and membranes, reducing sodium uptake by roots and improving salt tolerance. Calcium amendments, such as gypsum, can help displace sodium in the soil and improve soil structure.
   - *Magnesium (Mg):* Magnesium is essential for chlorophyll production and photosynthesis. Adequate magnesium can mitigate the negative effects of salt stress on chlorophyll content and photosynthetic efficiency.

- *Sulfur (S):* Sulfur aids in the synthesis of proteins and enzymes involved in stress responses. Ensuring sufficient sulfur levels can enhance the plant's ability to cope with salt stress.
- *Use of Organic Amendments*

– *Organic Matter:* Incorporating organic matter such as compost, manure, or green manure can improve soil structure, water retention, and nutrient availability, helping to reduce salt stress.
– *Biochar:* Adding biochar to the soil can improve soil pH, increase cation exchange capacity (CEC), and reduce the availability of sodium ions, thereby alleviating salt stress.

2. *Micronutrient Management*

- *Zinc (Zn):* Zinc is involved in various enzymatic functions and can improve the antioxidant defense system in plants under salt stress. Zinc fertilization can help mitigate oxidative damage caused by salinity.
- *Iron (Fe):* Iron is crucial for chlorophyll synthesis and can alleviate the chlorosis often associated with salt stress. Foliar or soil application of iron can be beneficial.
- *Manganese (Mn) and Copper (Cu):* Both are important for enzyme activation and the detoxification of reactive oxygen species (ROS) under salt stress.

3. *Use of Plant Growth Regulators*

- *Gibberellins (GA):* Gibberellins can promote plant growth under salt stress by enhancing cell elongation and division.
- *Salicylic Acid (SA):* Salicylic acid enhances plant tolerance to salt stress by regulating antioxidant enzyme activities and osmotic balance.
- *Abscisic Acid (ABA):* Abscisic acid plays a critical role in stomatal regulation and osmotic adjustment under stress conditions. Application of ABA can improve salt tolerance.

4. *Irrigation Management*

- *Leaching:* Adequate irrigation with non-saline water can help leach excess salts from the root zone. Careful management of water quantity and timing is essential to prevent waterlogging and further salt accumulation.
- *Use of Mulches:* Mulching helps conserve soil moisture, reduce evaporation, and maintain a favorable microenvironment around the roots, reducing the impact of salt stress.

5. *Soil Amendments*

- *Gypsum (calcium sulfate):* Its application can help displace sodium ions from soil particles, reducing sodium toxicity and improving soil structure.
- *Sulfur Compounds:* Elemental sulfur or sulfuric acid can be used to acidify alkaline soils, making sodium less available to plants and improving overall soil conditions.

6. *Genetic Approaches*
   - *Salt-Tolerant Varieties:* Planting salt-tolerant potato varieties is a fundamental strategy. These varieties are bred to withstand higher salt concentrations in the soil.
7. *Microbial Inoculants*
   - *Mycorrhizal Fungi:* Mycorrhizae can enhance nutrient uptake and improve plant tolerance to salt stress by altering the plant's hormonal balance and improving soil structure.
   - *Plant Growth-Promoting Rhizobacteria (PGPR):* PGPR can enhance plant growth under salt stress by producing phytohormones, solubilizing nutrients, and enhancing antioxidant defense mechanisms.
8. *Foliar Nutrition:* Foliar application of nutrients such as potassium, calcium, and micronutrients can provide immediate relief to plants under salt stress, as these nutrients can be directly absorbed by the leaves.
9. *Monitoring and Adjustment:* Regular soil and tissue testing can help monitor nutrient levels and adjust fertilization strategies accordingly to optimize nutrient availability and reduce salt stress.

Implementing these nutrient management strategies can significantly improve potato resilience to salt stress, leading to better growth, yield, and overall plant health.

Soil Acidity

Soil acidity is a major threat to land degradation. An excess of H^+ and Al^{3+} ions in the soil solution results in soil acidity. Acidic soil decreases the availability of essential minerals and exacerbates the negative impacts of pollutants. It reduces plant production; disrupts important biological processes in the soil, such as nitrogen fixation; and makes the soil more prone to erosion and structural degradation.

Nutrient Management Strategies to Mitigate Soil Acidity

Soil acidity greatly reduces the availability of nutrients to plants and interferes with vital microbial activities that break down organic materials and fix nitrogen. In highly acidic soils, the concentration of Fe and Al^{3+} in the soil solution significantly affects plant roots and growth.

The primary cause of the development of very acidic conditions in soils is the overuse of acidic fertilizers over time without the addition of lime (Rajasekharan et al. 2014). In some regions where soils are very acidic (acid sulfate soils), the use of dolomite, lime, or rice husk ash has proven effective in lowering acidity and increasing rice output; it is also effective for various other crops. Customized fertilization using lime, nitrogen, phosphorus, potassium, and magnesium, as well as foliar N-P-K treatments, have shown promise in several areas for a variety of crops. Additionally, the application of biochar raises the soil pH and decreases exchangeable H^+ and Al^{3+}, thereby reducing acidity. Applying arbuscular mycorrhizal fungi can significantly reduce Fe and Al concentrations, successfully offsetting their negative effects on growth and photosynthesis. Furthermore, adding rice husk biochar, which is rich in organic carbon, improves soil pH, lowers Al and Fe levels in acidic soil, and enhances both the biochemical characteristics of the soil and plant development.

Toxicity Caused by Various Metals

When plants are exposed to excessive amounts of essential and non-essential metals, they can experience various common toxic effects, ultimately resulting in plant death. These effects can include reduced biomass accumulation, chlorosis, inhibited growth and photosynthesis, disrupted water balance and nutrient assimilation, and senescence.

Toxicity Caused by Iron and Its Mitigation

Soils with limited cation exchange capacity and acid sulfate conditions are commonly susceptible to iron poisoning, a prevalent nutritional disease affecting all crops. These soils are often moderately to highly acidic and contain readily reducible active Fe, which severely impacts crop yields and soil fertility.

Increased polyphenol oxidase activity, resulting from Fe toxicity, leads to the production of oxidized polyphenols. This condition causes chlorosis, bronzing of the leaves, reduced root oxidation capacity, elongated roots, stunted growth, severely limited tillering, and iron plaque formation on the roots of rice and other crops. Iron toxicity in potato crops can manifest in several ways:

1. *Stunted Growth:* Plants may appear small and unhealthy.
2. *Bronzing of Leaves:* Leaves may turn brownish-red.
3. *Interveinal Chlorosis:* Young leaves may exhibit chlorosis between the veins.
4. *Buff Spots:* Buff-colored spots may appear on the leaves, eventually spreading to the entire interveinal area.
5. *Papery Leaves:* Leaves may become papery, dry, and brittle.

Iron toxicity can also lead to reduced root development, decreased water and nutrient uptake, and diminished photosynthetic activity.

To mitigate iron toxicity in potatoes, various approaches can be employed:

1. *Soil Testing and Management:* Regular testing of soil pH and iron levels is crucial. Based on these tests, lime can be applied to increase pH, which helps reduce Fe availability. Farmers should avoid blanket applications of fertilizers, as this practice can lead to yield losses in potatoes. Managing Fe-rich fertilizers can help control Fe toxicity to some extent.
2. *Improve Soil Drainage:* Waterlogged soils are susceptible to Fe toxicity. Enhancing drainage systems or preventing waterlogging can mitigate this issue. Raised beds or ridging can also help improve drainage and reduce anaerobic conditions.
3. *Use of Resistant Varieties:* Planting potato varieties that are more tolerant of high Fe levels or poor drainage conditions can be beneficial.
4. *Balanced Fertilization:* Ensure a balanced supply of nutrients, particularly phosphorus (P), potassium (K), and calcium (Ca), to prevent imbalances that may lead to increased Fe uptake. Additionally, the application of organic matter can

help buffer the effects of high Fe by improving soil structure and nutrient availability.

5. *Irrigation Management:* Avoid excessive irrigation that can lead to waterlogged conditions. Utilizing drip or sprinkler irrigation helps maintain optimal soil moisture levels without oversaturation, mitigating the effects of Fe toxicity.
6. *Chelating Agents:* Applying chelating agents that bind excess Fe can reduce its availability to plants. However, these should be used cautiously and under the guidance of soil or crop specialists.
7. *Foliar Applications:* Foliar sprays of micronutrients like manganese (Mn) or zinc (Zn) can help counteract the effects of Fe toxicity by promoting better nutrient balance within the plant.
8. *Crop Rotation and Cover Crops:* Implement crop rotation with species less sensitive to Fe toxicity and use cover crops that improve soil structure and organic matter content.

Other techniques, such as the addition of silicon, can enhance the activity of antioxidant enzymes (like catalase and ascorbate peroxidase) under mild Fe toxicity, decreasing lipid peroxidation and improving hydrogen peroxide detoxification. By lowering plant Fe levels and increasing antioxidant enzyme activity, the addition of silicon may mitigate the negative effects of iron toxicity.

Toxicity Caused by Manganese and Its Mitigation

Manganese toxicity often coexists with aluminum toxicity in acidic highland soils (pH less than 5.5). Potato crops sensitive to Mn can experience severe production losses and quality issues. While Mn is an essential element for plants, high concentrations can be harmful, particularly in acidic soils or specific environmental conditions.

Several factors contribute to Mn toxicity in potatoes:

- *Acidic Soil Conditions:* Lower pH increases Mn solubility and bioavailability, raising the risk of toxicity.
- *Poor Drainage and Waterlogging:* These conditions reduce oxygen availability in the soil, leading to increased Mn availability.
- *High Mn Content:* Soils with high natural Mn levels or those receiving Mn-rich fertilizers or amendments can lead to toxicity.
- *Soil Compaction:* Compacted soils limit root growth, increasing Mn accumulation in the root zone.

Symptoms of Mn toxicity in potatoes include:

- *Leaf Chlorosis:* Yellowing of leaves, often starting at the margins and tips.
- *Necrotic Spots:* Dark brown or black necrotic spots on older leaves, particularly along the veins.
- *Interveinal Chlorosis:* Yellowing between the veins, while the veins remain green.
- *Reduced Tuber Development:* Smaller tubers and poor overall quality.

To mitigate Mn toxicity in potatoes, consider the following strategies:

1. *Soil pH Management:* Apply lime to raise soil pH and reduce Mn solubility and toxicity. Regular soil testing is crucial.
2. *Improving Soil Drainage:* Ensure good drainage to prevent waterlogging. Use raised beds or ridging to promote better drainage.
3. *Balanced Fertilization:* Avoid over-application of Mn-rich fertilizers and ensure a balanced supply of essential nutrients, particularly calcium (Ca), which can compete with Mn uptake.
4. *Organic Matter Addition:* Incorporate organic matter to improve soil structure, increase microbial activity, and enhance nutrient balance, helping buffer high Mn levels.
5. *Use of Resistant Varieties:* Select potato varieties known to tolerate high Mn levels or acidic conditions.
6. *Crop Rotation:* Rotate potatoes with less sensitive crops to manage soil Mn levels over time.
7. *Chelation and Amendments:* Apply chelating agents to bind Mn and reduce its bioavailability, and use gypsum to add Ca without significantly affecting soil pH.
8. *Irrigation Management:* Avoid excessive irrigation that can lead to waterlogged conditions. Maintain optimal soil moisture without creating anaerobic conditions.

Manganese poisoning can reduce the quantities of chlorophyll, carotenoids, and net photosynthesis. Silicon has been shown to effectively increase Mn toxicity tolerance in rice by raising chlorophyll concentration and maintaining the structure of photosystems.

Toxicity Caused by Aluminum and Its Mitigation

Aluminum toxicity can impair a plant's ability to absorb, transport, and utilize essential nutrients such as phosphorus, potassium, calcium, magnesium, iron, molybdenum, and boron. This results in reduced nutrient and water absorption efficiency and possible suppression of microbial processes linked to limited root development.

Symptoms of Al toxicity in potatoes include:

1. *Root Inhibition:* Al toxicity primarily harms the root system, inhibiting root elongation and affecting the plant's ability to absorb water and nutrients.
2. *Nutrient Deficiency:* Excess Al in the soil can interfere with the uptake of essential nutrients, leading to symptoms like chlorosis and reduced tuber size.
3. *Reduced Photosynthesis:* Al toxicity can damage chloroplasts, reducing chlorophyll content and compromising energy production.
4. *Oxidative Stress:* High levels of Al generate reactive oxygen species, damaging cellular components and hindering growth.
5. *Increased Susceptibility to Diseases:* Weakened plants are more vulnerable to pathogens and pests, increasing the incidence of diseases like late blight.

To mitigate Al toxicity in potatoes, consider the following strategies:

1. *Soil pH Management:* Lime the soil to raise pH and reduce Al availability and toxicity. Regular soil testing is essential.
2. *Organic Matter Addition:* Incorporate organic matter to bind Al and improve soil structure, enhancing nutrient availability.
3. *Gypsum Application:* Apply gypsum to displace Al ions from the root zone and neutralize Al′s effects, promoting healthier root development.
4. *Use of Al-Tolerant Varieties:* Plant potato varieties that are naturally more tolerant to Al toxicity.
5. *Balanced Fertilization:* Apply balanced fertilizers, particularly those containing phosphorus, which forms insoluble complexes with Al, mitigating its toxic effects.
6. *Deep Tillage:* Break up compacted soil layers to improve root penetration and reduce exposure to toxic Al levels.
7. *Crop Rotation:* Rotate potatoes with less sensitive crops to manage soil conditions and reduce Al build-up.

Silicon can also reduce the effects of Al toxicity, improving plant development and reducing leaf chlorosis. Additionally, biochar can effectively reduce Al toxicity due to its properties, including electrical conductivity and cation exchange capacity. By substituting Al ions at soil exchange sites, biochar can promote the formation of more neutral Al hydroxides, enhancing soil pH.

Implementing these mitigation strategies can help potato producers minimize the detrimental effects of aluminum toxicity and improve crop yield and quality.

4.2 Strategies for Mitigating Abiotic Stresses Through Agronomic Practices

(a) *Mulch*

Mulch is an effective agricultural practice with numerous benefits, particularly in potato cultivation. Here's how mulch helps regulate soil temperature, reduce evapotranspiration, and maintain potato yield by alleviating the effects of heat stress:

1. *Regulating Soil Temperature*

- *Insulation:* Mulch acts as a natural insulator, moderating soil temperature by protecting it from extreme fluctuations. In hot weather, mulch prevents excessive soil warming by reflecting sunlight. During cooler periods, it retains soil warmth by trapping heat.
- *Consistent Conditions:* This stabilization creates a more consistent environment for potato roots, essential for their growth. Potatoes thrive in soil that is neither too hot nor too cold.

2. *Reducing Evapotranspiration*

- *Moisture Retention:* Mulch reduces water loss through evaporation by covering the soil surface, limiting direct exposure to sun and wind. This is especially beneficial in drought-prone areas.
- *Efficient Water Use:* By reducing evapotranspiration, mulch ensures more water remains available for the potato plants, leading to better hydration and reduced irrigation needs.

3. *Enhanced Soil Health:*
   Mulch prevents soil erosion, reduces weed competition, and gradually adds organic matter to the soil, creating a nutrient-rich environment for potatoes and promoting robust growth.
4. *Improved Plant Growth:*
   With regulated soil temperature and moisture levels, potatoes experience less stress and can focus on tuber development, resulting in higher yields and better-quality potatoes.
5. *Weed Suppression:*
   By minimizing weed growth, mulch reduces competition for nutrients and water, allowing potato plants to thrive.

(b) *Intercropping*
   Intercropping, the practice of growing two or more crops together, offers several benefits in potato cultivation, particularly in cooling soil temperature, conserving soil moisture, reducing irradiance, and maintaining potato tuber yield. It helps mitigate the negative impacts of heat stress on potato tubers:

1. *Cooling Soil Temperature*

- *Shade Creation:* Intercropping with taller or broader-leafed plants can provide shade to the soil surface, reducing direct sunlight exposure. This shading effect lowers soil temperatures, creating a cooler root zone, which is vital for optimal growth
- *Evapotranspiration Reduction*
- *Canopy Effect:* The canopy formed by intercrop plants helps reduce soil temperature by minimizing exposure to intense sunlight, which in turn decreases overall heat absorption by the soil.
- *Conserving Soil Moisture*

1. *Reduced Evaporation:* The shading effect from intercrops significantly lowers the rate of water evaporation from the soil surface. With less direct sunlight and wind reaching the soil, moisture loss is minimized, leading to better water conservation.
2. *Improved Soil Structure:* Some intercrops, particularly legumes, enhance soil structure and organic matter content, which improves the soil's ability to retain moisture. This consistency in soil moisture levels is crucial for potato tuber development.

- *Reducing Irradiance*

1. *Light Interception:* Intercropping systems effectively intercept sunlight before it reaches the soil surface, reducing the overall irradiance that affects potato plants and the soil. This protects them from excessive light exposure, which can lead to heat stress and increased water loss.
2. *Balanced Light Distribution:* Competition for light between intercrop and potato plants can lead to a more efficient use of available light, ensuring that both crops receive adequate light without the harmful effects of excessive irradiance.

Intercropping often leads to better utilization of available resources such as light, water, and nutrients. By optimizing these resources, potato plants can focus more energy on tuber formation, resulting in sustained or improved yields. Additionally, certain intercrops can act as natural pest repellents or reduce the spread of diseases, contributing to healthier potato plants and reducing the need for chemical inputs. Healthier plants are more likely to produce higher yields. Some intercrops, like legumes, fix nitrogen in the soil, improving soil fertility, which benefits potatoes through increased nutrient availability and enhanced growth and tuber production.

Hormone Treatments

Hormone treatments, such as paclobutrazol (PBZ), play a significant role in enhancing various physiological and biochemical processes in plants, including potatoes. PBZ increases chlorophyll content and boosts net photosynthesis, enhancing carbon fixation and resulting in higher yields while helping combat heat stress (Tekalign and Hammes 2005).

Increasing Chlorophyll a and b Content

1. *Inhibition of Gibberellin Biosynthesis:* Paclobutrazol is a plant growth regulator that inhibits gibberellin biosynthesis, a hormone responsible for stem elongation and leaf expansion. By suppressing gibberellin, PBZ promotes shorter, thicker, and more compact plants, which leads to an increase in chlorophyll concentration per unit leaf area.
2. *Enhanced Chlorophyll Synthesis:* PBZ treatment has been shown to upregulate the enzymes involved in chlorophyll biosynthesis, leading to higher concentrations of chlorophyll a and b. This increase enhances the plant's ability to capture light energy for photosynthesis, improving overall plant vigor and growth.

Increasing Net Photosynthesis

1. *Improved Light Absorption:* With increased chlorophyll content, potato plants have a greater capacity to absorb light, particularly in the blue and red wavelengths, which are crucial for photosynthesis. This leads to a more efficient conversion of light energy into chemical energy during photosynthesis.
2. *Stomatal Conductance:* PBZ treatment can improve stomatal regulation, balancing the opening and closing of stomata. This optimizes the intake of carbon dioxide (CO_2) for photosynthesis while minimizing water loss through transpiration, leading to improved photosynthetic efficiency.
3. *Enhanced Carbon Fixation:* The increased chlorophyll content, coupled with better stomatal control, results in higher rates of carbon fixation in the chloro-

plasts. This boosts net photosynthesis, providing more energy for plant growth and development, including tuber formation.

Maintaining Potato Tuber Yields

1. *Stronger and More Resilient Plants:* The application of PBZ results in more compact and robust plants with thicker leaves, leading to a more efficient distribution of resources. These structural changes enhance the plant's ability to support tuber growth, even under stress conditions like drought or nutrient deficiency.
2. *Optimized Source-Sink Relationship:* PBZ helps create a favorable source-sink relationship by enhancing photosynthesis (source) and directing more assimilates (photosynthates) toward the tubers (sink). This ensures that the energy produced is effectively utilized for tuber development, leading to stable or increased yields.
3. *Stress Tolerance:* PBZ-treated plants often show improved tolerance to environmental stresses, such as drought or salinity, which can negatively impact potato yields. By enhancing the plant's resilience, PBZ helps maintain consistent tuber production even under suboptimal conditions.

Irrigation Scheduling

Irrigation scheduling is crucial in potato cultivation, particularly for maintaining water use efficiency and combating drought stress. It helps maintain optimal moisture conditions in the potato root zone, even during drought, thus preserving yields and production (Kang et al. 2004; Kashyap and Panda 2003).

Maintaining Water Use Efficiency (WUE)

1. *Optimal Water Application:* Irrigation scheduling involves applying water at the right time and in appropriate amounts based on the crop's growth stage and environmental conditions. This precision ensures that water is used efficiently, maximizing the amount that contributes directly to potato growth.
2. *Minimizing Water Loss:* Through precision irrigation, farmers can reduce moisture losses due to evaporation, runoff, and deep percolation. This ensures a greater proportion of the applied water is available to potato roots, enhancing WUE.
3. *Preventing Water Deficit:* Potatoes are highly susceptible to water stress, which severely impacts tuber formation and yield. Proper irrigation scheduling ensures water is provided at critical growth stages, such as tuber initiation and bulking, to prevent water deficits that could harm the crop.
4. *Maintaining Soil Moisture:* Regular and well-timed irrigation helps maintain consistent soil moisture levels, vital for preventing drought stress. Keeping the soil adequately moist allows potato plants to continue physiological processes, such as photosynthesis and nutrient uptake, even during dry spells.
5. *Improving Plant Resilience:* Consistent soil moisture from well-scheduled irrigation reduces water stress, allowing plants to develop a stronger root system and more robust foliage. This increased resilience helps them better withstand drought periods, minimizing negative impacts on growth and yield.
6. *Efficient Use of Limited Water Resources:* In drought-prone areas, water resources are often limited. Irrigation scheduling allows farmers to optimize lim-

ited water supplies, prioritizing irrigation during critical periods to maintain crop health and productivity.

Drip Irrigation

Drip irrigation is an effective method for maintaining soil moisture, combating drought stress, and increasing potato yields.

Maintaining Soil Moisture

1. *Targeted Water Delivery:* Drip irrigation delivers water directly to the root zone of potato plants through a network of tubes and emitters. This precise delivery system ensures that water is applied exactly where needed, maintaining consistent soil moisture levels around the roots.
2. *Reduced Evaporation:* By applying water slowly and directly to the soil, drip irrigation minimizes evaporation losses compared to traditional methods like sprinklers or flood irrigation. This helps retain more water in the soil for plant use (Onder et al. 2005).
3. *Uniform Moisture Distribution:* Drip irrigation systems can be designed to provide uniform water distribution across the field, ensuring all potato plants receive an adequate and consistent supply of moisture. This uniformity is crucial for maintaining healthy growth and avoiding water stress.

Fighting Drought Stress

1. *Efficient Water Use:* During drought conditions, water is often scarce, making efficient use of available resources essential. Drip irrigation is highly efficient, using less water while providing sufficient moisture to the plants, conserving water, and extending its availability during dry periods (Kumari 2012).
2. *Reduced Water Stress:* By maintaining consistent soil moisture levels, drip irrigation reduces the likelihood of water stress in potato plants. This stability is crucial during drought, allowing the plants to continue physiological processes without interruption.
3. *Enhanced Root Development:* Consistent and adequate moisture from drip irrigation promotes deeper root growth in potato plants. A well-developed root system allows plants to access moisture from deeper soil layers, providing additional resilience against drought.
4. *Improved Plant Health:* By preventing water stress, drip irrigation helps maintain overall plant health, reducing the risk of diseases exacerbated by drought conditions. Healthier plants are better equipped to cope with environmental stressors and continue producing tubers.

Increasing Potato Yields

1. *Optimized Growing Conditions:* By ensuring that potato plants receive a consistent and adequate supply of water, drip irrigation creates an optimal environment for growth. This leads to better tuber formation, size, and quality, ultimately increasing overall yields.
2. *Reduced Competition with Weeds:* Drip irrigation delivers water directly to the crop's root zone, minimizing the amount available to weeds. This reduction in

weed growth decreases competition for nutrients and water, allowing potato plants to thrive and produce higher yields.

3. *Better Nutrient Uptake:* Drip irrigation systems can be integrated with fertigation, where nutrients are delivered along with the irrigation water. This ensures that nutrients are available where needed, enhancing uptake efficiency and contributing to higher potato yields.
4. *Minimized Waterlogging:* Unlike flood irrigation, which can lead to waterlogging and subsequent root rot, drip irrigation provides controlled water delivery, avoiding excess moisture in the soil. This balance prevents conditions that could negatively impact tuber development and yield.

Plant Growth-Promoting Rhizobacteria (PGPR)

Plant growth-promoting rhizobacteria (PGPR) are beneficial soil bacteria that colonize plant roots and enhance growth through various mechanisms. One key way PGPR helps potato crops combat drought stress is by enhancing the activity of reactive oxygen species (ROS) scavenging enzymes (Gururani et al. 2013).

Understanding ROS and Drought Stress

1. *ROS in Plants:* Reactive oxygen species (ROS) are highly reactive molecules, including hydrogen peroxide (H_2O_2), superoxide anion (O_2^-), and hydroxyl radical (OH^-), produced as byproducts of cellular metabolism. Under normal conditions, ROS play a role in cell signaling. However, during drought stress, ROS levels can rise dramatically, leading to oxidative stress and damage to proteins, lipids, and DNA in plant cells.
2. *Impact of Drought Stress:* Drought exacerbates ROS production as the plant's metabolic processes become imbalanced due to limited water availability. The excessive accumulation of ROS can cause significant cellular damage, leading to reduced growth, impaired photosynthesis, and ultimately lower crop yields.

Role of ROS Scavenging Enzymes

1. *ROS Scavenging Mechanism:* To counteract the harmful effects of ROS, plants produce ROS scavenging enzymes such as superoxide dismutase (SOD), catalase (CAT), and peroxidases (POD). These enzymes neutralize ROS by converting them into less harmful molecules, protecting cells from oxidative damage.
2. *Importance in Drought Tolerance:* During drought, the efficiency of these enzymes is critical for maintaining cellular homeostasis and protecting tissues. Enhanced enzyme activity helps plants withstand drought by reducing oxidative damage and supporting continued growth.

How PGPR Enhances ROS Scavenging Enzymes

1. *Induction of Enzyme Activity:* Certain strains of PGPR have been shown to induce the expression and activity of ROS scavenging enzymes in plants. By colonizing the root zone, these bacteria produce signaling molecules or trigger systemic resistance pathways that upregulate SOD, CAT, and POD in potato plants.
2. *Modulation of Stress Responses:* PGPR can influence the plant's hormonal balance, particularly levels of stress-related hormones like abscisic acid and sali-

cylic acid. These hormonal changes can lead to an enhanced antioxidant defense system, including the upregulation of ROS scavenging enzymes.
3. *Reduction of Oxidative Stress:* By enhancing ROS scavenging enzyme activity, PGPR helps reduce ROS accumulation during drought. This reduction in oxidative stress allows potato plants to maintain healthier cellular processes, minimizing damage to vital structures like chloroplasts, essential for photosynthesis.

Helping Potato Crops Fight Drought Stress

1. *Improved Water Use Efficiency:* PGPR-induced enhancement of ROS scavenging enzymes supports better cellular function, even under water-limited conditions. This improved cellular health allows plants to use available water more efficiently, maintaining growth despite drought stress.
2. *Enhanced Root Growth:* PGPR promotes better root growth and architecture, enabling potato plants to access moisture from deeper soil layers. Healthier roots, supported by reduced oxidative damage, can sustain plant functions better during drought, contributing to overall drought tolerance.
3. *Sustained Photosynthesis and Growth:* By mitigating oxidative stress, PGPR helps maintain the functionality of the photosynthetic machinery. This ensures that plants continue to produce the energy needed for growth, even under drought conditions, leading to more consistent and higher yields.

Physiological Disorders in Potato Tubers

Potato tubers can develop various physiological disorders due to environmental conditions, nutrient imbalances, and other non-pathogenic factors. These disorders can affect the quality, appearance, and marketability of the tubers. Below are some common physiological disorders of potato tubers, along with their symptoms, causes, and management practices (Table 1).

Physiological disorders in potato tubers can significantly impact the quality, appearance, and marketability of the crop. Managing these disorders involves understanding their causes, such as environmental stress and nutrient imbalances, and implementing practices to mitigate these factors, including proper irrigation, temperature control, and careful handling during harvest and storage.

5 Conclusion

Biotic and abiotic stress conditions, along with climate change, affect potato crop performance and yield, posing a serious threat to long-term potato production. Abiotic stresses, including drought, temperature extremes, salinity, and nutrient imbalances, present significant challenges to potato production worldwide. Effective management measures are required to reduce the negative impacts of these stresses and ensure long-term crop output.

Several methods can help reduce abiotic stress in potato production, including mulching, drip irrigation, plant growth-promoting rhizobacteria (PGPR), hormone therapies, and nutritional management. Proper nutrient management is essential for

Table 1 Physiological disorder of potato

Name of physiological disorder	Plant part affected	Symptom	Cause	Management
1. Black heart	Tubers-internal	*Dark, black or brown discoloration*: The affected areas of the tuber's internal flesh turn black or dark brown, often starting at the center and expanding outward. *No external signs*: In the early stages, the exterior of the potato appears healthy, and the discoloration is only visible when the tuber is cut open. *Soft or hollow center*: As the disease progresses, the blackened tissue may become soft, hollow, or shriveled. *Foul odor*: In advanced stages, the potato may develop a rotten smell due to tissue breakdown	*Poor ventilation*: Inadequate air circulation during storage or transportation reduces oxygen levels around the potatoes. *High storage temperatures*: Storing potatoes at temperatures that are too high can accelerate the depletion of oxygen in the tuber tissues. *Prolonged exposure to low oxygen environments*: When potatoes are stored in poorly ventilated containers, or packed too densely, they may experience oxygen deprivation. *Environmental stress*: Potatoes exposed to rapid temperature changes or physical damage during harvest can be more susceptible to black heart	*Proper storage conditions*: *Ventilation*: Ensure adequate airflow in storage facilities to maintain sufficient oxygen levels. Avoid tightly packed storage that restricts airflow. *Temperature control*: Store potatoes at a cool, stable temperature (ideally between 38–40 °F or 3–4 °C) to slow metabolic activity and prevent oxygen depletion. *Humidity control*: Maintain proper humidity levels (85–90%) to reduce moisture loss while preventing excessive condensation. *Harvest and handling*: *Avoid mechanical injury*: Minimize bruising and physical damage during harvesting, as injured tubers are more susceptible to oxygen deprivation. *Curing before storage*: Allow potatoes to cure for a few days at moderate temperatures and high humidity to heal any minor skin injuries, reducing the risk of black heart development. *Transport considerations*: *Adequate ventilation during transport*: Use well-ventilated containers during shipping to prevent oxygen depletion in the tubers. *Avoid prolonged exposure to high temperatures*: Transport potatoes in temperature-controlled conditions to prevent heat buildup, which can trigger black heart

2. Hollow heart	Tubers-internal	*Internal cavity formation*: A hollow space or cavity forms in the center of the potato tuber. The cavity is usually irregular in shape and not surrounded by rot. *No external signs*: Like brown core, the tuber's exterior remains normal in appearance, with no visible signs of the internal disorder. *Cavity without discoloration*: The hollow space may or may not be accompanied by slight brown discoloration at the cavity edges, but no rotting occurs. *Linked to rapid growth*: Hollow heart often occurs when potatoes grow too quickly due to environmental factors, such as excess moisture or nitrogen, after a period of slow growth.	*Rapid growth after slow development*: When potatoes experience a period of slow growth (due to factors like drought or cool temperatures) followed by a rapid growth spurt (due to sudden rain, excessive irrigation, or favorable weather conditions). This quick expansion of the tuber can result in internal cavities. *Temperature fluctuations*: Especially during early tuber development, they can disrupt the normal growth process, leading to the formation of hollow cavities in the center of the potato. *Plant spacing*: Wide plant spacing allows for larger tubers to develop quickly, which can increase the likelihood of hollow heart. *Nutrient imbalances*: Excess nitrogen or unbalanced fertilization can cause rapid vegetative growth, contributing to uneven tuber development. *Moisture stress*: Irregular water supply, such as alternating between dry and wet conditions, can cause uneven tuber expansion, increasing the chances of hollow heart. *Varietal susceptibility*: Certain potato varieties are more prone to hollow heart, especially those that produce large tubers. *Harvesting practices*: Harvesting of potato before tubers fully develop, may increase the incidence of hollow heart	*Consistent irrigation*: *Avoid moisture fluctuations*: Regular and consistent water supply can prevent stress. Drip irrigation is very effective. *Prevent water stress*: Ensure adequate watering during critical growth stages, particularly during tuber initiation and development. *Plant spacing*: *Optimal plant spacing*: Avoid wide plant spacing, which promotes the growth of larger tubers prone to hollow heart. *Use smaller tubers for planting*: Plant smaller seed tubers to help reduce the likelihood of oversized tubers. *Nutrient Management*: *Balanced fertilization*: Avoid excessive nitrogen application, as it can lead to rapid vegetative growth and uneven tuber development. Provide balanced fertilization adequate K&P. Always monitor soil nutrient levels and adjust fertilization accordingly to prevent nutrient imbalances. *Temperature control*: *Mulching*: Apply organic mulch to regulate soil temperature and prevent sudden fluctuations. *Timely planting*: Optimal time for planting, to avoid exposing tubers to extreme temperature fluctuations. *Choose resistant varieties*: *Varietal selection*: Choose potato varieties that are less susceptible to hollow heart *Gradual harvesting*: *Avoid early harvesting*: Ensure that tubers have fully matured before harvesting. *Avoid excessive irrigation after drought*: avoid over-watering immediately after drought, as the rapid intake of water can cause the tubers to expand too quickly, increasing the risk of hollow heart

(continued)

Table 1 (continued)

Name of physiological disorder	Plant part affected	Symptom	Cause	Management
3. Brown core	Tubers-internal	*Internal browning*: Brown or dark brown patches form inside the potato, usually around the core or vascular tissue. *No external symptoms*: The outer appearance of the potato may remain unaffected, making the condition hard to detect without cutting into the tuber. *Corky texture*: The affected internal areas often develop a corky, dry texture. *Triggered by stress*: Brown core is typically caused by environmental factors like rapid changes in temperature or water stress	*Temperature fluctuations*: Sudden changes in soil temperature, especially alternating periods of warm and cold, can lead to internal browning in the core of the potato. *Water stress*: Irregular watering, drought, or waterlogged soils can disrupt the potato's internal physiology, leading to brown core. Both excess moisture and water deficiency contribute to stress. *Nutrient imbalances*: Inadequate or unbalanced supply of essential nutrients like calcium, potassium, or phosphorus can increase susceptibility to brown core. *Oxygen deficiency*: Poor soil aeration or compacted soils can reduce oxygen levels available to the potato tuber, triggering brown core due to tissue damage. *Storage conditions*: Suboptimal storage conditions, especially high temperatures and poor ventilation, can exacerbate brown core symptoms in potatoes after harvest	*Optimize growing conditions*: *Maintain consistent soil moisture*: Avoid extreme fluctuations in soil moisture by providing regular and uniform irrigation. Drip irrigation is effective in ensuring consistent water delivery. *Improve soil drainage*: Ensure the soil is well-drained to prevent waterlogging, which can lead to oxygen deficiency in the tubers. *Avoid drought stress*: Monitor soil moisture levels closely, especially during tuber formation, to prevent drought stress. *Temperature control*: *Avoid temperature extremes*: Protect the crop from rapid temperature fluctuations, especially during critical growth stages. Mulching can help regulate soil temperatures. *Timely planting*: Plant potatoes at the optimal time for your region to avoid exposing them to extreme temperatures. *Nutrient management*: *Balanced fertilization*: Provide a balanced supply of nutrients, particularly calcium, potassium, and phosphorus, which are crucial for tuber development and reducing susceptibility to brown core. *Soil testing*: Regularly test soil to ensure proper nutrient levels and pH balance. *Soil aeration*: *Prevent soil compaction*: Avoid heavy machinery use in the field when the soil is wet, as this can compact the soil and reduce oxygen availability to the tubers. *Improve soil structure*: Incorporate organic matter to improve soil aeration and overall structure, allowing better root and tuber development. *Storage management*:

				Control storage temperature: Store potatoes in cool, well-ventilated environments, ideally between 38–40 °F (3–4 °C), to prevent brown core from developing post-harvest. *Proper ventilation*: Ensure adequate airflow in storage to avoid oxygen depletion, which can aggravate brown core
3. Cracking/ growth cracks	Tubers-external	Longitudinal or radial cracks in the potato tuber surface. The cracks can be shallow or deep and may extend along the length or around the tuber. The potato flesh exposed through the cracks may be discolored or may rot if secondary infections occur.	Growth cracks occur during varying soil moisture conditions Tubers crack when a dry period is followed by heavy rains or over-irrigation. Moisture uptake causes a quick increase in tuber moisture and growth and consequently an increase in tuber size Uneven plant population, over fertilization with nitrogen and nutrient imbalances contribute to the occurrence of growth crack Application of large amounts of nitrogen at one time after tuber formation	Avoid cultivars that are prone to crack if growth cracks regularly occur and cause loss in income Where possible ensure uniform growing conditions, even plant population, good irrigation scheduling and good fertilization practices
4. Enlarged lenticels	Tubers-external	Raised, white callus-like tissue on the surface of the tubers. The size may vary from inconspicuous to large	Enlarged lenticels are formed under conditions of oxygen shortage and excessive moisture in the soil surrounding the tubers, or in storage as the cells beneath the lenticels expand and the cell mass breaks through the corky layer	Avoid fields that are prone to water logging Reduce irrigation 2 weeks prior to harvesting by irrigating when the plant available water is 40–50%

(continued)

Table 1 (continued)

Name of physiological disorder	Plant part affected	Symptom	Cause	Management
5. Frost damage	Tubers-internal	Exposer for short periods of low temperature around 0 C to 2 C can result in gray or reddish tissue. The tissue can also turn dark gray or black.	Exposure of tubers to temperatures that vary from just above freezing point (2 °C) or below freezing point	Do not leave harvested tubers on the land overnight, especially if low temperatures are expected Avoid cultivars that tend to bear shallow in regions with low temperatures at the end of the growing season Do not store seed potatoes on the farm if the correct conditions are not available. Isolate stacks by covering it with grass and/or hessian bags, but preferably store it in a storeroom with temperature control and ventilation If seed potatoes must be transported in winter, the consignment must be properly covered and it must be done during the day Ensure that mechanisms are in place for proper temperature control during cold storage Harvest plantings with tubers close to the soil surface before the first cold front is expected
6. Greening	Tubers-external	Greening of the tissue immediately under the skin The intensity may differ and may also be accompanied by sunburn	Greening occurs when tubers are exposed to light (sun or artificial)	Avoid cultivars that are prone to bear shallowly Avoid high nitrogen application, especially with cultivars that are prone to form long stolons Plant in well prepared soils If possible, irrigate to prevent cracks occurring during dry weather conditions If tubers are exposed after foliage die-off, it can be ridged to cover it with soil Do not expose tubers to light for long periods after harvesting Use packaging materials that provide sufficient protection against light transmission if potatoes are exposed to light for more than a few days. White paper transmits more light than brown paper

7. Internal brown fleck	Tubers-internal	Irregular brown flecks that primarily occur in the vascular bundle ring	The primary cause of brown fleck is localized shortages in Calcium (Ca). Environmental factors that lead to brown fleck include air and soil temperature, soil types and soil moisture	Ensure that sufficient Ca is available for uptake where and when tubers are formed. Where granular fertilizer is used, the Ca must be available in the whole tuber zone Opting for planting times to avoid high temperatures during tuber formation and harvesting, reduces the risk associated with brown fleck Avoid stress conditions at the beginning of the growing season through proper fertilization and irrigation practices
8. Loose skin	Tubers-external	The skin is totally or partially removed from the tubers leading to the underlying tuber tissue being exposed	Excessive application of nitrogen (N) late in the season and wet soil may delay ripening and promote loose skin	Top-growth must be completely dead 14–21 days prior to harvesting to promote skin setting Ensure that the harvester is properly set to prevent tuber damage
9. Malformation	Tubers-external	Tubers form secondary tubers, some on short stolons and others directly on an eyelet	Malformed tubers form when an interruption in growth is experienced during the growing season as a result of high temperature, often accompanied by moisture stress and nutrient shortage	Avoid production in periods when heat waves generally occur Avoid cultivars that are not adaptable to high temperatures, or plant it during cooler times of the year Ensure uniform growing conditions
10. Netting	Tubers-external	Shallow fissures that give the skin a distinctive net like appearance	It is related to climatic conditions and physiological factors in the skin of young tubers. Moisture stress leads to an increased intensity of the disorder	Where possible tubers should not be exposed to varying soil moisture conditions
11. Sprouting	Tubers-external	Potatoes start to sprout when it is harvested, especially when tubers have been left in the soil for long.	Sprouting is promoted by high temperature before harvesting. Sprouting on the land is normally an indication that the cultivar is not adapted to the cultivation practices of a specific area	Evaluate new cultivars for at least 3 years in the production area and follow normal cultivation practices to identify cultivars that are not totally adapted

(continued)

Table 1 (continued)

Name of physiological disorder	Plant part affected	Symptom	Cause	Management
12. Thumb nail cracks	Tubers-external	Mechanical impact manifests often as thumb nail cracks on tubers	The cracks can develop after a slight impact or injury and initially only occur in the skin (periderm) of the tuber without bruising to the underlying tissue	To prevent the occurrence of thumb nail cracks, tubers should not be harvested or handled when it is cold. The temperature of tuber tissue should be >9 °C The crop should be fertilized optimally to ensure that sufficient Ca and Mg are present in the cell walls
13. Vascular browning	Tubers-internal	Vascular browning of variable intensity develops at the stolon end, but in serious cases it can extend throughout the whole vascular ring Vascular browning appears as speckled, light brown to brown and even dark brown stripes		

Source: Modified from Singh et al. (2014)

enhancing the resilience of potato plants to abiotic stress. Potatoes require a balanced supply of macronutrients (nitrogen, phosphorus, potassium) and micronutrients (calcium, magnesium, zinc) to support optimal growth and development. Nutrient deficiencies or imbalances can exacerbate the effects of abiotic stress, reducing the plant's ability to cope with unfavorable conditions.

Other agronomic practices also help combat abiotic stress. For instance, mulching is an effective agronomic practice that offers multiple benefits for managing abiotic stress in potato crops. Likewise, irrigation management—such as scheduling and drip irrigation—also helps manage drought stress in potato cultivation by providing a consistent and controlled water supply, thereby increasing water use efficiency (WUE).

Effective management of abiotic stress in potato cultivation requires an integrated approach that combines nutrient management, mulching, drip irrigation, PGPR, and hormonal treatments. Each of these strategies contributes to enhancing the resilience of potato plants to environmental stressors, supporting sustainable crop production in the face of climate change and resource limitations. By adopting these practices, farmers can mitigate the impacts of abiotic stress, ensuring stable and high-quality potato yields.

Physiological disorders in potato tubers can significantly impact the quality, appearance, and marketability of the crop. Managing these disorders involves understanding their causes, such as environmental stress and nutrient imbalances, and implementing practices to mitigate these factors, including proper irrigation, temperature control, and careful handling during harvest and storage.

References

Abid M, Tian Z, Ata-Ul-Karim ST, Cui Y, Liu Y, Zahoor R, Jiang D, Dai T (2016) Nitrogen nutrition improves the potential of wheat (*Triticum aestivum L.*) to alleviate the effects of drought stress during vegetative growth periods. Front Plant Sci 7:981

Aghaei K, Ehsanpour AA, Komatsu S (2009) Potato responds to salt stress by increased activity of antioxidant enzymes. J Integr Plant Biol 51:1095–1103

Ahmad P, Sarwat M, Bhat NA, Wani MR, Kazi AG, Tran LP (2015) Alleviation of cadmium toxicity in Brassica juncea L. (Czern. & Coss.) by calcium application involves various physiological and biochemical strategies. PLoS One 10:e0114571

Aien A, Chaturvedi AK, Bahuguna RN, Pal M (2017) Phenological sensitivity to high temperature stress determines dry matter partitioning and yield in potato. Indian J Plant Physiol 22(1):63–69

Akhtar SS, Anderson MN, Liu F (2015) Biochar mitigates salinity stress in potato. J Agron Crop Sci 201:368–378

Aldubai AA, Alsadon AA, Migdadi HH, Alghamdi SS, Al-Faifi SA, Afzal M (2022) Response of tomato (*Solanum lycopersicum* L.) genotypes to heat stress using morphological and expression study. Plan Theory 11(5):615

Ata-Ul-Karim ST, Liu X, Lu Z, Yuan Z, Zhu Y, Cao W (2016) In-season estimation of rice grain yield using critical nitrogen dilution curve. Field Crop Res 195:1–8

Awasthi LP, Verma HN (2017) Current status of viral diseases of potato and their eco-friendly management: a critical review. Virol Res Rev 1:1–16

Banks E, VanOostrum, M (2016) The impact of heat and drought on the 2016 Ontario potato crop. https://spudsmart.com/impact-heat-drought-2016-ontario-potato-crop/

Begum N, Ahanger MA, Zhang L (2020) AMF inoculation and phosphorus supplementation alleviates drought induced growth and photosynthetic decline in Nicotiana tabacum by up-regulating antioxidant metabolism and osmolyte accumulation. Environ Exp Bot 176:104088

Bello SK, Alayafi AH, AL-Solaimani SG, Abo-Elyousr KA. (2021) Mitigating soil salinity stress with gypsum and bio-organic amendments: a review. Agronomy 11(9):1735

Bendixen R, Gerendas J, Schirmer K, Sattelmacher B, Hansen UP (2001) Difference in zeaxanthin formation in nitrate-and ammonium-grown Phaseolus vulgaris. Physiol Plant 111:255–261

Binghua L, Liang C, Mingjun L, Dong L, Yangjun Z, Fengwang M (2012) Interactive effects of water and nitrogen supply on growth, biomass partitioning, and water-use efficiency of young apple trees. Afr J Agric Res 7:978–985

Busse JS, Wiberley-Bradford AE, Bethke PC (2019) Transient heat stress during tuber development alters post-harvest carbohydrate composition and decreases processing quality of chipping potatoes. J Sci Food Agric 99(5):2579–2588

Campos H, Ortiz O (2020) The potato crop. Its agricultural, nutritional and social contribution to humankind. Cham (Switzerland). Springer, Cham. ISBN 978-3-030-28683-5. 524 p

Cao MJ, Wang Z, Zhao Q, Mao JL, Speiser A, Wirtz M, Xiang CB (2014) Sulphate availability affects ABA levels and germination response to ABA and salt stress in Arabidopsis thaliana. Plant J 77:604–615

Chakrabarti SK, Sharma S, Shah MA (2022) Potato pests and diseases: a global perspective. In: Sustainable management of potato pests and diseases. Springer Singapore, Singapore, pp 1–23

Chourasia KN, Lal MK, Tiwari RK, Dev D, Kardile HB, Patil VU, Kumar A, Vanishree G, Kumar D, Bhardwaj V (2021). Salinity stress in potato: Understanding physiological, biochemical and molecular responses. Life, 11(6):545

CPRI Annual Report 2014–2015. CPRI Annual Report 2015. Printed: June 2015

Dahal K, Kane K, Sarhan F, Grodzinski B, Hüner NPA (2012) Cold acclimation inhibits CO2-dependent stimulation of photosynthesis in spring wheat and spring rye. Botany 90:433–444

Dahal K, Martyn GD, Vanlerberghe GC (2019) Improved photosynthetic performance during severe drought in Nicotiana tabacum overexpressing a non-energy conserving respiratory electron sink. New Phytol 208:382–395

Danial HF, Ewees MS, Moussa SA (2010) Significance of influence potassium on the tolerance to induce moisture stress and biological activity of some legume crops grown on a sandy soil Egypt. Egypt J Soil Sci 43:180–204

Duman F (2012) Uptake of mineral elements during abiotic stress. In: Abiotic stress responses in plants. Springer, New York, pp 267–281

Evers D, Lefevre I, Legay S, Lamoureux D, Hausman J-F, Rosales RO et al (2010) Identification of drought-responsive compounds in potato through acombined transcriptomic and targeted metabolite approach. J Exp Bot 61:2327–2343

Fahad S, Hussain S, Matloob A, Khan FA, Khaliq A, Saud S, Hassan S, Shan D, Khan F, Ullah N et al (2015) Phytohormones and plant responses to salinity stress: a review. Plant Growth Regul 75:391–404

FAO (2019) FAO statistical databases FAOSTAT. http://faostat3.fao.org

Fleisher DH, Timlin D (2006) Modeling expansion of individual leaves in the potato canopy. Agric For Meteorol 139(1–2):84–93

Fooladivanda Z, Zadehdelouei MH, Zarifinia N (2014) Effects of water stress and potassium on quantity traits of two varieties of mungbean (*Vigna radiata L.*). Cercet Agron Mold 47:107–114

Ge TD, Sun NB, Bai LP, Tong CL, Sui FG (2012) Effects of drought stress on phosphorus and potassium uptake dynamics in summer maize (*Zea mays*) throughout the growth cycle. Acta Physiol Plant 34:2179–2186

Ghosh SC, Asanuma K, Kusutani A, Toyota M (2001) Effect of salt stress on some chemical components, and yield of potato. Soil Sci Plant Nutr 47:467–475

Golovko TK, Tabalenkova GN (2019) Source-sink relationships in potato plants. Russ J Plant Physiol 66(4):664–671

Gonzalez-Real MM, Baille A (2000) Changes in leaf photosynthetic parameters with leaf position and nitrogen content within a rose plant canopy (*Rosa hybrida*). Plant Cell Environ 23:351–363

Gururani MA, Upadhyaya CP, Baskar V, Venkatesh J, Nookaraju A, Park SW (2013) Plant growth-promoting Rhizobacteria enhance abiotic stress tolerance in Solanum tuberosum through inducing changes in the expression of ROS-scavenging enzymes and improved photosynthetic performance. J Plant Growth Regul 32:245–258

Hailu B, Mehari H (2021) Impacts of soil salinity/sodicity on soil-water relations and plant growth in dry land areas: a review. J Nat Sci Res 12(3):1–10

Hancock RD, Morris WL, Ducreux LJM, Morris JA, Usman M, Verrall SR, Fuller J, Simpson CG, Zhang R, Hedley PE et al (2014) Physiological, biochemical and molecular responses of the potato (Solanum tuberosum L.) plant to moderately elevated temperature. Plant Cell Environ 37(2):439–450

Hassan MJ, Wang F, Ali S, Zhang G (2005) Toxic effects of cadmium on rice as affected by nitrogen fertilizer form. Plant Soil 277:359–365

Hastilestari BR, Lorenz J, Reid S, Hofmann J, Pscheidt D, Sonnewald U, Sonnewald S (2018) Deciphering source and sink responses of potato plants (Solanum tuberosum L.) to elevated temperatures. Plant Cell Environ 41(11):2600–2616

Havaux M (1993) Rapid photosynthetic adaptation to heat stress triggered in potato leaves by moderately elevated temperatures. Plant Cell Environ 16:461–467

Hermans C, Conn SJ, Chen J, Xiao Q, Verbruggen N (2013) An update on magnesium homeostasis mechanisms in plants. Metallomics 5:1170–1183

Hijmans RJ (2003) The effect of climate change on global potato production. Am J Potato Res 80:271–279

Hlavinka P, Trnka M, Semeradova D, Dubrovsky M, Zalud Z, Mozny M (2009) Effect of drought on yield variability of key crops in Czech Republic. Agric For Meteorol 149:431–442

Holden N, Brereton A, Fealy R, Sweeney J (2003) Possible change in Irish climate and its impact on barley and potato yields. Agric For Meteorol 116:181–196

Huang ZA, Jiang DA, Yang Y, Sun JW, Jin SH (2004) Effects of nitrogen deficiency on gas exchange, chlorophyll fluorescence, and antioxidant enzymes in leaves of rice plants. Photosynthetica 42:357–364

Islam S, Eusufzai TK, Ansarey FH, Hasan MM, Nahiyan ASM (2022) A breeding approach to enhance late blight resistance in potato. J Hortic Sci Biotechnol 97(6):719–729

Islam S, Jiana Li, Mohammad Ataur Rahman, FangruXie, Botao Song, BihuaNie (2024) Resistance to biotic and abioticstress in potato: the origin of the genesand corresponding molecular markers. Phytopathology Research, 6(4):3–14

Jaarsma R, de Vries RSM, de Boer AH (2013) Effect of salt stress on growth, NaCl accumulation and proline metabolism in potato (Solanum tuberosum) cultivars. PLoS One 8:e60183

Jensen KH, Berg-Sørensen K, Bruus, H, Holbrook, NM, Liesche J, Schulz A, Zwieniecki MA, Bohr T (2016) Sap flow and sugar transport in plants. Reviews of Modern Physics, 88(3):35007

Kang Y, Wang F-X, Liu H-J, Yuan B-Z (2004) Potato evapotranspiration and yields under different drip irrigation regimes. Irrig Sci 23:133–143

Kashyap PS, Panda RK (2003) Effect of irrigation scheduling on potato crop parameters under water stressed conditions. Agric Water Manag 59:49–66

Kumar A, Chandra S (2024) Chapter 7: Instability and growth scenario of potato in India: an empirical study. In: Futuristic trends in social sciences, IIP Series, vol 3., Book 14, Part 2. Iterative International Publishers (IIP). isbn:978-93-5747-431-3

Kumar D, Singh BP, Rawal S, Minhas JS, Pandey SK (2011) Rehabilitation of frosted potato crop through plant growth regulators. Potato J 38(1):18–25

Kumari S (2012) Influence of drip irrigation and mulch on leaf area maximization, water use efficiency and yield of potato (*Solanum tuberosum L.*). J Agric Sci 4:71–80

Lobanov AV, Hatfield DL, Gladyshev VN (2008) Reduced reliance on the trace element selenium during evolution of mammals. Genome Biol 9:R62

Majeed S, Akram M, Latif M, Ijaz M, Hussain M (2016) Mitigation of drought stress by foliar application of salicylic acid and potassium in mungbean (*Vigna radiata L.*). Legum Res 39:208–214

Marschner H (2012) Marschner's mineral nutrition of higher plants, 3rd edn. Academic, London, UK, pp 178–189
Meng X, Chen WW, Wang YY, Huang ZR, Ye X, Chen LS, Yang LT (2021) Effects of phosphorus deficiency on the absorption of mineral nutrients, photosynthetic system performance and antioxidant metabolism in Citrus grandis. PLoS One 16:e0246944
Munns R, Tester M (2008) Mechanisms of salinity tolerance. Annu Rev Plant Biol 59:651–681
Obidiegwu J, Bryan G, Jones G, Prashar A (2015) Coping with drought: stress and adaptive responses in potato and perspectives for improvement. Front Plant Sci 6:542
Obiero CO, Milroy SP, Bell RW (2020) Photosynthetic and respiratory response of potato leaves of different ages during and after an episode of high temperature. J Agron Crop Sci 206(3):352–362
Odemis B, Caliskan ME (2014) Photosynthetic response of potato plants to soil salinity. Turk J Agric Nat Sci 2:1429–1439
Oliveira ED, Bramley H, Siddique KHM, Henty S, Berger J, Palta JA (2012) Can elevated CO2 combined with high temperature ameliorate the effect of terminal drought in wheat. Funct. Plant Biol. 40:160–171
Onder S, Caliskan ME, Onder D, Caliskan S (2005) Different irrigation methods and water stress effects on potato yield and yield components. Agric Water Manag 73:73–86
Peck AW, McDonald GK (2010) Adequate zinc nutrition alleviates the adverse effects of heat stress in bread wheat. Plant Soil 337:355–374
Queirós F, Almeida D, Fidalgo F (2009) Ultrastructural aspects of a NaCl-adapted potato cell line. Microsc. Microanal. 15(Suppl. 3):41–42
Queirós F, Rodrigues JA, Almeida JM, Almeida D, Fidalgo F (2011) Differential responses of the antioxidant defense system and ultrastructure in a salt-adapted potato cell line. Plant Physiol Biochem 49:1410–1419
Rahman A, Mostofa MG, Nahar K, Hasanuzzaman M, Fujita M (2015) Exogenous calcium alleviates cadmium-induced oxidative stress in rice (Oryza sativa L.) seedlings by regulating the antioxidant defense and glyoxalase systems. Braz J Bot 39:393–407
Rajasekharan P, Nair KM, John KS, Kumar PS, Kutty MN, Nair AR (2014) Soil fertility related constraints to crop production in Kerala. Indian J Fert 10(11):56–62
Rykaczewska K (2013) The impact of high temperature during growing season on potato cultivars with different response to environmental stresses. Am J Plant Sci 4(12):2386–2393
Sadura I, Latowski D, Oklestkova J, Gruszka D, Chyc M, Janeczko A (2020) Molecular dynamics of chloroplast membranes isolated from wild-type barley and a brassinosteroid-deficient mutant acclimated to low and high temperatures. Biomol Ther 11(1):27
Sardans J, Penuelas J (2012) The role of plants in the effects of global change on nutrient availability and stoichiometry in the plant-soil system. Plant Physiol 160:1741–1761
Singh A, Siddappa S, Bhardwaj V et al (2015) Expression profiling of potato cultivars with contrasting tuberization at elevated tempera[1]ture using microarray analysis. Plant Physiol Biochem 97:108–116
Singh B, Singh DK, Pandey NK (2014) Physiological disorder in potato: causes and their management. Indian Farmer's Digest 47(5):40–42
Song J, Wang Y, Pan Y, Pang J, Zhang X, Fan J, Zhang Y (2019) The influence of nitrogen availability on anatomical and physiological responses of *Populus alba x P. glandulosa* to drought stress. BMC Plant Biol 19:1–12
Statista (2024) https://www.statista.com/statistics/1038959/india-production-of-potato
Tang R, Niu S, Zhang G, Chen G, Haroon M, Yang Q et al (2018) Physiological and growth responses of potato cultivars to heat stress. Botany 96:897–912
Tariq A, Pan K, Olatunji OA, Graciano C, Li Z, Sun F, Sun X, Song D, Chen W, Zhang A et al (2017) Phosphorous application improves drought tolerance of Phoebe zhennan. Front Plant Sci 8:1561
Tekalign T, Hammes PS (2005) Growth responses of potato (*Solanum tuberosum*) grown in a hot tropical lowland to applied paclobutrazol: 1. Shoot attributes, assimilate production and allocation. N Z J Crop Hortic Sci 33:35–42

Tiwari JK, Plett D, Garnett T, Chakrabarti SK, Singh RK (2018) Integrated genomics, physiology and breeding approaches for improving nitrogen use efficiency in potato: Translating knowledge from other crops. Functional Plant Biology, 45:587–605

Tiwari JK, Buckseth T, Singh RK, Kumar M, Kant S (2020) Prospects of improving nitrogen use efficiency in potato: lessons from transgenics to genome editing strategies in plants. Front. Plant Sci. 11:597481. https://doi.org/10.3389/fpls.2020.597481

Trapero-Mozos A, Morris WL, Ducreux LJM, McLean K, Stephens J, Torrance L, Bryan GJ, Hancock RD, Taylor MA (2018) Engineering heat tolerance in potato by temperature-dependent expression of a specific allele of HEAT-SHOCK COGNATE 70. Plant Biotechnol J 16(1):197–207

Tripura B, Ghosh, S (2017) Adoption constraints in use of True Potato Seeds (TPS) as planting materials in Tripura. International Journal of Research in Applied, Natural and Social Sciences, 6:95–100

Van Dam J, Kooman PL, Struik PC (1996) Effects of temperature and photoperiod on early growth and final number of tubers in potato (*Solanum tuberosum L*). Potato Res 39:51–62

Wang M, Zheng Q, Shen Q, Guo S (2013) The critical role of potassium in plant stress response. Int J Mol Sci 14:7370–7390

Wang Q-L, Chen J-H, He N-Y, Guo F-Q (2018) Metabolic reprogramming in chloroplasts under heat stress in plants. Int J Mol Sci 19(3):849

Wang-Pruski G, Schofield A (2012) Potato: improving crop productivity and abiotic stress tolerance. In: Tuteja N, Gill SS, Tiburcio AF, Tuteja R (eds) Improving crop resistance to abiotic stress. Wiley-VCH Verlag GmbH &Co. KGaA, Weinheim, pp 1121–1153

Waraich EA, Ahmad R, Ashraf MY, Saifullah Ahmad M (2011) Improving agricultural water use efficiency by nutrient management in crop plants. Acta Agric Scand B Soil Plant Sci 61:291–304

Wijesinha-Bettoni R, Mouillé B (2019) The Contribution of Potatoes to Global Food Security, Nutrition and Healthy Diets. American Journal of Potato Research, 96:139–149

Yadav SK, Srivastava AK (2015) A review on agronomical aspects of potato production in north-eastern region of India. International Journal of Applied and Pure Science and Agriculture, 1(6):26–34

Yang Y, Guo JY, Wang GX, Yang LD, Yang Y (2012) Effects of drought and nitrogen addition on photosynthetic characteristics and resource allocation of Abies fabri seedlings in eastern Tibetan plateau. New For 43:505–518

Yang J, Fan W, Zheng S (2019) Mechanisms and regulation of aluminium induced secretion of organic acid anions from plant roots. J Zhejiang Univ Sci B 20:513–527

Yuan L, Zhang L, Wu Y, Zheng Y, Nie L, Zhang S, Lan T, Zhao Y, Zhu S, Hou J et al (2021) Comparative transcriptome analysis reveals that chlorophyll metabolism contributes to leaf color changes in wucai (*Brassica campestris L.*) in response to cold. BMC Plant Biol 21(1):1–18

Zhang J, Wang Y, Zhao Y, Zhang Y, Zhang J, Ma H, et al. (2020) Transcriptome analysis reveals Nitrogen deficiency induced alterations in leaf and root of three cultivars of potato (Solanum tuberosum L.). PLoS One 15:e0240662

Zhu Z, Gerendas J, Bendixen R, Schinner K, Tabrizi H, Sattelmacher B, Hansen UP (2000) Different tolerance to light stress in NO3–and NH4+-grown Phaseolus vulgaris L. Plant Biol 2:558–570

Advances in Insect-Pest Management of Potatoes

B. C. Anu and Moti Lal Meena

Abstract Potatoes hold a distinctive position as a highly nutritious and easily digestible food source. Cultivated across diverse states in India, potatoes face persistent threats from various formidable insect pests. These pests, exhibiting high adaptability and prolific breeding, pose a significant risk of causing extensive damage to potato crops if left uncontrolled. In India, the cultivation of potatoes is confronted by a diverse range of insect pests that employ various mechanisms to inflict damage. The impact of climate change may result in the evolution of geographical range, resulting in increased population and, consequently, greater losses in both crop yields and post-harvest storage. The application of insect pest modeling in pest risk analysis has shown promise in providing insights and raising awareness about potential future pest risks under changing climatic conditions. In India, the insect pests of potatoes can be generally classified into three major categories: sucking pests, tuber- and root-damaging pests, and foliage feeders (defoliating pests). This classification reflects the diversity of pests affecting different parts of the potato plant and causing distinct types of damage. Notably, some potato pests in India have developed resistance to a wide array of chemicals, presenting a challenge for growers who have encountered situations where available chemicals fail to control these resistant pests. The implementation of integrated pest management (IPM) strategies is crucial for effectively managing potato pests. The success of IPM relies on interdisciplinary research and the adoption of nonchemical control methods tailored to the specific agroecology of potato crops. The goal is to minimize or eliminate the reliance on pesticides, promoting sustainable and environmentally friendly pest management practices. An in-depth understanding of individual potato production systems, including pest identification, understanding their biology, and recognition of infestation symptoms, is crucial for making decisions in the management of potato pests in an integrated manner.

B. C. Anu (✉) · M. L. Meena
Krishi Vigyan Kendra, Turki, Muzaffarpur, Dr. Rajendra Prasad Central Agricultural University, Pusa, Samastipur, Bihar, India

S. S. Solankey (ed.), *Advances in Research on Potato Production*,
Advances in Olericulture, https://doi.org/10.1007/978-3-031-82710-5_15

Keywords Agroecology · Foliage feeders · Insect pest modeling · IPM · Sucking pests · Symptoms

1 Introduction

Approximately more than 400 years ago, the potato was introduced to India, initially contributing to less than 1% of global production. However, with the development of high-yielding varieties and location-specific agronomic practices, India has emerged as the second largest global producer of potatoes and contributes approximately 12% to the total production. In the 2018–19 period, potato production reached 52.59 million tons from an area of 2.18 million hectares, making it the crucial food crop of the world alongside rice, wheat, and maize. The adaptability of potatoes to diverse agricultural environments is notable, as they are grown across various latitudes and elevations (Douches et al. 2004). Numerous biotic and abiotic factors, including insect pests, impact the production potential of potatoes significantly (Tiwari et al. 2020a, b). Insect pests are accountable for approximately 16% of global loss in the potato crop (Oerke et al. 1994), with variations depending on geographical locations and climatic conditions.

Potatoes are highly nutritious and easily digestible, providing carbohydrates, proteins, minerals, vitamins, and high-quality dietary fiber. A potato tuber is composed of 80% water and 20% dry matter, which includes 14% starch, 2% sugar, 2% protein, 1% minerals, 0.6% fiber, and 0.1% fat, along with adequate amounts of vitamins B and C. Consequently, potatoes offer superior nutrition compared to cereals and vegetables.

Given the challenges of diminishing arable land and a growing population in India, potatoes emerge as a viable substitute. This crop is particularly well-suited for cultivation in areas with limited land and abundant labor—a common scenario in developing countries. Especially in the most economically disadvantaged and underfed households, potatoes play a crucial role in ensuring dietary requirements are met reliably, even in conditions where other crops may face challenges (Lutaladio et al. 2009).

2 Geographical Distribution of Potato

Potatoes are cultivated across various states in India, encompassing a wide range of agroclimatic conditions. Recognizing the geographical variability and climatic distinctions, the cultivation of potato in various regions in India has been categorized into seven zones (Table 1).

Table 1 Potato growing agroclimatic zones in India

S. no.	Agroclimatic zones	Cropping season	States
1.	Northwestern plains	September–November, October–January/February, January–April/May	Punjab, Haryana, Rajasthan
2.	West-Central plains	October–January/February	Madhya Pradesh, West-Central Uttar Pradesh, Northwestern Gujarat
3.	Northeastern plains	November–February/ March	Assam, Bihar, Jharkhand, Orissa, West Bengal, Eastern Uttar Pradesh, Eastern Chhattisgarh, Northeastern Madhya Pradesh
4.	Plateau region	June/July–September/ October, November–January/February	Maharashtra, Karnataka, parts of Gujarat, Madhya Pradesh and Orissa
5.	Northwestern and Central hills	April–September, January–May	Jammu & Kashmir, Uttarakhand and Himachal Pradesh
6.	Northeastern hills	January–May, August–December	Meghalaya, Manipur, Mizoram, Tripura, Nagaland, Arunachal Pradesh
7.	Southern hills	April–September, August–December, January–May	Nilgiris and Palani hills of Tamil Nadu

Source: Chandel and Chandla (2003)

3 Potato and Its Insect Pests

In the global context, insect pests pose a significant challenge to the commercial production of potatoes. Dating back over 8000 years to its domestication in the central Andes, the potato has endured persistent infestations from several formidable insect pests (Radcliffe 1982). These pests encompass highly prolific and adaptable species, capable of causing complete destruction to affected crops if left unchecked. The intensification of worldwide trade has facilitated the dissemination of many insect herbivores, accompanying agricultural products across continents and emerging as severe pests in new areas (Carrillo et al. 2019).

Addressing the challenges posed by insect pests to potato cultivation requires attention to several unresolved issues:

- *Understanding the Insect Pests:* A thorough understanding of the ecology and biology of the insect pests of potato is imperative for developing effective countermeasures.
- *Tracking Insect Spread:* Examining the spread of invasive insects affecting potatoes is crucial for enhancing prevention measures and mitigating potential damage risks.
- *Understanding Interactions:* Investigating the occurrence of insects and plant pathogens in the fields of potato is essential for unraveling their interactions during crop damage.

- *Implementing Integrated Management Strategies:* The current reliance on synthetic insecticides in commercial potato production raises environmental concerns. Therefore, developing new strategies for more ecologically sustainable pest control is imperative.

In India, potato cultivation faces a diverse array of insect pests (Shah et al. 2020), which inflict damage through various means. This includes direct harm, such as the reduction of plant photosynthates through sap-sucking, removal of plant tissues, weakening of plants, inhibition of nutrient transport, and attacks on potato tubers (Chandel and Chandla 2003). Additionally, some insect pests indirectly contribute to losses through transmission of plant viruses from alternate hosts, weeds, and infected to healthy potato plants, leading to seed tuber degeneration. The economic impact of pest damage in India is substantial, with an annual loss of approximately 60 billion rupees, accounting for 10–20% of total potato production (Chandel et al. 2013). In India, the insect pests of potato can be generally classified into three major categories: sucking pests, tuber- and root-damaging pests, and foliage feeders (defoliating pests).

4 Impact of Changing Climatic on Insect Pests of Potato

Climate, particularly temperature, exerts an important and direct impact on the development and growth of pest populations. Herbivorous insects (like other arthropods) are exothermic organisms which are incapable of internally regulating their temperature. Their developmental processes are contingent upon the temperature prevailing in their environment. Changes in temperature associated with climate change may result in both increased and decreased pest development rates, impacting associated crop losses. Consequently, temperature variations can potentially influence the outbreaks and range expansion of various insect pests, including those affecting potatoes.

Novel modeling approaches, such as process-based phenology models which have climatic responses (Orlandini et al. 2017; Mujica et al. 2017), are being employed to assess the implications of temperature rise under global temperature changes for the year 2050 (Kroschel et al. 2016a). These models have been applied to a diverse range of potato pests, such as *Phthorimaea operculella* (Kroschel et al. 2016b) and the Whiteflies, *Bemisia tabaci* (Gennadius). The findings of these models unequivocally illustrate that insect pests will react to the change in climatic conditions by enhancing their geographical range, leading to increased population densities and consequently a greater loss in crop as well as post-harvest losses. However, various differences exist among various pest species. For instance, the damage potential of potato tuber moth is anticipated to gradually increase in areas where the pest currently prevails, with an expansion into temperate and tropical mountainous regions (Kroschel et al. 2016b). Global predictions suggest a slight to moderate reduction in the establishment potential of the pest in most tropical and subtropical potato growing areas, though it continues to pose significant pest risks. Furthermore, an expansion into temperate regions of Asia, North and South America, and Europe, along with the tropical and subtropical mountainous regions, is anticipated, accompanied by a slight increase in its establishment and damage potential (Mujica 2016).

The large-scale use of pesticides in potato cultivation raises significant concerns for both human and environmental health. Addressing these concerns necessitates the development and broader implementation of integrated pest management (IPM) approaches.

5 Major Insect Pests of Potato

A diverse array of insects poses threats to potato crops, either through direct damage, such as feeding on potato tubers and compromising the harvest, or indirectly by consuming foliage or stems and causing a number of diseases by transmitting a wide range of pathogens (Radcliffe 1982). Indirect damage, if severe, can result in diminished harvestable yield and compromised quality. Potato farmers encounter variable and intricate challenges due to insect pests. India, in particular, hosts various insects targeting potatoes. Few of these insects were introduced to new areas through seeds and tubers, while others were already present in areas where potatoes were introduced, resulting in the expansion of their host range to exploit the new environment. Given that the potato crops are propagated vegetatively from tubers, which carry pests and pathogens quite easily, several insect pest-related issues have accompanied potatoes to regions where they have been cultivated (Chandel et al. 2007). The impact of such pests on the crop includes feeding on green leaves, leading to reduced photosynthesis, invading stems, weakening plants, inhibiting the transport of nutrients, and targeting tubers designated for human use or use as seed (Chandel and Chandla 2003).

The classification of potato pests encompasses soil pests, foliage feeders, sap feeders, and storage pests (Table 2). In seed production, the primary concerns typically revolve around aphids, which are vectors of a wide range of potato viruses, especially *Myzus persicae* (Sulzer). The key pests may include insect pests that target potato tubers, viz. tuber moths, white grubs, and cutworms. Additionally, foliage feeders like noctuid moths and coccinellid beetles may assume significance in certain situations.

5.1 Tuber- and Root-Damaging Insect Pests

Prominent insect pests that feed on and damage potato tubers and roots include white grubs, cutworms, and potato tuber moths. These insect pests present formidable challenges to potato growers due to their persistent nature, the limited

Table 2 Insect pests of potato

S. no.	Common name	Insect species	Order	Family
Tuber- and root-damaging insect pests				
1.	Potato tuber moth	*Phthorimaea operculella* (Zeller)	Lepidoptera	Gelechiidae
2.	White grubs	*Holotrichia seticollis* Moser, *H. longipennis* (Blanchard), *Brahmina coriacea* (Hope), *H. serrata* (Fab.)	Coleoptera	Melolonthidae
3.	Cutworms	*Agrotis* spp.	Lepidoptera	Noctuidae
Foliage feeders or defoliating pests				
4.	Hadda beetles	*Epilachna dodecastigma* *Epilachna vigintioctopunctata*	Coleoptera	Coccinellidae
5.	Flea beetle	*Psyllodes plana* (Maulik)	Coleoptera	Chrysomelidae
	Colorado potato beetle	*Leptinotarsa decemlineata* (Say)	Coleoptera	Chrysomelidae
6.	Tobacco caterpillar	*Spodoptera litura* (Fabricius)	Lepidoptera	Noctuidae
7.	Gram pod borer	*Helicoverpa armigera* (Hubner)	Lepidoptera	Noctuidae
Sucking pests				
8.	Green peach aphid	*Myzus persicae* (Sulzer)	Hemiptera	Aphididae
9.	Whitefly	*Bemisia tabaci*	Hemiptera	Aleyrodidae
10.	Leafhoppers	*Empoasca devastans* (Distant)	Hemiptera	Cicadellidae
11.	Thrips	*Thrips palmi* (Karny), *Scirtothrips dorsalis* (Hood), *Frankliniella occidentalis* (Pergande)	Thysanoptera	Thripdae
Mites				
12.	Broad mite	*Polyphagotarsonemus latus* (Banks)	Acarina	Tarsonemidae
13.	Red spider mite	*Tetranychus neocaledonicus*	Acarina	Tetranychidae

effectiveness of insecticides, and their tendency to go unnoticed until they cause significant visible damage to the crop. The repercussions of these insect pests extend beyond a single field and season, resulting in substantial yield losses. The impact of these insects is not confined solely to direct tuber losses but also encompasses a reduction in the market value of potato tubers. Moreover, secondary infections on the tubers at damaged sites contribute to the overall economic losses associated with these pests. This chapter will delve into the specific management practices for insect pests of potatoes which feed on tubers and roots, addressing each pest individually.

5.1.1 Potato Tuber Moth

Scientific name *Phthorimaea operculella* (Zeller) (Lepidoptera: Gelechiidae)

Distribution The potato tuber moth (PTM), *P. operculella*, originated in South America's tropical mountainous regions. It exhibits a cosmopolitan distribution, being present in over 90 countries globally, and is found in all tropical and subtropical areas where potatoes are cultivated (Kroschel and Schaub 2013). The introduction of this pest into India occurred through Italy in 1906 via seed tubers, subsequently gaining economic significance in numerous potato-growing regions of the country (Chandel et al. 2020).

Host range It is an oligophagous pest primarily infesting the Solanaceous vegetable crops, including potato (*Solanum tuberosum* L.), tomato (*L. esculentum* Mill.), and tobacco (*Nicotiana tabacum* L.). Additionally, it targets many wild species, such as the notable black nightshade (*S. nigrum* L.), belonging to family Solanaceae. The overall host range encompasses 60 species. Furthermore, crops within the Chenopodiaceae family are susceptible to infestation, including eggplant (*Solanum melongena* L.), bell pepper (*Capsicum annuum* L.), Cape gooseberry (*Physalis peruviana* L.), and sugar beet (*Beta vulgaris* L.).

Damage symptoms The damaging stage of the potato tuber moth is larva that causes damage by tunneling into leaves, terminal shoots, and petioles, leading to plant wilting. Additionally, the adult females of PTM lay eggs on the eyes of exposed tubers, and upon hatching, the larva penetrate the tuber, creating galleries and tunnels. These tunnels are prone to secondary infections, rendering the tubers unfit for consumption and sale. Although the potato tuber moth is a pest in both field and storage conditions, its economic significance is particularly pronounced as a stored insect pest. The degree of tuber infestation varies across different geographic locations in India, ranging from 30% to 60% tuber infestation in the Kangra Valley of Himachal Pradesh (Singh et al. 1990) to 100% tuber infestations observed in particular regions of Karnataka (Trivedi et al. 1994).

Bionomics The female potato tuber moth deposits approximately 66 eggs on leaves, eye buds of tubers, and the soil surrounding the plants, with an incubation period of 4–7 days. Upon hatching, the larva makes tunnel into tubers, leaves, and, stems undergoing a developmental phase that lasts about 15–20 days until they reach full maturity. The lower developmental threshold and optimum temperature for the immature stages of the potato tuber moth range between 12.5–16.2 °C and 31.7–33.8 °C, respectively (Andreadis et al. 2017).

Pupation takes place in a cocoon amidst debris and clods of earth in the field, with a pupal period spanning 6–8 days. The entire life cycle of the potato tuber moth, under warm weather conditions, takes approximately 22–25 days. In the Kangra Valley of Himachal Pradesh, five to six generations are produced per year (Chandel et al. 2001). The climatic conditions in Assam, the northern plains, and plateau regions of India permit the completion of 11, 8–9, and 10–13 generations, respectively, for the potato tuber moth.

5.1.2 White Grubs

Scientific name *Holotrichia seticollis* Moser, *Holotrichia longipennis* (Blanchard), *Brahmina coriacea* (Hope), *Holotrichia serrata* (Fab.) (Coleoptera: Melolonthidae)

Distribution The Indian subcontinent hosts a diverse array of white grub species, with over 200 species described. Among these, 20 species have been documented on potatoes in India. Notably, *H. seticollis* Moser, *H. longipennis* (Blanchard), and *B. coriacea* (Hope) stand out as prevalent and destructive species in the hills of the northwest. In contrast, in Karnataka *H. serrata* (Fab.) has been identified as a species causing damage to potatoes (Misra and Chandel 2003).

Host range They exhibit polyphagous feeding habits, consuming over thousands (1000) of plant species. Among field crops, potato, sweet potato, maize, and soybean are susceptible to their feeding. Additionally, white grubs target fruits such as apples, grapes, and peaches along with trees and ornamental plants (Pathania and Chandel 2017).

Damage symptoms The larvae primarily inflict damage while residing in the soil. The damage is attributed to their feeding habits, which involve consuming roots of plants, stolons, underground stem parts, and tubers of plants. In the early instars, the grubs focus on feeding on the roots, potentially leading to the desiccation of plants. As tubers develop, the grubs start feeding on them, resulting in large holes in the affected tubers.

Bionomics The larval stages of white grubs share a similar shape and coloration. These larvae exhibit a C-shaped form with a brown head and well-developed thoracic legs. Adults typically emerge during specific periods from the soil, generally in May and June, although the timing of adult emergence can vary slightly based on geographical locations. White grubs predominantly spend the majority of their time in the larval stage within the soil. Throughout their life stages, apart from the adult beetles, which emerge during the evening hours from the soil, they remain in the soil and may defoliate various plant species.

5.1.3 Cutworms

Scientific name *Agrotis* spp. (*Agrotis segetum* Schiff., *Agrotis ipsilon* Hufn.) (Lepidoptera: Noctuidae)

Distribution *Agrotis* spp. exhibit a polyphagous nature and are distributed across numerous potato-growing regions in India. *A. segetum* is primarily observed in the hills, while *A. ispilon* is identified as causing damage to potatoes mainly in the plains. The activity of this insect varies in different geographic locations, with peak activity observed in May and June in Shimla, March–April in Punjab and Bihar, and August in peninsular India. The observed damage to the tuber recorded by this insect is 12.7% in Bihar and ranges from 9% to 16.4% in Himachal Pradesh.

Host range Cutworms, known for their polyphagous nature, have been identified as pests on a wide variety of crops, including mostly all vegetable crops and some cereals. Among the crops affected, cutworms are recognized as pests of potato, maize, clover, alfalfa, rice, sorghum, cotton, sugar beet, strawberry, and tobacco.

Damage symptoms Newly hatched cutworm larvae initially feed on leaves, causing defoliation of plants. However, as the larvae mature, particularly in later instars, they cause substantial plant losses by cutting them at the bases. During the night, these later instars of cutworms emerge from hiding places in cracks and crevices of the upper soil layer, cutting entire potato plants at the bases, sometimes necessitating replanting. Incidences of cutworms are more prevalent during dry periods, with reduced occurrence during the rainy season, although measurable losses still occur due to the creation of circular bore holes on potato tubers.

Cutworms further feed on potato tubers by creating deep irregular holes, significantly diminishing tuber quality and triggering infections by secondary pathogens. This renders affected tubers unsuitable for consumption and sale. The impact of cutworm losses is particularly pronounced in unirrigated fields with lighter soil, especially during hot and dry summers.

Bionomics The female moth deposits 600–800 eggs, either individually or in mass, on leaves, damp soil, or in fissures and crevices present in soil. These eggs are characterized by a dome shape and a creamy-white color and require approximately 4–7 days to hatch. The recently hatched larvae have a smooth, greasy, grayish-brown appearance and a plump look. The larvae undergo complete development within 22–30 days, attaining a length of approximately 4–5 cm when fully mature. Pupation takes place either in the soil or on folded leaves of potato plants, and the pupal stage lasts for 12–20 days. The male moth has a brief lifespan of 2–4 days, while the female moth survives for 5–8 days. Notably, cutworms can complete two generations within a single potato crop season.

5.2 Foliage Feeders or Defoliating Pests

5.2.1 Hadda Beetles

Scientific name *Epilachna dodecastigma* and *Epilachna vigintioctopunctata* (Fabricius) (Coleoptera: Coccinellidae)

Distribution The impact of *E. dodecastigma*, commonly known as the 12-spotted hadda beetle, is primarily confined to hilly regions. In contrast, the incidence of the 18-spotted hadda beetle, *E. vigintioctopunctata*, is observable in mid hills and plains.

Host range Hadda beetles exhibit a polyphagous nature with a broad host range, encompassing various plant species such as potato, tomato, brinjal, numerous other Solanum species, and, on occasion, cucurbits.

Damage symptoms Both hadda beetle grubs and adults contribute significantly to crop damage. They engage in voracious feeding on the leaves, actively scraping chlorophyll and leading to the skeletonization of leaves. The extensive damage to leaves manifests as a lace-like appearance, with severely affected leaves later transitioning to a brown, desiccated state and, in some instances, falling off. In favorable weather conditions, hadda beetles have the capacity to entirely defoliate potato crops before the maturation of tubers.

Bionomics The female beetle deposits approximately 500–700 eggs beneath the surface of the leaves, with a 3–4 days incubation period. The resulting grubs measure around 6 mm in length, displaying a yellow coloration and featuring six rows of branched spines on their dorsal surface. The larval phase spans 8–10 days, followed by pupation occurring on potato leaves, and the pupal stage lasting for 3–6 days (Trivedi and Rajagopal 1999). The entire life cycle gets completed within approximately 21–36 days. During the winter period, the adult beetles overwinter in weeds and grass.

5.2.2 Flea Beetle

Scientific name *Psyllodes plana* Maulik (Coleoptera: Chrysomelidae)

Distribution The distribution of the flea beetle spans across all the potato-growing regions of India.

Host range In addition to potatoes, the flea beetle also inflicts notable damage in other crops such as tomato, turnip, and brinjal.

Damage Symptoms Both flea beetle grubs and adults contribute to crop damage, with grubs primarily feeding on rootlets and adults targeting aerial parts, particularly leaves. The resulting damage persists throughout the crop season. Adult beetles create small and rounded holes in the leaves, giving rise to a distinctive symptom known as "Shot hole" (Chandla 1985). Upon emergence from the soil, adults commence their attack on the potato crop and continue causing damage until harvest.

Bionomics During the overwintering phase, adults seek refuge on the ventral surfaces of leaves, within grassy areas, and amidst residual crop remains, predominantly near the bunds of the field. Upon the cessation of adult diapause, these individuals initiate their feeding activities on weeds within and surrounding potato fields. The female beetle deposits approximately 50–80 eggs of a creamy white hue at the base of plants in the soil. The incubation period of the egg spans 6–7 days, leading to the emergence of whitish-colored grubs (Chandel et al. 2011). The larval development undergoes three molting, culminating in fully grown grubs measuring about 5 mm in length. These larvae (grubs) engage in soil penetration, subsequently feeding on crop rootlets for a duration of 2–3 weeks. The pupation within the soil occurs in earthen cells, resulting in pupae measuring approximately 0.5 mm in length. Adults exhibit variability in coloration, ranging from shiny black to black, with a prevailing occurrence of one to two generational cycles annually.

5.2.3 Colorado Potato Beetle

Scientific name *Leptinotarsa decemlineata* (Say) (Coleoptera: Chrysomelidae)

Distribution The indigenous habitat of *L. decemlineata* is Mexico, with its distribution having extended to Central America, a significant portion of the United States (excluding California, Alaska, and Nevada, Hawaii), and southern regions in Canada. Furthermore, the pest species was introduced into European territories as well as certain areas within the Asian continent (CABI 2017; Jacques and Fasulo 2015).

Host range This beetle is classified as an oligophagous pest, demonstrating a feeding preference for approximately 10 plant species within the family Solanaceae (Alyokhin et al. 2013). Noteworthy among these are potato, eggplant, and tomato.

Damage symptoms Both the adult and larval stages of the Colorado potato beetle exhibit voracious leaf-feeding behavior, capable of entirely defoliating the plants of potato. In exceptional instances, exposed tubers present at the soil surface may also become subject to consumption. The larvae and adults leave distinctive black and adhesive excreta on the stems and leaves.

Bionomics The reproductive capability of adult beetle is notable, with a single female laying a prolific range of 300–800 eggs. Adult beetles exhibit a facultative diapause overwintering within the soil. Emergence from diapause occurs in the early spring, facilitating copulation. The adult female deposits eggs on the undersides of host leaves in batches of approximately 24 over a 4–5 week period, resulting in a cumulative production of about 500 eggs/female. Subsequent to an incubation period of 4–9 days, the larvae emerge and promptly initiate feeding. The larval development encompasses four stages spanning 2–3 weeks, culminating in their entry into the soil, where they form spherical cells and undergo pupation. The emergence of new adults occurs within 5–10 days, initiating a repetition of the life cycle, leading to the development of a second generation.

5.2.4 Tobacco Cutworm

Scientific name *Spodoptera litura* (Fab.) (Lepidoptera: Noctuidae)

Distribution This pest is characterized by its sporadic occurrence, highly polyphagous feeding behavior, and extensive distribution across diverse regions worldwide.

Host range It is markedly polyphagous in nature, inflicting significant injury to prevalent crops such as tobacco, cotton, soybean, beet, cabbage, and chickpeas.

Damage symptoms Upon hatching, the early instars of the tobacco cutworm, *S. litura*, exhibit gregarious feeding behavior, collectively scraping the green matter from leaves. Subsequently, as the later instars progress, they disperse and engage in voracious feeding, capable of defoliating the entire plant (Butani and Jotwani 1984).

Bionomics The tobacco cutworm, exhibits nocturnal behavior and inflicts damage to crops during night. The female moth deposits approximately 400 eggs in clusters, with an incubation period ranging from 3 to 5 days. The caterpillar displays a pale brown coloration with a green or violet tinge, reaching a fully grown length of about 40–50 mm. The larval and pupal stages encompass periods of 20–28 days and 7–11 days, respectively. The complete life cycle of *S. litura* spans 30–40 days.

5.2.5 Gram Pod Borer

Scientific name *Helicoverpa armigera* (Hubner) (Lepidoptera: Noctuidae)

Distribution This pest is highly polyphagous in nature and distributed throughout India.

Host range In India, *H. armigera* (Hübner) is known to infest numerous economically significant crops, encompassing cotton, chickpea, sunflower, pigeon pea, chili, tomato, corn, and okra.

Damage symptoms The newly hatched larval instars exhibit gregarious feeding behavior by scraping leaves and causing leaf skeletonization. In contrast, mature larvae adopt solitary feeding habits. As the larvae progress to older instars, they tend to segregate and defoliate the leaves voraciously, causing absolute defoliation of plants. In instances where there is severe infestation, the entire crop may get defoliated overnight.

Bionomics The adult female deposits eggs singly on plants' leaves, and the eggs hatch within 3–7 days. The early larval instars engage in feeding by scraping the leaves, while the older instars are known to defoliate the plants. Fully grown larvae exhibit a length of approximately 35 mm, displaying a greenish to brown coloration with yellow colored dark stripes on the dorsal surface. The larval and pupal periods span 17–22 days and 6–12 days, respectively (Misra 1995). Pupation takes place in the soil, where the larvae overwinter in the pupal stage. The total life cycle is completed in approximately 4–6 weeks (Chandel et al. 2011), and the species can undergo five to eight generations within a year.

5.3 *Sucking Pests*

5.3.1 Green Peach Aphid

Scientific name *Myzus persicae* (Sulzer) (Hemiptera: Aphididae)

Distribution *M. persicae* (Sulzer) is recognized as the most significant pest of potato globally, with its believed origin traced to China. The green peach aphid demonstrates high adaptability and currently exhibits a cosmopolitan distribution.

This highly polyphagous species displays the capability to colonize over a hundred plant species from >40 (more than 40) different families (Blackman and Eastop 1994). Additionally, it serves as an efficient vector for more than 100 plant viruses.

Host range It exhibits an extraordinary degree of polyphagy, demonstrating the capability to feed on several hundred plant species.

Damage symptoms Aphids have the capacity to directly damage potatoes through sap-sucking activities, thereby weakening the plants. However, the indirect losses incurred due to the transmission of plant viruses by acting as vectors are more substantial (Naga et al. 2020). Common potato viruses transmitted by aphids encompass *Potato virus X* (PVX), *Potato virus Y* (PVY), *Potato virus A* (PVA), and *Potato leaf roll virus* (PLRV). While the direct losses resulting from aphids' sap-sucking behavior may be nonsignificant, the consequential losses due to the transmission of potato viruses are indeed significant.

Bionomics Adults of *M. persicae* (Sulzer) undergo asexual reproduction through parthenogenesis when infesting potatoes, resulting in the generation of genetically identical offspring. Apterous females exhibit variations in coloration, ranging from green to red, and possess an oval shape with a body length of 1.2–2.1 mm. Winged forms display a distinctive black dorsal abdominal patch. Under optimal temperature conditions, the lower and upper developmental thresholds of this pest are documented as 6.5 °C and 37.3 °C, respectively. Single female can lay approximately five offspring per week (Davis et al. 2006). However, under optimal fluctuating temperature conditions, this reproductive rate can increase to up to 12 offspring per week.

5.3.2 Whitefly

Scientific name *Bemisia tabaci* (Hemiptera: Aleyrodidae)

Distribution *Bemisia tabaci* is recognized as a polyphagous pest, extending its infestation beyond Solanaceae crops, including potato, tomato, and eggplant, to a broad spectrum of field crops and vegetables. Notably among these are sweet potato, cotton, cucumber, beans, soybean, melon, cassava, and numerous others.

Host range It is a sap-feeding insect prevalent in field crops located globally in tropical and subtropical regions. Being polyphagous, it targets plants across 63 families (Mound and Halsey 1978). The pest displays a preference for short-duration herbaceous plants, particularly various dicotyledonous crops in both agriculture and horticulture.

Damage symptoms Whitefly, due to polyphagous nature, causes notable direct losses to potatoes through the extraction of plant sap. However, its most significant economic impact lies in the transmission of plant viruses. Whiteflies act as vectors for the causative agent (ToLCNDVpotato) of apical leaf curl disease in potatoes (Naga et al. 2019). This disease has been detected in 40–70% of potato samples from the Indo-Gangetic plains (Venkatasalam et al. 2011). Occurrences of whitefly are more pronounced during the early growth phase of potatoes and are prevalent across all potato-growing areas in India, with an exception of the higher altitudes of the Himalayan region and the Western Ghats, where potatoes are cultivated during the summer season.

Bionomics The eggs of the whiteflies exhibit an ovoid or pyriform shape and are characterized by a peg-like extension of the chorion referred to as the pedicel. Adult females have the capacity to lay approximately 60 eggs under the surface of leaves, a capability influenced significantly by environmental conditions and the host plant. Upon hatching, the first instar crawlers emerge, displaying three segmented functional walking legs. These newly emerged crawlers traverse the leaf surface, seeking minor veins, and typically remain on the same leaf where the eggs were laid. The second and third instars of the whitefly are oval to elongate in shape and are sessile in nature. Additionally, the whitefly exhibits pupa-like instars that are flat and irregularly oval-shaped, featuring two to eight long dorsal setae. However, these setae are absent when the leaves possess a smooth surface. The complete life cycle of the whitefly, from egg to adult, spans approximately 15 days when infesting potatoes, a duration that may vary depending on the host plant (Fekrat and Shishehbor 2007).

5.3.3 Potato Leafhoppers

Scientific name *Empoasca devastans* Distant (Hemiptera: Cicadellidae)

Distribution The leafhopper *E. devastans* exhibits a widespread distribution. In India, they are prevalent in all potato-growing regions, including Punjab, Haryana, Gujarat, Maharashtra, Karnataka, Uttar Pradesh, and Bihar, where they pose a significant threat to potato production.

Host range It is a polyphagous pest, with primary hosts encompassing a range of crops. Noteworthy among these are potato, eggplant, alfalfa, celery, bean, cucumber, various cucurbits, melon, groundnut (*Arachis hypogaea* L.), rhubarb (*Rheum rhabarbarum* L.), tomato, strawberry, and sweet potato.

Damage symptoms Leafhoppers pose a notable threat through direct losses to crops resulting from their sap-sucking activities. Both the nymph and adult feed by extracting cell sap from the ventral surface of leaves. During this feeding process, leafhoppers inject toxic saliva, leading to symptoms such as yellowing, browning, cupping, and curling of leaves. In the field, characteristic V-shaped burns appear on the leaf tips of crop plants. The leaf margins become brittle and fragment into pieces upon crushing. Severe infestations present a scorched appearance referred to as "hopper burn." Additionally, leafhoppers, including *E. fabae*, contribute to the transmission of aster yellows mycoplasma-like organisms, resulting in the development of purple top disease in potatoes. The leafhopper exhibits year-round breeding potential but is particularly active from October to March, coinciding with the potato growing season. Infestations are more severe during the early growth phase.

The buildup of leafhopper populations is negatively impacted by heavy monsoon rains in hilly regions.

Bionomics The green leafhopper female lays approximately 200–300 fertilized eggs. These eggs are inserted into the leaf tissues or the stem of potatoes through the use of their ovipositor. The hatching period spans 4–10 days, leading to the emergence of light green, translucent, wingless nymphs that closely resemble adults in appearance, differing primarily in size and the absence of wings. Nymphs feed by extracting plant sap while residing on the undersurface of leaves. The process of development from nymph to adult takes approximately 15 days. Adults, on the basis of prevailing weather conditions, exhibit a lifespan ranging from 30 to 60 days. The green leafhopper has the capacity to complete 2–4 generations in a single year when infesting potatoes.

5.3.4 Thrips

Scientific name *Thrips palmi* Karny, *Scirtothrips dorsalis* Hood, *Frankliniella occidentalis* (Pergande) (Thysanoptera: Thripdae)

Distribution *T. palmi* stands as a predominant species recorded across the potato-growing states of India. These include Himachal Pradesh, Haryana, Punjab, Delhi, Madhya Pradesh (MP), Jammu and Kashmir, Chhattisgarh, Rajasthan, Maharashtra, Odisha, Karnataka, Tamil Nadu, Uttar Pradesh, and West Bengal.

Host range It is a polyphagous insect demonstrating the capability to feed on more than 200 plant species encompassing vegetables, fruits, and ornamental plants.

Damage symptoms The impact of thrips damage is particularly pronounced during dry climatic conditions. Both nymphs and adults feed on leaves, by concentrating on veins, and midribs while also targeting stems, particularly in the vicinity of the growing tip. Intensive feeding on potato leaves results in the development of silver or chlorotic dotting on the tissues, leading to deformation. Continuous feeding exacerbates the issue, causing the leaves' tips to wither, curl up, and eventually die. Additionally, they serve as vectors for the groundnut bud necrosis virus in potatoes, inducing stem necrosis disease (Raigond et al. 2017). Thrips infestations can cause a significant yield loss of approximately 15–30% in potato crops in northern Gujarat, parts of MP, and Rajasthan.

Bionomics *T. palmi* exhibits a pale yellow to whitish coloration with many dark setae on the body surface and possesses pale and slender fringed wings. Adult specimens have a body length ranging from 0.8 to 1.0 mm, with females slightly larger than males. The developmental period varies between 12 and 20 days, contingent upon prevailing temperature conditions. Parthenogenesis is a common reproductive mechanism in thrips, where a single female inserts approximately 50 eggs by making a slit cut into leaf tissues. The eggs take about 4.3–16 days to hatch into larvae. The larva closely resembles adult except for the size and presence of wing pads. The larvae progress to inactive and nonfeeding pre-pupal stages over a span of 5–15 days. Subsequently, the pre-pupae and pupae undergo a transformation into fully grown adults, a process that takes about 3–12 days, influenced by the prevailing temperature. The buildup of thrips populations is supported by elevated temperatures (30–35 °C) and low relative humidity. These environmental conditions typically coincide with the early potato production growth phase, occurring during September–October in plains.

5.4 Mites

Scientific name Broad mite, *Polyphagotarsonemus latus* Banks (Acarina: Tarsonemidae)

Red spider mite, *Tetranychus neocaledonicus* André (Acarina: Tetranychidae)

Distribution Broad mite infestations are notably more severe in potato crops cultivated during the Kharif season in Maharashtra and Karnataka. Reports also indicate instances of broad mite infestations affecting potatoes in Punjab, Gwalior (Madhya Pradesh), Kangra Valley (Himachal Pradesh), and Western Uttar Pradesh, particularly in early planted crops during the Rabi season.

Host range Mites, particularly broad mites, are characterized by a highly polyphagous nature, known to infest a diverse array of agriculture and horticulture crops. Among these are chili, tomato, brinjal, tea, cotton, beans, apple, mango, grapes, citrus, guava, papaya, pear, and coffee.

Damage symptoms Mites typically inhabit the upper portions of plants. Their feeding behavior involves extracting cell sap from the lower surface of young leaves and apical shoots. During the feeding process, mites inject toxic saliva, resulting in curling, bronzing, and discoloration of leaves. Plants that are severely damaged by *P. latus* exhibit stunted growth and do not successfully form tubers (Kreuze et al. 2020).

Bionomics The broad mite, *P. latus*, is exceptionally small, nearly microscopic, with the female measuring approximately 0.2 mm. They exhibit a light yellow to amber coloration, featuring a faint and indistinct median stripe that forks near the posterior end of the body. Males share a similar appearance with females, differing primarily in size and the absence of a distinct median stripe. The life cycle of a broad mite comprises four stages: egg, larva, nymph, and adult. Adult females lay around 30–76 eggs on the ventral surface of the leaves, and these eggs typically hatch within 2–3 days. The newly emerged larva is slow-moving and exhibits limited dispersal. After 2–3 days, the larva enters a quiescent larval (nymph) stage. The emergence of females from this quiescent stage is commonly referred to as the yellow mite stage.

6 Viruses of Potato

Fifty distinct viruses and one viroid have been documented to infect potatoes globally. While only a limited number of these viruses result in significant global losses, some exhibit local and/or temporary relevance, while others currently hold minor importance worldwide. Among the consequential viruses are PVY, PLRV, PVA, PVM, and, PVS, all of which are transmitted by aphids (Table 3). Except for PLRV, these viruses are nonpersistently transmitted by aphids, whereas PLRV is persistently transmitted. The efficacy of insecticides has traditionally been observed

Table 3 List of viruses of potato which are found worldwide

S. no.	Viruses	Genus, family	Transmission	Distribution
1.	*Potato virus Y* (PVY)	*Potyvirus*, Potyviridae	Aphids	Worldwide
2.	*Potato virus A* (PVA)			Worldwide
3.	*Potato virus X* (PVX)	*Potexvirus*, Alphaflexiviridae	Contact	Worldwide
4.	*Potato leaf roll virus* (PLRV)	*Polerovirus*	Aphids	Worldwide
5.	*Tomato spotted wilt virus* (TSWV)	*Tospovirus*, Bunyaviridae	Thrips	Worldwide
6.	*Tomato chlorotic spot virus* (TCSV)			South America
7.	*Groundnut bud necrosis virus* (GBNV)			India
8.	*Groundnut ringspot virus* (GRSV)			Americas
9.	*Tomato chlorosis virus* (ToCV)	*Crinivirus*, Closteroviridae	Whiteflies	Brazil, Spain, India
10.	*Tobacco streak virus* (TSV)	*Ilarvirus*, Bromoviridae	Pollen, thrips	Worldwide, reported in potato in Brazil
11.	*Alfalfa mosaic virus* (AlMV)	*Alfamovirus*, Bromoviridae	Aphids	Worldwide, sporadic in potato
12.	*Tomato leaf curl New Delhi virus* (ToLCNDV)	*Begomovirus*, Geminiviridae	Whiteflies	India
13.	*Beet curly top virus* (BCtV)	*Curtovirus*, Geminiviridae	Leaf hopper	Americas, Europe, Asia under dry conditions
14.	*Tobacco rattle virus* (TRV)	*Tobravirus*, Virgaviridae	Nematodes	Worldwide, common in cool climates, or Australia
15.	*Potato spindle tuber viroid* (PSTVd)	*Pospiviroid*, Pospiviroidae	Contact, aphids (when infecting along with PLRV)	Worldwide

Source: Kreuze et al. (2020)

against persistently transmitted viruses, likely contributing to the decline in the prevalence of PLRV over the past five decades in both developed and emerging economies.

Viruses transmitted by thrips and whiteflies persist to incite outbursts in potatoes within climatic regions which are warmer. The frequency of outbursts associated with viruses transmitted by thrips and whiteflies is steadily increasing because of global climate warming. Notably, one of these viruses, *Tomato Leaf Curl New Delhi Virus* (ToLCNDV), has recently emerged as a major potato pathogen in India (Jeevalatha et al. 2017).

7 Integrated Pest Management in Potato

While integrated pest management (IPM) has gained widespread attention from researchers, policymakers, and agricultural practitioners, its universal adoption remains a distant goal. Nonetheless, there have been substantial shifts toward incorporating IPM into mainstream agricultural production. The success of IPM hinges on interdisciplinary research and the implementation of effective nonchemical control methods tailored to the specific agroecology of a given crop, aiming to minimize or eliminate pesticide usage. Despite these aspirations, many researches focused on IPM and tend to concentrate on plant protection methods individually, neglecting systematic examinations of the optimization and compatibility of concurrently applied techniques of pest management (Stenberg 2017). The International Potato Center (CIP) has pioneered an extensive working framework, for potato pest management research and development, which has practical applications for farmers. This holistic way has led to the establishment of IPM strategies, tailored to diverse agroecologies for potatoes in Peru, particularly under production for smallholders. These strategies prioritize ecological, environmental, and economic benefits (Mujica and Kroschel 2018).

Although individual IPM programs of potato exhibit variability based on specific agroecologies and socioeconomic surroundings, they share key components that set them apart from conventional approaches. Unlike calendar-based pesticide applications, IPM programs emphasize a multifaceted and context-specific approach to manage pests. The implementation of IPM involves several key components and practices to achieve effective and sustainable pest control in potato cultivation:

- *Suitable Cultural Practices:* This encompasses the utilization of seed that is healthy, appropriate intercropping, and crop rotation systems. These practices not only contribute to crop health but also support mechanisms of natural and biological control.
- *Correct and Timely Pest Identification:* Accurate identification and monitoring of pest populations on a regular basis are essential. Monitoring tools, viz. pheromone or physical traps like yellow sticky traps, aid in this process. Educated pest

management leverages specific aspects of pest biology to diminish populations preventively.
- *Economic Thresholds Application:* Rather than attempting to complete the elimination of pests, IPM advocates for control actions when the cost of pest damage exceeds the expense of control measures. While control thresholds exist for various potato pests, they require verification and adjustment based on specific locations and the conditions of the used potato variety.
- *Diversified Control Tactics:* An amalgamation of a controlled set of strategies, which include regulatory, cultural, biological, and pesticides, should be employed in a complementary or synergistic manner. Pesticides are considered a last resort and are integrated with other controls to form a comprehensive and unified approach against pest populations.
- *Evaluation of Results:* Continuous monitoring and assessment of the effectiveness of IPM control programs are crucial. Regular evaluations allow for necessary adjustments, considering economic, ecological, and environmental factors.

These principles contribute to a holistic and sustainable approach to pest management in potato cultivation, aligning them with the goals of IPM.

7.1 Cultural Control

- *Deep Tilling:* Utilizing deep tillage practices results in the exposure of white grubs to predators or adverse climatic conditions such as sunlight or frost.
- *Crop Rotation:* Implementing crop rotation of main crop with nonhost or less susceptible crops, like alfalfa, cowpea, pigeon peas (deep-rooted legumes), onions, garlic (*Allium* crops), and cruciferous crops for white grub, potato tuber moth, and cutworms, respectively. Additionally, employing the African marigold (*Tagetes* sp.), castor, and sunflower as trap crops.
- Ensuring field sanitation and balanced fertilizers, complete elimination of plant debris after harvest, planting seed tubers at a depth of 10 cm, and earthing up 6–7 weeks after planting to reduce damage by potato tuber moth.
- *Cutworm Control:* For cutworms, implementing earthing up to prevent tuber exposure to cutworm larvae. Flooding potato fields before planting to kill cutworms in the soil. Using light traps for mass collection and destruction of adult cutworm moths.
- *Deep Summer Ploughing:* Utilizing deep summer ploughing to eliminate overwintering populations.
- *Irrigation:* Implementing irrigation in infested crops to minimize hadda beetle populations.
- *Intercropping:* Employing intercropping with garlic, onion, or coriander to negatively impact aphid populations. Using high-quality disease-free seed tubers and removing diseased potato plants to reduce the spread of potato viruses by aphids.

- *Varietal Tolerance:* Cultivating potato varieties like cv. Kufri Bahar, which are tolerant to whitefly, and destroying alternate hosts to mitigate whitefly incidence.
- *Thrips Management:* Avoiding intercropping with onion, tomato, corn, artichoke, and pumpkin, as these crops support high thrips populations. Opting for late planting of potatoes to reduce thrips incidence, avoiding early planting.
- *Mite Infestation Control:* Delaying planting to minimize mite infestation. Utilizing sprinkler irrigation to restrict the incidence of mites by increasing humidity on plant leaves. Ensuring proper field sanitation by removing infested plants and volunteer crops to check mite populations.
- *Weed Removal:* Implementing weed removal as an effective strategy to control various insect pests.

7.2 Monitoring

- *Pheromone Traps:* Employing pheromone traps for both monitoring and mass trapping. Installation of a large number of traps to reduce the adult male moth population. For monitoring purposes, installing five traps/hectare.
- *Sex Pheromone Traps:* Using yellow and white traps containing sex pheromones to predict potential attacks at the time of spring. Larval populations can also be monitored by baited traps in the field.
- *Yellow Sticky Traps:* Utilizing these sticky traps as a valuable tool for monitoring aphids, whiteflies, and leafhoppers. These traps aid in the timely detection and documentation of relative abundance of whiteflies over time.
- *Haulm Cutting:* Recommending haulm cutting for seed potato crops when aphid populations surpass the economic threshold level of 20 aphids per 100 compound leaves. Cultivating seed potatoes during the winter season, in a low aphid or aphid-free window, is known as the seed plot technique.
- *Leaf Hopper Monitoring:* Monitoring nymphs by observing the underside of potato leaves, especially middle leaves. The use of entomological nets for monitoring hopper populations. The economic threshold for leaf hoppers in potatoes is ten or more than 10 nymphs/100 leaves, or 10% of nymph-infested leaves (Cook et al. 2004).
- *Monitoring of Thrips Population:* Regularly monitoring thrips populations using blue sticky traps and taking prompt action upon the appearance of thrips or their damage symptoms.
- *Monitoring of Mite Infestation:* Like aphids and whiteflies, infestations of mite typically start at the edges of potato fields, especially the ones which are adjacent to other fields. Dry weather conditions favor mite buildup, emphasizing the importance of constant monitoring and early detection to manage mite incidence.

7.3 Mechanical Control

- *Light Trap Installation:* Deploying light traps at night during May–June for mass trapping of adult beetles. Subsequently, immersing the trapped adults in kerosene or chemically treated water to effectively reduce the incidences of white grub in potato crops.
- *Mechanical Collection:* Employing mechanical methods in order to collect and destroy the grubs and various life stages of the pests. This approach is recommended to effectively mitigate pest populations.
- *Egg Mass and Larval Collection:* Identifying visible egg masses and early gregarious larvae, especially in the case of *H. armigera*. Utilizing mechanical means to collect and destroy these early stage caterpillars of *S. litura*, effectively managing their populations.

7.4 Biological Control

- Various immature stages, particularly nymphs, are susceptible to parasitization by entomopathogenic fungi (EPF), including *Verticillium lecanii* Zimmerman, *Paecilomyces fomosoroseus* Wise, and *P. farinosus* Holmsk. These fungi have been harnessed for development as biopesticides.
- The conservation of Anthocorid bugs, specifically *Orius maxidentex* and *Carayonocoris indicus*, holds significant potential in naturally controlling thrips populations.
- The predatory mite *Amblyseius barkeri* Hughes (Acarina: Phytoseiidae) is recognized for effectively regulating the population of broad mites. Therefore, emphasis should be placed on the conservation of these predatory mites.

7.5 Microbial Control

- The application of the entomopathogenic fungus *Metarrhizium anisopliae* at a concentration of 5 gm/liter of water demonstrates effective control over the incidence of white grubs.
- Utilizing *Bacillus thuringiensis* ssp. *kurstaki* (*Btk*) provides protection against potato tuber moth. Treatment involves the direct application (powdered) to potato tubers immediately after harvest, followed by storage. Additionally, a granulosis virus-based biopesticide, PhopGV, formulated in powder form, can also be employed to safeguard stored potatoes.

7.6 *Chemical Control*

7.6.1 White Grub Control

- Drenching of the ridges with Chlorpyrifos 20 EC at 2.5 ml/l of water at the time of appearance of adult beetles of white grub.
- Spraying the host trees with Chlorpyrifos 20 EC at 2 ml/l of water immediately after the first monsoon shower.

7.6.2 General Insecticide Application for Potato Crop

- Spraying Deltamethrin 2.8EC (1350 ml) or Cypermethrin 25EC (450 ml) in 750 liters of water/ha on 30-day-old crop.
- Soil application of Chlorpyrifos 20 EC at 2 ml/l water at 2% of plant damage.
- Need-based sprays of Indoxacarb 15.8 EC at 0.06%, Flubendiamide 20 WDG at 0.06%, or Emamectin benzoate 5% SG at 220 g/ha to control caterpillars.
- Azadirachtin 0.03% WSP at 2500–5000 g/ha and *Bt* serovar *kurstaki* 5% WP at 1000–1250 g/ha can also be effective against caterpillars.

7.6.3 Aphid and Virus Management

- Application of mineral oils at 1–3% has been observed to protect the plants of potatoes from aphids as well as the viral diseases transmitted by aphids.
- Avoiding broad-spectrum insecticides to prevent a flare-up of *Myzus persicae* population.
- Avoiding synthetic pyrethroids to prevent whitefly resurgence.
- Foliar application of Imidacloprid 17.8 SL at 0.03% at 75% crop emergence followed by the spray of Thiamethoxam 25WG at 0.05% after 15 days for managing aphids, whiteflies, and leaf hoppers. Repeat as needed.

7.6.4 Mite Control

- Application of wettable sulfur at 3 g/l water for the prevention of damage by mite.
- Application of Spiromesifen 22.9 SC (40–75 ml/100 liters) for effective control of mite infestation; repeat the spray after 7 days if necessary.

7.7 *Host Plant Resistance (HPR)*

HPR has a long history in agriculture, dating back at least 200 years. The increasing significance of host-plant resistance is evident in the context of modern agriculture, marked by the ongoing integration of agricultural landscapes. Before the green revolution, the crops were characterized by diversity and typically found in landscapes that were heterogeneous, which included trees, woodlands, and many other naturally found habitats. However, the shift to monocultures has led to a reduction in landscape diversity, resulting in the loss of habitats for natural enemies. This reduction in diversity has heightened the susceptibility of the crops to pests (Meehan et al. 2011). Factors such as expansion of agriculture, decline in agricultural diversity, and diminishing of genetic diversity of crops underscore the need for a greater emphasis on intrinsic or innate plant protection. In the 2000s, research into HPR, both conventional and transgenic techniques, emerged as a dominant focus in plant protection strategies. This trend overshadowed other management approaches, including research on natural enemies of pests and different methods of biocontrol. Notably, in the case of tuber moths, substantial progress has been achieved in utilizing pheromones for pest management.

7.7.1 Native and Crop Potatoes as Sources of Resistance

Solanum tuberosum, commonly referred to as the crop potato, is cultivated worldwide, alongside eight other diploid and tetraploid species of potato which are generally known as native potatoes. These native potatoes, such as *S. pureja* Juz. and Bukasov, *S. goniocalyx* Juz. and Bukasov, and *S. tuberosum andigena*, have been domesticated, particularly in Andean regions. *S. t. andigena* is suggested to be the precursor of cultivated potato crops (National Research Council 1989). While the majority of potato varieties that are edible, including the crop and native potatoes, are vulnerable to a number of pests (Jansky et al. 2009), there are significant variations in the pest performance observed among a number of crops and native cultivars of potatoes. Many varieties have the ability to mitigate the damage caused by insect pests in the field as well as storage conditions. Notably, in Peru, Horgan et al. (2012) identified the levels of high resistance (up to 55%) against the two potato tuber moth species in native purple potatoes (*S. t. andigena*). This finding suggests that the mechanisms of resistance are suitably broad-spectrum and general. Moreover, even in some cases of breeding materials where resistance is initially low, careful attention during breeding programs for the resistant traits can enhance the levels of resistance.

7.7.2 Wild Potatoes: Sources of Resistance

In the era of breeding of potatoes for enhanced insect- and disease-resistance, around 230 tuber-producing wild *Solanum* species play a crucial role (Flanders et al. 1999). These wild potato species are indigenous to regions extending from the southwestern United States, through Central America and Mexico, along South America, and into the plains of Argentina, Uruguay, and Paraguay. While nearly all species exhibit a restricted geographical range determined by particular ecogeographic attributes, a few are distributed widely. Various wild *Solanum* species provide valuable genetic materials for potato breeders. Latest cellular techniques, like somatic hybridization, offer solutions to reduce reproductive barriers between incompatible and compatible species of potato crop (Chen et al. 2008). Two remarkable wild species which are studied extensively for their resistance properties are *S. berthaultii* Hawkes and *S. chacoense* Bitt. These species showcase distinct modes of action, with *S. berthaultii* relying on glandular trichomes, and *S. chacoense* employing defensive glycoalkaloids. Nevertheless, evidence shows that many other mechanisms of resistance may contribute to the defense mechanisms in wild species of potatoes and their respective interspecific hybrids.

7.7.3 Utilizing Genetic Transformation Methods

Advancements in tissue culture and genetic transformation have paved the way for innovative strategies for pest control centered around the utilization of resistant "transgenic" plants. These methods involve the incorporation of resistance factors which encode new genes into potato crop's genome. These genes can be sourced from various origins, which include wild *Solanum* species (Sagredo et al. 2006), various other crops (Lecardonnel et al. 1999), and many other bacteria (Mohammed et al. 2000). To enhance the effectiveness of resistance factors, promoters that are specific to green tissue or tubers may be employed, restricting the expression of transgene to either above or below ground plant tissues. This not only reduces costs to the plant but also enhances the efficacy of factors resulting in direct resistance, addressing the concerns of consumers about the potential negative impacts of these novel proteins (Kumar et al. 2010). Consequently, as conventional breeding methods are limited to the introgression of the gene from those species that are closely related, genetic transformation significantly expands the potential or ability to discover the latest resistance sources.

8 Storage Management

Control measures for potato tuber moth (PTM) must be implemented both in fields and during storage. Infestations are common in potato stores that are managed by farmers in developing countries, particularly when conditions are conducive to rapid population growth, and storage persists for several months. To mitigate infestations, storage facilities should undergo thorough cleaning before storing potato tubers. The installation of fine nets in the windows can prevent the entry of adult potato tuber moths into the storage facilities. Selection of the healthy tubers should be done for storage, and any infested ones must be promptly destroyed. Despite these precautions, initial infestations may not be easily detectable, emphasizing the importance of implementing preventive measures. Water traps with sex pheromones in it, or Delta traps (funnel traps) can be employed to monitor the moths. In regions where PTM is known to occur, treatment of potatoes should be done before storing them.

9 Conclusion and Future Thrust

Given the global significance of potatoes as staple food and their susceptibility to a wide range of pests, effective pest management strategies are crucial to ensure sustainable, resilient, and economically viable potato production. While the popularity of potatoes is increasing in developing countries due to their nutritional and agronomic traits, the overreliance on synthetic insecticides poses a huge risk to the health of humans as well as the environment. Therefore, there is an urgent need to adopt IPM approaches to address these challenges.

The development, popularization, adaptation, and implementation of IPM strategies will be a key focus of future research in potato production worldwide. Biological control or biocontrol, as a component of IPM, plays an important part in regulating insect populations without the negative consequences associated with synthetic insecticides. Emphasizing the identification and utilization of locally adapted biological control agents is essential in reducing dependence on chemical interventions. This approach not only minimizes the risk of insecticidal residues in food and the environment but also helps to mitigate the chances of development of resistance in pest populations due to insecticides.

References

Alyokhin A, Chen YH, Udalov M, Benkovskaya G, Lindström L (2013) Evolutionary considerations in potato pest management. In: Giordanengo P, Vincent C, Alyokhin A (eds) Insect pests of potato: global perspectives on biology and management. Academic Press, Oxford, pp 543–571

Andreadis SS, Spanoudis CG, Zakka G, Aslanidou B, Noukari S, Savopoulou-Soultani M (2017) Effect of temperature on rate of development, survival and adult longevity of Phthorimaea operculella (Lepidoptera: Gelechiidae). Eur. J. Entomol. 114:35–41

Blackman RL, Eastop VF (1994) Aphids on the world's trees: an identification and information guide. In: Aphids on the world's trees: an identification and information guide. https://doi.org/10.1080/00305316.2001.10417292

Butani DK, Jotwani MG (1984) Insects in vegetables. Colour Publications, Mumbai

CABI (2017) Leptinotarsa decemlineata (Colorado potato beetle) data sheet. In: Invasive species compendium. CAB International, Wallingford. http://www.cabi.org/isc/datasheet/30380. Accessed 25 Feb 2018

Carrillo CC, Fu Z, Burckhardt D (2019) First record of the tomato potato psyllid *Bactericera cockerelli* from South America. Bull Insectology 72:85–91

Chandel RS, Chandla VK (2003) Managing tuber damaging pests of potato. Indian Hortic 48:15–17

Chandel RS, Kumar R, Kashyap NP (2001) Bioecology of potato tuber moth, *Phthorimaea operculella* Zeller in mid hills of Himachal Pradesh. J Entomol Res 25:195–203

Chandel RS, Dhiman KR, Chandla VK, Kumar V (2007) Integrated pest management in potato. In: Jain PC, Bhargava MC (eds) Entomology: novel approaches in entomology. New India Publishing Agency, New Delhi, pp 377–398

Chandel RS, Sharma PC, Verma KS, Mehta PK, Vinod K (2011) Insect pests of potato—III: leaf eating and defoliating insects. Pestology 35:60–66

Chandel RS, Chandla VK, Verma KS, Pathania M (2013) Insect pests of potato in India: biology and management. In: Giordanengo P, Alyokhin A, Vincent C (eds) Insect pests of potato. Global perspectives on biology and management. Elsevier, Waltham, pp 227–265

Chandel RS, Vashisth S, Soni S, Kumar R, Kumar V (2020) The Potato Tuber Moth, *Phthorimaea operculella* (Zeller), in India: Biology, Ecology, and Control. Potato Res 63:15–39. https://doi.org/10.1007/s11540-019-09426-z

Chandla VK (1985) Potato pests and their management. Indian Farming 34:31–32

Chen Q, Li HY, Shi YZ, Beasley D, Bizimungu B, Goettel MS (2008) Development of an effective protoplast fusion system for production of new potatoes with disease and insect resistance using Mexican wild potato species as gene pools. Can J Plant Sci 88:611–619

Cook KA, Ratcliffe ST, Gray ME, Steffey KL (2004) Potato leafhopper (*Empoasca fabae* (Harris)). Insect Fact Sheet, University of Illinois, Department of Crop Sciences. https://ipm.illinois.edu/fieldcrops/insects/potato_leafhopper.pdf

Davis JA, Radcliffe EB, Ragsdale DW (2006) Effects of high and fluctuating temperatures on *Myzus persicae* (Hemiptera: Aphididae). Environ Entomol 35(6):1461–1468

Douches DS, Pett W, Santos F, Coombs J, Grafius E, Li W, Metry EA, El-din TN, Madkour M (2004) Field and storage testing *Bt* potatoes for resistance to potato tuberworm (Lepidoptera: Gelichiidae). J Econ Entomol 97:1425–1431

Fekrat L, Shishehbor P (2007) Some biological features of cotton whitefly, *Bemisia tabaci* (Homoptera: Aleyrodidae) on various host plants. Pak J Biol Sci. https://doi.org/10.3923/pjbs.2007.3180.3184

Flanders KL, Arnone S, Radcliffe EB (1999) The potato: genetic resources and insect resistance. In: Clement SL, Quinsberry SS (eds) Global plant genetic resources for insect-resistance crops. CRC Press, Boca Raton, pp 207–239

Horgan FG, Quiring DT, Lagnaoui A, Pelletier Y (2012) Life-histories and fitness of two tuber moth species feeding on native Andean potatoes. Neotrop Entomol. in press

Jacques RL, Fasulo TR (2015) Colorado potato beetle *Leptinotarsa decemlineata*. DPI Entomology Circular 271. Florida Department of Agriculture and Consumer Services, Division of Plant Industry and University of Florida. http://entnemdept.ufl.edu/creatures/veg/leaf/potato_beetles.htm

Jansky SH, Jin LP, Xie KY, Xie CH, Spooner DM (2009) Potato production and 793 breeding in China. Potato Research 52:57–65

Jeevalatha A, Chakrabarti SK, Sharma S, Sagar V, Malik K, Raigond B, Singh BP (2017) Diversity analysis of Tomato leaf curl New Delhi virus–[potato], causing apical leaf curl disease of potato in India. Phytoparasitica 45:33–43. https://doi.org/10.1007/s12600-017-0563

Kreuze JF, Souza-Dias JAC, Jeevalatha A, Figueira AR, Valkonen JPT, Jones RAC (2020) Viral diseases in potato. In: The potato crop. https://doi.org/10.1007/978-3-030-28683-5_11

Kroschel J, Schaub B (2013) Biology and ecology of potato tuber moths as major pests of potato. In: Giordanengo P, Vincent C, Alyokhin A (eds) Insect pests of potato: global perspectives on biology and management. Academic, Oxford, pp 165–192

Kroschel J, Mujica N, Carhuapoma P, Sporleder M (2016a) Pest Distribution and Risk Atlas for Africa—potential global and regional distribution and abundance of agricultural and horticultural pests and associated biocontrol agents under current and future climates. International Potato Center (CIP), Lima, p 650. https://doi.org/10.4160/9789290604761. (online publication: http://cipotato.org/riskatlasforafrica/)

Kroschel J, Sporleder M, Carhuapoma P (2016b) Potato tuber moth, *Phthorimaea operculella* (Zeller). In: Kroschel J, Mujica N, Carhuapoma P, Sporleder M (eds) Pest distribution and risk atlas for Africa—potential global and regional distribution and abundance of agricultural and horticultural pests and associated biocontrol agents under current and future climates. International Potato Center, Lima, pp 7–23. https://doi.org/10.4160/9789290604761. (online publication: http://cipotato.org/riskatlasforafrica/)

Kumar M, Chimote V, Singh R, Mishra GP, Nail PS, Pandey SK, Chakrabarti SK (2010) Development of *Bt* transgenic potatoes for effective control of potato tuber moth by using cry1Ab gene regulated by GBSS promoter. Crop Prot 29:121–127

Lecardonnel A, Chauvin L, Jouanin L, Beaujean A, Prévost G, Sangwann-Norreel B (1999) Effects of rice cystatin I espression in trasngenic potato on Colorado potato beetle larvae. Plant Sci 140:71–79

Lutaladio N, Ortiz O, Haverkort A, Caldiz D (2009) Sustainable potato production; guidelines for developing countries. FAO, Rome

Meehan TD, Werling BP, Landis DA, Gratton C (2011) Agricultural landscape simplification and insecticide use in the Midwestern United States. Proc Natl Acad Sci 108:11500–11505

Misra SS (1995) Potato pests and their management, Technical Bulletin No. 45. CPRI, Shimla

Misra SS, Chandel RS (2003) Potato white grubs in India, Technical Bulletin No. 60. CPRI, Shimla

Mohammed A, Douches DS, Pett W, Grafius E, Coombs J, Liswidowati L, Li W, Madkour MA (2000) Evaluation of potato tuber moth (Lepidoptera: Gelechiidae) resistance in tubers of Bt-cry5 transgenic potato lines. J Econ Entomol 93:472–476

Mound LA, Halsey SH (1978) Whitefly of the World. A Systematic Catalogue of the Aleyrodidae (Homoptera) with Host Plant and Natural Enemy Data. British Museum of Natural History and John Wiley & Sons, London, pp 340

Mujica N (2016) Ecological approaches to manage the leafminer fly *Liriomyza huidobrensis* (Blanchard) in potato-based agroecosystems of Peru. In: Kroschel J (ed) Tropical agriculture 21–Advances in crop research, vol 11. Margraf Verlag, Weikersheim, p 255

Mujica N, Kroschel J (2018) Ecological, economic and environmental assessments of Integrated Pest Management technologies: case study for potato lowland systems in the Cañete valley. Food Energy Secur 8:e00153. https://doi.org/10.1002/fes3.153

Mujica N, Sporleder M, Carhuapoma P, Kroschel J (2017) A temperature-dependent phenology model for the leafminer fly Liriomyza huidobrensis (Blanchard) (Diptera: Agromyzidae). J Econ Entomol 110:1333–1344

Naga KC, Buckseth T, Kumar R, Subhash S, Sundaresha S, Shah MA, Chakrabarti SK (2019) Whitefly mediated PALCV transmission in potato and its endosymbionts profile. In: International conference on plant protection in horticulture crops: advances and challenges, 24–27 July, ICAR-IIHR, Bengaluru, India, p 109

Naga KC, Bairwa A, Verma G, Kumar R (2020) Transmission efficiency of potato leaf roll virus (Plrv) by potato aphid *Aulacorthum solani* and green peach aphid *Myzus persicae*. https://doi.org/10.5958/0974-8172.2020.00017.6

National Research Council (1989) The crops of the Incas: little-known plants of the Andes and the promise for worldwide cultivation. National Academy Press, Washington, DC, p 415

Oerke EC, Weber A, Dehne DW, Schönbeck F (1994) Crop production and crop protection—estimated losses in major food and cash crops. Elsevier, Amsterdam, p 742

Orlandini S, Magarey RD, Woo Park E, Sporleder M, Kroschel J (2017) Methods of agroclimatology: modeling approaches for pests and diseases. In: Hatfield JL, MVK S, Prueger JH (eds) Agroclimatology. Linking agriculture to climate, Agron Monogr 60. ASA, CSSA, and SSSA, Madison. https://doi.org/10.2134/agronmonogr60.2016.0027

Pathania M, Chandel RS (2017) Life history strategy and behaviour of white grub, *Brahmina coriacea* (Hope) (Coleoptera: Scarabaeidae: Melolonthinae) an invasive pest of potato and apple agro-ecosystem in northwestern India. Orient Insects 51:46–69. https://doi.org/10.1080/00305316.2016.1247756

Radcliffe EB (1982) Insect pests of potato. Annu Rev Entomol 27:173–204

Raigond B, Sharma P, Kochhar T, Roach S, Verma A, Jeevalatha A, Verma G, Sharma S, Chakrabarti SK (2017) Occurrence of Groundnut bud necrosis virus on potato in North Western hills of India. Indian Phytopathol 70:478–482. https://doi.org/10.24838/ip.2017.v70.i4.76993

Sagredo B, Lafta A, Casper H, Lorenzen J (2006) Mapping of genes associated with leptine content of tetraploid potato. Theor Appl Genet 114:131–142

Shah MA, Bairwa A, Naga KC, Subhash S, Raghavendra KV, Mhatre PH, Bhatnagar A, Malik K, Venkatasalam EP, Sharma S (2020) Important potato pests and their management. In: Singh AK. Chakrabarti SK, Singh B, Sharma J, Dua VK (eds) Potato science and technology for subtropics. New India Publishing Agency, New Delhi, pp 295–326

Singh MB, Bhagat RM, Sharma DC (1990) Life history and host range of potato tuber moth (*Phthorimaea operculella* Zeller). Himachal J Agric Res 16:59–62

Tiwari RK, Lal MK, Naga KC, Kumar R, Chourasia KN, Subhash S, Kumar D, Sharma S (2020a) Emerging roles of melatonin in mitigating abiotic and biotic stresses of horticultural crops. Sci Hortic 272:109592. https://doi.org/10.1016/j.scienta.2020.109592

Tiwari RK, Kumar R, Sharma S, Naga KC, Subhash S, Sagar V (2020b) Continuous and emerging challenges of silver scurf disease in potato. https://doi.org/10.1080/09670874.2020.1795302

Trivedi TP, Rajagopal D (1999) Integrated pest management in potato. In: Upadhayay RK, Mukeriji KG, Dubey OP (eds) IPM system in agriculture cash crops, vol 6. Aditya Books Pvt. Ltd., New Delhi, pp 299–313

Trivedi TP, Rajagopal D, Tandon PL (1994) Environmental correlates of the potato tuber moth, *Phthorimaea operculella* (Zeller) (Lepidoptera: Gelechiidae). Int J Pest Manag 40:305–308

Venkatasalam EP, Singh S, Sivalingam PN, Malathi VG, Das Garg I (2011) Polymerase chain reaction and nucleic acid spot hybridisation detection of begomovirus(es) associated with apical leaf curl disease of potato. Arch Phytopathol Plant Protect. https://doi.org/10.1080/03235401003633741

Advances in Disease Management of Potatoes

Santosh Kumar, Mahendra Singh, Vinit Pratap Singh, Ram Chandra, and S. Vinod Kumar

Abstract Potato popularly known as vegetable king is globally one of the utmost essential economic staple crops after rice and is cultivated in more than 100 countries. The underground part of the stem is the edible part of potato, which is known as tuber. Due to their short growing period, this crop is most fitted for catch cropping, intercropping, and relay cropping. Potato tuber is rich in starch and other carbohydrates. Potato is extensively cultivated in the Indian states including, Uttar Pradesh, Himachal Pradesh, Punjab, West Bengal, Bihar, Assam, Madhya Pradesh, Gujarat, Maharashtra, and Karnataka. Potato crops suffer many biotic and abiotic constraints, which reduces production and productivity. Various biotic factors reduce the quality and quantity of the crop. High relative humidity, medium pH, temperatures ranging from 10 to 25 °C, etc. are highly conducive to the development of potato diseases. Altered cultural practices including more than 4 years of crop rotation, timely irrigation and fertilization application, an adapted delay between haulm killing and harvest, need-based application of prophylactic chemicals, and storage of potato tuber under dry and cool conditions are good ways to control potato diseases.

Keywords Potato disease · Symptomatology · Pathogen ecology · Advanced management

---

S. Kumar (✉) · R. Chandra · S. V. K
Department of Mycology and Plant Pathology, Institute of Agricultural Sciences, Banaras Hindu University, Varanasi, Uttar Pradesh, India

M. Singh
Department of Soil Science and Agriculture Chemistry, Acharya Narendra Deva University of Agriculture & Technology, Kumarganj, Ayodhya, Uttar Pradesh, India

V. P. Singh
College of Agriculture Campus, Acharya Narendra Deva University of Agriculture & Technology, Kumarganj, Ayodhya, Kotwa, Azamgarh, Uttar Pradesh, India

S. S. Solankey (ed.), *Advances in Research on Potato Production*, Advances in Olericulture, https://doi.org/10.1007/978-3-031-82710-5_16

1 Introduction

Potato (*Solanum tuberosum* L.) is the fourth highest food crop for mankind in the world next to rice, wheat, and maize (FAO 2008). Globally, potato crops are popularized as the population's diet because it is consumed throughout the year in most countries. Potato originated from Peru (South America) and is presumed to have been introduced to India by the Portuguese in the early seventeenth century. Potatoes are used in industries for manufacturing starch, alcoholic beverages, and other processed products. Currently, India accounts for 14.93% of the total world production of potatoes with 376.12 million tons in a cultivable area of 18.13 million hectares. The productivity in India is 24.24 tons per hectare which is higher than the world average productivity of 20.75 tons per hectare (Singh et al. 2023). Major potato-producing states of India include Uttar Pradesh, Bihar, West Bengal, Madhya Pradesh, Punjab, Haryana, and Karnataka. It has the potential to feed hungry populations and is a good source of carbohydrates, vitamins, and essential minerals. Even though the production of potatoes is encouraged by the high-yielding varieties and increasing cultivable area, various plant pathogens greatly affect potato production. This chapter incorporates information on the major potato diseases and recent advances in their disease management.

2 Major Diseases in Potato

In India, a dozen potato diseases acknowledged to be incited by fungi, bacteria viruses, and nematodes which have a negative influence on both the agricultural industry and potato growers (Khurana et al. 1998; Thyagharajan and Kiruba Raji 2019; Abbas and Hameed 2012; Gul et al. 2013; Ashraf et al. 2012) as presented in Table 1.

2.1 Late Blight of Potato

The late blight disease has countless significance, particularly in ancient plant pathology. During 1844–1845, whole potato plantation was slew early across Europe, particularly in Ireland, leading to the "Irish Potato Famine" (Large 1940). One million people died due to hunger and another one million fled to the USA and other parts of the world. In India, this disease was first documented between 1870 and 1880 in the Nilgiri hills (Butler 1918). Late blight is a highly disastrous and most important disease of potato with a maximum yield loss of up to 85%. However, Ahmed et al. (2015) reported that 100% yield loss of potato crops is due to late blight in Pakistan. According to the Indian scenario, depending upon weather conditions, 5–90% decrease in potato production occurs with an average of 15% across

Table 1 List of major diseases of potato

S. no.	Disease	Causal organism
Fungal diseases		
1.	Late blight	*Phytophthora infestans*
2.	Early blight	*Alternaria solani*
3.	Potato scab	*Streptomyces scabies*
4.	Stem canker and black scurf	*Rhizoctonia solani* Kuhn
5.	Wart	*Synchytrium endobioticum* (Schilb) Perc.
6.	Powdery scab	*Spongospora subterranea* f.sp. *subterranea*
7.	Black dot	*Colletotrichum coccodes*
8.	Charcoal rot	*Macrophomina phaseolina*
9.	Fusarium wilt and dry rot	*Fusarium* sp.
10.	Verticillium wilt	*Verticillium alboatrum* Reinke & Berth.
11.	White mold or stem rot	*Sclerotinia sclerotiorum* (Lib.) de Bary
Bacterial diseases		
12.	Bacterial soft rot	*Erwinia carotovora* subsp. *caratovora*
13	Brown rot or bangle blight	*Ralstonia solanacearum*
Viral and viroid diseases		
14.	Potato leaf roll	Potato leaf roll virus (PLRV)
15.	Potato mop-top	Potato mop-top virus (PMTV)
16	Potato spindle tuber	Potato spindle tuber viroid (PSTVd)

the country (Collins 2000). It was the first plant disease that led to the discovery that microorganism causes plant disease and the emergence of plant pathology as a science. Since this disease causes more damage in medium or late-maturing cultivars, it is known as late blight. The foliage and stems of potato plants die at any time during the growing season due to infestation. It also attacks potato tubers in the field and storage (Agrios 2005). In India, late blight was first observed in epiphytotic form in Nilgiri Hills and later appeared every year in the Himalayas, Nilgiris, and Indo-Gangetic plains between 1870 and 1880 (Butler 1918). Since 1961, it has attained an annual feature infecting both summer and autumn crops (Srivastava 1962). In the northern parts, late blight first came to notice in 1883 at Darjeeling and advanced rapidly to the nearby hills (Butler 1903; Dastur 1917). It was observed in Khasi Hills in 1885, Kumaon Hills in 1897, and Shimla Hills in 1902 (Butler 1903; Dastur 1917). In the plains, it was first observed in 1898–1900 in the Hooghly district of West Bengal and persisted in severe form for two consequent years, followed by many areas of Assam and Bihar in 2013 (Butler 1903; Basu 1913; Dastur 1917; Lal 1949; Woodhouse and Dutt 1913). It not only caused reduction in yield but also severely affected the quality of the crop.

Late blight of potato was first reported to be caused by *Phytopthora infestans* by Anton de Bary in 1876. The pathogen *Phytophthora* is a member of the *Oomycota*, order *Peronosporales*, and family *Peronosporaceae*. Hyphae of this pathogen are hyaline, coenocytic, endophytic, and intercellular with double club-shaped haustoria. *P. infestans* produces indeterminate, well-branched sporangiophores bearing hyaline, lemon-shaped sporangium with a characteristic papilla. The sporangia are dispersed by rain splash or wind-borne rainfall. In cool, wet weather conditions, the sporangium releases reniform, laterally biflagellate zoospores with heterokont flagellar pattern. The zoospores swim in the thin film of waters toward suitable host infection sites and then encysts. Later, it produces the infection peg and causes infection in the potato plant. In hot weather conditions, sporangia may germinate directly. Sexual reproduction occurs in the presence of both mating types A1 and A2. A nucleus is released from the male gametangia, i.e., antheridium, and enters the female gametangia-oogonium. Following karyogamy, a thick-walled, diploid, oospore is produced.

2.1.1 Symptomatology

All the parts of plants such as leaves, stems, and tubers get affected by this disease. Initial symptoms include small, water-soaked blackish/brown lesions on the leaf tips, margin, or any other part of the leaf (Fig. 1). It expands rapidly and the entire leaf becomes necrotic. In humid (wet) weather conditions, the spots expand quickly and turn into dark brown or black lesions. The lower surface of the leaf is covered

Fig. 1 Water-soaked blackish/brown lesions on the leaf tips and margin

with sporangia and sporangiophores. All the lesions coalesce together, and the entire plant is destroyed in a few days after the advent of primary lesions. Under dry conditions, water-soaked areas are dry and turn brown. The infection reaches potato tubers in the field when sporangia are washed from the foliage part into the soil. Initially infections commonly start from cracking tuber, eyes, or lenticels. Infected tuber tissues turn copper brown, reddish, or purplish in color.

2.1.2 Pathogen Ecology, Survival, Dispersal/Transmission

The pathogen, *P. infestans*, endures between potato crops as mycelium in infected tubers in the field as well as in the storage acts as a source of primary infection. The pathogen over summers inside the infected tubers in the cold storage. When the infected tuber is planted or left behind in the fields, the fungus sporulates and produces sporangia and spreads to the nearby plants/fields by rain splashes/wind currents. The sporangia also move in the soil with rain water or dew and infect the nearby healthy tubers. With the advancement of the disease, both mating types (A1 and A2) interact with each other resulting in the development of thick-walled oospores which persist in the soil or plant tissues for a longer period. The diseased tubers are primarily accountable for the continuous occurrence of the disease from season to season. When the infected tuber is stored during off-season and used as seed materials during the following season, the fungus is returned to the soil and causes further infection, thus the disease cycle is maintained.

2.1.3 Epidemiology

There are specific requirements of weather conditions such as temperature, humidity, dew, and rain for the initiation and further buildup of late blight. Requirements of temperature based on growth stages such as for mycelial growth (16–20 °C), spore germination (10–20 °C), and spore production (18–22 °C) and for infection and disease progress (10–22 °C & 18 ± 1 °C). Hundred percent relative humidity is required for spore germination and infection, while lower humidity (<75%) results in the killing of the spores. Cloudiness favors disease development. Hot dry conditions may temporarily stop the spread of the disease.

Various forecasting models and warning services have been framed by the scientists for the prediction of late blight of potato and toward optimizing the application of fungicides without endangering the crop and human health. A prediction model for Shimla, Shillong, and Ootacamund was developed by Bhattacharyya et al. (1983) based on the following parameters:

(i) Seven-day moving precipitation such as for Shimla, 30 mm; for Ootacamund, 28.9 mm; and for Shillong, 38.5 mm lines linked with a mean temperature of 23.9 °C or less continues for seven consecutive days, late blight would appear within 3 weeks.

(ii) If the hourly temperature ranges from 10 to 20 °C linked with high RH (80% or more) for a constant 18 h for two successive days, the blight would appear within a week. This model helped to provide alerts every year since 1978 for Shimla Hills through All India Radio, Shimla. Van Everdingen (1926) established using "Dutch Rules" for the late blight prediction in Holland based on weather parameters, viz. RH >90%, temperature of 10–25 °C, night temperature of 10 °C, cloudiness on the next day, and rainfall at least 0.1 mm. These forecasting models are not common in all areas but have impetus for several workers.

2.1.4 Management Strategies

Late blight management has always been challenging. For effective management, the primary sources of inoculums must be slowed down, especially in cool weather conditions. The most effective and advanced control method is comprehensive disease deterrence to be attained through consistent application (usually weekly) with a combination of preventive and systemic fungicides. Removal of waste heaps, infected tubers, and volunteer plants can also help in the control of primary sources of inoculum. Application of broad-spectrum fungicides with a diverse mode of mechanism permits the plant to resist late blight more comprehensively and, ultimately, reduces the amount of chemicals being used. Sensitivity of *P. infestans* should be monitored at provincial ancestries to the useful chemicals periodically to check resistance development and disease outbreaks. Prophylactic spray of propineb (0.2%)/mancozeb (0.2%)/chlorothalonil (0.2%) followed by cymoxanil + mancozeb (0.3%)/dimethomorph/fenamidone-based fungicides followed by one more spray of contact or systemic fungicides depending upon disease severity is mostly advised. First spray with ridomil MZ 63.5% WP at a rate of 2 kg ha^{-1} followed by 2–3 sprays of Dithane M-45 at a rate of 3 kg ha^{-1} were found to be most effective in controlling late blight disease. Combined application of Metalaxyl + Mancozeb was reported to be highly effective against this disease (Thind et al. 2001). To manage tuber-borne diseases, tubers are dipped in boric acid (3%) for 30 minutes. Cultural practices such as the use of disease-free seed tuber, field sanitation, demolition of diseased crop debris, proper earthing up, and elusion collateral hosts are considerably highly effective. Adapting sound crop rotation at least 2–3 years reduces the risk of soil-borne infections. Avoiding irrigation during cloudy weather conditions and adjustment of sowing date reduce the risk of the disease. Minimized injuries of tubers during harvest time and storage are highly essential. The development of resistant varieties has a significant role in the management of late blight (Joseph et al. 2011; Bhardwaj et al. 2013). The resistant varieties recommended for cultivation include Kufri Naveen, Kufri Jeevan, Kufri Alenkar, Kufri Khasi Garo, and Kufri Moti.

2.2 *Early Blight*

Early blight disease is one of the most substantial fungal diseases which mainly affects the early maturing cultivars (Abuley and Nielsen 2017). This disease hampers vegetative development with yield losses of 20–54% under field conditions if not properly and timely controlled (Srikantaiaya 1962; Kadam et al. 1974). This disease was first described in 1882 in the USA. Early blight is problematic in many potato-growing areas, where it not only affects potatoes but also tomatoes and other solanaceous crops (Chester 1892). Early blight is caused by *Alternaria solani* (earlier named *Macrosporium solani* by Ellis and Martin 1882), an *Ascomycota* fungus. *Alternaria* sp. is a significant group of pathogenic fungi causing a number of diseases in several crop plants. The hyphae of the fungus are haploid and septate, becoming darkly pigmented with age. The asexual spore called conidia is borne singly or in chains on distinct, brown, septate, geniculate conidiophores produced in groups or single. Conidia are bottle-shaped, brown, and muriform, with beak, and normally possess 9–11 transverse septa, and there may or may not be a few longitudinal septa.

2.2.1 Symptomatology

The disease occurs on foliage and sometimes also on tubers. On the older leaves, dark brown, more or less circular lesions, with concentric rings resembling a target board, are observed (Rotem 1994) (Fig. 2). Depending on environmental conditions and the crop variety, lesions enlarge from 0.5 to 2.0 cm in diameter and become

Fig. 2 Dark brown, more or less circular lesions on potato leaf

associated with chlorotic leaf area around and between the lesions. Under dry conditions, lesions may drop out leaving a hole (shot-hole appearance). The disease generally leads to defoliation, but sometimes dry leaves remain hanging on the plants. Similar lesions also occur on petioles and stems. Stem infection may cause breaking of stems and death of the uninfected plant portions above the lesion (Zachmann 1982). Symptoms in the tuber include circular to irregular, brown, and depressed lesions. The underneath flesh turns dry, brown, and corky. During storage, the lesions tend to enlarge, and the affected tubers become shriveled. Yield reduction caused by foliage attack reaches more than 25%. In stored potatoes, losses caused by *A. solani* may be substantial and reach 80% affected by early blight lesions.

2.2.2 Ecology of Pathogen, Survival, Dispersal/Transmission, and Epidemiology

A. solani survives as conidia and mycelium in the soil and crop debris. Besides during storage, they also survive in the infected tubers. The primary source of inoculum for early blight is the infected tubers. The conidia germinate within 30 minutes when high moisture and temperature between 24 and 34 °C. The developing germ tube penetrates leaves through the epidermis and stomata. Septate and branched hyphae expand within the leaves. Infection usually starts from the lower leaves near the soil line resulting in infection appearing in young immature tubers that are open from the soil. Secondary dispersal is through the conidia being disseminated primarily by wind and seldom by splashing rain or overhead irrigation. Moderate temperatures ranging from 22 to 26 °C, high humidity with intermittent rain, and poor manuring favor the disease development. A long moist period is essential for sporulation, but it can also happen under conditions of alternating wet and dry periods. Light and temperatures between 19 and 23 °C stimulate the development of conidiophores (Zachmann 1982).

2.2.3 Management Strategies

Early blight diseases of potato can be managed by the efficient application of cultural practices (Schultz and French 2009). Use of disease-free seed tubers, removal and destruction of infected plant debris and haulms of the affected potato crop, following 3–5 years crop rotation with nonhost crops (millets, grain crops like maize, soybean, and forage crops), avoiding excess dose of nitrogenous fertilizers, avoiding cultivation of solanaceous crops nearby potato field, and avoiding overhead irrigation aid in the management of the disease. Foliage killing at least 2–3 weeks before harvesting prevents tuber infection. Early harvesting under wet conditions may be avoided so as not to bruise the tubers during harvesting and sorting. Application of alternate contact and systemic fungicides is most practical for the control of this disease early in the growing season (Schultz and French 2009). Early spraying of potato crops with chlorothalonil (0.2%), mancozeb (0.2%), and copper oxychloride (0.2%) and repeating it every 15–20 days give effective management. Application of bio-pesticides may also play an important role in the management of disease. Soaking potato with Beejamrutha (5 l/10 kg tubers) for 30 min followed by two foliar applications with Jeevamrutha at 5% at the interval of disease onset evidenced to be the most effective and eco-friendly management of early blight under field conditions (Patel and Gohel 2023).

2.3 *Stem Canker and Black Scurf*

Black scurf is also a most damaging disease of potatoes and is distributed in different regions in low to severe form. The infection is much more severe in the fields where potato is grown continuously. Stem canker and black scurf are caused by the fungus *Rhizoctonia solani* Kuhn. (Teleomorph: *Thanatephorus cucumeris* (Frank) Donk). *Rhizoctonia solani* causing infection in potatoes was first time described by J. G. Kuhn in 1858. The AG-3, anastomosis group of *R. solani*, is responsible for black scurf of potatoes (Wharton et al. 2007). The fungus produces superficial, irregular, scab-like sclerotia on the infected tissues (Arora 2012).

2.3.1 Symptomatology

Disease symptoms appear on both above- and belowground parts of the plant. The fungus causes black scurf on the mature tubers, stems, underground stems, and stolons. In the case of black scurf, the fungus produces dark brown to black, hard, masses of sclerotia (a resting body of fungus) on the surface of tubers. Black scurf (encrustation) forms in the tuber when these sclerotia get dumped endlessly on the surface of the tuber. Young sprouts are infected in the early season. Initial symptoms occur as brown lesions which coalesce and kill the tuber resulting in poor and uneven stands of plants. Besides, gridling of the stem is observed along with curling of the foliage. The foliage turns pinkish to purplish color. Due to starch translocation, aerial tubers are produced.

2.3.2 Ecology of Pathogen, Survival, Dispersal/Transmission, and Epidemiology

Seed tubers play an important role as the chief source of the disease (Lal et al. 2014). In offseason, the pathogen infects alternate crops or over summers in resting stage. The fungus actively grows at an optimum temperature of 25–30 °C. Sclerotial germination is observed at 23 °C. Moderately cool temperatures of 16–23 °C, moist conditions, and dry light soils are favorable for the development and spread of pathogens.

2.3.3 Management Strategies

Management of this disease is very tedious owing to its soil and seed-borne nature, high level of survivability, and wide host range. Planting sclerotia-free seed tubers in warm soil and covering them with little soil is excellent management. Long crop rotation reduces inoculum that can cause cankers. Treating seed tubers with *T. harzianum* at 10 gm/seed or with suitable fungicide is effective against seed-borne scurf (Lal et al. 2014). To manage tuber-borne diseases, tubers are dipped in boric acid (3%) for 30 minutes. Cattle manure compost was also shown effective against *R. solani* in soil mainly due to enhanced activities of antagonistic microorganisms (Kuter et al. 1983).

2.4 Potato Wart Disease

Wart is a worldwide popular potato disease and is named by various names like black wart, scab, tumor, canker, and many other descriptive terms (Franc 2007). This disease reduces quality and yield of potatoes, which may be unmarketable. In India, potato wart was introduced from Denmark to Darjeeling hills of West Bengal in 1953 (Ganguly and Paul 1953; Mehta 1963). Now this disease has been reported in various parts of Europe, Asia, Africa, North America, and South America (Obidiegwu et al. 2014).

Potato wart disease is incited by the fungus *Synchytrium endobioticum* (Schilberszky) and has worldwide importance as a quarantine plant pathogen. It is a chytrid fungus recognized by the presence of short-lived swarm (summer) sporangial stage by uniflagellate zoospores and a resting (winter) sporangial stage by conjugated zoospores (biflagellate zygotes). Resting sporangia are golden brown and spheroidal with prominent exterior ridges in the wall (Franc 2007).

2.4.1 Symptomatology

This disease of potato is not recognized until the tubers are dug out. Gall (prominent warty protuberances) formation as outgrowth on stem base, stolon buds, and tuber eyes is the main diagnostic feature of this disease. Belowground galls are white to brown and turn black as they decay, while aboveground galls are green to brown, turning black at maturity, and are prone to decay. Infected plants may develop small, greenish, warty growths on the stalks.

2.4.2 Ecology of Pathogen, Survival, Dispersal/Transmission, and Epidemiology

Wart disease is both tuber- and soil-borne. *S. endobioticum* can endure up to 30 years in the soil. Seed tubers, infected soil, or planting materials carrying infected soil initiate the primary infection. Soil contaminated with sclerotia plays an important role in the dissemination of disease. The disease more commonly occurs in cold and humid climates. The disease is worse in the wet season.

2.4.3 Management Strategies

Intercropping with maize and crop rotation with radish or bean will be effective in the management of potato warts. Cultural practices such as the use of disease-free tubers, weed management, and destruction of infected plants/debris reduce inoculum. Soil solarization by plastic mulching is an effective and eco-friendly way to manage wart disease. Seed treatment with *Trichoderma viride*, a bio-formulation developed by the Central Potato Research Institute, was reported to be effective (Arora and Somani 2001). To manage tuber-borne diseases, tubers are dipped in boric acid (3%) for 30 minutes.

2.5 Bacterial Brown Rot or Wilt Disease

Bacterial brown rot or wilt is the most damaging disease of potatoes in Asia, Africa, and Central and South America (Charkowski et al. 2020), which is incited by the bacterium *Ralstonia solanacearum* (Smith). This disease was first observed in Egypt from potato tubers producing rot symptoms (Sabet 1961). *R. solanacearum* is gram-negative, rod-shaped, motile bacteria with one or several polar flagella.

2.5.1 Symptomatology

The pathogen causes infection in both the above- and underground parts of the crop. During hot sunny days, the initial symptoms include wilting in the leaves in the top branches. The affected leaves droop due to loss of turgidity and result in total unrecoverable wilt. Upon squeezing the infected sample, dull white ooze is released from the cut ends of the stem. Two types of symptoms are produced on the tuber surface: vascular rot and pitted lesion. In vascular rot, the tuber tissues express water-soaked, brown, dirty white, sticky drops. In severe conditions, bacterial ooze is in the eyes which results in the attachment of soil glued with the bacterial ooze. Initially, water-soaked lesions appear in the lenticels which further enlarge in the form of pitted lesions. In storage, the tubers may not rot or may not show any vascular browning (Sagar 2015).

2.5.2 Ecology of Pathogen, Survival, Dispersal/Transmission, and Epidemiology

The pathogen over winters in the infected seed tubers and plant debris in the soil. The bacteria enter the plant through various means such as mechanical wounds, natural openings, cracks, or root tips, and then colonize the cortex (sub-epidermal tissue) and infect the xylem vessels. As the bacterium spreads in the xylem, it blocks the plant's water transport channel leading to wilting and eventually fatal to the plant. Potato tubers carry the infection in three ways: vascular tissues, tuber surfaces, and lenticels. Latent infection is the most important for the transmission of the disease through tubers. The survival of the pathogen is affected by high soil moisture, temperature, and oxygen stress. Hingorani et al. (1956) reported that the disease in potatoes is favored by relatively high temperatures (21–38 °C) and high soil moisture (50–100% water-holding capacity of soil). Pathogen is dispersed from one location to another by means of infected planting materials, irrigation or rainwater, and farm equipment.

2.5.3 Management Strategies

Owing to the seed- and soil-borne nature of bacteria, complete control of the disease is very tedious. Use of fresh and healthy vigorous planting material can reduce almost 80% of bacterial wilt problems. Obtaining seeds from disease-free areas for propagation might reduce the possibility of infection and spread of the disease. Fallow crop rotation for at least 2–3 years with nonhost crops such as wheat, maize, oat, sun hemp, finger millet, and vegetables like garlic, onion, and cabbage will reduce the inoculum. Avoiding other solanaceous vegetables like brinjal, tomato, chili, etc. will have a positive impact. Infected tubers should be disinfected with hot air treatment at 112 °F for 30 minutes (Tsang and Shintaku 1998). Application of bleaching powder at 12 kg/ha at the time of potato planting in a furrow along with fertilizer reduces the pathogen population from the field by up to 80% and also improves the yield (Shekhawat and Gadewar 1988).

2.6 Potato Leaf Roll Virus (PLRV) Disease

More than 50 viruses have been reported to infect potatoes (Kreuze et al. 2020). Among all viral diseases, PLRV is devastating and more prevalent in most potato-growing areas worldwide and can severely reduce yields and quality (Peters and Jones 1981; Harrison 1984; Storch and Manzer 1985). Globally, the annual loss in potato production by this viral disease is more than 20 million tons (Kumari et al. 2020; Patton et al. 2020). Potato leaf roll is incited by *potato leaf roll virus* (PLRV), also known as potato virus 1 or *Solanum* virus 14 (Mehrotra and Aggrarwal 2003). Its genus is *Polerovirus* and family *Solemoviridae.* The genome consists of 5.3–5.7 kb long positive sense (+), monopartite, ssRNA genome (LaTourrette et al. 2021; Walker et al. 2021).

PLRV is also associated with other viruses such as potato virus (PVY), potato virus x (PVX) causing extra losses to potato production (Valkonen 2015; Okeyo 2017; Byarugaba et al. 2020). The incidence of this disease has been estimated as high as 60–70% and annual loss of about 20 million tons of tubers worldwide (Tripathi 2007).

2.6.1 Symptomatology

Typical symptoms are characterized by the inward rolling of lower leaflets, extending ultimately to the upper leaves (Fig. 3). The infected leaves become dry and hard, chlorotic, and often show purple discoloration. Necrosis of phloem tissue is found in the stems and tubers. Infected plants produce small tubers, which are not used for eating purposes and marketing due to deterioration in quality.

Fig. 3 Inward rolling of leaflets

2.6.2 Infection and Transmission

The nature and severity of this viral disease depend on the virus strain, potato cultivar, weather conditions, time, and infection source. Initial infection of PLRV is introduced into the potato field by infected seed tubers or by aphids. The amount and movement of insect vector has a countless influence on the spatial and temporal patterns of viral infections. In nature, PLRV is only transmitted by aphids in a persistent manner. Luteoviruses contain nucleic acid as RNA and are phloem-limited viruses that are transmitted by aphids (Casper 1988). *Myzus persicae* (green peach aphids) is the competent aphid vector, but *Aulacorthum solani* and *Macrosiphum euphorbiae* may also transmit PLRV. Aphid feeding introduces potato leaf roll virus into the phloem tissue where the virus multiplies, spreads, and initiates the disease. Humid soil promotes the infection.

2.6.3 Management Strategies

Avoidance of infection sources, management of vectors, cultivation of resistant varieties, and use of transgenic plants are most important toward management. Burn or bury infected plants to avoid further infection. Use of suitable insecticides is generally recommended to break the transmission of viral diseases by vectors. Spray with imidacloprid (17.8 SL) at 1 ml/3 liter of water should be started from the early stage of plant growth at 10–15 days interval.

3 Conclusion

Potato is the third most economically important global food crop and a source of energy at low cost. It is versatile, delicious, and a staple in many cuisines. It is a temperate crop grown under subtropical conditions in India. Late blight, early blight, wart, stem canker and black scurf, black scurf, bacterial brown rot or wilt disease, potato leaf roll disease, etc. are the most common diseases of potatoes that can greatly reduce their yield and tuber quality. In this chapter, major emphasis is given to the economic significance, etiology, symptomatology, ecology, survival, dispersal/transmission, and epidemiology of pathogen and management strategies of major diseases of potato. A holistic approach is necessary to manage the diseases for sustainable potatoes because most of the farmers use chemical fungicides, which are not economically feasible in many low-input farming systems. Therefore, integrated disease management is a holistic approach to the sustainable management of potato diseases. Agronomic/cultural practices such as crop rotation, optimum spacing, inter/mixed cropping, change in date of sowing, bio-fungicides, botanicals, host plant resistance (HPR), and fungicides should be potentially viable parts for IDM. However, the efficacy of control strategies is dependent on the proper identification and diagnosis of the causal pathogen of the disease. In addition, the identification of suitable biocontrol strains and biotechnological approaches such as genetic engineering, marker-assisted selection, tissue culture, and hybridization are modern techniques to develop resistance cultivars against a wide range of diseases. Disease modeling, decision support systems, and application of remote sensing to refine up-scale and disseminate IDM technologies are emphasized.

References

Abbas MF, Hameed S (2012) Identification of disease-free potato germplasm against potato viruses and PCR amplification of potato virus X. Int J Biol Biotechnol 9:335–339

Abuley IK, Nielsen BJ (2017) Evaluation of models to control potato early blight (*Alternaria solani*) in Denmark. Crop Prot 102:118–128

Agrios GN (2005) Plant pathology, 5th edn. Academic, London, New York, p 922

Ahmed N, Khan MA, Khan NA, Ali MA (2015) Prediction of potato late blight disease based upon environmental factors in Faisalabad, Pakistan. J Plant Pathol Microbiol S3:008

Arora RK (2012) Eco-friendly management of soil and tuber borne diseases of potato. Indian Phytopathol 65(2):116–121

Arora RK, Somani AK (2001) A bioformulation for control of soil and tuber borne diseases of potato. J Indian Potato Assoc 28:88–89

Ashraf A, Rauf CA, Abbas MF, Rehman R (2012) Isolation and identification of Verticillium dahliae causes wilt on potato in Pakistan. Pak J Phytopath 24:112–116

Basu SK (1913) The late blight of potato. Agric J Bihar Orissa 1:142–149

Bhardwaj V, Srivastava AK, Sharma S, Kumar V, Kaushik SK, Singh BP (2013) Efficiency of different potato (*Solanum tuberosum* L) cross combinations in late blight resistance breeding. Int J Innov Hort 2(1):63–69

Bhattacharyya SK, Phadatare SG, Khanna RN, Srivastava DS, Prasad B (1983) Efficacy of some fungicides in controlling late blight of potato in India. Indian J Agric Sci 53:153–157

Butler K (1903) Potato disease of India. Agric Ledger Crop Dis Pest Ser 8:87–124

Butler EJ (1918) Fungi and disease in plants. Thacker Spink and Co., Calcutta

Byarugaba AA, Mukasa SB, Barekye A, Rubaihayo PR (2020) Interactive effects of potato virus Y and potato leafroll virus infection on potato yields in Uganda. Open Agric 5:726–739

Casper R (1988) Luteoviruses. In: Koenig R (ed) The plant viruses, polyhedral virions with monopartite RNA genomes. Plenum, New York, pp 235–258

Charkowski A, Sharma K, Parker ML, Secor GA, Elphinstone J (2020) The potato crop. In: Bacterial diseases of potato, pp 351–388

Chester FD (1892) Diseases of the round potato and their treatment. Annu Rep Del Coll Agric Exp Stn 5:67–70

Collins WW (2000) The global initiative on late blight- alliance for the future. In: Khurana SMP, Shekhawat GS, Singh BP, Pandey SK (eds) Potato global research and development, vol I. Indian Potato Association, CPRI, Shimla, pp 513–524

Dastur JF (1917) Conditions influencing the distribution of potato blight in India. Indian J Agric Res (Special Indian Congress) 12:90–96

Ellis JB, Martin GB (1882) *Macrosporium solani* E&M. Am Nat 16:1003

FAO (Food and Agriculture Organization) (2008) Potato world: Africa-international year of the potato

Franc G (2007) Potato wart. Online. APS*net* Features. https://doi.org/10.1094/APSnetFeature-2007-0607

Ganguly A, Paul DN (1953) Wart disease of potato in India. Sci Cult 18:605–606

Gul Z, Khan AA, Khan AUR, Khan ZU (2013) Incidence of potato viruses in different districts of Khyber Pakhtunkhawa. Pak Sci J Plant Pathol 2:32–36

Harrison BD (1984) Potato leafroll virus. In: CMI/AAB descriptions of plant viruses, vol 291
Hingorani MK, Mehta PP, Singh NJ (1956) Bacterial brown rot of potato India. Indian Phytopathol 9:67–71
Joseph TA, Singh BP, Kaushik SK, Bhardwaj V, Pandey SK, Singh PH, Singh GJ, Bhat MN, Gupta VK (2011) Kufri Girdhari: a medium maturing, late blight resistant potato variety for cultivation in Indian hills. Potato J 38(1):26–31
Kadam VC, Sarode MS, Bendre NJ, Shingte VV, Khot SB, Lokhande SB (1974) Late blight of potato *Phytophthora infestans* (Mont.) de Bary. Curr Sci 43:260
Khurana SM, Paul Garg ID, Singh BP, Gadewar AV (1998) Major diseases of potato and their management. In: Upadhyay RK et al (eds) Integrated pest and disease management. A.P.H. Publishing Corporation, New Delhi, pp 11–64
Kreuze F, Sousa-Dias J, Jeevalatha A, Figueria A, Valkonen J, Jones R (2020) Viral diseases in potato. In: Campos H, Ortiz O (eds) The potato crop. Its agricultural, nutritional and social contribution to humankind. Springer, Cham, pp 389–430
Kumari P, Kumar J, Kumar RR, Ansar M, Rajani K, Kumar S et al (2020) Inhibition of potato leafroll virus multiplication and systemic translocation by Sirna constructs against putative *Atpase* fold of movement protein. Sci Rep 10:1–11
Kuter GA, Nelson EB, Hoitink HAJ, Madden LV (1983) Fungal population in container media amended with composted hardwood bark suppressive and conducive to Rhizoctonia damping-off. Phytopathology 73:1450–1456
Lal TB (1949) Occurrence of late blight in the plains of India. Indian Phytopathol 1:164–181
Lal M, Sharma S, Yadav S, Kaushik SK (2014) Bioefficacy of new molecule: penflufen 240 FS against black scurf of potato. Int J Agric Stat Sci 10(1):63–66
Large EC (1940) The advance of the fungi. Jonathan Cape, London, p 488
LaTourrette K, Holste NM, Rodriguez-Peña R, Leme RA, Garcia-Ruiz H (2021) Genome-wide variation in betacoronaviruses. J Virol 95:e0049621
Mehrotra RS, Aggrarwal A (2003) Viruses, viroids and their diseases. In: Plant pathology, 2nd edn. Tata McGraw-Hill Publishing Company Limited, New Delhi, pp 703–749
Mehta PR (1963) Biological basis of domestic quarantine against the wart disease. Indian Potato J 5:76–79
Obidiegwu JE, Flath K, Gebhardt C (2014) Managing potato wart: a review of present research status and future perspective. Theor Appl Genet 127:763–780
Okeyo OG (2017) Response of potato genotypes to virus infection and effectiveness of positive selection in Management of Seed Borne Potato Viruses. J Agric Sci 10:71
Patel JV, Gohel NM (2023) Eco-safe management of potato early blight caused by *Alternaria solani* (Ellis and Martin) Jones and Grout. Potato Res 66:1023–1032
Patton MF, Bak A, Sayre JM, Heck ML, Casteel CL (2020) A Polerovirus, potato leafroll virus, alters plant–vector interactions using three viral proteins. Plant Cell Environ 43:387–399
Peters D, Jones RAC (1981) Potato leafroll virus. In: Hooker WJ (ed) Compendium of potato diseases. American Phytopathological Society, St Paul, pp 68–70
Rotem J (1994) The genus Alternaria: biology, epidemiology and pathogenicity. American Phytopathological Society, St. Paul, p 326
Sabet AH (1961) Geology and mineral deposits of Gebel El-Sibai area, Red Sea Hills, Egypt, U.A.R., Unpublished Ph.D. Thesis, I.T.C. Delfet, Holland, p 189
Sagar V (2015) Bacterial wilt and brown rot. In: Singh BP, Nagesh M, Sharma S, Sagar V, Jeevvlatha A, Sridhar J (eds) A manual on diseases and pest of potato, Technical Bulletin No. 101. ICAR-Central Potato Research Institute, Shimla, pp 20–23
Schultz D, French RD (2009) Early blight of potatoes and tomatoes. Texas AgriLife Extension Service; The Texas A&M System: PLPA-Pot 009-01
Shekhawat GS, Gadewar AV (1988) Bacterial wilt of potato. In: Khurana SMP (ed) Pathological problems of economics crop plants and their management. Scientific Publishers, Jodhpur, pp 511–534

Singh HP, Singh B, Kumar R (2023) Approaches to sustaining potato production in India, Technical Bulletin 2023. Confederation of Horticulture Associations of India, New Delhi, p 18
Srikantaiaya M (1962) Late blight of potato in Mysore State. Indian Potato J 4:49–50
Srivastava SNS (1962) Epidemic of late blight of potato in South India. Plant Prot Bull 14:26–28
Storch RH, Manzer EE (1985) Effect of time and date of inoculation, plant age, and temperature on translocation of potato leafroll virus into potato tubers. Am Potato J 62:137–143
Thind TS, Lakhvir S, Mohan C, Paul J (2001) Monitoring for metalxayl resistance in populations of *Phytophthora infestans* and their characterstics in Punjab. Indian Phytopathol 54:91–97
Thyagharajan K, Kiruba Raji I (2019) A review of visual descriptors and classification techniques used in leaf species identification. Arch Comput Methods Eng 26:933–960
Tripathi DP (2007) Introductory plant virology, 2nd edn. Kalyani Publishers, New Delhi, p 532
Tsang MMC, Shintaku M (1998) Hot air treatment for control of bacterial wilt in ginger root. Appl Eng Agric 14:159—163
Valkonen JP (2015) Elucidation of virus-host interactions to enhance resistance breeding for control of virus diseases in potato. Breed Sci 65:69–76
Van Everdingen E (1926) Het.verband tusschen de weergesteldhied en de aarolppelziekte, Phytopthora infestans (the relation between weather conditions and potato blight, Phytophthora infestans) Tijdschr. Plan Theory 32:129–140
Walker PJ, Siddell SG, Lefkowitz EJ, Mushegian AR, Adriaenssens EM, Alfenas-Zerbini P et al (2021) Changes to virus taxonomy and to the international code of virus classification and nomenclature ratified by the international committee on taxonomy of viruses (2021). Arch Virol 166:2633–2648
Wharton P, Driscoll J, Douches D, Hammerschmidt R, Kirk W (2007) Common scab of potato. Extension Bulletin E-2990 of Michigan State University, p 4
Woodhouse EJ, Dutt NL (1913) Potato disease in Bihar. Agric J India 1:15–38
Zachmann R (1982) Early blight of potato: *Alternaria solani*, Technical Bulletin 17. International Potato Centre, Lima, p 13

Exploring Indian Potato Sector: From Farm to Table

Arindam Nag, Krishna D. K., Kevin Christopher, Aditya Sinha, Anukul Chandra, and Bholanath Saha

Abstract Worldwide potato is an important staple food crop after cereals. With the ever-growing population, food and nutritional security remains a challenge, and potato could be an important food choice for this. Today, globally India ranks second in potato production, but productivity remains a point of worry. Moreover, challenges like degeneration, storage in summer season, dependence on informal seed potato system, and limited processing prevail in potato business. Changing consumption preferences aggravates demand for ready-to-eat potatoes; however, only 7–8% of the total production is processed in India. Climate change is a global threat and also a significant threat to potato production. Rising temperatures, altered precipitation patterns, and increased frequency of extreme weather events can negatively impact yield and quality. Climate change is not only affecting the frequency of pest and disease occurrences in potato but also creating supply chain disruptions. Greater participation of stakeholders (public and private enterprise) in collaborative research, extension, and business may result in greater adoption of climate-resilient practices, securing profitability in potato farming business for small and marginal farmers. In this chapter we will try to understand various pertinent issues in Indian potato sector.

A. Nag (✉) · K. Christopher
Department of Extension Education, Dr. Kalam Agricultural College, Kishanganj, Bihar Agricultural University, Sabour, Bhagalpur, Bihar, India
e-mail: arindam.apdj@gmail.com

K. D. K.
Department of Extension Education, Advance Centre on Sericulture, Kishanganj, Bihar Agricultural University, Sabour, Bhagalpur, Bihar, India

A. Sinha · A. Chandra
Department of Extension Education, Bihar Agricultural College, Bihar Agricultural University, Sabour, Bhagalpur, Bihar, India

B. Saha
Department of Soil Science, Dr. Kalam Agricultural College, Kishanganj, Bihar Agricultural University, Sabour, Bhagalpur, Bihar, India

S. S. Solankey (ed.), *Advances in Research on Potato Production*,
Advances in Olericulture, https://doi.org/10.1007/978-3-031-82710-5_17

Keywords Indian potato sector · Potato farming situation · Potato growers · Issues in Indian potato

1 Introduction

One of the most significant noncereal food crops in the world, potatoes (*Solanum tuberosum* L.) are native to Peru and Bolivia in the Andes region of South America. It has high-quality protein, a range of minerals, rich vitamins, and trace elements in addition to carbohydrates. Potatoes are low in fat (between 0.1% and 1.1%) and high in dietary fiber (Dongyu and Director-general 2022).

The Food and Agriculture Organization of the United Nations has strongly suggested potatoes as a food security crop as the world faces issues with food supply due to population growth. Potatoes are the third most significant food crop in terms of worldwide consumption (Devaux et al. 2014). Potatoes' ability to adapt to a variety of growth environments strengthens their contribution to food security. Potatoes are a sustainable crop in areas with difficult agricultural conditions since they can flourish in a variety of temperatures and soil types. This flexibility contributes to the stability of food availability for populations around the world by guaranteeing a steady supply of food, even in locations vulnerable to environmental changes.

Approximately two-thirds of the global population eats potatoes as a staple meal, about half of potatoes are used as vegetables or household staples, and in high-income countries, the average percentage of potatoes processed is more than 40% (Dongyu and Director-general 2022). When potatoes are harvested, they can be utilized in many different ways, such as a fresh vegetable for home cooking, a raw ingredient for processing into food items, fodder for animals, food ingredients, starch, and alcohol, as well as seed tubers for planting the crop for the following season. More and more potatoes are being processed to satisfy the growing need for ready-to-eat foods and snacks as consumer demand shifts away from fresh tubers and toward processed items globally. Growing urban populations, rising wages, dietary diversity, and lifestyles that do not allow for as much time for preparing fresh food for consumption are the main factors driving this development. While the majority of potatoes consumed in developing nations are still fresh potatoes, fast food and high-value leisure consumption are steadily rising in China, India, Russia, and other nations. While consumption dropped in European and American nations, it surged dramatically in Asian nations. In the world, almost 40% of potatoes are consumed in China, India, and Bangladesh.

Because of its widely varied distribution pattern, present cultivation, and demand—particularly in developing nations with high rates of poverty, hunger, and malnutrition—the potato crop, among other food sources, can help meet food security globally (Devaux et al. 2021).

1.1 Potato Production Recent Trend

There is a gradual shift of potato cultivation from west to east. Now the developing countries in the Asia and Africa region are showing rapid growth in potato production. Between 1961 and 2020, the combined output of potatoes in Asia, Africa, and Latin America climbed from 32 to 226 million tons, a rise of more than six times, with an average annual growth rate of 3.37% (Dongyu and Director-general 2022). According to FAO's database (FAOSTAT), the world produced about 375 million tons of potatoes in 2022, with China producing 95.5 million tons and India producing 56 million tons (Pieterse 2024).

It is believed that the Portuguese brought potatoes to India in the early seventeenth century from Europe. It began as a significant cool-season crop in both the valleys and the mountains. However, it is currently grown in practically every Indian state under various agro-climatic circumstances. India is the world's second-largest producer of potatoes, right after China. Potato is cultivated as one of the main crops in the country and is grown in 23 states. Uttar Pradesh, West Bengal, Bihar, Gujarat, Madhya Pradesh, Punjab, Assam, Haryana, Jharkhand, and Chhattisgarh are the major potato-producing states in India (Table 1). Over the past 15 years, India's potato production has increased significantly; nevertheless, area expansion rather than yield enhancement has driven most of this development (Rana and Anwer 2018). Despite increases in area, production, and productivity, the average productivity is still significantly lower than the highest productivity in the globe (Sahu et al. 2024).

Table 1 State wise distribution of potato area, production, and yield during 2021–2022

S. no.	States	Area (000′ hectares)	Area share in % of all India	Production 000′ tons	Production share in % of all India	Yield (kg/ha)
1	Uttar Pradesh	622.5	28	16161.97	30	25,963
2	West Bengal	447.45	20	12403.13	23	27,720
3	Bihar	330	15	9125.8	17	27,654
4	Gujarat	127.8	6	3700	7	28,951
5	Madhya Pradesh	158.4	7	3587.9	7	22,651
6	Punjab	110	5	2850	5	25,909
7	Haryana	29.54	1	782.5	1	26,490
8	Assam	103.44	5	761.84	1	7365
9	Jharkhand	49.28	2	698.99	1	14,184
10	Chhattisgarh	42.54	2	651.48	1	15,315
11	Others	181.2	8	2663.74	5	14,701
	All India	2202.15	100	53387.35	100	24,243

Source: Agricultural Statistics at a Glance (2022)

Box 1 Some Important Potato Varieties Cultivated Across India

- *Kufri Badshah*—Cultivated in Jammu Kashmir, Haryana, Punjab, Gujarat, Uttar Pradesh, and Madhya Pradesh. The variety has higher productivity.
- *Kufri Sutlej*—Mostly grown in Bihar, Uttar Pradesh, Haryana, Punjab, and Madhya Pradesh. This variety is moderately susceptible to late blight disease.
- *Kufri Chandramukhi*—Can be found in Bihar, Gujarat, Haryana, Punjab, Madhya Pradesh, Himachal Pradesh, Uttar Pradesh, Orissa, and West Bengal. This variety is suitable for chips making.
- *Kufri Jawahar*—This variety is grown in Haryana, Punjab, Gujarat, Madhya Pradesh, and Karnataka. This variety is moderately resistant to late blight disease, suitable for cooking/table purpose.
- *Kufri Sindhuri*—It is mainly cultivated in Bihar, Gujarat, Maharashtra, Punjab, Uttar Pradesh, Karnataka, and Jammu Kashmir. It is grown in the regions having high temperature and water stress and used for preparing different processed food.
- *Kufri Bahar*—Major growing states of this variety are Haryana, Uttar Pradesh, Himachal Pradesh, and Jammu Kashmir.
- *Kufri Ashoka*—Cultivated in Bihar, Punjab, Haryana, Uttar Pradesh, and West Bengal.
- *Kufri Lauvkar*—Maharashtra, Madhya Pradesh, and Karnataka are the main growing states of this variety. This variety is good for chips preparation.
- *Kufri Lalima*—This variety is mainly cultivated in Uttar Pradesh and Bihar moderately resistant reaction to early blight, not good for processing.
- *Kufri Jyoti*—Bihar, Maharashtra, Gujarat, Haryana, Punjab, Uttar Pradesh, Karnataka, and West Bengal. This variety is moderately resistant to early and late blight disease.
- *Kufri Pukhraj*— Important for Bihar, Punjab, Haryana, Uttar Pradesh, Maharashtra, Himachal Pradesh, Madhya Pradesh, Orissa, and West Bengal. This variety shows good resistance against early bight as well as moderate resistance against late blight disease.

1.2 Export of Potato from India

Indian potatoes have a huge market for exports. Due to cheaper labor, shorter crop duration, and reduced manufacturing costs, it is more economical than its European equivalent. The majority of Indian potatoes are harvested between January and March, a time when fresh potatoes are scarce in most of the Northern Hemisphere.

It opens up a significant market for Indian table potatoes to export to the EU and other nations (Ranjan 2017). India is the second largest exporter of potatoes, exporting to more than 30 countries. Varieties that are mostly exported are Kufri Pukhraj, Kufri Jyoti, Kufri Bahar, Bhura aloo, etc. India's top export destinations of potatoes are Nepal, the United States, the United Arab Emirates, Bhutan, Sri Lanka, Japan, Qatar, Maldives, Malaysia, Thailand, Vietnam, and Ghana.

The demand for high-quality potatoes for processing has increased in line with the rise in potato processing in India in recent years. The Indian potato processing industry has experienced phenomenal growth as a result of growing urbanization and consumers' preference for quick, ready-to-eat meals, especially French fries, which are among the best-selling foods. In India 7% of the crop gets processed (in contrast to over 80% in industrialized nations). There may be room for future growth in the Indian potato-processing business given current global trends in shifting food consumption preferences (Kaur et al. 2023). India is progressively growing into a significant supplier of French fries to nations such as Singapore, Malaysia, Bangladesh, UAE, Sri Lanka, and Mauritius. India exports dried potato products to the UK, Sri Lanka, Maldives, UAE, and South Africa. These items include potato flour, starch, and dehydrated chips and flakes (Ranjan 2017).

In order to increase potato production, guarantee positive yield growth, and maintain global exports of potato products, it is necessary to shift research priorities, concentrate on stable markets and products driven by consumer demand, and fortify the relationships between policy planners, extension workers, and researchers.

2 Emerging Issues of Potato Sector in India

The Indo-Gangetic Plains (IGP), namely Uttar Pradesh, Bihar, and West Bengal, are the primary locations for potato cultivation in India, accounting for around 85% of the nation's total output. With their easily accessible irrigation and rich soils, these areas are perfect for growing potatoes. However, optimizing yields is hampered by the brief winter season. Farmers that live in hilly areas double crop their potatoes in order to have two harvests a year.

Potato produces comparatively higher economic yield per unit area and unit time and, therefore, has become more important where cultivable lands are scarce. That is why even the small and marginal farmers of the country have taken up potato cultivation. But, over and again the potential yield of potato is limited by various biotic and abiotic factors in hills, plains, and plateau regions of our country. Among the various factors related to quality seed and ware potato production, farmers are often forced to face water scarcity, changes in rainfall patterns and its intensity, frost and fog, post-harvest losses, pests and diseases, issues related to processing and entrepreneurship, etc.

2.1 Issues Related to Potato Cultivation

Pandey et al. (2005) summarized major obstacles to successful potato production and development.

I. First, there were no potato cultivars or types that could be adapted to the long days of India. Since the winter growing season in subtropical plains produces over 80% of potatoes under conditions very different from those seen in temperate European countries.
II. Second, these cultivars were found to degenerate rapidly under aphid (vector)-infested areas of Indian subtropical plains because potatoes are a vegetative propagated crop maintained over generations through tubers wherein a variety of systemic viruses or virus-like agents accumulate progressively.
III. Third, in contrast to European settings when the potato harvest is followed by severe summer heat waves, the occurrence of hot summers soon after harvest made post-harvest potato storage a major issue.

Apart from this, challenges posed by small and marginal land holding, capital-intensive nature of farming, labor availability front, short days and reduced sunshine in growing period as compared to temperate regions, post-harvest losses, cold storage, transportation mechanism, etc. are listed by Singh and Rana (2014).

2.2 Potato Research in India

About 55 years ago, indigenous potato research and development programs were started in India in response to these initial obstacles. Over these years, focused breeding research along with persistent efforts to improve the nation's germplasm collection for potential uses led to the development of 72 high-yielding indigenous varieties including 9 processing varieties and 2 heat-tolerant varieties suitable for growing under different agro-climatic conditions (Singh and Dutt 2024).

In 1941, the Indian subcontinent accounted for less than 1% of the world's total potato area and production. The main cause of this was the lack of regionally appropriate potato cultivating techniques and varieties in subtropical climates. In order to address these issues, the Imperial Agricultural Research Institute, New Delhi, opened three breeding and seed production sites at Shimla, Kufri (both Shimla hills), and Bhowali (Kumaon hills) on April 1, 1935, as part of a systematic research program on potatoes.

In August 1949, the Central Potato Research Institute (CPRI) of the Indian Council of Agricultural Research (ICAR) was founded in Patna, Bihar, India, following the suggestion of Sir Herbert Steward, the Government of India's Agricultural Advisor. The institute was eventually relocated to Shimla, Himachal Pradesh, in 1956 in order to preserve the health of seed potatoes and promote hybridization work in potato breeding.

The true difficulty lies in figuring out how to best incorporate potatoes as a commodity into the country's agricultural output in order to link millions of marginalized and small-scale farmers in places like India to global markets. Along with significant policy changes aimed at favorable techno-economic feasibility combined with good governance, other measures include: (i) developing and disseminating appropriate yield-enhancing technology for potatoes, as processing varieties yield less than table/ware potatoes; (ii) developing suitable temperature-insensitive processing varieties suitable for growing in cooler regions of the northwest and central plains; and (iii) developing and disseminating appropriate low-cost technologies for the production of a variety of processed potato products affordable for low-income consumers and small-scale rural entrepreneurs to move up at commercial scale (Pandey et al. 2005).

3 Potato Processing Sector and Value Addition in India

3.1 Innovations in Value Addition

Potatoes are prone to spoilage and undergo undesirable quality changes when stored fresh for extended periods. To combat this issue, minimize post-harvest losses, and ensure a year-round potato supply, the creation of value-added processed products is a beneficial solution. Value addition and processing of potatoes not only extend their shelf life but also provide consumers with safe, nutritious, and convenient food options (Table 2). This approach also benefits farmers by increasing the profitability of their potato produce (Marwaha et al. 2010).

Potatoes offer a wide range of processed products, such as cooked potatoes prepared through methods like blanching or boiling. Par-fried potato strips, partially fried for convenience, are used in various dishes, including French fries. French fries, a popular snack, are appreciated for their taste and convenience. Potato chips, thinly sliced and either deep-fried or baked, create a crispy snack. Potato starch, extracted from potatoes, serves as a thickening agent. Potato granules and flakes, dehydrated forms, can be reconstituted for mashed potatoes or used in recipes. Dehydrated diced potatoes are convenient for soups and stews. Additionally, potatoes can be utilized in bioethanol, organic acids, and glucose syrup production through fermentation and hydrolysis. The quality of these processed products depends on raw potato quality, meeting specific criteria like tuber morphology, low reducing sugar content ($\leq$0.1%), and high dry matter content ($\geq$20%). Not all potato varieties suit processing, and selection depends on intended techniques. The highly industrialized and market-driven potato processing industry provides diverse value-added products, meeting consumer preferences for taste, convenience, and nutritional value, thus playing a crucial role in the global food industry.

Table 2 Value additions of potato and its versatility

Value addition	Description
Fresh produce	Potatoes are sold as raw vegetables in markets
Potato chips	Sliced, fried, and seasoned for snacks
French fries	Cut and fried to make a popular fast food
Mashed potatoes	Boiled and mashed, often served as a side dish
Potato starch	Extracted and used in food processing
Potato flour	Ground into flour for baking and thickening
Potato vodka	Fermented and distilled into alcoholic drinks
Potato salad	Mixed with various ingredients for salads
Hash browns	Shredded and pan-fried as a breakfast item
Potato skins	Baked or fried potato skins as appetizers
Dehydrated potatoes	Dried for long-term storage or instant meals
Potato pizza	Pizza crust made with grated potatoes
Potato wedges	Thick-cut potato slices, often seasoned
Scalloped potatoes	Sliced potatoes baked with cheese and cream
Potato gnocchi	Dumplings made from potato, flour, and egg
Potato croquettes	Mashed potatoes formed into patties and fried
Potato pancakes	Grated potatoes mixed with other ingredients
Potato leek soup	A creamy soup made from potatoes and leeks
Potato dumplings	Boiled or steamed dumplings made from potato
Potato bread	Bread made with potato flour or mashed potatoes
Potato crisps	Thinly sliced and fried potato chips

3.2 *Economic Impact and Market Dynamics of Processed Potato Products*

- *Overview of Global Processed Potato Market*
- The potato market size is estimated at USD 115.74 billion in 2024 and is expected to reach USD 137.46 billion by 2029 (Anonymous 2024). Key product types include frozen items, chips and snack pellets, and dehydrated potatoes, finding applications in snacks, ready-to-cook items, and prepared meals. Current global potato production stands at 329 million tons, covering over 19 million hectares. China leads as the top potato producer with 75 million tons, followed by the Russian Federation (37 million tons) and India (25 million tons).
- *Potato Consumption Trends and Patterns in India*
- In India, potatoes are considered a vegetable, unlike in developed countries where they serve as a staple food. Positive correlation was observed between income levels and potato consumption (Scott et al. 2000). Changing consumption patterns over the past decade, influenced by population growth and urbanization, especially among double-income families. Despite changing consumption patterns, per capita potato supply in developing nations, including

India, lags significantly behind developed countries. India's per capita supply is at 17 kg/year, considerably lower than China (35.3 kg/year) and the Russian Federation (122.0 kg/year) (Pandey et al. 2006). According to FAOSTAT data, in 2020 per capita consumption of potato in India was reported 24.95 kg/year, in China 47.24 kg/year, and in Russian Federation 87.16 kg/year.

- *Urgency for Quality-Enhancing Technologies in Indian Potato Processing*
- India urged to expedite the development and adoption of quality-enhancing technologies in processing to compete in the predominantly European-dominated global potato trade. Emphasis shall be given to the necessity for institutional innovations alongside technological advancements to secure a leading role in the international potato processing trade.
- *Significance of Potato Processing Growth for India's Self-Reliance*
- The growth of the potato processing industry is considered vital for India's journey toward self-reliance in food and nutrition. Diversifying processing options, including inexpensive dehydrated products, is crucial for sustaining domestic markets, especially among the low-income population. The expansion of the potato processing sector in India is poised to trigger a "crispy revolution," reflecting a significant transformation and expansion in the industry (Pandey et al. 2006).

4 Potato Marketing System

In India, considering Triennium average export quantity based on APEDA for the year 2022–23 shows, 401,801.88 tons of table potatoes; 351.15 tons of potato seeds; 156,506.36 tons of processed potatoes were exported from India. India's potato export grew at a rate of 14.1% (CAGR) from 2003 to 2022 (Kharumnuid et al. 2024).

4.1 Table Potato Marketing System

Table potatoes, commonly referred to as fresh potatoes, constitute a significant segment of the agricultural industry, representing those potatoes consumed directly by consumers without undergoing substantial processing. The marketing system for table potatoes encompasses a series of interconnected stages, spanning from production to consumption, each playing a pivotal role in delivering quality produce to the end consumer.

At the production stage, the cultivation of table potatoes involves considerations such as the geographical location and scale of farms, as well as the diverse varieties grown to meet varying consumer preferences.

4.2 Seed Potato Marketing System

The seed potato marketing system is a critical component of the agricultural supply chain. In potato seed systems, informal and formal subsystems play an important role (Thiele 1999). Informal subsystem is predominant in resource-poor places and involves a complex network. In contrast, well-off farmers go for formal seed subsystems.

Seed producers play a pivotal role in multiplying and distributing certified seed potatoes to farmers. Education and extension services disseminate knowledge about best practices, technological advancements, and sustainable farming methods, empowering farmers to make informed decisions. Apart from this, global and regional factors, including climate conditions, trade policies, and geopolitical dynamics, influence the dynamics of the seed potato market. Regulatory frameworks, such as phytosanitary standards and certification programs, provide guidelines for quality control and ensure the integrity of seed potatoes in national and international trade.

- *Informal Seed Potato Subsystem*
- This subsystem includes sources like own preserved seeds, seeds preserved by fellow farmers of the same village, and seed preserved by neighboring villages. Monitoring of seed potato quality through institutional mechanisms is largely absent, and degenerated produce is also sold for use as seed in small and marginal farmers' scenarios (National Academy of Agricultural Sciences (NAAS 2021). Farmers are often compelled to use it year after year causing reduction in yield also (Ali et al. 2013).
- *Formal Seed Potato Subsystem*
- Only 10% of seed potato demand is met by certified seeds (Buckseth et al. 2020). Major sources of formal seed potato subsystem are research institutes for providing basic seed and public and private agencies involved in the multiplication of foundation and certified seeds.

Elango et al. (2019) discuss strategies for developing a sustainable potato seed system in the Indian state of Assam where they suggested the following measures. (1) Adoption of slowly degenerating varieties for farmers as they mostly use their own potato seeds. So, the strategy of introduction of new genotypes, participatory trial of varieties, and subsequent release, dissemination, and adoption studies advocated. (2) Strategy of on-Farm practices for improving and maintenance of seed quality. This includes adaptation of farm practices, training material development, training of farmers and extension agents, etc. (3) Strategically use of certified seeds. It may be done through a participatory market chain and selection of suitable areas (by observing vector population) for seed production.

Public-private partnerships and collaborations with governmental bodies are essential for establishing and enforcing regulatory frameworks. These partnerships contribute to the creation of a supportive environment that encourages compliance with quality standards, ensuring the integrity of the seed potato supply chain.

4.3 Processed Potato Marketing System

The processed potato marketing system encompasses a comprehensive approach to promoting, distributing, and selling a variety of potato-based products that have undergone processing. This system involves several key components, including product development, branding, distribution channels, retail presence, and adaptation to market challenges.

Due to changes in consumer preference, processing industries invest in R&D to introduce diverse processed potato products with flavors, healthier options, taste, texture, etc. Over the next 40 years, there is demand for French fries (11.6% ACGR) followed by potato flakes/powder (7.6%) and potato chips (4.5%). A list containing various processed potato products is mentioned in Table 2, and demand is expected to grow from 2.8 million tons in 2010 to 25 million tons during the year 2050 at an CAGR of 5.61%. Presently only 7–8% of total potato produce is processed in India, so, high growth opportunity exists in processed potato market in India (Singh and Dutt 2024). Various leading private players like FMEL Agrifriz Pvt. Ltd. (Falcon group), Agristo India (Wave Agro), JK Singhania, JR Simplot Foods Private Limited, Farmton, Chillfill Foods, and SK International have already ventured into potato processing, and many more are about to come in this sector (Gupta et al. 2024).

5 Future Trends and Challenges in the Potato Sector

The potato (*Solanum tuberosum* L.) is a vital crop with a global production exceeding 368 million metric tons in 2021 (FAOSTAT 2021). Despite its importance, the potato sector confronts various challenges that necessitate strategic planning and collaborative efforts. As per the Food and Agriculture Organization (FAO) in 2022, there is potential for potatoes to make a more significant impact on global history and food security. The FAO suggests that the overall production of this essential crop has the capacity to double within the next decade.

This section discusses the future trends and challenges in the potato sector, focusing on sustainable growth and the importance of collaborative research initiatives.

5.1 Future Trends in Potato Sector

- *Climate Change Impact*
- Climate change poses a significant threat to potato production. Rising temperatures, altered precipitation patterns, and increased frequency of extreme weather events can negatively impact yield and quality (CIP 2018). Adaptation strategies, including the development of climate-resilient potato varieties, are essential for the sector's sustainability. Potatoes and sweet potatoes, being robust and fast-

growing crops, hold significance in climate-smart agriculture. Utilizing next-generation technologies, there is potential to create climate-smart varieties that exhibit rapid maturation, resilience against pests and diseases, and tolerance to heat, drought, or soil salinity. This can contribute to the production of more nutritious food and increased incomes in challenging climates. The International Potato Center (CIP) has been at the forefront, introducing innovative agronomic techniques aimed at enhancing the quality of planting material.
- *Shifting Consumer Preferences*
- Changing dietary patterns and consumer preferences influence the demand for different potato products. As consumers increasingly seek healthier and more sustainable options, the potato sector must innovate to meet these evolving demands (WRI 2016).
- *Technological Advancements*
- Technological innovations, such as precision agriculture, genetic engineering, and data analytics, offer opportunities to enhance productivity and sustainability in the potato sector (Devaux et al. 2021). Integration of these technologies requires investment and collaborative efforts to ensure widespread adoption.

5.2 Challenges in the Potato Sector

- *Pest and Disease Management*
- Potatoes face numerous pests and diseases that threaten crop yields. Sustainable pest and disease management practices, including integrated pest management (IPM) and the development of resistant varieties, are critical for mitigating these challenges (Hameed et al. 2018).
- *Resource Constraints*
- Limited water availability, land degradation, and resource-intensive cultivation practices contribute to the strain on potato production systems. Sustainable practices, including efficient water use and soil conservation, are essential for long-term viability (Singh 2011).
- *Global Supply Chain Disruptions*
- Globalization exposes the potato sector to supply chain disruptions, as witnessed during the COVID-19 pandemic. Strengthening local and regional supply chains and implementing resilient production systems are essential for mitigating such risks (FAO 2020).

5.3 Recommendations for Sustainable Growth

- *Promotion of Climate-Resilient Varieties*
- Investments in breeding programs focusing on climate-resilient potato varieties can enhance the sector's ability to withstand changing climatic conditions (CIP 2020).

- *Adoption of Sustainable Agricultural Practices*
- Encouraging the adoption of sustainable agricultural practices, including precision farming, organic cultivation, and agroecological approaches, can contribute to resource efficiency and environmental conservation (Cassani and Gomez-Zavaglia 2022).
- *Public-Private Partnerships*
- Fostering collaboration between government agencies, private sector entities, and research institutions can accelerate the development and dissemination of innovative technologies and practices. In industrialized nations, potato storage facilities primarily function as a guarantee for year-round tuber availability for the processing industry. Research activities among storage facility providers focus on sustainable technology. There is a need to facilitate knowledge exchange to support the development of small-scale potato storage facilities in rural and low-income countries through public-private partnerships.

5.4 Collaborative Research Initiatives

- *International Research Networks*
- Establishing international research networks can facilitate knowledge exchange, collaborative studies, and the development of best practices for potato production (Duncan et al. 2021).
- *Data Sharing and Analytics*
- Encouraging data sharing among stakeholders and leveraging advanced analytics can enhance decision-making processes and enable evidence-based solutions for the challenges faced by the potato sector (CIP 2021).
- *Capacity Building and Extension Services*
- Investing in capacity building programs and extension services can empower farmers with the knowledge and skills needed to adopt sustainable practices and navigate challenges effectively (Oyesola et al. 2021).

6 Conclusion

Potato remains a crucial staple food crop globally, particularly in India, where it ranks second in production. However, key challenges such as low productivity, seed degeneration, inadequate storage infrastructure, and limited processing capacity continue to hinder its full potential. The shifting consumption preferences toward ready-to-eat products further strain the sector, with only a small percentage of the total production being processed. Additionally, climate change presents a significant threat to potato cultivation, affecting not only yields but also the frequency of pest and disease outbreaks and supply chain stability.

To overcome these challenges, it is imperative to encourage greater participation from all stakeholders—public and private—along the entire value chain, from farm to table, to ensure success in potato cultivation and facilitate its journey to consumers. Collaborative research, extension, and innovative business practices must be at the forefront of developing climate-resilient strategies that ensure the long-term profitability of potato farming, particularly for small and marginal farmers. By addressing these issues collectively, the potato sector in India has the potential to achieve sustainable growth, contributing to both food and nutritional security amidst the pressures of a growing population and climate uncertainties.

References

Agricultural Statistics at a Glance (2022) Economics & Statistics Division, Department of Agriculture & Farmers Welfare, Ministry of Agriculture & Farmers Welfare, Government of India

Ali S, Kadian MS, Ortiz O, Singh BP, Chandla VK, Masood AKH (2013) Degeneration of potato seed in Meghalaya and Nagaland states in northeastern hills of India. Potato J 40(2):122–127

Anonymous (2024). https://www.mordorintelligence.com/industry-reports/potato-market. Accessed 20 May 2024

Buckseth T, Singh RK, Tiwari JK, Sharma AK, Singh S, Chakrabarti SK (2020) A novel sustainable aeroponic system for healthy seed production in India-an update. Indian J Agric Sci 90(2):243–248

Cassani L, Gomez-Zavaglia A (2022) Sustainable food systems in fruits and vegetables food supply chains. Front Nutr 9:829061

CIP (2018) Climate-resilient potatoes for the future. Retrieved from https://cipotato.org/research/climate-change/

CIP (2020) International potato center annual report 2019. Retrieved from https://cipotato.org/publications/reports-documents/annual-reports/

CIP (2021) Harnessing the power of data to transform potato agriculture. Retrieved from https://cipotato.org/research/data-transform-potato-agriculture/

Devaux A, Kromann P, Ortiz O (2014) Potatoes for sustainable global food security. Potato Res 57:185–199

Devaux A, Goffart JP, Kromann P, Andrade-Piedra J, Polar V, Hareau G (2021) The potato of the future: opportunities and challenges in sustainable agri-food systems. Potato Res 64(4):681–720

Dongyu Q, Director-general FAO (2022) Role and potential of potato in global food security. The potato of the future: opportunities and challenges in sustainable agri-food systems. Potato Res 64(4):681–720

Duncan E, Abdulai AR, Fraser ED (2021) Modernizing agriculture through digital technologies: prospects and challenges. In: Handbook on the human impact of agriculture, Cheltenham: Edward Elgar Publishing; pp 138–161

Elango D, Vikram P, Kreuze JF, Andrade-Piedra JL (2019) Strategic plan for developing a sustainable potato seed system in Assam, India. Lima (Peru). CGIAR Research Program on Roots, Tubers and Bananas (RTB). RTB Working Paper. No. 2019-1. Available online at: www.rtb.cgiar.org

FAO (2020) The impact of COVID-19 on food security and nutrition: an interim report. Retrieved from http://www.fao.org/3/ca8666en/ca8666en.pdf

FAOSTAT (2021). www.fao.org/faostat/en/#data/QC. Accessed 11 Apr 2024

Gupta VK, Sood S, Mangal V, Luthra SK, Chaudhary B (2024) Potato processing scenario and future prospects in India. Indian Farming 74(05):35–38

Hameed A, Zaidi SSEA, Shakir S, Mansoor S (2018) Applications of new breeding technologies for potato improvement. Front Plant Sci 9:925
Kaur S, Aggarwal P, Babbar N (2023) Evaluating progress of Indian potato processing industry: an updated review. Potato Res 66(4):945–964
Kharumnuid P, Kumar A, Singh B (2024) Indian marketing and exports of potato and its processed products. Indian Farming 74(5):48–51
Marwaha RS, Pandey SK, Kumar D, Singh SV, Kumar P (2010) Potato processing scenario in India: industrial constraints, future projections, challenges ahead and remedies—a review. J Food Sci Technol 47:137–156
NAAS (2021) Innovations in potato seed production. Strategy Paper No. 14, National Academy of Agricultural Sciences, New Delhi, p 20
Oyesola OL, Aworunse OS, Oniha MI, Obiazikwor OH, Bello O, Atolagbe OM et al (2021) Impact and management of diseases of Solanum tuberosum. In: Solanum tuberosum: a promising crop for starvation problem; IntechOpen, London, UK
Pandey SK, Khurana SMP, Singh SV, Kumar D, Kumar P (2005) Evaluation of Indian and exotic potato varieties for sustaining processing industries in north western plains of India. Indian J Hort Sci 62:155–159
Pandey SK, Sarkar D, Singh SV (2006) Potato processing in India: today and tomorrow. Potato J 33(1–2):11–19
Pieterse L (2024) Global potato production: insights from the FAO's latest data. Retrieved on 18 Jan 2024 from https://www.potatonewstoday.com/2024/01/06/global-potato-production-insights-from-the-faos-latest-data/
Rana RK, Anwer MD (2018) Potato production scenario and analysis of its total factor productivity in India. Indian J Agric Sci 88(9):1354–1361
Ranjan N (2017) Marketing and export of potato in India. In: Summer school, p 179
Sahu PK, Das M, Sarkar B, Adarsh VS, Dey S, Narasimhaiah L, Raghav YS (2024) Potato production in India: a critical appraisal on sustainability, forecasting, price and export behaviour. Potato Res:1–37
Scott GJ, Rosegrant MW, Ringler C (2000) Roots and tubers for the 21st century: trends, projections, and policy options. Food, agriculture, and the environment discussion paper 31. International Food Policy Research Institute, Washington, USA.
Singh S (2011) Farm to fork supply chain of potato in the state of Bihar in India. Available at SSRN 2015255
Singh B, Dutt S (2024) Seventy five years of potato in India An inspiring success story. Indian Farming 74(05):03–08
Singh BP, Rana RK (2014, November) History of potato and its emerging problems in India. In: Souvenir–National seminar on emerging problems in potato. Indian Potato Association, Shimla, pp 7–21
Thiele G (1999) Informal potato seed systems in the Andes: why are they important and what should we do with them? World Dev 27(1):83–99
WRI (World Resources Institute) (2016) Shifting diets for a sustainable food future. https://doi.org/10.13140/RG.2.1.3808.2961

Basics of Economics and Marketing Practices in Potato Production

Rajkumari Asha Devi, Anupam Adarsh, Avinash Kumar, and Aniruddh Pratap Singh

Abstract Potato production plays a vital role in the global agricultural economy, offering significant opportunities for income generation and food security. Understanding the basics of economics and marketing practices in potato production is crucial for maximizing profitability and ensuring sustainable operations. The economic aspects of potato farming involve analyzing production costs, yield, and profitability across different scales of farming, from small-scale to large-scale operations. Key factors such as seed quality, input costs, labor, and harvesting techniques directly impact the overall efficiency and output of the production process. Marketing practices in potato production are equally important, determining the accessibility of markets and the returns farmers can secure for their produce. Farmers utilize various marketing channels, including direct sales, wholesale markets, cooperatives, contract farming, and export markets. Each channel offers distinct advantages and challenges, influenced by factors such as farm scale, market access, and product quality standards. Small-scale farmers often rely on direct sales and local markets, while larger-scale producers have better access to wholesale markets and export opportunities. These foundation elements of potato production economics and marketing emphasize the need for strategic planning and market linkages to enhance profitability and competitiveness. By adopting innovative marketing practices and optimizing production economics, potato producers can improve their market position and contribute to the broader agricultural value chain.

Keywords Potato · Economics · Marketing · Profitability · Farmers

R. A. Devi (✉)
Department of Horticulture (Vegetable Science), Bihar Agricultural College, Bihar Agricultural University, Sabour, Bhagalpur, Bihar, India

A. Adarsh · A. P. Singh
Department of Horticulture (Vegetable Science), Nalanda College of Horticulture, Noorsarai, Nalanda, Bihar Agricultural University, Sabour, Bhagalpur, Bihar, India

A. Kumar
Department of Agricultural Economics, Bihar Agricultural College, Sabour, Bihar Agricultural University, Sabour, Bhagalpur, Bihar, India

S. S. Solankey (ed.), *Advances in Research on Potato Production*, Advances in Olericulture, https://doi.org/10.1007/978-3-031-82710-5_18

1 Introduction

Potato (*Solanum tuberosum* L.) is one of the important cash crops from the family Solanaceae. It was introduced to India in the seventeenth century by Portuguese traders or the Britishers from Peru-Bolivian region in the Andes (South America). It is one of the major vegetable crops of India, occupies an important position among food crops, and provides staple foodstuff for millions of people in many parts of the world. In India, there is a huge demand for and supply of this crop, which is 47.5 mt and 51.3 mt, respectively, thereby it contributes a considerable share in the exports (395748.12 mt) and generate revenue of (41182.5 lacs) as compared to other food crops. The largest importer of the potato from India is Nepal, importing 2.74 mt of value 20,301.37 lacs. India contributes 11–12% of the world's potato production and is the second largest producer after China. India produces about 43.8 mt of potato from an area of 2.13 mha. Among the states, Uttar Pradesh is the leading state with an annual production of 15.55 mt from 0.61 mha area, but the productivity is highest in West Bengal (29.9 mt/ha) (Anon., 2018). Bihar is producing 7.7 mt from 0.3 mha area. In recognition of virtues the importance of potato as a staple food the FAO head declared 2008 as the "International Year of the Potato" and has rightly identified it as the "food of future" due to its high nutritive value content of energy (77 Kcal), carbohydrates (17.49 g), protein (2.05 g), total fat (0.10 g), cholesterol (0 mg), and dietary fiber (2.1 g) [USDA National Nutrient data base]. The potato has a high potential to generate revenue for the farmers, and being high mechanized crop requires high capital. Apart from India, the other major producers of potato in the world are China, the Russian Federal, Poland, the USA, Ukraine, Germany, the Netherlands, and the UK.

2 Origin and Distribution

Potato was introduced to India in the seventeenth century by Portuguese traders or the Britishers from the Peru-Bolivian region in the Andes (South America). Cultivation of potato in northern hills started later than plains but became as cash crop much earlier. Cultivation of potato began first in the Nilgiri hills in 1822. It is grown in India in almost all the states except Kerala.

3 Climate and Soil

Potato is basically a cool season crop. It is grown in winter in the plains of India. However, in northern hills, it is grown as a summer season crop. It is a long day plant but cultivated as day plant. It requires favorable environmental conditions such as low temperature and short-day conditions at the time of tuberization for rapid bulking rate, which requires an optimum temperature of 25 °C for germination and 17–19 °C for its growth. The best temperature for vegetative growth is 20 °C. Highest tuber formation occurs at 20 °C Day and 14 °C night temperature. At temperature above 24 °C tuber formation stops completely. At higher temperatures, the respiration rate increases, and the carbohydrates produced by photosynthesis are consumed rather than stored in tuber. High temperatures at any part of the growing period affect the size of leaflets, thereby reducing the tuber formation. It grows best under long day sunshine condition along with cooler nights are essential for reducing the spread of diseases.

Grown in all types of soil except saline and alkaline soil (causes scab disease). Best suited in well-drained sandy loam to clay loam soil with a pH of 5.5–7.5. Light soil is preferred, because they tend to promote more uniform soil temperatures and make harvesting of the crop easier.

4 The Potato: A Global Staple

The potato ranks as the third most important food crop in the world, following only rice and wheat. This versatile tuber is a dietary staple for over a billion people worldwide, with annual production exceeding 300 million metric tons. Its global importance is rooted not only in its nutritional value but also in its remarkable adaptability to various climates and soil conditions. The potato's efficiency in producing more food per unit of water than any other major crop highlights its crucial role in global food security.

5 The Journey of the Potato: From the Andes to the World

The story of the potato begins in the Andean region of South America, where it was first domesticated by the pre-Inca people around 8000 years ago. In the sixteenth century, Spanish conquistadors encountered this humble tuber and introduced it to Europe. Over the centuries, the potato spread across the globe, becoming a vital food source due to its nutritional value and versatility. From its ancient origins in the highlands of Peru and Bolivia to its status as a global staple, the potato's journey is a testament to its enduring significance in the world's agricultural and culinary history.

6 The Nutritional Power of Potatoes

Potatoes are not only a dietary staple but also a nutritional powerhouse. Rich in vitamins C and B6, potassium, and fiber, they contribute to heart health and aid digestion. Their high carbohydrate content provides essential energy, making them

Table 1 Year-wise export of potatoes by India

Year	Qty (MT)	Value (in crore)
2018	320,000	6.88
2019	350,000	7.55
2020	375,000	7.97
2021	400,000	8.81
2022	425,000	9.23
2023	450,000	10.07
2024	460,000	10.91

Source: APEDA (2018–2024)

a key player in global food security. Potatoes are also laden with antioxidants, which help combat chronic diseases and support overall health. Their adaptability to diverse climates and soil conditions makes them a primary crop in many countries, further emphasizing their importance in both nutrition and agriculture.

Potato cultivation is a cornerstone of Indian agriculture, vital both as a staple food and an economic driver. Across the country, a diverse array of potato varieties flourishes, each with distinct colors—ranging from white to purple—and textures, whether waxy or starchy. Remarkably, potatoes yield more per acre than many other staple crops, with a single plant often producing 5–10 tubers, depending on the variety and growing conditions. This narrative explores India's leading potato-producing states, each contributing significantly to the nation's agricultural wealth.

7 India's Potatoes on the Global Stage

India has carved out a significant role in the global potato market, emerging as a major exporter of this essential crop. In 2022, India's potato exports reached $108 million, earning the country the rank of 12th largest potato exporter in the world. Indian potatoes find their way to neighboring countries like Nepal, as well as to more distant markets such as Oman, Saudi Arabia, Indonesia, and Malaysia. Despite challenges like crop diseases and supply chain disruptions, the export of certain potato varieties—particularly those used for processing—experienced a notable surge in early 2024. This growth is driven by global demand amidst tight supplies, underscoring India's potential to leverage its agricultural strengths on the international stage, even in the face of adversity.

The export of potatoes from India saw a significant increase between 2018 and 2019. In 2018, the export value was ₹432.84 crore for a quantity of 375,572 metric tons (MT). By 2019, the export value rose to ₹547.14 crore, corresponding to a quantity of 432,895 MT. This growth reflects a rising global demand for Indian potatoes, with an 18% increase in quantity and a 26% increase in value within a year. Table 1 highlights the strengthening position of India in the international potato market, driven by both higher volumes and improved pricing.

8 India's Love Affair with the Potato

In India, the potato is more than just food—it is a culinary chameleon, adapting to a myriad of dishes across the country's diverse kitchens. The nation's per capita potato consumption reached 25.4 kg in 2021, a slight increase from the previous year, and a far cry from the modest 3.45 kg recorded in 1962. This steady rise mirrors the potato's growing significance in Indian diets, driven by its versatility and the increasing demand from urban, higher-income populations. As India urbanizes and its dietary patterns evolve, potato remains a beloved ingredient, seamlessly bridging the gap between traditional culinary practices and modern tastes.

8.1 Uttar Pradesh: The Titan of Tuber Cultivation

Leading the charge is Uttar Pradesh, with a monumental output of 15.89 million tons. The state's fertile plains, graced with a favorable climate, provide an ideal setting for potato farming. Agra, Farrukhabad, and Meerut stand out as the crown jewels, their high yields reinforcing UP's dominance in the nation's potato production.

8.2 West Bengal: The Robust Contributor

Close on Uttar Pradesh's heels is West Bengal, boasting an impressive 12.6 million tons of potatoes. The state's alluvial soil, coupled with its conducive climate, particularly in Hooghly, Burdwan, and Bankura, fuels its strong presence in the potato market.

8.3 Bihar: The Rising Star

Bihar secures the third spot with a robust 9.13 million tons. The state's embrace of advanced farming techniques and high-yield varieties has sparked a significant surge in potato production, with Nalanda and Purnia at the forefront of this agricultural renaissance.

8.4 Gujarat: The Innovator

Gujarat contributes a solid 3.78 million tons to India's potato basket. The state's innovative irrigation strategies and adoption of modern farming technologies have transformed districts like Banaskantha into potato-growing powerhouses.

8.5 Madhya Pradesh: The Diverse Producer

Madhya Pradesh follows with 3.58 million tons, its varied agro-climatic zones providing a prolonged cropping season. Indore and Ujjain emerge as key players, benefiting from the state's diverse agricultural landscape.

8.6 Punjab: The Granary's Bounty

Known as the "Granary of India," Punjab excels in potato farming, contributing 2.85 million tons. The state's progressive farming practices, combined with its fertile soil, secure its place as a significant player in potato production.

8.7 Haryana: The Consistent Cultivator

In Haryana, potatoes are more than just a crop—they are a staple, with the state producing 765,900 tons. Sirsa and Hisar lead the charge, their innovative farming methods ensuring consistent and high-quality yields.

8.8 Assam: The Unique Contributor

Assam adds 761,840 tons to the national tally. The state's distinctive topography and climate create a perfect environment for potato cultivation, with Tinsukia and Dibrugarh being key contributors.

8.9 Jharkhand: The Emerging Force

Jharkhand, producing 733,770 tons, is quickly rising in the ranks. The state's focus on agricultural development has spurred an increase in potato farming, particularly in Ranchi and East Singhbhum.

8.10 Chhattisgarh: The Agricultural Diversifier

Rounding out the list is Chhattisgarh, with a production of 623,650 tons. The state's commitment to agricultural diversification has led to a rise in potato cultivation, with Raipur and Durg at the helm.

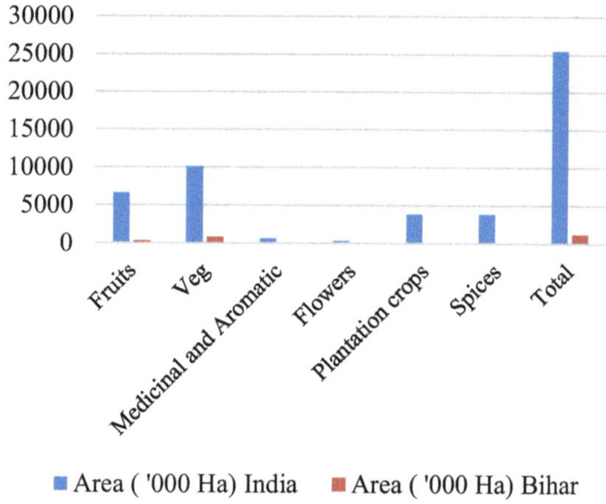

Fig. 1 Area of horticulture crops (2023–24). (Source: APEDA (2023–24))

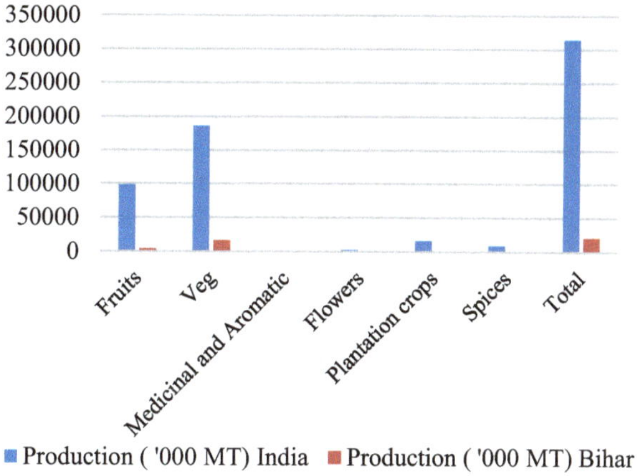

Fig. 2 Production of horticulture crops (2023–24). (Source: APEDA (2023–24))

Each of these states plays a crucial role in feeding the nation, their collective efforts ensuring that the humble potato continues to thrive as a vital component of India's agricultural landscape.

Table 1 and Figs. 1 and 2 indicate area and production of horticulture crops. During the 2018–2019 period, India's horticulture sector demonstrated significant contributions to the agricultural landscape, covering a total area of 25,492 thousand hectares and yielding a massive production of 3,13,730 thousand metric tons (MT). Bihar, a key agricultural state, contributed notably to this national output, with a horticultural area of 1211.77 thousand hectares and a production of 21,149.95 thousand MT. Fruits formed a substantial portion of the horticultural sector, with India dedicating 6,648,000 hectares to fruit cultivation, yielding 98,579,000 MT. Bihar, on a smaller scale but still significant, contributed 313.95 thousand hectares and produced 4384.46 thousand MT. Vegetables were another dominant category, with

Table 2 Area and production of horticulture crops (2023–2024)

Crops	Area ('000 Ha)		Production ('000 MT)	
	India	Bihar	India	Bihar
Fruits	6648	313.95	11,263	4346
Vegetables	10,100	872.55	20,496	16,684
Medicinal and aromatic	656	6.79	819	268
Flowers	313	1.41	2865	1052
Plantation crops	3880	8.79	16,368	4012
Spices	3895	8.28	9216	1233
Total	**25,492**	**1211.77**	**313,730**	**21149.95**

Source: APEDA (2024)

the largest area and production figures. India devoted 10,100,000 hectares to vegetable cultivation, resulting in a production of 185,883,000 MT (Table 2). Bihar's share in this was considerable, with 872.55 thousand hectares yielding 16,699.84 thousand MT, highlighting the state's robust vegetable production capabilities. Medicinal and aromatic plants occupied a smaller area but still played a crucial role, with India utilizing 656,000 hectares for these crops, producing 819,000 MT. Bihar's contribution was modest, with 6.79 thousand hectares and a production of 2.68 thousand MT. Flowers, while covering only 313,000 hectares across India, produced were 2,865,000 MT. Bihar's involvement in floriculture was minimal, with 1.41 thousand hectares contributing 10.52 thousand MT. Plantation crops and spices also formed vital segments of horticulture. India allocated 3,880,000 hectares to plantation crops, yielding 16,368,000 MT, and 3,895,000 hectares to spices, with a production of 9,216,000 MT. Bihar's participation was relatively minor in these areas, with 8.79 thousand hectares and 40.12 thousand MT for plantation crops, and 8.28 thousand hectares and 12.33 thousand MT for spices (Table 2).

Overall, Bihar's contribution to India's horticulture is significant, particularly in vegetables, reflecting the state's strong agricultural practices.

8.11 *Storing Potatoes in India: A Blend of Tradition and Modernity*

In India, the journey of the potato continues long after it leaves the field. To preserve their longevity and quality, potatoes are stored using a combination of age-old wisdom and contemporary techniques. Traditionally, potatoes find refuge in cool, dry rooms with ample ventilation, often resting on bamboo racks or nestled in pits. These methods, passed down through generations, ensure that the tubers remain fresh and viable for months.

In modern agricultural practices, large-scale cold storage facilities have become the cornerstone of potato preservation. Here, potatoes are kept at temperatures between 8 and 12 °C, a sweet spot that prevents sprouting and spoilage. Seed potatoes, destined for future harvests, are stored at even lower temperatures of 2–4 °C. To further extend their shelf life, some potatoes undergo treatment with sprout suppressants. These combined efforts safeguard the quality of potatoes, ensuring they continue to grace Indian tables throughout the year.

Storing potatoes in cold storage is a vital aspect of post-harvest management, ensuring that the tubers remain in optimal condition for extended periods. It is important for maintaining the supply chain and minimizing losses due to spoilage. It is a critical component of the potato supply chain, offering numerous benefits in terms of extending shelf life, preserving quality, and reducing post-harvest losses. However, it also presents challenges, including high energy costs and the risk of disease. As research and innovation continue, the future of cold storage for potatoes looks promising, with potential for even greater efficiency and sustainability.

1. *Importance of Cold Storage for Potatoes*

   Potatoes are a staple food in many parts of the world, and their year-round availability is crucial for both consumers and industries. Cold storage plays a significant role in extending the shelf life of potatoes by slowing down their metabolic processes. According to Harris (2021), the respiration rate of potatoes decreases significantly at lower temperatures, which helps in preserving their nutritional content and reducing the risk of spoilage.

   Cold storage is essential for:

   - *Maintaining Quality*: Potatoes stored at optimal temperatures retain their firmness, flavor, and nutritional value for longer periods.
   - *Preventing Sprouting*: The sprouting of potatoes is a common issue during storage. Cold temperatures inhibit sprout development by slowing down physiological processes (Jones & Armstrong, 2019).
   - *Minimizing Weight Loss*: Moisture loss during storage can lead to weight loss in potatoes. Cold storage helps in maintaining moisture content, thereby reducing weight loss (Brown et al., 2018).
   - *Reducing Post-Harvest Losses*: Proper cold storage techniques can reduce post-harvest losses, which are a significant concern for farmers and distributors (Kumar et al., 2020).

2. *Optimal Storage Conditions*

   To achieve the benefits of cold storage, specific environmental conditions must be maintained. These include temperature, humidity, ventilation, and light exposure.

   (a) *Temperature*

   The most critical factor in storing potatoes is temperature. Potatoes should ideally be stored at temperatures between 4 °C and 10 °C (39 °F to 50 °F). Temperatures below 4 °C can cause the starches in potatoes to convert into sugars, leading to a sweet taste and increased browning during cooking. This phenomenon, known as "cold-induced sweetening," is undesirable for both consumers and processors.

   On the other hand, temperatures above 10 °C can accelerate sprouting and increase the risk of disease. Therefore, maintaining a stable temperature within the recommended range is crucial.

(b) *Humidity*

Relative humidity in the storage environment should be kept between 90% and 95% to minimize moisture loss (Burton, 1989). Low humidity can cause excessive dehydration, leading to shrinkage and weight loss. However, too high humidity may encourage the growth of mold and other pathogens, which can lead to rot.

(c) *Ventilation*

Proper ventilation is essential to prevent the buildup of carbon dioxide (CO_2) and ethylene, both of which can negatively impact the quality of stored potatoes. According to, high levels of CO_2 can lead to the accumulation of undesirable compounds, such as glycoalkaloids, which are toxic in large quantities. Ventilation also helps in maintaining uniform temperature and humidity levels throughout the storage facility.

(d) *Light Exposure*

Potatoes should be stored in darkness to prevent the development of chlorophyll, which causes greening. Greening is often accompanied by the formation of solanine, a toxic compound that can make potatoes unsafe for consumption (Friedman, 2006). Therefore, storage facilities should be designed to minimize light exposure.

3. *Challenges in Cold Storage*

While cold storage offers numerous benefits, it also presents several challenges that must be addressed to ensure the successful preservation of potatoes.

(a) *Energy Costs*

Maintaining the optimal temperature and humidity levels in a cold storage facility requires significant energy. According to a study by, energy costs can account for up to 60% of the total operating costs of a cold storage facility. As energy prices fluctuate, this can impact the cost-effectiveness of cold storage.

(b) *Disease Management*

Even under optimal conditions, stored potatoes can be susceptible to diseases, such as late blight (*Phytophthora infestans*) and soft rot (*Pectobacterium* spp.). These diseases can spread rapidly in a cold storage environment if not properly managed. Regular monitoring and the use of fungicides are necessary to prevent outbreaks (Morris & Taylor, 2010).

(c) *Technological Requirements*

Modern cold storage facilities require sophisticated technology to monitor and control environmental conditions. This includes temperature and humidity sensors, ventilation systems, and backup power supplies. The initial investment in such technology can be high, and ongoing maintenance is required to ensure its effectiveness.

4. *Best Practices for Cold Storage of Potatoes*

(a) *Pre-storage Handling*

Before placing potatoes in cold storage, they should be cured to heal any wounds or bruises sustained during harvest. Curing involves holding the potatoes at a temperature of 15–20 °C (59–68 °F) and a high relative humidity of 85–95% for about 10–14 days. This process helps in developing a protective skin that reduces moisture loss and disease incidence during storage.

(b) *Monitoring and Control*

Continuous monitoring of temperature, humidity, and CO_2 levels is essential to maintain optimal storage conditions. Automated control systems can help in adjusting environmental parameters in real time, reducing the risk of human error.

(c) *Regular Inspection*

Stored potatoes should be inspected regularly for signs of spoilage, disease, or sprouting. Early detection of issues allows for timely intervention, such as the removal of affected tubers to prevent the spread of disease (Storey, 2007).

(d) *Rotational Storage*

To ensure that potatoes are used within their optimal storage period, a rotational storage system should be implemented. This involves regularly cycling the oldest stock out of storage for processing or sale, while newer stock is stored.

5. *Future Trends and Innovations*

The field of cold storage is continuously evolving, with new technologies and practices being developed to improve efficiency and reduce costs. Some emerging trends include:

(a) *Renewable Energy Integration*

To address the challenge of high energy costs, there is growing interest in integrating renewable energy sources, such as solar and wind power, into cold storage facilities. This can help in reducing the environmental impact and operational costs of cold storage (Gustavsson et al., 2020).

(b) *Smart Storage Systems*

Advancements in the Internet of Things (IoT) and artificial intelligence (AI) are enabling the development of "smart" cold storage systems. These systems can predict and respond to changes in storage conditions, optimizing energy use and reducing the risk of spoilage (Kumar et al., 2021).

(c) *Biological Control Methods*

Research is ongoing into the use of biological control methods to manage diseases in stored potatoes. For example, certain beneficial bacteria and fungi can be used to inhibit the growth of pathogens, reducing the need for chemical fungicides (Fravel, 2005).

9 Importance of Potato Production

Potatoes are grown in over 150 countries, and they provide a significant source of nutrition, especially in developing regions. According to the Food and Agriculture Organization (FAO, 2020), global potato production reached 370 million tons in

2019. Potatoes are a rich source of carbohydrates, vitamins, and minerals, making them an essential part of diets worldwide.

The economic importance of potato production is multifaceted:

- *Income Generation*: Potatoes are a high-value crop that provides income for millions of smallholder farmers, particularly in regions like South Asia, Africa, and Latin America (Scott et al., 2000).
- *Employment*: Potato production and processing create substantial employment opportunities in rural areas, contributing to poverty alleviation and rural development (Horton & Anderson, 1992).
- *Food Security*: Potatoes are a vital crop for food security due to their adaptability to various agro-climatic conditions and their ability to produce high yields in a relatively short period (FAO, 2014).

9.1 Economics of Potato Production

Potatoes are semi-perishable and bulky, and therefore, their transport to long distances is a problem. The economics and marketing of potato production are crucial for understanding the dynamics of this crop, which plays a vital role in food security, income generation, and rural development. Often potatoes rot during transit because of high temperatures at the time of transport. The problem is compounded further due to shortage of transporting wagons.

(a) *Production Costs*

The cost of potato production varies widely depending on factors such as location, scale of operation, and input prices. Key components of production costs include:

- *Land Preparation*: Costs associated with land preparation include plowing, harrowing, and leveling. These costs can vary depending on the type of machinery used and the condition of the land.
- *Seed*: Seed potatoes are one of the most significant costs in potato production. Certified seed potatoes are preferred for their disease resistance and higher yield potential, but they can be expensive (Haverkort et al., 2009).
- *Fertilizers and Pesticides*: Potatoes require a balanced supply of nutrients, which necessitates the use of fertilizers. Additionally, the crop is susceptible to various pests and diseases, leading to expenditures on pesticides (Westermann, 2005).
- *Irrigation*: In regions where rainfall is insufficient, irrigation is necessary. The cost of irrigation depends on the method used (e.g., drip, sprinkler) and the availability of water.
- *Labor*: Labor costs can be significant, particularly in regions where mechanization is limited. Labor is required for planting, weeding, irrigation, pest control, and harvesting.

- *Harvesting and Post-Harvest Handling*: The harvesting process can be labor-intensive if not mechanized. Post-harvest handling, including grading, sorting, and packaging, also incurs costs (Storey, 2007).

  The total cost of production can range from \$2000 to \$5000 per hectare, depending on the aforementioned factors (Haverkort & Struik, 2015). Understanding these costs is essential for farmers to make informed decisions about resource allocation and to optimize profitability.

(b) *Yield and Returns*

Potato yields can vary significantly based on the variety, growing conditions, and management practices. In developing countries, average yields are around 10–20 tons per hectare, while in developed countries, yields can exceed 40 tons per hectare (FAO, 2020). The revenue generated from potato production depends on the yield and market price.

If a farmer achieves a yield of 20 tons per hectare and sells the potatoes at ₹25,000 per ton, the gross revenue would be ₹5,00,000 per hectare. Subtracting the production costs of ₹3, 33,000 per hectare, the net return would be ₹1,67,000 per hectare. However, market prices can fluctuate significantly based on supply and demand dynamics, affecting profitability (Scott et al., 2000).

(c) *Profitability Analysis*

A comparative study of potato production economics is presented in Table 3. Profitability in potato production is influenced by several factors:

Table 3 Comparative table for potato production economics

Parameter	Small-scale farms	Medium-scale farms	Large-scale farms
Land preparation costs	₹8300–16,600 per hectare	₹6640–12,450 per hectare	₹4150–8300 per hectare
Seed costs	₹41,500–83,000 per hectare	₹33,200–66,400 per hectare	₹24,900–49,800 per hectare
Fertilizers and pesticides	₹24,900–49,800 per hectare	₹20,750–41,500 per hectare	₹16,600–33,200 per hectare
Labor costs	₹33,200–99,600 per hectare	₹24,900–83,000 per hectare	₹16,600–66,400 per hectare
Irrigation costs	₹8300–33,200 per hectare	₹6640–24,900 per hectare	₹4980–16,600 per hectare
Harvesting costs	₹16,600–41,500 per hectare	₹12,450–33,200 per hectare	₹8300–24,900 per hectare
Total production costs	₹1,32,800–3,23,700 per hectare	₹1,04,580–2,61,450 per hectare	₹75,530–1,99,200 per hectare
Average yield	10–20 tons per hectare	20–30 tons per hectare	30–40 tons per hectare
Gross revenue	₹2,49,000–4,98,000 per hectare	₹4,98,000–7,47,000 per hectare	₹7,47,000–9,96,000 per hectare
Net profit	₹91,300–1,74,300 per hectare	₹2,36,550–4,85,550 per hectare	₹5,47,020–7,96,020 per hectare

Source: Adopted from FAO (2022), USDA (2022)

- *Input Costs*: Efficient use of inputs, such as seeds, fertilizers, and pesticides, is crucial for maximizing returns. Overuse or misuse can lead to increased costs and reduced profitability.
- *Yield*: Higher yields generally lead to increased profitability, but they also require optimal growing conditions and effective management practices.
- *Market Price*: Potato prices can be volatile due to factors such as seasonal variations, supply chain disruptions, and market demand. Farmers need to carefully monitor market trends to time their sales for maximum profit.
- *Scale of Operation*: Larger-scale operations often benefit from economies of scale, leading to lower per-unit costs and higher profitability (Haverkort et al., 2009).

9.2 Marketing of Potatoes

The marketing of potatoes involves the movement of the crop from the farm to the consumer, including various intermediaries such as wholesalers, retailers, and processors. The efficiency of the marketing system has a direct impact on the returns received by farmers and the prices paid by consumers. The markets of potato-producing belts in northern India are also not properly integrated with major consuming markets of metropolitan cities in other regions. Because of these factors marketing of potato is a complicated process and potato prices are prone to high fluctuations and often glut situations. The National Agricultural Co-operative Marketing Federation of India Limited (NAFED) is the central body dealing with procurement, distribution, and export and import of selected agricultural commodities. For potato, it is the nodal agency, in cooperation with state agencies, for implementing market intervention schemes. At the state level, State Agricultural Marketing Boards and Directorates of Agricultural Marketing develop and regulate markets in the organized sector and collect and disseminate market information under the Agricultural Produce Markets (Regulation) Acts. State Cooperative Marketing Federations implement market intervention schemes, while seed certification agencies certify potato in Bihar, Haryana, Himachal Pradesh, Karnataka, Punjab, Tamil Nadu, Uttar Pradesh, and West Bengal.

Potatoes are marketed through several channels (Table 4), including:

- *Direct Sales*: Farmers may sell potatoes directly to consumers at local markets or through farm-gate sales (Table 4). This method can offer higher prices but requires more effort in terms of marketing and distribution (Mendoza, 1995).
- *Wholesale Markets*: In many countries, potatoes are sold in bulk to wholesalers who then distribute them to retailers, processors, or exporters (Table 4). This channel provides access to larger markets but often results in lower prices for farmers due to the involvement of intermediaries (Timmer, 2017).
- *Contract Farming*: Some farmers engage in contract farming agreements with processors or retailers. Under these agreements, farmers grow potatoes according

Table 4 Comparative table for different marketing channels of potato

Marketing channel	Small-scale farmers	Medium-scale farmers	Large-scale farmers
Direct sales	High prices, local markets, limited reach	Moderate prices, regional markets	Lower prices, limited reach
Wholesale markets	Moderate prices, wider reach	Moderate prices, regional reach	Higher prices, extensive reach
Contract farming	Rare, due to scale	Common, access to processors	Very common, large-scale contracts
Cooperatives	Limited participation	Moderate participation	High participation
Export markets	Rare, limited by quality	Occasional, requires quality standards	Frequent, large-scale and quality-focused

Source: FAO (2021), World Bank (2021)

to specific quality and quantity requirements, often in exchange for a guaranteed price (Reardon et al., 2009).

- *Cooperatives*: Farmer cooperatives can play a significant role in the marketing of potatoes by aggregating produce, negotiating better prices and providing access to larger markets (Bijman and Ton, 2008).

(a) *Price Determinants*

Several factors influence the price of potatoes in the market:

- *Supply and Demand*: The most significant factor affecting potato prices is the balance between supply and demand. Seasonal variations in production can lead to price fluctuations, with prices typically higher when supply is low (Rashid et al., 2005).
- *Quality*: The quality of potatoes, including size, appearance, and absence of defects, plays a critical role in determining market prices. Premium-quality potatoes command higher prices in both domestic and export markets (FAO, 2014).
- *Processing Requirements*: Potatoes intended for processing into products such as chips, fries, and starch have specific quality requirements. Farmers who meet these standards can often secure higher prices through contracts with processors (Haverkort et al., 2009).
- *Transportation and Storage Costs*: The cost of transporting and storing potatoes can significantly affect their final price. Efficient logistics and storage systems are essential to minimize losses and maintain profitability (Van der Vorst et al., 2009).

(b) *Challenges in Potato Marketing*

Potato marketing faces several challenges that can impact the profitability of farmers:0

- *Price Volatility*: Potato prices can be highly volatile, leading to uncertainty for farmers. Price crashes during periods of oversupply can result in significant losses (Minot, 2010).

- *Post-Harvest Losses*: Post-harvest losses due to inadequate storage, transportation, and handling can be substantial. In some regions, losses can exceed 20% of the total production, affecting both market availability and farmer income (Kitinoja & Kader, 2003).
- *Market Access*: Smallholder farmers often face difficulties accessing larger markets due to a lack of infrastructure, information, and bargaining power. This can result in lower prices and limited market opportunities (Poulton et al., 2006).
- *Global Competition*: The potato market is increasingly globalized, with competition from both domestic and international producers. Farmers need to focus on quality, efficiency, and innovation to remain competitive (Scott et al., 2000).

10 Opportunities and Innovations in Potato Marketing

Despite the challenges, several opportunities exist for improving the economics and marketing of potato production:

(a) *Value Addition*
Processing potatoes into value-added products such as chips, fries, and dehydrated products can significantly increase their market value. Value addition not only provides higher returns for farmers but also creates employment opportunities in rural areas (Reardon et al., 2009).

(b) *Export Markets*
Expanding into export markets offers significant growth potential for potato producers. However, this requires meeting stringent quality standards, understanding international market dynamics, and navigating trade regulations (Haverkort & Struik, 2015).

(c) *Digital Marketing*
The rise of digital platforms offers new opportunities for farmers to market their produce directly to consumers and retailers. E-commerce platforms and mobile applications can help farmers access larger markets, reduce transaction costs, and improve price transparency (Schuster & Maertens, 2016).

(d) *Sustainable Practices*
Sustainable potato production practices, such as integrated pest management (IPM), organic farming, and water-efficient irrigation, can lead to higher yields, lower production costs, and access to premium markets that value environmentally friendly products (FAO, 2014).

(e) *Cooperative Models*
Strengthening farmer cooperatives can enhance the bargaining power of smallholder farmers, improve market access, and facilitate investment in storage, transportation, and processing infrastructure (Bijman & Ton, 2008).

11 Conclusion

The potato (*Solanum tuberosum* L.), introduced to India in the seventeenth century by Portuguese traders or the British, is a key cash crop and a staple food worldwide. As the second-largest potato producer after China, India produces around 43.8 million tons annually, with Uttar Pradesh leading in production and West Bengal having the highest productivity. Potatoes are highly nutritious, offering essential energy, carbohydrates, protein, and dietary fiber. Their adaptability to various climates and soil conditions makes them vital for global food security, as they yield more food per unit of water than any other major crop. Originating in the Andean region of South America over 8000 years ago, potatoes were first domesticated by pre-Inca people and later spread globally by Spanish conquistadors in the sixteenth century. Today, they rank as the third most important food crop globally, after rice and wheat, due to their nutritional richness, including vitamins C and B6, potassium, and antioxidants. In India, potatoes face challenges like transportation and storage issues due to their semi-perishable nature, leading to price fluctuations and marketing complications. However, India has become a significant player in the global potato market, ranking as the 12th largest exporter with $108 million in exports in 2022. The potato's role in Indian cuisine is also growing, with per capita consumption rising to 25.4 kg by 2021, reflecting its increasing importance in the country's diet and economy. The economics and marketing of potato production are complex, involving numerous factors that influence profitability and market dynamics. While potato production offers significant economic benefits, farmers face challenges related to price volatility, post-harvest losses, and market access. However, opportunities exist for improving profitability through value addition, export markets, digital marketing, and sustainable practices. By adopting innovative approaches and strengthening market linkages, potato producers can enhance their competitiveness and secure better returns in an increasingly globalized market.

References

APEDA (2024) Comparative statement for export of agri and processed food products. DGCIS Principal commodities data 2024–25

Bijman J, Ton G (2008) Farmer cooperatives and value chain integration. J Agric Econ 59(2):257–274

Brown P, Allen E, Griffiths D (2018) Postharvest management of potatoes. J Agric Sci 154(3):523–534

Burton WG (1989) The potato: evolution, biodiversity, and genetic resources. Cambridge University Press

FAO (2014) Potatoes and food security. Food and Agriculture Organization of the United Nations

FAO (2020) Global potato production. Food and Agriculture Organization of the United Nations

FAO (2021) Potato value chain analysis and market trends. Food and Agriculture Organization of the United Nations. Available at: www.fao.org

FAO (2022) Global potato production and value chain overview. Food and Agriculture Organization of the United Nations. Available at: www.fao.org
Fravel D (2005) Commercialization and implementation of biocontrol. Annu Rev Phytopathol 43:337–359
Friedman M (2006) Potato glycoalkaloids and metabolites: roles in the plant and in the diet. J Agric Food Chem 54(23):8655–8681
Gustavsson J, Cederberg C, Sonesson U, Van Otterdijk R, Meybeck A (2020) Global food losses and food waste: extent, causes and prevention. FAO
Harris P (2021) The potato crop: the scientific basis for improvement. Springer
Haverkort AJ, Struik PC (2015) Yield levels of potato crops: recent achievements and future prospects. Field Crop Res 182:76–85
Haverkort AJ, Van Koesveld MJ, Schepers HTAM, Wijnands JHM, Wustman R (2009) The potato atlas: description and analysis of potato production systems in the World. Plant Research International
Horton DE, Anderson JL (1992) The potato as a world food crop: current status and trends. Am Potato J 69(4):195–209
Jones P, Armstrong S (2019) The role of cold storage in reducing food losses. J Food Preserv 43(2):410–423
Kitinoja L, Kader AA (2003) Small-scale postharvest handling practices: a manual for horticultural crops. University of California, Davis, Davis
Kumar S, Patel M, Singh R (2020) Post-harvest management of root and tuber crops. Indian J Agric Res 54(4):450–461
Kumar V, Singh S, Reddy B (2021) Smart cold chain management. Trends Food Sci Technol 108:77–89
Mendoza G (1995) A primer on marketing channels and margins. Loyola University
Minot N (2010) Staple food prices in Tanzania. International Food Policy Research Institute
Morris J, Taylor M (2010) Potato diseases: the Grower's guide. CAB International
Poulton C, Dorward A, Kydd J (2006) The future of small farms: new directions for services, institutions, and intermediation. World Dev 34(10):1724–1737
Rashid S, Gulati A, Cummings R (2005) Grain marketing parastatals in Asia: results from six case studies. World Dev 33(3):475–487
Reardon T, Barrett CB, Berdegué JA, Swinnen JFM (2009) Agrifood industry transformation and small farmers in developing countries. World Dev 37(11):1717–1727
Schuster M, Maertens M (2016) Do private standards create exclusive supply chains? New evidence from the Peruvian asparagus export sector. Food Policy 58:14–26
Scott GJ, Rosegrant MW, Ringler C (2000) Global projections for root and tuber crops to the year 2020. Food Policy 25(5):561–597
Storey RMJ (2007) The harvest and storage of potatoes. In: Dick Vreugdenhil, John Bradshaw, Christiane Gebhardt, Francine Govers, Donald K.L. Mackerron, Mark A. Taylor, Heather A. Ross (eds) The potato crop: the scientific basis for improvement. Potato Biology and Biotechnology, Elsevier Science B.V, pp 611–675
Timmer CP (2017) Food security, structural transformation, markets, and government policy. Asia Pac Policy Stud 4(1):4–19
USDA (2022) Potato production and market trends report. United States Department of Agriculture. Available at: www.usda.gov
Van der Vorst JGAJ, Tromp S-O, Zee D-JVD (2009) Simulation modelling for food supply chain redesign; integrated decision making on product quality, sustainability and logistics. Int J Prod Res 47(7):1755–1777
Westermann DT (2005) Nutritional requirements of potatoes. Am J Potato Res 82(4):301–307
World Bank (2021) Agricultural marketing and potato supply chain report. World Bank Group. Available at: www.worldbank.org

Index

S. S. Solankey (ed.), *Advances in Research on Potato Production*, Advances in Olericulture, https://doi.org/10.1007/978-3-031-82710-5

Q

R

MIX
Papier aus verantwortungsvollen Quellen
Paper from responsible sources
FSC® C105338

If you have any concerns about our products,
you can contact us on
ProductSafety@springernature.com

In case Publisher is established outside the EU,
the EU authorized representative is:
Springer Nature Customer Service Center GmbH
Europaplatz 3, 69115 Heidelberg, Germany

Printed by Libri Plureos GmbH
in Hamburg, Germany